Handbook of Silicon Carbide Materials and Devices

This handbook presents the key properties of silicon carbide (SiC), the power semiconductor for the 21st century. It describes related technologies, reports the rapid developments and achievements in recent years, and discusses the remaining challenging issues in the field.

The book consists of 15 chapters, beginning with a chapter by Professor W. J. Choyke, the leading authority in the field, and is divided into four sections. The topics include presolar SiC history, vapor-liquid-solid growth, spectroscopic investigations of 3C-SiC/Si, developments and challenges in the 21st century; CVD principles and techniques, homoepitaxy of 4H-SiC, cubic SiC grown on 4H-SiC, SiC thermal oxidation processes and MOS interface, Raman scattering, NIR luminescent studies, Mueller matrix ellipsometry, Raman microscopy and imaging, 4H-SiC UV photodiodes, radiation detectors, and short wavelength and synchrotron X-ray diffraction.

This comprehensive work provides a strong contribution to the engineering, materials, and basic science knowledge of the 21st century, and will be of interest to material growers, designers, engineers, scientists, postgraduate students, and entrepreneurs.

Series in Materials Science and Engineering

The series publishes cutting edge monographs and foundational textbooks for interdisciplinary materials science and engineering. It is aimed at undergraduate and graduate level students, as well as practicing scientists and engineers. Its purpose is to address the connections between properties, structure, synthesis, processing, characterization, and performance of materials.

Conductive Polymers: Electrical Interactions in Cell Biology and Medicine
Ze Zhang, Mahmoud Rouabhia, Simon E. Moulton, Eds.

Silicon Nanomaterials Sourcebook, Two-Volume Set
Klaus D. Sattler, Ed.

Advanced Thermoelectrics: Materials, Contacts, Devices, and Systems
Zhifeng Ren, Yucheng Lan, Qinyong Zhang

Fundamentals of Ceramics, Second Edition
Michel Barsoum

Flame Retardant Polymeric Materials, A Handbook
Xin Wang and Yuan Hu

2D Materials for Infrared and Terahertz Detectors
Antoni Rogalski

Fundamentals of Fibre Reinforced Composite Materials
A. R Bunsell. S. Joannes, A. Thionnet

Fundamentals of Low Dimensional Magnets
Ram K Gupta, Sanjay R Mishra, Tuan Anh Nguyen

Emerging Applications of Low Dimensional Magnets
Ram K Gupta, Sanjay R Mishra, Tuan Anh Nguyen

Handbook of Silicon Carbide Materials and Devices
Zhe Chuan Feng, Ed.

Bioelectronics: Materials, Technologies, and Emerging Applications
Ram K. Gupta and Anuj Kumar, Eds.

Series Preface

The series publishes cutting edge monographs and foundational textbooks for interdisciplinary materials science and engineering.

Its purpose is to address the connections between properties, structure, synthesis, processing, characterization, and performance of materials. The subject matter of individual volumes spans fundamental theory, computational modeling, and experimental methods used for design, modeling, and practical applications. The series encompasses thin films, surfaces, and interfaces, and the full spectrum of material types, including biomaterials, energy materials, metals, semiconductors, optoelectronic materials, ceramics, magnetic materials, superconductors, nanomaterials, composites, and polymers.

It is aimed at undergraduate and graduate level students, as well as practicing scientists and engineers.

Proposals for new volumes in the series may be directed to Carolina Antunes, Commissioning Editor at CRC Press, Taylor & Francis Group (Carolina.Antunes@tandf.co.uk).

Handbook of Silicon Carbide Materials and Devices

Edited by
Zhe Chuan Feng

CRC Press
Taylor & Francis Group
Boca Raton London New York

CRC Press is an imprint of the
Taylor & Francis Group, an **informa** business

First edition published 2023
by CRC Press
6000 Broken Sound Parkway NW, Suite 300, Boca Raton, FL 33487-2742

and by CRC Press
4 Park Square, Milton Park, Abingdon, Oxon, OX14 4RN

CRC Press is an imprint of Taylor & Francis Group, LLC

Library of Congress Cataloging-in-Publication Data
Names: Feng, Zhe Chuan, editor.
Title: Handbook of silicon carbide materials and devices / edited by Zhe Chuan Feng.
Description: First edition. | Boca Raton : CRC Press, [2023] |
Series: Series in materials science & engineering |
Includes bibliographical references and index.
Identifiers: LCCN 2022025660 (print) | LCCN 2022025661 (ebook) |
ISBN 9780367188269 (hbk) | ISBN 9781032383576 (pbk) |
ISBN 9780429198540 (ebk)
Subjects: LCSH: Electronics–Materials. |
Silicon carbide–Electric properties. | Semiconductor films. | Epitaxy.
Classification: LCC TK7871.15.S56 H36 2023 (print) |
LCC TK7871.15.S56 (ebook) | DDC 621.3815/2–dc23/eng/20220826
LC record available at https://lccn.loc.gov/2022025660
LC ebook record available at https://lccn.loc.gov/2022025661

ISBN: 9780367188269 (hbk)
ISBN: 9781032383576 (pbk)
ISBN: 9780429198540 (ebk)

DOI: 10.1201/9780429198540

Typeset in Palatino
by Newgen Publishing UK

Contents

Part I   General

Part III   SiC Materials Studies and Characterization

Preface

Silicon carbide (SiC) has been an important wide bandgap (WBG) semiconductor material comparable or superior to traditional semiconductors. It exhibits interesting mechanical and electrical properties, such as high thermal conductivity, high critical electric breakdown field, excellent temperature stability, and good chemical inertness. These characteristics make SiC a promising material for a wide variety of high-power and high-temperature electronic device applications in harsh environment where traditional semiconductors lack sufficient durability. SiC, a IV-IV semiconductor with Si-to-C tetrahedral structure, forms close-packed planes stacking in different sequences and exhibits a large number (>250) of polytypes. Their different electronic energy band structures and invariant chemical composition of these polytypes allow a novel type band gap engineering for scientific and technological applications.

Recent industrial and economic developments have moved the technology focus towards minimizing cost on the over-use, and elimination of harmful and possibly destructive energy sources. With the development and utilization of fossil energy, human society is facing global energy resource shortage, environmental pollution, climate change, and other problems. These have raised severe concerns at a global level for all the countries including those lacking or rich in energy sources. To tackle these problems, it is essential to promote new technology development on energy saving. Among various development directions, silicon carbide semiconductors are favored by capital and industry attention. In the capital market, silicon carbide semiconductors are extremely active. Silicon carbide semiconductor concept-based industry stocks have become daily limit hot stocks. Research and development (R&D) on SiC have been very active continually for several decades. Together with other WBG materials of III-Nitrides (GaN, AlN, InN, and alloys) and oxides (ZnO, GaO, and others), SiC is becoming a major research and industrial factor in the world and this growth will continue for a long time into the future.

Tremendous research and industry achievements and developments have been made in recent years. The current *Handbook of Silicon Carbide Materials and Devices* reviews many of these significant results and progresses, covers the basic and critical aspects, and serves as an up-to-date reference for professors, scientists, engineers, students, entrepreneurs, and industrialists in the field. The handbook is organized for a wide range of audiences and covers each of the basic and critical aspects of SiC science and technology. Each chapter, written by experts in the field, reviews the important topics and achievements in recent years, discusses progresses made by different groups, and suggests the needed further works. This book provides useful information about SiC material growth/processing, characterization, devices and developments.

This handbook consists of 15 well-written review chapters, lead by the authorized Prof. W. J. Choyke. These chapters are divided into four sections. The topics of the book include, *Part-I General*: (1) Silicon Carbide: Presolar SiC Stardust Grains and the Human History of SiC from 1824 to 1974, (2) Recent Progresses in Vapor-Liquid-Solid Growth of High-Quality SiC Single Crystal Films and Related Techniques, (3) Spectroscopic Investigations for the Dynamical Properties of Defects in Bulk and Epitaxially Grown 3C-SiC/Si (100), (4) SiC Materials, Devices and Applications: A Review of Developments and Challenges in the 21st Century; *Part-II SiC Materials Growth and Processing*: (5) CVD of SiC Epilayers – Basic Principles and Techniques, (6) Homo-Epitaxy of Thick Crystalline 4H-SiC Structural

Materials and Applications in an Electric Power System, (7) Epitaxial Growth and Structural Studies of Cubic SiC Thin Films Grown on Si-face and C-face 4H-SiC Substrates, (8) SiC Thermal Oxidation Process and MOS Interface Characterizations: From Carrier Transportation to Single-Photon Source; *Part-III SiC Materials Studies and Characterization*: (9) Multiple Raman Scattering Spectroscopic Studies of Crystalline Hexagonal SiC Crystals, (10) Near-Infrared Luminescent Centers in Silicon Carbide, (11) SiC Substrate and its Epitaxial Layers' Anaysis by Spectroscopic Ellipsometry, (12) Raman Microscopy and Imaging of Semiconductor Films Grown on SiC Hybrid Substrate Fabricated by the Method of Coordinated Substitution of Atoms on Silicon; *Part-IV SiC Devices and Developments*: (13) 4H-SiC-Based Photodiodes for Ultraviolet Light Detection, (14) SiC Radiation Detector Based on Metal-Insulator-Semiconductor Structures, (15) Internal Atomic Distortion and Crystalline Characteristics of Epitaxial SiC Thin Films Studied by Short Wavelength and Synchrotron X-ray Diffraction.

The handbook presents the key properties of SiC, describes related technologies and discusses the remaining challenging issues in SiC for further R&D in the 21st century. This book can serve material growers and evaluators, device design and processing engineers, potential users and newcomers, postgraduate students, engineers, scientists, and entrepreneurs well. In 2003–2004, I edited two books on SiC, *SiC Power Materials – Devices and Applications, published by Springer*, and *Silicon Carbide: Materials, Processings and Devices*, with Jian H. Zhao, published by Taylor & Francis also.

Developments on SiC materials and devices are moving quickly. It is important to stay up-to-date and publish on these developments. This new book will cover the rapid new developments and achievements in the field, especially those made after entering the 21st century and within recent years. It will serve and add to the engineering, materials, and basic science knowledge base.

Editor Biography

Professor Zhe Chuan Feng earned his PhD in condensed matter physics from University of Pittsburgh in 1987, and, earlier, BS (1962–68) and MS (1978–81) from the Department of Physics at Peking University. He has worked at Emory University (1988–92), National University of Singapore (1992–94), Georgia Tech (1994–95), EMCORE Corporation (1995–97), Institute of Materials Research & Engineering, Singapore (1998–2001), Axcel Photonics (2001–02), Georgia Tech (2002–03), National Taiwan University (NTU) (2003–2015) as a professor at the Graduate Institute of Photonics & Optoelectronics and the Department of Electrical Engineering; and Guangxi University (GXU) (2015–2020) as a distinguished professor at the School of Physical Science and Technology. After retiring from NTU and GXU and moving back to Georgia, USA, he established the Science Exploring Lab and in January 2022 joined Kennesaw State University as an Adjunct Professor in the Department of Electrical and Computer Engineering, Southern Polytechnic College of Engineering and Engineering Technology.

He has long been devoted to materials research and growth of III-V and II-VI compounds, LED, III-nitrides, SiC, ZnO, GaO and other semiconductors and oxides. Professor Feng has edited 12 review books on compound semiconductors and microstructures, porous Si, SiC and III-nitrides, ZnO devices, and nanoengineering, especially in the 21st century on WBGs: *SiC Power Materials: Devices and Applications*, Springer (2004); *III-Nitride Semiconductor Materials*, Imperia College Press (2006); *III-Nitride Devices and Nanoengineering*, Imperia College Press (2008); *Handbook of Zinc Oxides and Related Materials: Volume 1) Materials, and Volume 2) Devices and Nano-Engineering*, T&F/CRC (2012); *Handbook of Solid-State Lighting and LEDs*, T&F/CRC (2017); and *III-Nitride Materials, Devices and Nanostructures*, World Scientific Publishing (2017).

He has authored and co-authored more than 570 scientific papers with more than 420 indexed by Science Citation Index (SCI) and cited more than 6600 times, with h-index:**40** and i10-index:**152**. Among these, he has published more than 50 journal papers and more than 70 conference papers on SiC as well as three review books on SiC. He has been a symposium organizer and invited speaker at numerous international conferences and universities. He has served as a guest editor for special journal issues and has been a visiting or guest professor at Sichuan University, Nanjing Tech University, South China Normal University, Huazhong University of Science & Technology, Nankai University, and Tianjin Normal University. Professor Feng has been a fellow of SPIE since 2013. More details on his academic contributions can be found at https://scholar.google.com/citations?hl=en&user=vdyXZpEAAAAJ and www.ee.ntu.edu.tw/profile1.php?teacher_id=941011&p=5.

Contributors

Chong Chen
State Key Laboratory of Luminescence and Applications, Changchun Institute of Optics, Fine Mechanics and Physics, Chinese Academy of Sciences, China

Wolfgang J. Choyke
Department of Physics and Astronomy University of Pittsburgh, Pittsburgh, Pennsylvania, USA

Zhe Chuan Feng
Department of Electrical and Computer Engineering, Southern Polytechnic College of Engineering and Engineering Technology, Kennesaw State University, Kennesaw, Marietta, Georgia USA

ChangCai Cui
National and Local Joint Engineering Research Center for Intelligent Manufacturing Technology of Brittle Material Products,
Huaqiao University, Xiamen, China
and
Institute of Manufacturing Engineering, Huaqiao University, Xiamen, China

Roman Drachev
ON Semiconductor,
Hudson, New Hampshire, USA

Ian T. Ferguson
Southern Polytechnic College of Engineering and Engineering Technology, Kennesaw State University, Kennesaw, Marietta, Georgia USA

Yasuto Hijikata
Graduate School of Science and Engineering, Saitama University, Japan

Rongxiang Hu
Department of Physics, Auburn University, Alabama, USA

Ivan G. Ivanov
Department of Physics, Chemistry and Biology, Linköping University, Linköping, Sweden

Yuping Jia
State Key Laboratory of Luminescence and Applications, Changchun Institute of Optics, Fine Mechanics and Physics, Chinese Academy of Sciences, China

Benjamin Klein
Southern Polytechnic College of Engineering and Engineering Technology, Kennesaw State University, Marietta, Georgia, USA

Sergey A. Kukushkin
Institute for Problems in Mechanical Engineering of the Russian Academy of Sciences, Saint Petersburg, Russia

Dabing Li
State Key Laboratory of Luminescence and Applications, Changchun Institute of Optics, Fine Mechanics and Physics, Chinese Academy of Sciences, China

HuiHui Li
National and Local Joint Engineering Research Center for Intelligent Manufacturing Technology of Brittle Material Products,
Huaqiao University, Xiamen, China
and
Institute of Manufacturing Engineering, Huaqiao University, Xiamen, China

Hao-Hsiung Lin
Graduate Institute of Electronics
   Engineering, Graduate Institute of
   Photonics & Optoelectronics
and
Department of Electrical Engineering,
   National Taiwan University, Taipei,
   Taiwan, ROC

Chengling Lu
Beijing University of Posts and
   Telecommunications, Beijing, China,

Min Lu
Compound Semiconductor China (CSC)

Yuji Matsumoto
Department of Applied Chemistry, School
   of Engineering, Tohoku University,
   Sendai, Japan

Yu-ichiro Matsushita
Tokyo Tech Academy for Convergence
   of Materials and Informatics, Tokyo
   Institute of Technology, Japan

Yingxi Niu
Institute of Semiconductors, Chinese
   Academy of Sciences, Beijing, China
and
Wuhu Advanced Semiconductor
   Manufacturing Co., LTD, Anhui, China

Takeshi Ohshima
Quantum Beam Science Research
   Directorate, National Institutes for
   Quantum and Radiological Science and
   Technology, Japan

Andrey V. Osipov
Institute for Problems in Mechanical
   Engineering of the Russian Academy of
   Sciences, Saint Petersburg, Russia

Tatiana S. Perova
Department of Electronic and Electrical
   Engineering, Trinity College Dublin,
   The University of Dublin, Dublin,
   Ireland

Zhi Ren Qiu
State Key Laboratory of Optoelectronic
   Materials and Technologies and School
   of Physics, Sun Yat-Sen University,
   Guangzhou, China

Vishal Saravade
Southern Polytechnic College of
   Engineering and Engineering
   Technology, Kennesaw State
   University, Marietta, Georgia, USA

Nguyen T. Son
Department of Physics, Chemistry
   and Biology, Linköping University,
   Linköping, Sweden

Xiaojuan Sun
State Key Laboratory of Luminescence
   and Applications, Changchun Institute
   of Optics, Fine Mechanics and Physics,
   Chinese Academy of Sciences, China

Devki N. Talwar
Department of Physics, University
   of North Florida, Jacksonville,
   Florida, USA,
and
Department of Physics, Indiana
   University of Pennsylvania, Indiana,
   Pennsylvania, USA

Chin-Che Tin
Department of Physics, Auburn University,
   Alabama, USA

Lingyu Wan
Center on Nano-Energy Research,
   Laboratory of Optoelectronic Materials
   & Detection Technology, Guangxi
   Key Laboratory for the Relativistic
   Astrophysics, School of Physical
   Science & Technology, Guangxi
   University, Nanning, China

Zhengyun Wu
Department of Physics, School of Physical
   Science and Engineering, Xiamen
   University, Xiamen, China

Bin Xin
Department of Materials Science and
   Engineering, King Abdullah University
   of Science and Technology, Jeddah,
   Saudi Arabia

Gu Xu
Department of Materials Science and
   Engineering, McMaster University,
   Hamilton, Ontario, Canada

Alireza Yaghoubi
Center for High Impact Research,
University of Malaya, Kuala Lumpur,
   Malaysia

Weifeng Yang
Department of Microelectronics and
   Integrated Circuit, School of Electronic
   Science and Engineering (National
   Model Microelectronics College),
   Xiamen University, Xiamen, China

Jeffrey Yiin
Southern Polytechnic College of
   Engineering and Engineering
   Technology, Kennesaw State University,
   Marietta, Georgia, USA

Part I

General

1

Silicon Carbide: Presolar SiC Stardust Grains and the Human History of SiC from 1824 to 1974

Wolfgang J. Choyke

Department of Physics and Astronomy
University of Pittsburgh, Pittsburgh, USA

1 Introduction

The rest of the articles in this Handbook deal with SiC crystal growth, characterization, and device technologies. As such, the readers of this volume will most likely be condensed matter scientists or electrical engineers. Alas, this community will not be familiar with journals such as the *Astrophysical Journal*, the *Annual Review of Astronomy and Astrophysics*, *Meteoritics and Planetary Science* as well as *Geochimica et Cosmochimica*. Why is this interesting? In the last 33 years these journals have been reporting on experimental findings on presolar SiC stardust grains and the importance of this work to astrophysics. In the following section we will attempt to sketch the progress in this field up to the present and show the importance of SiC to modern astrophysics.

SiC is becoming a major industrial factor in the world and this growth will continue for a long time into the future. Our industrial effort on SiC is completely based on synthetic SiC. In the formation of our solar system the conditions were not favorable for the growth of "natural" SiC. Under these conditions how did we stumble on this wonderful material and who were some of the scientists who developed the field and led it to our current technological maturity? In the final section of this article we will briefly cover some of the exciting developments from roughly 1824 to 1974, which we might call "the formative stage of SiC".

2 Presolar SiC Stardust Particles

In the second half of the 1940s George Gamow, then a professor at George Washington University in Washington DC, started to develop a theoretical model leading to what is currently called the "big bang" theory of the expanding universe [1-2]. Gamow at first suggested that all the elements could be created in the enormously high temperature and density at the start of our universe. However, in 1954, Fred Hoyle, the Plumian Professor of Astronomy and Experimental Philosophy at Cambridge University, published an

DOI: 10.1201/9780429198540-2

immensely important paper entitled "Synthesis of the elements from carbon to nickel". This paper [3] predicted and calculated the nucleosynthesis of the elements from carbon to nickel and the associated increase of the elements heavier than hydrogen and helium in the universe. This is what is now termed "primary nucleosynthesis". Astronomers generally term all elements heavier than hydrogen and helium as metals. Importantly, Hoyle also discussed verbally an equation for the rate of growth for the heavy element mass in the interstellar medium. Once Gamow became aware of the results described in Hoyle's 1954 paper, he changed his mind and also believed that all elements heavier than helium are produced by thermonuclear reactions in stars. Only two years later (1956), three famous astronomers and a famous theoretical physicist got together at Cal Tech to write another momentous paper [4] entitled "Synthesis of the elements in stars". This *Review of Modern Physics* paper is one of the most cited works in astrophysics. In this paper the authors primarily describe the secondary processes of nucleosynthesis. Secondary nucleosynthesis describes the process by which one heavy nucleus might change into another within a star, but which does not contribute to increasing the metallicity of the galaxy in time. Of particular interest to our later discussion of SiC stardust is the treatment of modes of element synthesis. We will frequently come across the term "s process" slow neutron capture and "r process" for rapid neutron capture. The former is the process of neutron capture with gamma ray emission. The neutron capture occurs on a time scale of ~100 years to 10^5 years. The "s" stands for slow rate for neutron capture compared to the beta decays. The "r" process of neutron capture occurs at a very short time scale, 0.01 to 10 seconds compared to the beta decays. In all, they proposed eight different nucleosynthetic processes, which take place in different stars. In the same year (1957) A.G.W. Cameron also published another important paper for the general theory of nucleosynthesis in stars [5] entitled "Stellar evolution, nuclear astrophysics, and nucleogenesis".

Our solar system formed roughly 4.6 billion years ago. It consisted of a cloud of interstellar gas and dust, which had its origin in the spewing off of dust and gas into the interstellar medium by aging stars and nebulae. This interstellar cloud collapsed forming the solar nebula. Due to gravitational attraction, a very dense center core formed leading to the birth of our sun. Farther away from the early sun, gravity also caused the collection of clumps of matter, which in turn collided with each other ultimately forming the planets, the asteroid belt, and meteorites. In 1960 it was believed that all the matter making up the solar system was a complete homogenization of the dust and gas of the solar nebula. What this meant was that if you looked at the isotopic ratios of any of the elements from specimens taken from meteors, they would always be the same and identical to the elements found on earth. This meant that only astronomically based spectroscopy would ever give us an insight into the makeup of celestial objects. Fortunately, Nature is kind to us and this is not our fate.

In 1960 J.H. Reynolds at Berkeley published a *Physical Review Letter* [6] in which he showed the mass spectrum of xenon, which was extracted from the chondritic (stone like) meteorite Richardton. He observed a substantial enrichment of Xe^{129} compared to terrestrial Xe. This isotope had to be formed from the radioactive decay of $I^{129.}$ This isotope is now extinct on earth, but it was not extinct at the time of the formation of the stellar dust that became part of the meteorite during the formation of the solar system. This paper was a wakeup call and started a new field of examining the isotopic nature of noble gases trapped in a type of meteorite called a carbonaceous chondrite.

Fast forward 15 years to 1975 and quite a lot of new work has been done on extinct radioactive nuclei. At this point theoretical physicist Donald D. Clayton, then a Professor at Rice University in Houston, presented a new picture [7] for the origins of the extinct radioactive nuclei such as ^{129}I and ^{244}Pu. He suggested that these radioactive nuclei were precipitated in grains forming in the rapidly cooling ejecta of the outer atmospheres of exploding old stars. This implies that the radioactive decay occurred in the interstellar grains rather than in meteorites. This is an important hint that perhaps not all matter was homogenized in the formation of the meteorites as the solar system formed. Professor Clayton, early in his career worked with Fred Hoyle and through his long career has been a very important guide to the emerging scientific community studying interstellar grains.

In 1978 a paper was published by two scientists at the Enrico Fermi Institute of the University of Chicago entitled "Noble gases in the Murchison meteorite: Possible relics of s-process nucleosynthesis" [8]. From here on in our story the Murchison meteorite will play a major role. The Murchison meteorite fell on 28 September 1968 near Murchison, Victoria in Australia. Many fragments were found, and the total weight of the meteorite exceeded 100 kilograms. The Murchison belongs to the CM group of carbonaceous chondrites, which are rich in carbon and are among the most chemically primitive meteorites. The Chicago group used stepwise heating and chemical techniques for the extraction of gases from Murchison specimens. They found a new type of xenon component, enriched by up to 50% in five xenon isotopes, ^{128}Xe, ^{129}Xe, ^{130}Xe, ^{131} Xe, and ^{132}Xe. In krypton they find a smaller enrichment for ^{80}Kr and ^{82}Kr. Helium and neon are also found to be highly anomalous with high enrichments for ^{3}He and ^{22}Ne. They point out that these patterns are suggestive of three nuclear processes believed to take place in stars termed red giants. Process one is the s process where neutron capture is on a "s" slow time scale. Process two refers to nuclear reactions in the shell of a star filled with He and process three pertains to nuclear reactions proceeding in the shell of a star filled with hydrogen. Note that astronomers refer to these nuclear reactions as "burning". Srinivasan and Anders suggest that their results tend to show that primitive meteorites can contain **presolar material** in the form of dust grains ejected from red giant stars. The four decades old belief that the early solar system was isotopically and chemically uniform does not stand up to the discovery that presolar matter appears to survive in a meteor such as the Murchison.

We have one more section of background before we can hope to fully appreciate how SiC presolar grain research got its start mainly at Washington University in St. Louis. In 2007 Kevan D. McKeegan, a professor at UCLA and an early member of the Washington University team, wrote a very informative paper entitled "Ernst Zinner, lithic astronomer" [9]. This paper was meant to be a tribute to Professor Ernst K. Zinner on his 70th birthday. It is a very human description of Zinner's great contributions to a new era of space exploration and we will only touch on a few items to help us understand how SiC presolar grains are an important aspect to this new field. Ernst Zinner had come from Austria to Washington University in St. Louis to get a PhD in experimental particle physics. He submitted his dissertation in 1972. At about the same time Robert M. Walker had been appointed McDonnell Professor of Physics to develop the new Laboratory for Space Physics. He asked Zinner to join his group as a postdoc. At the time, Walker's group was focused on understanding and exploiting radiation damage in crystal in part to understand what happens under a radiation environment in space. Walker must have been very persuasive to convince Zinner to move from particle physics to a new area where one would hope to analyze what happens to matter in space. It was the start of a very fruitful collaboration. Zinner's first project in space science was measuring the abundance of heavy (iron group) ions in the solar wind, which was part of an Apollo 17 experiment. He recognized rapidly that there was a great need for better microanalytical methods to make progress with analyzing small interplanetary dust particles. New novel surface science techniques and instrument development became a priority at the McDonnel Center for the Space Sciences at Washington University in St. Louis. Zinner was central to the development and use of secondary ion mass spectroscopy (SIMS) in the study of micron sized dust particles from space. Especially noteworthy are his innovative developments on the Cameca 3f micro-probe and more recently the development of the Cameca NanoSIMS machine. By the end of the eighties the Zinner-Walker Laboratory was well prepared for performing innovative research on microparticles obtained from carbonaceous chondrites such as the Murchison meteorite. At this point, another interesting wrinkle must be mentioned. The close scientific collaboration between the Zinner group in St. Louis and the Anders, Lewis, Ming group at the University of Chicago turns out to have an important influence on the dramatic events ready to unfold in 1987. At the 1987 Lunar and Planetary Science Conference, Roy Lewis made the spectacular announcement that he was carrying a small vial of white powder in his pocket, derived from the Murray C2 chondrite meteorite, which contained inter-stellar diamond. Ed Anders, Roy Lewis, and Tang Ming announced that they had found a large concentration of Ne-E and exotic (s-process) Xe in the powder, and x-ray powder patterns had indicated the presence of nanoscale diamonds. Ernst Zinner's wife, Brigitte Wopenka, was at the conference and she had available to her, at Washington University, micro-Raman spectroscopy. Lewis asked Dr. Wopenka whether she might have a look at this white powder at Washington University and confirm that the powder was indeed nanodiamonds. The little vial flew to St. Louis with Brigitte Wopenka. Unfortunately, her experiments were not successful, but she asked Lewis and Anders whether they might permit their material to be studied by the ion probe group at Washington University. They agreed, and a whole new era of astrophysics was set into motion. On November 27, 1987 two papers were accepted, by *Nature*, which gave strong evidence of SiC from a carbon-aceous meteorite. These experiments were the result of the collaboration of the groups at Washington University in St. Louis and the Enrico Fermi Institute at the University of Chicago. The two papers were published one after the other in *Nature*, December 1987 [10,11]. In the first paper with lead author, Thomas Bernatowicz, specimens were obtained from the University of Chicago group [12,13,14] and were used to provide evidence for interstellar SiC in the Murray carbonaceous meteorite. A bulk sample from this meteorite was subjected to HF-HCL, oxidized with $Cr_2O_7^{2-}$ and boiling $HCLO_4$ for removal of organic and graphitic carbon. The grain sizes were separated into five fractions by centrifuging and the obtained sizes were 0.3μm to 2.0μm. This collection of grains was called CF. Another Murray sample was treated in a similar fashion but sample sizes were separated by filtra-tion and this residue was called CJ. CF has a very high concentrations of Xe-S and CJ has a very high concentration of Ne-E(H). Sample CJ under energy dispersive x-ray exam-ination (EDX) revealed to have substantial amounts of silicon. This was an indication of the presence of SiC. The samples were studied by scanning electron microscopy (SEM), ion microscope, transmission electron microscope (TEM), and micro Raman spectroscopy. Silicon carbide was positively identified by the micro-Raman spectroscopy in CJ, which had been subjected to an oxygen plasma discharge for 30 minutes and then called CJa. The SiC (TO) Raman peak was clearly observed. Further experiments also showed that the sep-arate, CF, actually contained crystals of SiC. They found irregular twinned and also faulted SiC crystals from several hundred Ångstroms to several thousand Ångstroms Å in size. Electron diffraction patterns revealed that all their SiC specimens appeared to be cubic SiC and twinned along multiple [111] planes. Some crystals showed severe stacking disorder. TEM images from both CJ and CJa showed the presence of SiC. Especially interesting is that they observed several SiC crystals up to 1μm in size. This report is clearly the first definitive identification of stardust SiC in meteorites. The story continues in the following paper [11] of Vol. 330 of *Nature*. The title of the paper is "Large isotopic anomalies of Si, C,

N and noble gases in interstellar silicon carbide from the Murray meteorite" and its lead author is Ernst Zinner. Isotope measurements were made on CJa using the Cameca IMS 3F Washington University ion probe. Agglomerates of CJa residues were mounted on gold foil. Silicon, carbon, and nitrogen isotopes were then measured on 5 to 10µm spots, as secondary ions, produced by the Cs^+ bombardment. Nitrogen as CN^- was measured with a mass resolving power (MRP) of 6500. This was a high enough resolution to separate $^{13}C^{14}N^-$ from $^{12}C^{13}N^-$ and $^{11}B^{16}O^-$. During the nitrogen measurements, carbon was measured as C_2^- and during silicon measurements as C^- and at an MRP of 3500. All isotopic compositions are normalized for instrumental mass fractionation and are given as (delta) values relative to the normal terrestrial standards.

Delta (δ) is defined as follows:

$$\delta\ ^{29}Si = (1000) \times ((^{29}Si/^{28}Si)_{sample} / (^{29}Si/^{28}Si)_{standard})$$

The authors conclude that SiC is responsible for the isotopically heavy carbon seen in C1-C2 chondrites upon stepped combustion at $\geq 900°C$. At this high temperature SiC is more resistant to combustion than diamond. Tentatively they believed that both Xe-S and Ne-E(H) could be contained in SiC. They conclude that several types of SiC could be circumstellar condensates from highly evolved stars where C/O is larger than unity. The heavy carbon and light nitrogen are qualitatively consistent with production in the CNO cycle (Bethe-Weizsäcker cycle). The CNO cycle and the p-p cycle are the two sets of fusion reaction responsible for the conversion of hydrogen to helium in stars. The pp-chain reaction starts at about $4 \times 10^6 K$ and is the dominant source in small stars including the sun. The CNO chain reaction starts at about $15 \times 10^6 K$ and becomes the dominant source of energy near $17 \times 10^6 K$. The CNO fusion reaction is believed to be dominant in stars exceeding 1.3 times the mass of our sun. Since Si is not involved in the hydrogen nuclear reaction (hydrogen burning) the anomalous Si isotope readings must be caused by later stages of nucleosynthesis. Xe-S requires the s-process for formation on the other hand Ne-E, presumable derived from ^{22}Na (half-life 2.6 years) requires an explosive process such as found in novaes or supernovaes. It is pointed out that the evidence seems to indicate that the SiC grains may be relatively young. The SiC grains were incorporated in the meteorite not long after their formation. This is suggested by the fact that the ^{22}Ne in the CJ group gives a cosmic ray exposure age of less than 60 million years for the SiC. Secondly, in a sister sample from CF they observed that the size distribution of the SiC is as expected for a primary crystal growth distribution that has not been altered by erosion or fragmentation. The authors noted that the isotopic anomalies and noble gases in their SiC stardust grains were 10 to 10,000 times larger than in most of the other elements found in meteorites. They pointed out that the solar system has a relatively low C/O ratio of about 0.6 and so SiC could not have formed in the stars that provided the material for the solar nebula. It is generally believed that SiC forms only in the outer envelopes of stars when the C/O ratio exceeds unity.

At this point I have to make a confession. I have been an avid researcher of SiC from the point of view of solid state physics since 1955 and have tried to keep a sharp eye on the literature but I was not aware of papers [10] and [11] until I saw an article by Bernatowicz and Walker in the December 1997 *Physics Today* [15]. This is a well written short review of this new field up to 1997 and I recommend it as reading to all.

It is now 33 years later, and since 1987 a vast number of experiments and theoretical modeling has been done. This short historical look cannot do this work justice. Fortunately there have been excellent reviews every couple of years by the leaders in the field and

I shall reference just three of these [16], [17], and [18] for those readers interested in a summary of the developments in the field as well as ample references.

From the point of view of the likely readers of this SiC Handbook it would be remiss if I do not mention extensive TEM work that has gone on in studying the presolar SiC grain polytype distribution. First, we report on a paper in 2003 by a group from the Argonne National Laboratory, the Naval Research Laboratory, Washington University in St. Louis and the Enrico Fermi Institute at the University of Chicago [19]. In this paper the authors start out by giving a listing (up to 2002) of papers in which isotopic anomalies, which characterize stardust SiC, are presented:

1. Isotopic anomalies in C and Si, Xe and Ne as well as a trace element such as N, [11], [20]
2. Size distribution of SiC measured in the range 0.2 to 6.0μm [21]
3. Analysis of He, Ne, Ar, Kr and Xe [22]
4. ^{26}Mg daughters of extinct ^{26}Al [23]
5. Ca [24]
6. Ti [25]
7. Sr [26]
8. Zr [27]
9. Mo [28]
10. Ba [29]
11. s-process Ba, Nd, and Sm [30]
12. Dy [31]

A second point to mention at this stage is the observation of SiC by means of astronomical spectroscopy. Molecular equilibrium models have predicted SiC particles and other particle species in the atmospheres of AGB carbon stars. In 1974 Treffers and Cohen [32] observed an IR band at about 11.3μm, which they attributed to SiC. Later attempts to obtain crystallographic features of the SiC from details of 11.3μm spectra have led to much controversy but relatively little believable additional insight. Hence looking at SiC stardust in the laboratory with the latest electron microscopy and diffraction techniques is a very compelling thing to do. Daulton and coworkers use TEM to determine the polytype distribution in stardust SiC obtained via acid dissolution from the Murchison CM2 carbonaceous meteorite. A total of 508 SiC presolar grains were studied. In [21] the size distribution of the SiC presolar grains from the Murchison CM2 has been measured from 0.2 to 6μm. The size fractions are labeled KJA to KJH. In this study the 508 specimens all come from KJB, which has a range of diameters from approximately 0.32 to 0.70μm, with a mean diameter of 0.49μm. It represents 70% of the Murchison SiC grains. The KJB size fraction was chosen because it most closely resembles the SiC population in the Murchison meteorite. The presolar SiC population in primitive meteorites is now believed to be a reasonably representative sample of SiC stardust produced in stellar outflows. The TEM measurements were made with a JEOL JEM-3010 instrument operating at 300keV and using a LaB$_6$ filament. They used a measured point-to-point resolution of 2.1Å for structural microcharacterization. The electron optical magnifications were between 600K and 1000K. Polytype determinations are made by analysis of selected area electron diffraction (SAED), convergent beam electron diffraction (CBED), and HR-TEM lattice images. Lattice images and SAED patterns show that only two types of SiC polytypes were found, namely cubic or 3C SiC, and hexagonal or 2H SiC. A number of disordered SiC grains were found but they could not be classified as distinct polytypes. The 3C SiC population is roughly 79.4%

of the total, 2H SiC is approximately 2.7% of the total, and 17.1% represents intergrowths of 3C and 2H. A small number of one-dimensionally disordered SiC grains were found as well as a category called "all other polytypes" with less than 0.20%. In laboratory growth studies on earth, 2H is the lowest temperature SiC polytype formed and 3C develops next. Finding 2H in the Murchison meteorite is the first evidence that 2H grows in nature. In this paper it is also concluded that the measured polytype distribution is a measure of the original growth abundances in stellar atmospheres and not an alteration by thermal transformation during the formation of the solar system. The formation of the 2H SiC and 3C SiC stardust grains are the results of specific physical conditions in AGB circumstellar atmospheres. For example, the total gas pressures in the extended stellar atmospheres are quite low (<100dyne/cm^2), which consequently leads to the formation of 2H SiC and 3C SiC at temperatures below 2000K. One proposed scheme is that 3C SiC first condenses at small radii in AGB atmospheres where temperatures are relatively higher compared to those at larger radii where 2H forms. At intermediate radii 2H/3C intergrowth grains are likely to form.

By 2021 more than 19,000 SiC stardust particles have been isolated from meteorites and studied. Characterization technology keeps improving and micro-Raman spectroscopy has been used for rapid selection of specimens. As we might expect, theoretical models are also improving and so the interpretation of the stardust grains is constantly being refined. Currently, a rough breakdown of SiC presolar grains is as follows: MS (mainstream) grains are the most common isotopic group constituting about 90% of all SiC stardust. Next come AB grains at ~5%, Y grains at ~1–5%, Z grains at ~1–5%, X grains at ~1–2%, C grains at <1%, U grains at <1%, and PNG grains at <1%. The current interpretation of the origin of SiC presolar grains in stars, novae, and supernovae is found in many of the papers, which have already been cited.

A very recent TEM study looks at eight SiC stardust grains in great detail both in isotopic characterization and with the latest TEM technology. Finer structural details are found as compared to past studies posing new challenges for the interpretation. The report comes from a group at the US Naval Research Laboratory, Washington University in St. Louis, and the Carnegie Institution of Washington, DC [32]. The detailed TEM analysis is confined to eight carefully characterized presolar SiC samples. There are four mainstream MS samples, one Y sample, and three X samples. The authors found new subtle structural variations, which are a sign that we need to consider more complex histories than homogeneous gas to dust condensation in the monotonically cooling circumstellar envelopes of stars. It is found that the SiC grains have a diversity of structures. There are single crystal 3C and 2H domains but there are also complex intergrowth domains in which higher order hexagonal polytypes are found. Examples are 2H-4H in an MS grain, 3C-8H and 3C-2H-10H-14H in X grains. They found small traces of elements such as Al and Mg and subgrain phases of AlN, TiC, TiN, ZrC, CaS, and Fe metal carbides and silicides. Many voids were found. An estimate of 0.98 to > 1.03 was made for the C/O ratio in the region where the SiC presumably formed in the atmospheres of the stars. This was based on the texture and composition of the SiC grains. The authors point out that studying just eight samples is very limiting. Remedying this short coming is left for the future.

Finally, we want to touch on the question as to how old these SiC stardust grains might actually be? Dating of stardust by means of astronomical techniques, such as spectroscopy, appears at present not to be possible. Similarly, the well-known and accurate dating methods of the decay of radioactive nuclides are also presently not useful because the initial isotopic compositions of the decaying elements in the parent stars are presently unknown. A third method, which is currently most attractive, is used in a paper published in 2020 [33] and determines the ages of 40 SiC stardust grains by means of a technique based on cosmogenic ^{21}Ne. This work was done by a team of scientists from the United States, Switzerland, and Australia. Large SiC stardust grains from the original "LS + LU" Murchison meteorite separation were characterized with electron microscopy and NanoSIMS was used to classify them. He and neon isotopes emanating from the presolar SiC grains, after heating samples with an IR laser, were analyzed with an ultrahigh sensitivity, noble gas, mass spectrometer at ETH in Zürich. In the paper the group reports on a laboratory-based determination of the interstellar lifetimes of individual large SiC stardust particles from the "LS + LU" Murchison separation. If one is able to determine the time it takes for a SiC particle to travel from its parent star to the solar system one just has to add on the well-known age of the solar system (4.6 billion years), to get the total age of the specimen. As we already mentioned these grains were studied for determination of their large isotopic anomalies with NanoSIMS to exclude any possibility that they originated in our solar system. In addition, these SiC grains survived processing in the interstellar medium (ISM) as well as during the formation of the solar system. The trick is to use noble gas mass spectroscopy to determine the abundance of He^3 and ^{21}Ne, produced by spallation reactions in the SiC stardust particle by high energy protons and alpha particles, so called galactic cosmic rays (GCR), during the time that the SiC particle was traveling from its native star to the early solar system. For example, the total number of ^{3}He or ^{21}Ne particles that accumulate during the flight in interstellar space is proportional to the time the SiC grain was bombarded by the GCR. Importantly, the authors are also able to estimate that the effect, if any, on presolar SiC grains by the particle flux due to the early active sun is likely not too significant. To calculate an internally consistent set of GCR exposure times they used the latest cosmogenic nuclide production rates as well as new estimates for nuclear recoil corrections. The Ne component (^{21}Ne) produced by the GCR during flight in the SiC grain can be easily separated by the ETH noble gas mass spectrometer from the nucleosynthetic neon termed (Ne-G) formed in the star and any neon adsorbed from the atmosphere. Finally, upper exposure times were determined for the ^{3}He and ^{21}Ne produced in the SiC grains by the GCR during the flight from the star to the solar system. Since recoil corrections for the ^{3}He are larger than for the ^{21}Ne it is reasonable to assume that the neon values are more reliable. For the full details of the analysis of the interstellar exposure times and relevance to astrophysics the reader is urged to read [33] and many of the references therein. In summary, the range of times for the SiC stardust grains prior to the beginning of the solar system are estimated to be 3.9×10^6 years +/- 1.6×10^6 years to about 3×10^9 years +/- 2×10^9 years. Add to that the start of the solar system, at about 4.6×10^9 years, then the oldest SiC stardust in hand at the present time is about seven billion years. It is impressive to think that the chapters in this book are all on a substance some of whose history goes back billions of years before the formation of our solar system.

3 SiC History from 1824 to 1974

Jöns Jakob Berzelius, one of Sweden's great scientists, was born near Linköping, Sweden and received his secondary education in that city. He then went on to Uppsala University (1796 to 1802) where he studied medicine. From 1807 to 1832 he was at the Karolinska Institute where he was a professor of medicine and pharmacy. Soon after his arrival at the Karolinska Institute, he shifted his interests to inorganic chemistry and he is now credited with discovering the element Si. This is where our story begins. Description of the preparation of elementary Si by reducing SiF_4 with potassium were first reported by Berzelius in Vetenskaps Academiens Handlinger for 1823 but then published in the Annalen der Physik (1824) [34]. Doing careful chemistry of the whole process he discovered carbon was given off and he wondered where the carbon came from. He then added, "Where does this carbon come from, how can it be chemically bonded with the Si?" It is man's first recognition that a Si-C compound might be possible. It is also interesting to note that in recent years the Linköping University has developed one of the most vigorous research programs on SiC.

In 1885 Eugene H. and Alfred C. Cowles reported in the *Scientific American* [35] on the development of an electric furnace for smelting metal ores. After the cooling of the furnace they observed, in the reacted mass, the formation of hard, shiny crystals, which they believed to be artificial sapphire. No analysis of these crystals was ever made and unfortunately a fire destroyed their furnace and so this discovery was never pursued.

At about the same time that the Cowles brothers were trying to develop furnaces for smelting, Edward Goodrich Acheson of Mononghahela, Pennsylvania near Pittsburgh was concerned with producing cheap abrasives for industry. He was hoping that he could convert carbon into "crystalline carbon" and have a product with the stiffness and hardness of diamond. Acheson had evolved a scheme, "To cause carbon to be dissolved in melted silicate of alumina and by cooling the same to the point of solidification, cause the contained carbon to crystallize". The first experiment used an iron bowl, which was lined with carbon and the interior of which was filled with a mixture of carbon and clay [36,37]. A substantial current was passed through the mixture, fusing the mass. After cooling the central mass was removed and carefully broken apart. When Acheson examined the broken material, he discovered a few bright crystals with a blue color around the carbon electrodes, which appeared to be extremely hard. After the first experiments it became clear that the crystals were not diamond and a name had to be found for the new product. The blue sapphire color and hardness suggested that the material was composed of carbon and alumina. At this point no chemical analysis of the blue crystals had been made. Hence the supposed combination of carbon and corundum suggested the name **carborundum.** It was clear to Acheson that the production of carborundum would be a lucrative business and a company was organized to make and market this product. Right at the start of this company, the Carborundum Corporation, a chemical laboratory was built and equipped. Dr. Otto Mulhaeuser was hired and he started an exhaustive analytical study of carborundum. The end result of Dr. Mulhaeuser's careful work revealed that carborundum was actually the compound **silicon carbide** (SiC) with a specific gravity of 3.22. The erroneous name carborundum has nevertheless held on to this very day.

Neither the Carborundum Corporation nor Acheson made any attempts to study the crystallography of their blue slender crystals of SiC. However, Acheson sent some of the crystals to Professors Frazier and Richards at Lehigh University. In an Appendix to Acheson's paper in 1893 [37], B.W. Frazier gave a short report dated June 21, 1893 and entitled "Report of an examination of crystals furniched by Mr. E.G|. Acheson, president of the Carborundum Company". Goniometry and polarization studies showed that some crystals were rhombohedral and others hexagonal. An important observation was that the blue crystals (likely 6H SiC) were hexagonal whereas the yellow green crystals were rhombohedral (likely 15R SiC). In addition, Professor Richards determined the specific gravity by the "specific gravity bottle" at 25°C to be 3.123.

Things moved rapidly thereafter. In July of 1894 Professor Gintl, a professor of chemistry at the German Technical Hochschule in Prague, gave F. Becke in Prague crystals of

SiC obtained from a factory near Prague and produced by means of the Acheson method. A crystallographic analysis was made, and it led to a comparison with B.W. Frazier's results as well as a measure of the ordinary index of refraction, $n = 2.786$ under Na light. Becke also studied the double refraction in the hexagonal SiC crystals. An estimate of the extra-ordinary index of refraction under Na light using his ordinary index value of $n = 2.786$, Becke obtained a value of $\varepsilon = 2.832$. These results were published in the Zeitschrift für Kristallographie [38] in 1895.

In 1905 William Weber Coblentz, known as the founder of astronomical infrared spec-troscopy, joined the National Bureau of Standards and soon afterwards measured the reflection from a SiC crystal, on its basal plane, in the infrared from about 1.0μm to 15μm [39]. The polytype of the sample was not given. He observed a fairly constant reflectivity roughly between 15% and 18% from 1μm to 9μm then a dip to about 6% at 10.2μm and a huge broad peak centered at 12μm with a reflectivity of about 97%. At 14.5μm the reflect-ivity levels off to about 52%. This peak is now known as the famous SiC Restrahlen, or residual ray peak.

Two years later, 1907, a remarkable British engineer, Henry Joseph Round, was working for the Marconi Company in New York when he was studying the "unsymmetrical passage of current through a contact of carborundum". On applying a voltage of 10 volts between two points on a crystal of carborundum he observed yellowish light coming from the crystal. By increasing the voltage, he soon found many luminescing samples with yellow, green, or blue light being given off. In all cases that he tested the glow appeared to come from the negative pole and at the positive pole there appeared a blue green spark. To our knowledge this is the first report of what we now call luminescence from a light emitting diode [40]. In both World Wars, Captain Round made important technical contributions to the British war effort.

In two papers [41,42] in 1912 and 1915, respectively, H. Baumhauer of Freiburg in Switzerland, reviews the goniometric measurements made by B.W. Frazier [37] and F. Becke [38] as well as presenting his own measurements on SiC crystals. He concludes that there are three main types of SiC crystals. More importantly he makes the important discovery of polytypism in SiC.

In 1915, Oskar Weigel published a trailblazing paper on the optical properties of SiC [43]. At the time he was a Privatdozent at the University of Göttingen but later that year he was appointed "ausserordentlicher" Professor for Minerology and Petrology and Director of the Institute of Mineralogical Petrology at the Philipps University Marburg. Oskar Weigel studied natural sciences with a major in Minerology at the University of Göttingen and received his PhD under Professor Theodor Liebisch in 1905. Weigel used large, almost transparent, hexagonal, likely 6H SiC, flat single crystal plates, for his optical experiments. Exactly where these SiC plates were produced is not clear. However, on a visit to the Mineralogical Museum of the Philipps – University in Marburg, I saw Weigel's collection of SiC crystals and there was a card, which implied that they came from the Lonza Works in Waldshut, on the Swiss border. Several slender prisms were fabricated from these crystals and the ordinary ω index and extraordinary ε index of refraction were measured from about 430nm to 710nm. ω and ε were determined from 20C to 1140C. A table with precise data is given at 471.33nm, 501.57nm, and 667.84nm at temperatures of 22°C, 456°C, 585°C, and 758°C. He terminated the table at that point because absorption sets in. He also points out that the UV absorption edge moves toward longer wavelengths with rising tempera-ture. At the time (1915), semiconductors and the concept of the variation of the band gap with temperature was a good 30 years in the future. Careful absorption measurements were also carried out on thick transparent plates that were also likely 6H SiC. The absorption coefficient α (cm^{-1}) is given from about 410nm to 530nm at temperatures from 79°C to 984.8°C. It is a lovely exhibition of the change of the band gap with temperature. He arbitrarily determined the absorption edge for his SiC to be at an extinction coefficient of 0.109 or a transmission of 0.452. It wasn't until the understanding of indirect optical transitions in semiconductors, in the mid-50s, that the field could move forward. Oskar Weigel had died in 1944.

In June of 1912, Sommerfeld at the Physikalische Gesellschadft of Göttingen reported that Max von Laue, Paul Knipping, and Walter Friedrich in Munich had demonstrated the diffraction of x-rays. Two years later Max von Laue was given the Nobel Prize for this work. Just a few years later in 1918 C.L. Burdick and E.A. Owen used x-ray diffraction to determine the atomic structure of silicon carbide [44]. They thought that determining the atomic structure of SiC was an important problem since the crystallographic relationships between diamond and SiC were extremely close and their physical properties were also closely related. This work got its start in the laboratory of Professor W.H. Bragg at the University of London and then was continued at MIT where Dr. A.A. Noyes facilitated the research. Measurements were made of the angles of reflection of palladium x-rays from the principal planes of a crystal of SiC. Measurements of the intensities of reflection of the different orders were made and using von Laue theory these results were interpreted. In their language, they found that the silicon atoms and carbon atoms are each arranged on face-centered rhombohedral lattices displaced with respect to one another along the hexagonal axis a distance equal to 0.36 of the basal plane spacing. From these x-ray measurements the distance between atomic planes was determined. This enabled them to calculate the density of SiC to be 3.11, which compared well with the value of the specific gravity of 3.123 obtained by Professor Richards at Lehigh. The determining of the atomic structure of SiC by means of x-rays was a pioneering achievement.

Oleg Vladimirovich Losev was born in May 1903 in the city of Tver known in Soviet times as Kalinin. Tver is located about 110 miles northwest of Moscow. Losev graduated from secondary school in 1920 but this was only three years after the Bolshevik Revolution and the subsequent civil war and since he came from a noble family, he was barred from a University education and career advancement. He went to work as a technician at the Nizhny Novorod Radio Laboratory (NNRL) under Professor Vladimir Lebedinsky. Losev was a truly self tought scientist and original thinker who never benefitted from the support of colleagues or being in a well-known research team. He published 43 papers and had 16 patents (Inventor Certificates) to his name. Sadly, he died at 38 from starvation during the siege of Leningrad, (St. Petersburg). Much of Losev's work is published in Russian and I rely on a review article by Egon E. Loebner [45] for references to Losev's publications. Egon Loebner himself, was a pioneer in the development of LED's and at the time of writing this article, he was posted to the Embassy of the United States in Moscow. As we have already seen, light emission from a SiC diode was discovered by Henry Joseph Round in 1907 [40] but it was rediscovered by Losev in Novgorod in 1922. Losev's diode was from an array of high frequency oscillating and amplifying SiC detector diodes that he was working on. This important re-discovery led to 16 publications and ten patents. In March 1923 Losev reported seeing green light emission from a reverse biased point contact SiC detector. The steel point was negative with respect to the SiC. Applying positive bias, he observed a greater current flow but no light emission. In September 1924 Losev published an extended discussion of "glow" from contacts on SiC crystals. He got microphotographs and measured the area that was emitting light to be 700µm^2. He obtained a spectrum of the emitting light and determined the threshold for light emission to be 0.1ma or a current density of 15A/cm^2. Losev was distracted from SiC for three years by his development of

ZnO crystal detectors. The study of LED's was continued in 1927 on the strength of the argument that his low voltage light source could be modulated to high frequencies and in that sense it was the first "true electronic light generator". He was encouraged by top Soviet physicists to consider his LED work for communications applications. In 1927 he published that light emission with forward bias is quite distinct from light emission in reverse bias. He gave a detailed study of their *I-V*, *I*-φ characteristics, spectral and spatial characteristics. He found that forward biased emission shifted toward the blue as the voltage was increased. He demonstrated that the light emission did not depend on temperature nor the surface conditions of his crystal but was a bulk property. Losev was many decades before his time and his untimely death ended an amazing career in midstream.

We have now come to the mid-50s where Ge is king in the semiconductor world. However, both industry and the military are concerned that Ge cannot handle either high temperature applications or yield devices that are sufficiently radiation resistant. This was a driver for the Electronic Research Directorate at the Air Force Cambridge Research Center in Bedford, Massachusetts to start SiC programs in universities and industry in the United States. Westinghouse already had an interest in SiC through their production of SiC lightning arrestors but they sensed a much bigger market for power and radiation resistant devices was out there. A sizable program was developed at the Research Laboratories, the Penn Avenue Plant and at the Youngwood plant, all in the Pittsburgh district. By the end of the fifties just the Research Laboratories and Penn Avenue had at least 16 professionals engaged in fundamental research, device development, and crystal growth. Of course, there were also many programs not only in the States but in Europe and Japan. I will confine myself to the Westighouse effort because it was one of the major actors and more importantly I was part of this effort and I can report on some of the achievements with some confidence. Even having pared things down to Westinghouse there is not space to discuss more than the major results that came out of the fundamental studies. During the period from roughly 1956 to 1974 the fundamental effort was carried out mainly by W.J. Choyke, Lyle Patrick, and Don Hamilton. The mid-50s were an exciting time for the theoretical and experimental development of semiconductors. Many of the things that we now take for granted such as electronic band structure, indirect transitions and excitons etc. were concepts under heated debate. In that light, we were not even sure that SiC was an indirect semiconductor. It was only in 1954, that Bardeen, Blatt, and Hall [46] presented a theoretical treatment for indirect interband transitions, which gave us an idea of how to think about them. At that time it was supposed that in 3C SiC the band minima were expected to be along [100] axes as in Si with the valence-band maximum at $k = 0$. Unfortunately, we had no good quality cubic SiC to test out this supposition. We had a transparent plate of hexagonal SiC, likely 6H SiC, with a thickness of 95μm and used it for transmission measurements [47]. We also used several *p-n* junctions as absorbers and detectors. A known quantity of radiation, from a chopped light beam coming from the monochromator, fell on the junction surface and the developed photovoltage was measured. This photovoltage should be proportional to the absorption coefficient for small photovoltages. At that time, the energy band structure of any of the hexagonal polytypes was unknown. However, our optical absorption and photovoltaic measurements were able to be interpreted according to the scheme used by Macfarlane and Roberts [48] for Si, which showed that indirect transitions were being observed. We were able to fit the data with one phonon with an energy of about 90meV. This was a puzzle since we knew that 4H SiC should have 24 phonons participating in the absorption process and 6H SiC should have 36 phonons participating. It was so early in the game that we felt the future would take care of it. Indeed, about 60 years later, with much better material and much higher optical resolution, in our two latest papers on 6H and 4H SiC [49,50], we resolved 21 out of 36 phonons in 6HSiC and 20 out of 24 phonons in 4H SiC.

Having done previous experiments on the photoemission of CsSb and RbSb, it seemed interesting to see whether reverse biased SiC junctions might act as electron emitters. The large band gap of SiC ensures high energy electrons at breakdown and it was estimated at the time that for impact ionization electron energies of 3/2 the band gap would be necessary. With band gaps of about 3eV in hexagonal polytypes one expected conduction-band electrons of about 4.5eV, which exceeded the 4.0eV estimated electron affinity given by H.R. Philipp [51]. We measured [52] electron emission currents from 20 reversed bias SiC *p-n* junctions. The maximum emissions were measured to be between 10^{-12} and 10^{-6} amp. Measurements were made in high vacuum and the best results were obtained heating the samples at 270°C for several hours in vacuum.

By 1961 Don Hamilton had designed and built several modifications to the Lely crystal furnace as well as carrying out experiments to reduce impurity content of the grown crystals. Each successful crystal run was monitored by obtaining photoemission spectra at low temperatures as well as by conducting transport measurements. In 1961 we had in hand many high quality single crystals of 6H and 15R SiC, grown by Hamilton, as well as a few cubic SiC samples. Exciton and interband absorption measurements were made [53] on these samples. The next step was to do lattice absorption measurements on these high purity single crystals [54] at room temperature. Using four phonons it was possible to explain, as summation bands, ten absorption bands lying between 130 and 300 millivolts. Phonons close to three of the four, used in the analysis, were obtained from our indirect interband absorption measurements of 6H SiC. It is important to note that no significant differences in energy were found for lattice absorption measurements of 6H, 15R, and 3C SiC.

By September 1962 a detailed low temperature photoluminescence study of 6H SiC was published in *Physical Review* [55]. A 6K photoluminescence spectrum consisting of approximately 50 lines was observed. The spectrum was attributed to exciton recombination at unionized nitrogen atoms. Since the unit cell of 6H SiC has six carbon atoms and six silicon atoms it provides three inequivalent sites for nitrogen and 36 phonon branches. Seeing 50 lines in the luminescence spectrum is thus no surprise. At the time, 17 phonon replicas were identified and plotted in a plausible way in an appropriate extended **k** space. Free exciton luminescence was also observed. The temperature dependence of the indirect exciton energy gap was given from 6K to about 180K. It was obtained by measuring the temperature dependence of the sharp break in the absorption at LA in the $E\perp c$ absorption curve. The value of E_{GX} was determined to be 3.023eV.

The optical properties of 15R SiC were next [56]. 6H and 15R results from optical experiments were compared. The indirect, exciton producing transitions in 15R SiC, across the exciton energy gap, yielded E_{GX} to be 2.986eV at 6K. 15R SiC has five inequivalent sites but only four no-phonon lines were found. At 6K, 18 phonon conserving energies were found out of a possible 30.

The comparison of the phonon energies of 6H and 15R SiC was very close. Since cubic (3C) SiC has the zincblende structure, with only one C and one Si atom per unit cell the absorption and luminescence spectra are much simpler than 6H SiC or 15R SiC spectra. Unfortunately, growing good single crystals of 3C SiC in a Lely furnace was difficult and more of a challenge than 6H or 15R. However, Hamilton, in time, managed to grow 3C SiC crystals from the vapor starting with relatively pure carbon and silicon. Most 3C SiC crystals were grown between 1950°C and 2000°C. In addition, 3C crystals were also grown during the cooling period after growth of 6HSiC or 15RSiC at temperatures above 2250°C.

Absorption measurements [57] of 3C SiC at 4.2K showed that the absorption edge was due to indirect, exciton creating transitions. The exciton energy gap, E_{GX}, was found to be 2.309eV. The conserving phonons were found to have energies of 46, 79, 94, and 103meV from 6K, four-particle, nitrogen exciton complexes, luminescence spectra. The 3C SiC results are compared with other polytypes. By 1964, we already knew the indirect exciton gaps for seven polytypes and so an interesting linear relationship was noticed between the E_{GX} of these seven polytypes, 3C, 8H, 21R, 15R, 33R, 6H, and 4H SiC and the so-called "percent hexagonal". 3C SiC has the smallest E_{GX}, being zero hexagonal or (k) in the Jagodzinski notation and 4H SiC, which has the largest E_{GX} of these seven polytypes being 50 percent hexagonal or (hk) in the Jagodzinski notation.

In that same year (1964) we also presented data for the absorption edge of 4H SiC as well as preliminary data on the luminescence of 4H SiC [58]. E_{GX} for 4H SiC was found to be 3.265eV and the E_{GX} temperature dependence of 4H SiC is given from 4.2K to about 190K. There are 4C and 4 Si atoms per unit cell in 4H SiC but only two inequivalent C and Si sites. The two no-phonon lines were found to be at 7 and 20meV.

In another paper [59], the 4H SiC, 6K, luminescence of Lampert complexes is given. From the spectral analysis it was possible to obtain 18 phonon energies. Recall that for radiative recombination, having the conduction-band minimum for 4H SiC at the M point on the 4H SiC Brillouin zone boundary, we should be able to see 24 phonons. As mentioned previously, in our latest 4H paper [50] we were able to resolve 20 phonon replicas. This 1965 paper on 4H SiC also gives an interesting discussion of symmetry and the large zone relevant to the polytypes of SiC.

Next, we ventured on to the rhombohedral polytype 21R or hkkhkkk in the Jagodzinski notation. Absorption measurements are taken from 4.2K to 300K. The shape of the edge shows that 21R SiC, as well as all other polytypes measured up to this time, is due to phonon assisted transitions and the creation of excitons. At 4.2K the exciton energy gap is found to be 2.853eV. The temperature dependence of the 21R exciton energy gap is given from 4.2K to 300K. The 21R SiC photoluminescence spectrum of the four particle nitrogen complexes taken at 6K shows six out of a possible seven no-phonon lines. Exciton biding energies of these sites are obtained by subtracting the measured energy of the lines from $E_{GX} = 2.853$. In principle, for each of the seven no-phonon lines there could be 42 phonon crystal momentum conserving phonons since there are 14 atoms per unit cell, but in fact far fewer lines are seen. 21R SiC crystals are not readily available and to our knowledge, no modern high-resolution emission spectrum is found in the literature.

The longest polytype that we were able to grow is 33R SiC (hkkhkkhkkhk) in the Jagodzinski notation or $(3332)_3$ in the Zhdanov notation. It has a slender rhombohedral cell, 83 Å long, along the trigonal axis. It belongs to space group $R3m$ and its unit cell contain 11 carbon atoms and 11 silicon atoms. The crystals used for these measurements were grown in a Lely furnace and are all from furnace runs, which produced mostly 6HSiC. In the paper [60] the absorption edge is given at 4.2K for light polarized $E \perp c$. The structure is characteristic of indirect exciton-creating transitions. At 4.2 K only the phonon emission structure is observed. The structure is relatively poorly resolved, and it is due to only a small number of the 66 (3 × 22) crystal momentum conserving phonons. It is similar to the absorption edges of 6H and 15R SiC. The 33R SiC absorption strength is everywhere intermediate between 6H SiC and 15R SiC. The four particle photoluminescence spectrum of 33R SiC, at 6K, has ten no-phonon lines due to complexes at ten of the 11 inequivalent nitrogen sites. In combination with the absorption spectrum the indirect exciton energy gap is found to be $E_{GX} = 3.003$eV. The missing no-phonon line, just as in 15R and 21R SiC, is likely to have too small a value of E_{GX} and is not stable even at liquid He temperatures.

Arguments are given that this 33R site is probably the same site missing in 15R SiC as in 21R SiC. Since there are 22 atoms in the 33R SiC rhombohedral unit cell each no-phonon line could have 66 momentum conserving lines associated with it. The observed intense emission lines in the spectrum involve only a few "principal" phonons, which once again suggests that we analyze the situation in a *large* zone. In the large zone the number of phonon branches is reduced to six. The six "principal" phonons all have the same k value as that of the conduction-band minima in the *large* zone. These six intense lines can be easily recognized because the energies of these lines, in many polytypes, is very nearly the same. The two no-phonon lines with the smallest E_{4X} generate most of the phonon replicas. The donor sites with the smallest E_{4X} have the largest capture cross sections for excitons. This is also observed for 4H, 6H, 15R, and 21R SiC. A comparison of the E_{4X} values of the principal phonons in 6HSiC, 33RSiC, and 15R SiC shows that the energy values are almost the same except in the case of the LA branch phonon where the 33R SiC phonon is an average between 6H and 15R. The no-phonon lines of 6H, 33R, and 15R are compared and certain of the 33R lines are associated with 6H and the remainder with 15R. A code is formulated for the three inequivalent donor sites of 6H SiC. Nitrogen donors in 6H SiC substitute on the carbon lattice sites. In the code, distances are measured from a particular donor to nearest like planes of Si atoms. By "like" is meant the letters A, B, or C in the Ramsdell notation. Nearest neighbors are always at 3 for all sites so they are eliminated from the code. The distances for each donor constitute the numbers of the code. For example, in 6H SiC we get the code numbers [9,15], [11,13], and [5,19]. By comparison of the codes for 6H, 33R, and 15R it is shown which 33R SiC no-phonon lines appear to be associated with 6H or 15RSiC. This simple code appears to work but it is not based on fundamental theory. Finally, it has to be pointed out that the so-called three-particle spectrum has been purposely left out of all our discussions. The data is fine, but the interpretation has been superceded by our discovery that this spectrum is actually due to titanium centers. The titanium spectrum in SiC will be discussed later on.

Contrary to the situation in stars the polytype, 2H SiC is not one of the common polytypes grown on earth by humans. However, in the sixties, Don Hamilton built special growth facilities at Westinghouse, in the fashion of Adamsky and Merz [61], and grew relatively pure, single crystals of 2H SiC. The largest of these slender crystals was about 3mm parallel to the c axis but with a cross section of no more than 0.3mm. The Adamsky-Merz crystals were black but of about the same size as the Westinghouse crystals. The Hamilton crystals on the other hand were completely clear and the level of the impurities was greatly reduced. Due to the geometry of these 2H SiC crystals only absorption measurements with $E{\perp}c$ polarization was feasible but for luminescence both polarizations were possible. In the case of optical absorption, one had to thread the incoming light through the length of the slender 2H SiC sample, mounted in a liquid He Dewar, and then focus the light coming out onto the entrance slit of the monochromator without unacceptable light fluctuations. This required the construction of a unique feedback apparatus, which would keep the light stably focused on the entrance slit of the monochromator. Results from this 2H SiC experiment can be found in *Physical Review* in March 1966 [62]. Absorption and luminescence measurements were carried out from about 2K to about 8K and they show that 2H SiC, in agreement with all other polytypes thus far measured, is an indirect semiconductor. The indirect exciton energy gap E_{GX} was found to be 3330eV, making it the largest of the SiC exciton energy gaps measured. The gap of 4H SiC, previously reported, is 3.265eV and very close to the gap of 2H SiC. This finding does not fit the previously discussed empirical relationship of polytype exciton energy band gaps as a function of hexagonality since 4H SiC is only 50% hexagonal and 2H SiC is 100% hexagonal. This strongly suggested that 2H SiC does not have the conduction-band minima at or very near the *M* points, on the Brillouin zone boundary, as do the other hexagonal polytypes. A detailed group theoretical analysis was made of the polarized luminescence spectra and selection rules were derived for the observed transitions. The observed spectrum was found to be consistent with the selection rules provided the conduction-band minima of 2H SiC are at the *K* points of the Brillouin zone. Modern band-structure calculations confirm that the conduction-band minima for 2H SiC are indeed at the *K* points of the Brillouin zone.

As we have seen, our 6H SiC absorption measurements have revealed an indirect interband edge near 3.0eV. Is there additional structure in the absorption edge? We were able to prepare samples by grinding and polishing to a thickness down to about 1.8μm. In the reported measurements [63], as grown crystals, with thicknesses ranging from 3.6μm to 200μm were used as well as the thinner ground and polished plates of 1.8μm. Absorption measurements were possible to almost 4.9eV where the absorption coefficient is estimated to be 4.6×10^4 cm^{-1}. The geometry of the 6H SiC samples made only $E \perp c$ measurements possible. The results show three successive indirect absorption edges at approximately 3.0eV, 3.7eV, and 4.1eV. Another edge is seen at 4.6eV but it was not positively identified as a direct or indirect transition.

During the sixties Westinghouse had a large program on laser development. Argon ion lasers were being built in house and Don Feldman and Jim Parker were part of this laser development program. Prior to lasers, Raman scattering was done with lamp sources and for SiC it proved to be very difficult to see more than a few Raman lines. Feldman and Parker had an argon ion laser working controllably at 4880Å with a power output of about 400mW. Two high quality 6H SiC crystals were cut and polished to 3 × 9 × 0.9mm and 3 × 6 × 0.8mm with the smallest dimension parallel to the c-axis. The Raman scattering measurements were all done at room temperature and reported in *Physical Review* [64]. Fifteen phonon lines were observed in the first-order Raman spectrum of 6H SiC. Polarized light was used to identify the mode symmetries. A large zone analysis was used to classify the modes. Plotting the results in this way one obtains what looks like dispersion curves. All the narrow lines in this Raman spectrum are consistent with the given interpretation. Only two of all the expected lines were not seen. Accurate values of the discontinuities in the large zone were shown to be 4–8cm^{-1} and were obtained by measuring the doublets in the Raman spectrum.

This work was quickly followed up by measuring the Raman scattering dispersion curves for 3C, 4H, 15R, and 21R SiC [65]. As already explained for 6H SiC, a new method for obtaining phonon dispersion curves is used, which is based on the property of polytypes. The property assumed to be common to all polytypes is the phonon spectrum in the axial direction derived from the first-order Raman spectra. In 4H SiC we observe nine one-phonon lines, in 15 SiC we observe 16 lines and in 21R SiC 14 lines. Polarization analysis was used to determine the symmetry type of each mode. The modes were further classified by the use of a standard large zone. The 4H, 15R, and 21R one-phonon lines together with the 15 lines already reported for 6H SiC were assigned to their positions in a single large zone. This yields six SiC dispersion curves in the axial direction, which look very much like those obtained much more recently by neutron diffraction or electron synchrotron inelastic scattering. From these dispersion curves it is possible to obtain transverse and longitudinal acoustic velocities, which are found to be in excellent agreement with measured sound velocities in SiC.

In 1969 we returned to cubic SiC to study several properties. Cubic SiC due to its simplicity is the easiest of the polytypes in which to arrive at interpretations. Unfortunately, growing cubic SiC crystals is a major challenge. In the 51 years that have elapsed since then much effort has been put into growing high quality 3C SiC but only with limited success. We compared the optical absorption of *n*-type samples of 3C SiC with that of relatively pure samples [66]. Measurements were carried out on samples that were vapor grown overgrowths of cubic layers on hexagonal crystals of SiC. These 3C overgrowths were deposited on the hexagonal substrates during the cool down phase of the Lely furnace. The cubic samples were prepared by grinding and polishing off the hexagonal layers. Various pressures of nitrogen gas were used in the Lely furnace during crystal growth. No electrical measurements could be made on the small cubic overgrowths but measurements on the hexagonal substrates show donor densities of $10^{17}cm^{-3}$ at the lowest nitrogen pressure and $10^{19}cm^{-3}$ at the highest nitrogen pressure. No sign of donor doping could be seen on layers doped at the $10^{17}cm^3$ nitrogen level. The absorption coefficient α (cm^{-1}) at 300K was obtained from a photon energy of about 0.7eV to 3.5eV. However, measurements were also made at 4.2K. The heavily doped cubic SiC showed a free carrier absorption with a $\lambda^{2.8}$ wavelength dependence. An additional narrow band was seen near 3.1eV and shown to be a direct transition from the X_{1c} conduction-band minimum to a higher X_{3c} conduction-band level. These transitions are now known as Biedermann band transitions. Higher indirect absorption edges in cubic SiC were also measured at this time [67]. In earlier measurements on cubic SiC we identified the first indirect transition at 2.39eV as X_{1c}-Γ_{15v}. and now a second indirect transition absorption edge is found at 3.55eV. Band calculations by Herman, Van Dyke, and Kortum that same year [68] suggest that this transition is $X_{1c} - L_{3V}$. Another absorption edge is found at 4.2eV but theory suggested an indirect edge at 5.6eV, so the 4.2eV absorption edge was not identified. The absorption coefficient for cubic SiC at 5.0eV was determined to be about 2.4×10^4 cm^{-1}.

Donor acceptor pairs in 3C SiC were first observed by us in 1963 and were attributed to N-Al pairs [57]. However, at the time it was seemingly impossible to give a satisfactory analysis of the data. Donor acceptor (DA) pair luminescence consists of many sharp lines and this is well illustrated by the Bell Labs work on DA pairs in GaP [69]. More importantly, the theoretical interpretation of DA pairs given by John Hopfield was critical for our understanding of the 3C SiC DA pair spectrum reported in 1970 [70]. In 1963 we used a mercury arc source to excite the 3C SiC crystals but in 1970 we had the benefit of an argon ion laser and that yielded a 3C SiC DA spectrum, at 1.8K, quite comparable to the ones obtained by the Bell group in GaP. The 3C SiC spectrum obtained is type II and attributed to N-Al pairs. Lines were sharp enough, so they could be resolved up to the 80th shell. van der Waals and multipole interactions were evaluated quite similarly as was done for the pair spectra in GaP. The limiting photon energy for distant pairs in 3C SiC was found to be $h\nu_\infty$ = 2.9034eV. Interestingly enough, some 40 years later Professor Ivanov in Linköping, Sweden, asked me whether I still had the 3C SiC crystal on which the N-Al DA spectra were obtained. He wanted to evaluate the binding energy of phosphorus in 3C SiC by using donor-acceptor pairs. Fortunately, I was able to locate the sample and send it to him. He obtained $h\nu_\infty$ by fitting the spectrum with a computer algorithm. To our pleasure he also obtained exactly the same value for $h\nu_\infty$, namely 2.9034eV.

We have seen donor-acceptor pair spectra in 4H, 6H, and 15R SiC but evaluating the spectra in these polytypes is a greater challenge.

In 1954 William Shockley submitted a patent application on ion implantation in semiconductors [71]. However, industry showed little interest in ion implantation until after 1972 and the introduction of the hand-held calculator HP-35 by Hewlett-Packard. The HP-35 used ion implanted semiconductor circuit elements and proved to be a huge success. We were interested in the effect of implanting ions in SiC as early as 1960 with a special focus on lattice damage and annealing. An ion implanter was built in the mid-60s out of available parts. Since the Westinghouse Si semiconductor engineers were not interested in ion implantation, we were able to apply our implanter strictly to SiC. Our first publication on radiation defects in cubic SiC or 3C was published in *Physical Review* in 1971 [71]. Cubic SiC was bombarded with 120keV He $^+$ ions (10^{14} cm^{-2} fluence) and a van de Graaff accelerator was used for electron bombardments (6×10^{17} cm^{-2} fluence at 1MeV). Annealing was done in an argon atmosphere. After an anneal at about 1000^{0}C a strong new spectrum was observed and later named D$_1$ to distinguish it from another intrinsic defect spectrum found to appear after a 1300°C anneal. It was named D$_2$. We will discuss the D$_2$ spectrum in the following paragraph. Both D$_1$ and D$_2$ spectra were found to persist to at least 1700°C. The luminescence spectrum of D$_1$, at 1.5^{0}K, shows a new vibrionic spectrum with a number of resonant modes, and a strong localized mode in the gap between the acoustic and optic phonon branches. It does not depend on the implant species and is likely a pure defect complex. The spectrum has an unusual temperature dependence with two abrupt changes below 13^{0}K. The changes between 1.3^{0}K (6279.8Å) and 13^{0}K (6269.5Å) were attributed to the Jahn-Teller effect and the distortion is accounted for by coupling with the strong 66.5meV resonant mode.

As already mentioned, after the implanted cubic samples were annealed to a temperature of 1300°C another new intrinsic defect spectrum D$_2$ appeared [72]. Both D$_1$ and D$_2$ spectra are much stronger in samples irradiated with ions rather than electrons. This is likely due to the fact that the much higher defect concentration formed by the bombarding ions favors the formation of centers with two or more defects. Both spectra survived a 1700°C anneal and both spectra are independent of the ions used for bombardment. Thus, these luminescent centers are termed intrinsic. These centers are formed from Si or C vacancies or interstitials. The high temperature of formation indicates that they are likely to be complexes. The simplest intrinsic complexes are di-vacancies and di-interstitials. A striking feature of the D$_2$ luminescence is the number and strengths of vibrational luminescent lines with frequencies above the lattice limit of about 120meV. The highest phonon replica found is 164.7meV. Interestingly enough this is very close to the highest lattice mode in diamond. The acoustical sidebands of this spectrum are very weak but there is more strength in the optical sidebands and the maximum intensity is reached in the high energy localized modes. This intensity pattern of D$_2$ has also been seen in a number of other polytypes. Beyond the 164.7meV localized mode of D$_2$ there is no sharp structure until the beginning of the two-phonon region at about 200meV. In thinking about a model, we note that high energy localized modes are generated by light atoms and require the lattice force constants to be stiffened by interstitial atoms rather than be weakened by the presence of vacancies. This led to the suggestion that the D$_2$ center is the carbon di-interstitial.

The next step was to see what the nature of the D$_1$ center would be like in a lower symmetry environment such as in the 6H polytype [73]. Radiation effects were introduced in the same manner as in the previously discussed measurements on 3C SiC [71]. The created defects in 6H SiC produce a low temperature luminescence that is independent of the implanted ion and one sees a new D$_1$ spectrum, which persists beyond a 1700°C anneal. A comparison of irradiation damage creation by ion and electron bombardment shows that the spectral intensity is strongly dependent on the defect concentration. This suggests that the defect center is an intrinsic defect and possibly a divacancy. In 6H SiC one sees three no-phonon lines because 6H SiC has three inequivalent sites. There is also a strong vibrionic structure with localized and resonant modes. The 6H SiC temperature dependence is different from that found in 3C SiC. We have a 1.4^{0}K low temperature D$_1$ spectrum, which is quenched while the high temperature 77^{0}K spectrum is activated. A model suggests that the abrupt change of the spectrum is due to a low temperature lattice distortion.

The D_1 and D_2 spectra were also investigated in 15R SiC and reported at the Defects in Semiconductor Conference in Reading, England [74]. In 15R SiC there are five inequivalent Si or C lattice sites and the non-axial divacancy model proposed in [73] suggests that there should be five L spectra and five H spectra. In fact, at 1.3^0K only four L spectra are observed. Recall that in rhombohedral polytypes one observes one fewer no-phonon lines than the number of inequivalent lattice sites. Not seeing the fifth L component of D_1 in 15R SiC is likely due to the fact that the exciton binding energy of the fifth site is simply too small to bind the exciton. The exciton binding energies vary from 361meV to 429meV. At 43^0K most of the L spectra have quenched and they are replaced by three H spectra. L_1 is quite weak at 1.3^0K and possibly H_1 is not seen for that reason. The 15R SiC, L or H, one-phonon sidebands are similar to the sidebands seen in 6H SiC. In 15R SiC the D_2 spectrum is similar to what was observed for 3C SiC [73] after an anneal of 1550°C. However, looking at the spectrum at 1.3^0K, after an 1550°C anneal, we see only two no-phonon lines with relatively small binding energies of 52meV and 59meV. Looking at the spectrum where the no-phonon line has an exciton binding energy of 59meV, the high energy localized modes are once again the dominant spectral features. The D_2 phonon energies in 15R SiC are similar to those in 3C SiC. In 3C SiC the identification of the one-phonon lines was confirmed by the observation of a well resolved two phonon spectrum. The very large exciton binding energies found for D_1 and the rather moderate exciton binding energies found for D_2 are seen to be comparable in all the polytypes that have been studied thus far. One finds that the characteristic D_1 and D_2 phonon structure, in all polytypes studied, is characteristic of the particular center and practically independent of polytype.

In the early 1970s it was a great puzzle why hydrogen had not been detected in covalent semiconductors. After all, hydrogen was used in some way in all growth procedures. We decided to implant hydrogen and deuterium into SiC single crystals and clarify the situation in SiC. H and D atoms could produce very high frequency local modes. Our first report on such implantations was on 6H SiC [75]. Implantation of H (protons) into 6H SiC leads to luminescence centers having vibrational modes of 369meV. This is more than three times the SiC lattice limit of 120.5meV. In the case of deuterium implants we got vibrational modes of 274meV. For H and D, respectively, the energies are seen to be those of the CH and CD vibrational stretching modes well-known from the infrared spectra of organic molecules. Annealing the samples to 800°C increases the luminescence intensity. At the higher annealing temperatures, H and D diffuse to vacant Si sites where H and D can bond with one of the neighboring carbon sites to form the CH or CD centers.

As a further check on this idea we grew 6H SiC in a hydrogen atmosphere. The samples were then implanted with other than H or D ions to produce vacancies and after annealing to 800°C a CH bond-stretching mode was observed.

We next implanted 15R SiC with 150keV protons and deuterons with a fluence of 3×10^{14} H or D / cm^2 [76]. Quite like in 6H SiC, an 800°C anneal produces a spectrum at 1.3^0K whose vibronic part of the spectrum has sharp peaks corresponding to momentum conserving phonons and far beyond the lattice limit of 120.5meV there is a strong line displaced from the no-phonon line by 369meV. Implantation by deuterons yields a high frequency mode at 273meV. As in 6H SiC we recognize these high frequency modes to be CH and CD stretching mode frequencies. In 6H SiC we had three CH and CD spectra because 6H has three inequivalent sites.

However, in 15R SiC we have five inequivalent sites, but we see only two CH and CD spectra. In 15R SiC the excitons are bound to the two visible centers by 4.0 and 43meV and most likely the exciton cannot bind to the center in the remaining three sites. The lower-frequency bond bending frequencies were not seen, again quite like in 6H SiC. After a

1200°C anneal the CH and CD spectra are no longer observed in the 1.3⁰K luminescence and in fact the spectrum now seen is the one normally seen for preimplanted 15R SiC samples. It is reasonable to assume that at 1200°C the H and D atoms are unable to stabilize the single vacancies and that is why the observed CH and CD spectra vanish with a 1200°C anneal.

After reporting preliminary accounts of the CH and CD spectra in 6H and 15R SiC it was time for a more detailed look at the 6H SiC H and D spectra [77]. A model is derived from an annealing study and by analysis of three components of the phonon spectrum (a) 370meV stretching mode, (b) two Si vacancy localized modes, and (c) many momentum conserving modes. When looking at the CD spectrum one sees that there is a large change in the stretching mode from 370meV to 274meV and the change is small for the localized modes and nearly zero for the momentum conserving phonons. Thermally excited states were seen at various temperatures up to 26⁰K. The H and D luminescence at 1.3⁰K is by far the most efficient luminescence we have observed in SiC and about 100 times more efficient than the luminescence from neutral-donor N centers. The extra electron present in the excited state of the nitrogen donor center leads to radiation-less Auger recombination. The absence of an extra electron in the H and D center is the principal factor for achieving the high efficiency. Finally, it should be noted that by this time H and D luminescence had been observed in 6H,15R 4H but not in 3C SiC. The usual defect is likely formed in 3C SiC, but the center does not appear to bind an exciton. In addition, high resolution studies of the CH and CD spectra showed two components in each no-phonon line. Each no-phonon line in the H and D spectra is now observed to have two components, with intensities approximately in the three to one ratio of non-axial to axial centers. The spectral separation between H_1, H_2, and H_3 varies from 0.18 to 0.36meV. This discovery forced a modification of the model, which is given in [78]. The dominant spectrum at 1.3⁰K is now renamed the primary spectrum. The other spectrum, strong between 3⁰K and 30⁰K, formerly called axial is now termed the secondary spectrum. Since both spectra depend on annealing in the same way suggests, that they should be assigned to two different forms of the same center. Now having found two spectra under high resolution and a model predicting two forms of the center, symmetry evidence is used to match spectra and luminescence centers. A possible mechanism that can account for the difference between the primary and secondary spectrum is that we are dealing with different charge states of the same center.

The last polytype in which we studied H and D implantation is 4H SiC, $E_{GX} = 3.265eV$ [79]. In 4H SiC only the primary spectrum was found, which has a phonon structure characteristic of exciton recombination at a neutral center. 4H SiC has two crystal-graphically inequivalent centers with large binding energies of 110 and 126meV. A striking new feature was the discovery of rapid quenching upon UV irradiation in one of the centers of the H-implanted 4H SiC. This quenching was *not* found for deuteron implanted samples. In the hydrogen implanted samples, at 4.2⁰K, one of the centers is efficient during the first minute of UV exposure and then fades out completely in the next hour. The sample can be completely rejuvenated by warming it up to room temperature. The quenching mechanism is believed to be due to a change in the charge state of this center. In other ways, the 4H SiC luminescence spectrum is similar in many ways to the primary spectrum of H- and D-implanted 6H SiC. The very large exciton binding energies in 4H SiC do lead to some significant differences with the results found for H- and D-implanted 6H SiC.

To top off the CH and CD measurements in 6H, 15R, 4H, and 3C SiC we did high resolution magneto-optical measurements on H-implanted 6H SiC [80]. Magneto-optical measurements were made at 1.5, 4.2, and 27⁰K on the primary and secondary spectra of the hydrogen implanted and annealed 6H SiC. The magnetic splitting found for the secondary spectra are of a kind predicted by effective mass theory as developed by Thomas and Hopfield [81] and giving g_h = 2.8 and g_e = 1.8. The primary spectrum, as we have already discussed, is due to the recombination of an exciton strongly bound to the same center but in a neutral state. The observed magnetic splitting can be explained by a strong-exchange model, with the singlet-exciton level 6.5meV above the triplet level. The triplet level is found to have a zero-field splitting of 0.2meV. This is due to a preferred spin alignment along the axial direction. Changing the angular position of the 34kG magnetic field causes strong variation in the intensity of different spectral lines associated with the center. The results appear to be explainable by use of the model. It has now also been found that in 6H SiC as well as 4H SiC we have quenching of the luminescence when irradiating with UV light. However, the speed at which centers quench in 6H SiC, as compared to the one center in 4H SiC, is very much reduced. It is likely that a third charge center is responsible for the slow quenching in 6H SiC. A two-meter Baird atomic grating spectrometer, operated in second order, and using a 1200 line/mm grating, blazed at 1.0μm, was used to obtain the high-resolution spectra required for these magnetic measurements. Such high resolution enabled us to resolve the axial-nonaxial pairs that are predicted by the model and actually at 4.2K we clearly resolve the doublet structure in each of the seven no-phonon doublets.

A very bright and prominent part of the 6H luminescence spectrum with deep no-phonon lines was named the ABC spectrum in 1963 [82]. A similar spectrum was also found in polytypes 4H, 15R, and 33R but not in 21R or 3C SiC. Originally the ABC spectrum was assigned to ionized nitrogen donors [82].

The earliest doubts about the role of nitrogen in the ABC spectrum followed our study of the 4H SiC low temperature luminescence [58]. We deduced from the data that the nitrogen ionization energies in 4H SiC, which the model generated, were simply too large to believe. However, it was clear that the spectrum was due to the same center as the ABC spectrum in 6H SiC [82]. In 1972, Dean and Hartman [83] published a magneto-optical study of the ABC spectrum in 6H SiC. As interpreted by Patrick [84], the Dean and Hartman data showed that the luminescence could be attributed to a substitutional atom but clearly something other than nitrogen. As part of the Dean and Hartman experiment their use of high-resolution equipment enabled them to observe two satellite lines on either side of the A, B, and C lines. They concluded that the satellite lines arose from *isotope* shifts of the no-phonon lines. It is important to recall that Ti has five isotopes whose abundances are consistent with the intensities of the main line and the four satellite lines seen by Dean and Hartman. Other evidence to point away from associating the ABC spectrum with the nitrogen donor were the electrical measurements on *n*-type 6H SiC by Hagen and Kapteyns [85]. They found the ionization energy associated with the nitrogen donor in 6H SiC to be less than 100meV, which is a good bit smaller than any of the three binding energies attributed to nitrogen in [82]. In 1973 Hagen and van Kemanade [86] grew 6H SiC crystals at Philips with the isotope ^{15}N and found that the ABC spectrum was unchanged. Finally, Hopfield had shown in 1964, in his theory of exciton binding [87], that an exciton bound to an ionized donor was possible only when the electron-hole mass ratio is small, and then only with very small binding energy. No other spectrum has been found in SiC that can be attributed to the ionized nitrogen donor and so it seems likely that the bound state of the ionized nitrogen donor in SiC does not exist. All this evidence and the fact that Ti is a common impurity in SiC convinced us at that point to attribute the ABC lines to the presence of the five Ti isotopes. In addition, in 1974, van Kemenade and Hagen [88] grew a sample of 6H SiC, which was intentionally doped with the Ti isotope, ^{46}Ti, and they recorded its luminescent spectrum, which further confirmed the assignment of Ti to the

ABC spectrum. We summarized [89] the Ti photoluminescence in 4H (E_{GX}=3.265eV), 6H (E_{GX}=3.023eV), 15R (E_{GX}=2.986eV), and 33R SiC (E_{GX}=3.003eV) and used Patrick's model of the isoelectronic Ti center to explain why all the no-phonon lines of the ABC spectrum fall in the narrow energy range of 2.79 to 2.86 despite much greater variation in poly-type exciton energy gaps. The absence of the ABC spectrum in 3C (E_{GX}=2.390eV) and 21R (E_{GX}=2.853eV) SiC is also consistent with Patrick's model. To further confirm the presence of Ti in 4H SiC high resolution, 4.2K, photoluminescence measurements were carried out on the no-phonon line A at 4350Å, using a Jarrell-Ash spectrometer with a dispersion of 0.4mm/Å and using 20µm slit widths. The weak ^{49}Ti line was not resolved. The identification of the isotope lines follows the rules established for 6H [88], that lower mass lines fall at lower energies. The separation between successive isotopes is 0.13meV. The relative line intensities could not be measured accurately on a 103aF photographic plate, but they appear to be in accord with the natural isotope abundances of Ti.

4 Final Thoughts

The reader may well wonder why we stopped at the end of 1974? There are a number of factors that converged at this time to basically stop SiC research. The mid-70s were a time of double-digit inflation, which prompted industry management to argue, wrongly, that a time to make a profit from fundamental studies was over. Furthermore, crystal production of semiconductor purity SiC was still limited to the use of Lely furnaces and that was not a mass production technology. Si technology had reached maturity and the semiconductor wise men argued that Si could handle anything in the future. Finally, the US Government agencies no longer felt that SiC was a new and exciting field and felt that industry should now carry the load. It was clearly the end of the initial phase of SiC research and development in this country.

Luckily new methods for large SiC wafer growth came along in the late seventies and homo-epitaxial growth of SiC made substantial progress in the mid-80s. In addition, Si could not handle all the problems of the Universe, and new spirits in Universities, the US Government and a few industries saw new possibilities for high temperature power devices and substrates for LED lamps. Now in 2021, we are approaching a new peak for a very impressive and worldwide industrial development of SiC. It is precisely what the exciting articles in this Handbook are meant to promote. Happy reading!

References

1. George Gamow, "Expanding Universe and the Origin of Elements", Physical Review 70, 572–573 (1946).
2. R.A. Alpher, H. Bethe, and G. Gamow, "The Origin of Chemical Elements" Physical Review 73, No.7, 803–804, (1948).
3. F. Hoyle, "Synthesis of the Elements from Carbon to Nickel", Astrophysical Journal (Suppl. 1), 121–146 (1954).
4. E. Margaret Burbidge, G.R. Burbidge, William A. Fowler, and Fred Hoyle, "Synthesis of the Elements in Stars", Reviews of Modern Physics, 29, No. 4, 547 (1957).

5. A.G.W. Cameron, "Stellar Evolution, Nuclear Astrophysics, and Nucleogenesis", CRL 41, Chalk River, Ontario, (1957).

6. J.H. Reynolds, "Determination of the Age of the Elements", Physical Review Letters, 4, No. 1, 8 (1960).

7. Donald D. Clayton, "22Ne, Ne-E, extinct radioactive anomalies and unsupported 40Ar", Nature 257, 36–37 (1975).

8. B. Srinivasan and Edward Anders, "Noble Gases in the Murchison Meteorite: Possible Relics of s-Process Nucleosynthesis", Science 201, (4350), 51–56, (1978).

9. Kevin D. McKeegan, "Ernst Zinner, lithic astronomer", Meteoritics & Planetary Science 42, No. 7/8, 1045–1054, (2007).

10. Thomas Bernatowicz, Gail Fraundorf, Tang Ming, Edward Anders, Brigitte Wopenka, Ernst Zinner, and Phil Fraundorf, "Evidence for Interstellar SiC in the Murray Carbonaceous Meteorite", Nature 330, 24/31, 728–730, (1987).

11. Ernst Zinner, Tang Ming, and Edward Anders, "Large Isotopic Anomalies of Si,C,N and Noble Gases in Interstellar Silicon Carbide from the Murray Meteorite", Nature 330, 24/31, 730–732, (1987).

12. Tang Ming, Roy S. Lewis, Edward Anders, M.M. Grady, I.P. Wright and C.T. Pillenger, "Isotopic Anomalies of Ne, Xe, and C in Meteorites. I. Separation of Carriers by Density and Chemical Resistance", Geochimica et Cosmochimica Acta 52, No. 5, 1221–1234, (1988).

13. Tang Ming and Edward Anders, "Isotopic Anomalies of Ne, Xe, and C in meteorites. II. Interstellar diamond and SiC: Carriers of exotic noble gases" Geochimica et Cosmochimica Acta 52, No. 5, 1235–1244, (1988).

14. Tang Ming and Edward Anders, "Isotopic Anomalies of Ne, Xe, and C in Meteorites. III. Local and Exotic Noble Gas Components and their Interrelations", Geochimica et Cosmochimica Acta, 52, No. 5, 1245–1254, (1988).

15. Thomas J. Bernatowicz and Robert M. Walker, "Ancient Stardust in the Laboratory", Physics Today, 50, No. 12, 26–32 (1997).

16. Edward Anders and Ernst Zinner, "Interstellar Grains in Primitive Meteorites: Diamond, Silicon Carbide and Graphite, Meteoritics 28, 490–514, (1993).

17. Ernst Zinner, "Stellar Nucleosynthesis and the Isotropic Composition of Presolar Grains from Primitive Meteorites", Annual Review of Earth and Planetary Science, 26, 147–188, (1998).

18. Larry R. Nittler and Fred Ciesla, "Astrophysics with Extraterrestrial Materials", The Annual Review of Astronomy and Astrophysics, 54, 53–93, (2016).

19. T.L. Daulton, T.J. Bernatowicz, R.S. Lewis, S. Messenger, F.J. Stadermann and S. Amari, "Polytype Distribution of Circumstellar Silicon Carbide: Microstructural Characterization by Transmission Electron Microscopy", Geochimica et Cosmochimica Acta, 67, No. 24, 4743–4767, (2003).

20. E. Zinner, M.Tang, and E. Anders, "Interstellar SiC in the Murchison and Murray Meteorites: Isotopic Compositions of Ne,Xe, Si, C, and N", Geochimica et Cosmochimica Acta 53, 3273–3290, (1989).

21. Sachiko Amari, Roy S. Lewis, and Edward Anders, "Interstellar Grains in Meteorites: I. Isolation of SiC, Graphite and Diamond; Size Distributions of SiC and Graphite", Geochimicha et Cosmochimica Acta 58, 459–470, (1994).

22. Roy S. Lewis, Sachiko Amari, and Edward Anders, "Interstellar Grains in Meteorites: II. SiC and its Noble Gases", Geochimica et Cosmochimica Acta 58, 471–494, (1994).

23. E. Zinner, S. Amari, E. Anders, and R. Lewis, "Large Amount of Extinct 26Al in Interstellar Grains from the Murchison Meteorite", Nature 349, 51–54, (1991).

24. S. Amari, E. Zinner, and R.S.Lewis, "Isotopic Compositions of Different Presolar SiC Size Fractions from the Murchison Meteorite", Meteor. Planet. Sci. 35, 997–1014, (2000).

25. T.R. Ireland, E.K. Zinner, and S.Amari, "Isotopically Anomalous Ti in Presolar SiC from the Murchison Meteorite", Astrophysical Journal, 376, L53-L56, (1991).

26. F.A. Podosek, C.A. Pombo, S. Amari, and R.S. Lewis, "s-Process Sr Isotopic Compositions in Presolar SiC from the Murchison Meteorite", The Astrophysical Journal, 605, 960–965, (2004).

27. G.K. Nicolussi, A.M. Davis, M.J. Pellin, R.S. Lewis, R.N. Clayton, and S. Amari, "s-Process Zr in Presolar SiC Grains", Science 277, 1281–1283, (1997).

28. G.K. Nicolussi, M.J. Pellin, R.S. Lewis, R.N. Clayton, and S. Amari, "Mo Isotopic Composition of Individual Presolar SiC Grains from the Murchison Meteorite", Geochimica et Cosmochimicha Acta 62, 1093–1104, (1998).

29. U. Ott and F. Begemann, "Discovery of s-Process Barium in the Murchison Meteorite", Astrophysical Journal 353, L57–L60, (1990).

30. E. Zinner, S. Amari, and R.S.Lewis, "s-Process Ba, Nd, and Sm in Presolar SiC from the Murchison Meteorite", Astrophysical Journal 382, L47–L50, (1991).

31. P. Hoppe and U. Ott, "Mainstream SiC Grains from Meteorites. In *Astrophysical Implications of the Laboratory Study of Presolar Materials* (eds. T.J. Bernatowicz and E.K. Zinner), pp. 27–58, AIP, New York (1997).

32. S.A. Singerling, N. Liu, L.R. Nittler, C.M. O'D Alexander, and R.M. Stroud, "TEM Analysis of Unusual Pre-solar Silicon Carbide: Insights into the Range of Circumstellar Dust Condensation Conditions", The Astrophysical Journal, 913, 90 (65pp), (2021).

33. Phillip R. Heck, Jennika Greer, Levke Kööp, Reto Tappitsch, Frank Gynyard, Henner Busemann, Colin Maden, Janaina N. Ávila, Andrew M. Davis and Rainer Wieler, "Lifetimes of Interstellar Dust from Cosmic Ray Exposure Ages of Presolar Silicon Carbide", PNAS 117, No.4, 1884–1889, (2020).

34. J.J. Berzelius, "Untersuchungen über Flussspathsaüre und deren merkwurdingsten Verbindungen" Annalen der Physik, B. 77. St. 1, p. 205 (1824). ST.5.

35. Eugene H. Cowles, Alfred H. Cowles, and Charles F.Mabery, "On the Electrical Furnace and the Reduction of Oxides of Boron, Silicon, Aluminum and other Metals by Carbon", Scientific American, September 26, 1885.

36. E.G. Acheson, "Production of Artificial Crystalline Carbonaceous Materials", U.S. Patent #492,767, Patented Feb 28, 1893.

37. E.G. Acheson: "Carborundum: its History, Manufacture and Uses" Jour. Franklin Institute, 194–203, Sept. 1893 and 279–289, Oct. 1893.

38. F. Becke, "XXIV. Beitrag Zur Kenntniss der Carborundum-krystalle CSi.", Z. Kristallogaphie, 24, 537 (1895).

39. William W. Coblentz, "Investigations of Infra-red Spectra, Part IV-Infra-Red Reflection Spectra", Carnegie Institution of Washington, pp. 94–95, Dec. 1906.

40. Henry Joseph Round, "A Note on Carborundum", Electrical World, p309, February 9, 1907.

41. Heinrich Baumhauer, "VII. Über die Krystalle des Carborundums", Zeitschrift für Kristallographie 50, 33–39, (1912).

42. H. Baumhauer, "XII. Über die verschiedenen Modifikationen des Carborundums und die Erscheinung der Polytypie" Zeitschrift für Kristallographie 55, 249–259, (1915).

43. Oskar Weigel, "Über einige physikalische Eigenschaften des Carborands", Nachrichten der Königlichen Gesellschaft der Wissenschaften zu Göttingen, Math.-phys Klasse, 264–298, (1915).

44. C.L. Burdick and E.O. Owen, "The Atomic Structure of Carborandum Determined by X-Rays", The Journal of the American Chemical Society, Vol. XL, No. 12, 1699–1759, (1918).

45. Egon E. Loebner, "Sub-histories of the Light Emitting Diode", IEEE Transactions on Electron Devices, Vol. ED-23, No. 7, 675, (1976).

46. John Bardeen, F. J. Blatt and L. H. Hall, in Proceedings of the Conference on Photoconductivity, Atlantic City, November 4–6 (1954), edited by R.G. Breckenridge, B.R. Russel, and E.E. Hahn (John Wiley and Sons, Inc., New York, (1956), p. 146.

47. W.J. Choyke and Lyle Patrick, "Absorption of Light in Alpha SiC near the Band Edge", Phys. Rev. 105, No. 6, 1721–1723, (1957).

48. G.C. Macfarlane and V. Roberts, Phys. Rev. 97, 1714 (1955); 98, 1865 (1955).

49. Walter M. Klahold, Wolfgang J. Choyke, and Robert P. Devaty, "Newly Resolved Phonon-Assisted Transitions and Fine Structure in the Low Temperature Wavelength Modulated

Absorption and Photoluminescence spectra of 6H SiC", Material Science Forum, 963, 341–345, (2019).

50. W.H. Klahold, W.J. Choyke, and R.P. Devaty, "Band Structure Properties, Phonons, and Exciton Fine Structure in 4H SiC Measured by Wavelength-Modulated Absorption and Low-Temperature Photoluminescence", Phys. Rev. B 102, 205203, (2020).

51. Heckrert R. Philipp, "Intrinsic Optical Absorption in Single Crystal Silicon Carbide", Phys. Rev. 111, 440, (1958).

52. Lyle Patrick and W.J. Choyke, "Electron Emission from Breakdown Regions in Sic p-n Junctions", Phys. Rev. Lett. 2, No. 2, 1–2, (1959).

53. W.J. Choyke and Lyle Patrick, "Exciton and Interband Absorption in SiC", Proceedings of the International Conference on Semiconductor Physics, Prague, 1960 (Czechoslovakian Academy of Sciences, Prague, 1961) p. 432.

54. Lyle Patrick and W.J. Choyke, "Lattice Absorption Bands in SiC", Phys. Rev. 123, No. 3, 813–815, (1961).

55. W.J. Choyke and Lyle Patrick, "Exciton Recombination Radiation and Phonon Spectrum of 6H SiC", Phys. Rev., 127, No. 6, 1868–1877, (1962).

56. Lyle Patrick, D.R. Hamilton, and W.J. Choyke, "Optical Properties of 15R SiC: "Luminescence of Nitrogen-Exciton Complexes, and Interband Absorption", Phys. Rev., 132, No. 5, 2023–2031, (1963).

57. W.J. Choyke, D.R. Hamilton, and Lyle Patrick, "Optical Properties of Cubic SiC: Luminescence of Nitrogen-Exciton Complexes, and Interband Absorption", Phys. Rev. 133, No 4A, A1163-A1166, (1964).

58. W.J. Choyke, L. Patrick and D.R. Hamilton, "Optical Properties of 4H SiC: Absorption and Luminescence", Proc. 7th Int. Conf. on Physics of Semiconductors, Dunod, Paris. (1964) pp. 751–758.

59. Lyle Patrick, W.J. Choyke, and D.R. Hamilton, "Luminescence of 4H SiC, and Location of Conduction-Band Minima in SiC Polytypes", Phys. Rev. 137, No. 5A, (1965).

60. W.J. Choyke, D.R. Hamilton and Lyle Patrick, "Exciton Complexes and Donor Sites in 33R SiC", Phys. Rev. 139, No. 4A, A1262–A1274, (1965).

61. R.F. Adamsky and K.M. Merz, Z. Krist. 111, 5 (1959).

62. Lyle Patrick, D.R. Hamilton, and W.J. Choyke, "Growth, Luminescence, Selection Rules, and Lattice Sums of SiC with Wurtzite Structure", Phys. Rev. 143, No. 2, 526–536, (1966).

63. W.J. Choyke and Lyle Patrick, "Higher Absorption Edges in 6H SiC", Phys. Rev. 172, No. 3, 769–772, (1968).

64. D.W. Feldman, James H. Parker, Jr., W.J. Choyke, and Lyle Patrick, "Raman Scattering in 6H SiC", Phys. Rev. 170, No. 3, 698–704, (1968).

65. D.W. Feldman, James H. Parker, Jr., W.J. Choyke, and Lyle Patrick, "Phonon Dispersion Curves by Raman Scattering in SiC, Polytypes 3C, 4H, 6H, 15R, and 21R", Phys. Rev. 173, No. 3, 787–793, (1968).

66. Lyle Patrick and W.J. Choyke, "Optical Absorption in n-Type Cubic SiC", Phys. Rev. 186, No. 3, 775–777, (1969).

67. W.J. Choyke and Lyle Patrick, "Higher Absorption Edges in Cubic SiC", Phys. Rev. 187, No. 3, 1041–1043, 1969.

68. F. Herman, J.P. Van Dyke, and R.L. Kortum, Mater. Res. Bull. 4, S167 (1969).

69. J.J. Hopfield, D.G. Thomas, and M. Gershenson, "Pair Spectra in GaP", Phys. Rev. Letters 10, 162 (1963).

70. W.J. Choyke and Lyle Patrick, "Luminescence of Donor Acceptor Pairs in Cubic SiC", Phys. Rev. B, 2, No. 12, 4959–4965, (1970).

71. W.J. Choyke and Lyle Patrick, "Photoluminescence of Radiation Defects in Cubic SiC: Localized Modes and Jahn-Teller Effect", Phys. Rev. B, 4, No. 6, 1843–1847, (1971).

72. Lyle Patrick and W.J. Choyke, "Localized Vibrational Modes of a Persistent Defect in Ion Implanted SiC", J. Phys. Chem. Solids, 34, 565–567, (1973).

73. Lyle Patrick and W.J. Choyke, "Photoluminescence of Radiation Defects in Ion-Implanted 6H SiC", Phys. Rev. B, 5, No. 8, 3253–3259, (1972).

74. W.J. Choyke and Lyle Patrick, "Intrinsic Luminescent Defects in 15R SiC Induced by Ion Bombardment and Annealing", Proc. Int. Conference on Defects in Semiconductors. Reading, UK, 218–222, (1972).

75. W.J. Choyke and Lyle Patrick, "CH and CD Bond Stretching Modes in the Luminescence of H-and D-Implanted SiC", Phys. Rev. Lett., 29, No. 6, 355–356, (1972).

76. W.J. Choyke and Lyle Patrick, "Recombination Radiation of Excitons in H or D implanted 15R SiC", Proc. 11th Int. Conf. on the Physics of Semiconductors (PWN – Polish Scientific Publishers, Warsaw (1972)), 177–182.

77. Lyle Patrick and W.J. Choyke, "Efficient Luminescence Centers in H and D Implanted 6H SiC", Phys. Rev. B 8, No. 4, 1660–1669, (1973).

78. Lyle Patrick and W.J. Choyke, "Luminescence Centers in H- and D- Implanted 6H SiC", Phys Rev B 9, No. 4, 1997, (1974).

79. W.J. Choyke and Lyle Patrick, "Photoluminescence of H- and D-implanted 4H SiC", Phys. Rev. B 9, No. 8, 3214–3219, (1974).

80. W.J. Choyke, Lyle Patrick, and P.J. Dean "Magneto-Optical Measurements on H-Implanted 6H SiC", Phys. Rev. B 10, No. 6, 2554–2565, (1974).

81. D.G. Thomas and J.J. Hopfield, "Optical Properties of Bound Exciton Complexes in Cadmium Sulfide", Phys. Rev. 128, No. 5, 2135–2148, (1962).

82. D.R. Hamilton, W.J. Choyke, and Lyle Patrick, "Photoluminescence of Nitrogen-Exciton Complexes in 6HSiC", Phys. Rev. 131, No. 1, 127–133, (1973).

83. P.J. Dean and R.L. Hartman, "Magneto -Optical Properties of the Dominant Bound Excitons in Undoped 6H SiC", Phys. Rev. B 5, No. 12, 4911–4924, (1972).

84. Lyle Patrick, "Interpretation of Dean and Hartman's 6H SiC Magneto Optical Data", Phys. Rev. B 5, No. 7, 1719–1721, (1973).

85. S.H. Hagen and C.J. Kapteyns, Philips Research Reports 25, 1, (1970).

86. S.H. Hagen and A.W.C. van Kemanade, J. Luminescence 3, 131, (1973).

87. J.J. Hopfield, "The Quantum Chemistry of Bound Exciton Complexes", Proceedings of the Seventh International Conference on the Physics of Semiconductors, (Dunod, Paris, 1964) p. 725.

88. A.W.C. van Kemenade and S.H. Hagen, "Proof of the Involvement of Ti in the Low Temperature ABC Luminescence spectrum of 6H SiC", Solid State Communications 14, 1331–1333, (1974).

89. Lyle Patrick and W.J. Choyke, "Phololuminescence of Ti in Four SiC Polytypes", Phys. Rev. B 10, No. 12, 5091–5094, (1974).

2

Recent Progresses in Vapor-Liquid-Solid Growth of High-Quality SiC Single Crystal Films and Related Techniques

Yuji Matsumoto

Department of Applied Chemistry, School of Engineering, Tohoku University, Sendai, Japan

1 Introduction

Among many kinds of polytypes of SiC, 4H-SiC, which has a high breakdown electric field strength, a high saturation drift velocity, and high thermal stability [1-3], is expected as one of the most promising candidates for power device applications [4-6]. On the other hand, 3C-SiC, which is the only cubic polytype with no crystallographic anisotropy, has high electron mobility and shows excellent metal oxide semiconductor (MOS) properties; it is expected to be used as field-effect transistors and capacitors based on the MOS structure. [7-9]. Depending on their crystal structures and properties, it is hence necessary to develop process techniques for the selective growth of each SiC polytype. For example, there have been many intensive efforts so far to develop growth techniques for high-quality 4H-SiC bulk single crystals for wafer use as well as epitaxial films. In the mass-production of the 4H-SiC bulk single crystal wafers, only the existing process is the sublimation process, while for epitaxial films on the bulk single crystal wafers, chemical vapor deposition (CVD) is one of the processes commercially available. However, high-temperature chemical vapor deposition (HT-CVD) [10], halide CVD [11], and gas-source [12] methods have also been rapidly developed, taken as being the closest to their commercialization as new bulk and film production processes.

Figure 1 is a schematic mapping of the processes for SiC bulk crystals and films in terms of their growth temperature and growth rate, displaying the potential usability of each growth technique. In general, the growth rate increases with an increase of the growth temperature in vapor deposition, including the sublimation process. In addition, there is reportedly a threshold temperature lying around 1500°C, above or below which the 4H- or 3C-SiC polytype is preferentially obtained. In contrast, a vapor-liquid-solid (VLS) mechanism is among the vapor deposition processes, but 4H-SiC can be obtained in the homoepitaxial growth even at temperature lower than the threshold temperature. This is because, as will be explained in the next section, the VLS process has some features characteristic of the solution growth, in which thin film crystals are grown via liquid flux. Furthermore, the growth rate in the VLS process potentially increases over several hundred μm per hour, as expected from the similarity to conventional solution growth

DOI: 10.1201/9780429198540-3

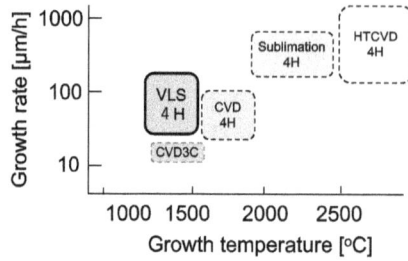

FIGURE 1

Growth rate-temperature map for vapor deposition for SiC bulk and film crystals.

TABLE 1

Comparison of vapor deposition processes for 4H-SiC

	VLS	CVD	Sublimation	HTCVD
Process temperature [°C] (energy saving)	1300	1600–1700	2000	2500
Growth rate [μm/hour]	40–100	30–100	200–300	1000
Susceptor crucible	no use	carbon	carbon	carbon
Purity	excellent	excellent	poor	average
Impurity (carrier) density [cm^{-3}]	1×10^{13}–1×10^{20}	1×10^{13}–1×10^{20}	1×10^{18}–1×10^{19}	1×10^{15}–1×10^{19}
Point defects	good	good	average	poor
Source	Gas	Gas	SiC powder	Gas

processes. In Table 1, the merits and demerits of each growth technique in addition to the growth temperature and growth rate are summarized. In all the processes except for VLS, some of inside parts, such as susceptor and crucible are usually made of carbon, a cost-effective heat-resistant material. However, even when high-purity carbon products with impurity levels less than ppm order are used, a possibility of the grown SiC crystals and films being contaminated from such carbon parts could not be completely ruled out owing to the nature of their high process temperature over 1600°C, at which impurities coming out from the inside bulk carbon can be emitted from the wall of the carbon parts. In view of the potential contamination from carbon parts through the process, the VLS process is advantageous over any other processes listed in Table 1 because of not using any carbon parts and its low process temperature in the VLS process. On the other hand, as has been so far pointed out, there is another potential contamination from liquid flux in the VLS process, making it difficult to control the residual carrier density in VLS-grown SiC crystals and films, which should be overcome for the VLS technique to be adopted in a commercial production process of SiC. In fact, as will be later shown in more detail, when Pt is used as an impurity flux in VLS growth of 4H-SiC films, almost no Pt is detectable by secondary ion mass spectroscopy (SIMS) in the SiC films, because Pt has a relatively large atomic radius for the lattice sites in crystal SiC.

This chapter presents our recent progresses in developing an innovative, vapor-liquid-solid (VLS) mechanism-based thin film processes for fabricating SiC films with single crystal quality, including our originally developed pulsed laser deposition (PLD)-based process chambers and an *in situ* visualization technique of solution growth interfaces by means of confocal laser scanning microscopy (CLSM). Accordingly, we demonstrate advanced VLS processes with novel additives such as Ni, Al, and Pt for high-quality

4H- and 3C-SiC films. The basis for our claim of "advanced VLS processes" is that our VLS processes take the most advantage of potential catalytic-like effects of additives in liquid flux on the thin film growth, beyond the scope of thermodynamical merit characteristic to solution growth processes, even in vapor deposition. All these examples presented in this review illustrate the potential of the SiC VLS growth being applied to industrial mass-production in the future.

2 Vapor-Liquid-Solid (VLS) Growth Mechanism

2.1 Brief History of the VLS Growth Mechanism Toward Single Crystal Films

"VLS" is the abbreviation of "vapor-liquid-solid", a growth mechanism of crystals via liquid in vapor deposition, as already mentioned in the previous section. The term was first used by R.S. Wagner and coworkers in their work published in *Applied Physics Letters* in 1964 [13], in which they reported on the growth of single crystalline Si whiskers via Au droplets in CVD. At that time, the VLS mechanism had drawn little attention for the first 30 years, but in the 1990s, especially since the 2000s there has been a rapid growing interest in the VLS process, which might be partly related to the so-called National Nano-technology Initiative (NNI) launched by the US in 2001. Since the VLS processes are, in most cases, employed for fabricating whiskers and nanowires of various materials, it has now become one of the most representative nano-technologies in materials science.

On the other hand, a flux growth method, one of the solution-growth processes [14], is often employed for bulk single crystal growth, and its derivative method for single crystal films is the liquid phase epitaxy [15, 16]. The VLS process, as shown in Figure 2, which is performed in vacuum by continuously feeding the source materials into a liquid flux layer on a single crystal substrate from the gas phase, can be regarded as an exquisite extension of these solution-growth methods to the vapor deposition process. The feature is to include both the merits of liquid phase epitaxy or flux growth and vapor deposition, i.e.,

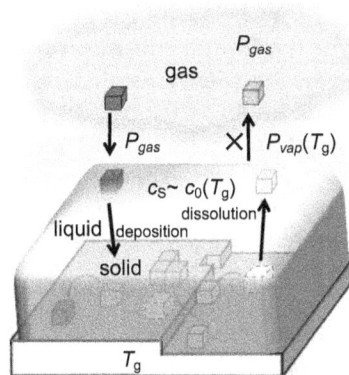

FIGURE 2
Schematic of the VLS process.

the nanoscale controllability of a film thickness in a similar way to the gas phase supply in the vapor deposition and high-crystallinity owing to its thermodynamically quasi-equilibrium growth conditions via liquid. Accordingly, these kinds of VLS applications are often termed "tri-phase epitaxy (TPE)" or "flux-mediated epitaxy (FME)" to differentiate from the conventional VLS processes for whiskers and nanowires. Applications of VLS processes based on CVD and PLD, often termed CVD-VLS and PLD-VLS, respectively, to the fabrication of single crystal films have been reported since the 2000s. In PLD-VLS, it is not necessary to care about possible complex decomposition reactions of gas source materials, which is the process inevitable in CVD-VLS, though the growth rate in the CVD-VLS is much higher than those in the PLD-VLS. Typical examples of the PLD-VLS growth [17] in the early years of the 2000s were limited to oxide materials such as super-conducting $NdBa_2Cu_3O_{7-\delta}$ (Nd123) grown with a Ba–Cu–O self-flux [18] and ferroelectric $Bi_4Ti_3O_{12}$ with a Cu-stabilized BiO_x flux available in vacuum [19]. On the other hand, it was in 2002 that G. Ferro and his coworkers first reported the successful fabrication of SiC single crystal films with a Si melt [20] by the CVD-VLS process, since then, it has been followed by those with a multi-component Si-based flux as Si-Ge [21], Si-Al, Si-Ni and Si-Co, Si-Fe [22] and Si-Cu [23]. Since the late years of the 2000s, there has been a revival of VLS growth of SiC single crystal films by the PLD-VLS, which is the main contents that will be introduced in this chapter. Incidentally, more recently, the VLS process is being applied to the growth of organic crystals and films of pentacene [24], fullerene (C_{60}) [25], and a charge-transfer complex of DBTTF-TCNQ [26] with ionic liquids, which are a new class of organic solvents stable in liquid even in vacuum [27].

2.2 Chemical Engineering Aspects in the VLS Growth Mechanism

As already explained, there are three steps involved in the VLS mechanism: (1) feeding of source material precursors from the gas phase into a liquid flux layer, (2) diffusion of the precursors in the liquid flux, and (3) crystallization of the precursors on a substrate. In this section, the first two steps characteristic of the VLS mechanism, which are related to the mass transfer of the source material precursors, are discussed from the viewpoint of chemical engineering. To simplify the mass transfer process in the VLS mechanism, a one-dimensional diffusion model, as shown in Figure 3(a), is employed, where an infinite 2D liquid flux layer covers a substrate with a uniform thickness of d and the liquid surface is homogeneously exposed to source material precursors in gas phase. When starting to supply the source material precursors from the gas phase, the precursors penetrate the liquid surface into the interior of the liquid, diffuse in the liquid along the x axis, and reach the substrate, being removed by their crystallization. After a while, the situation will reach a steady-state condition, under which the supply rate of the precursors from the gas phase v_{Dep}, their diffusion rate v_{Diff} in the liquid and the crystal growth rate v_{Epg} all become balanced to the same value. In order to confirm this steady-state model, the time-development of the concentration profile $c(x, t)$ of the precursors in the liquid was numerically simulated. For this, the differential equations with the boundary conditions at the substrate-liquid interface of $x=0$, and at the liquid-vacuum interface of $x=d$ are described as follows:

$$\frac{\partial c}{\partial t} = D\frac{\partial^2 c}{\partial x^2} \quad \text{for all } x \tag{1}$$

$$D\frac{\partial c}{\partial x} = \alpha v_{Dep} \quad \text{at } x = d \tag{2}$$

$$D\frac{\partial c}{\partial x} = kc \quad \text{at } x = 0 \tag{3}$$

where D is the diffusion coefficient of the precursors in the liquid in eqn (1)–(3), α is just a coefficient to ensure that the dimensions are equal on both sides of eqn (2), and k is the rate constant in eqn (3), assuming the first-order reaction kinetics for crystal growth.

The simulation result of a time-development of the concentration profile of the precursors in the liquid just after starting to supply the source material precursors from the gas phase is displayed in Figure 3(b) [28]. The concentration first starts to increase at the liquid-vacuum interface ($x/d \sim 1$), and then the concentration increase gradually proceeds toward the interior of the liquid layer. After enough time has passed (i.e., at time [a.u.] = 4 in the simulation), an almost-linear concentration gradient is finally established throughout the liquid layer. This is the steady-state condition that was mentioned above. Under the assumption of the crystal growth following the first-order reaction kinetics in this simplified model, the crystal growth rate on the substrate v_{Epg} is proportional to the concentration of the precursors at the substrate-liquid interface ($x/d = 0$) $c = c_S$. For different values of the supply rate v_{Dep}, different c_S values are obtained under the steady-state conditions, respectively, and the plot of $v_{Epg} = k\,c_S$ against v_{Dep} results in a good linear relationship between them (inset). This numerical simulation reconfirms that the supply rate v_{Dep}, the diffusion rate v_{Diff} and the crystal growth rate v_{Epg} all become balanced to the same value under the steady-state condition. Consequently, the crystal growth rate under the steady-state condition is represented by the diffusion rate as follows (eqn (4)):

$$D\frac{\partial c}{\partial x} = D\frac{c(x=d)-c(x=0)}{d} \leq D\frac{C_{max}}{d} \tag{4}$$

where C_{max} is the solubility of the source material precursors in the liquid. In the steady-state condition, $c(x=d)$ should be equal or less than C_{max}, and therefore the maximum crystal growth rate attainable in the VLS mechanism is, if there is no limit of the supply rate of the precursors, $\sim C_{max} D / d$. Since the diffusion constant D and the solubility C_{max} are

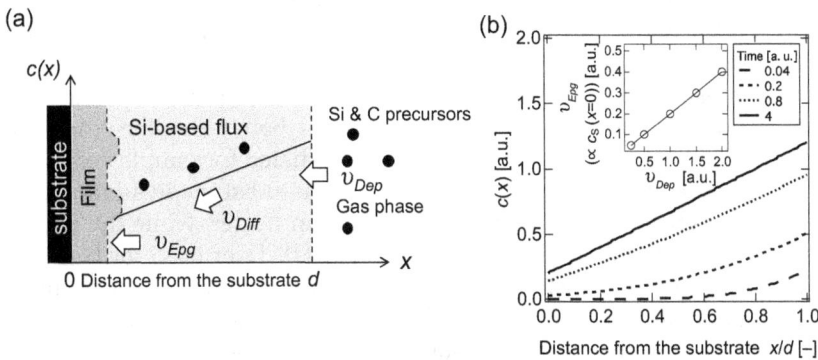

FIGURE 3

(a) Schematic of a simple one-dimensional diffusion model of the VLS process, assuming first-order reaction kinetics. (b) Numerical simulation results of the time-development of the concentration profile $c(x)$ in a liquid flux, approaching a constant gradient. After a while, the epitaxial growth rate v_{Epg} (corresponding to the concentration at the growth interface $c(0)=c_S$) becomes constant, which is well proportional to the deposition rate v_{Dep} (inset) . Reproduced with permission from [28]. Copyright 2020, The Royal Society of Chemistry.

the parameters pre-determined by the choice of the material/liquid combination, the variable process parameter is only the liquid thickness d; accordingly, the thinner the liquid layer is, the larger the achievable crystal growth rate is. In fact, G. Ferro and coworkers also show and discuss this expected tendency experimentally in the selective epitaxial growth of SiC by the VLS mechanism [22]. Another important simulation result is that the time required for reaching the steady-state condition is expressed by eqn (5), proportional to the square of d [28]:

$$\tau \approx d^2/D \tag{5}$$

This result indicates that the thinner the liquid layer is, the smaller the time constant is; the thinner liquid layer is more desirable for precision control of the VLS-grown film thickness on the nanometer scale.

3 Experiment

3.1 Pulsed Laser Deposition (PLD)-Based VLS

In a PLD process, by irradiating a high-power ultra-violet (UV) laser on a material target its constituent atomic/molecular clusters are produced in gas phase through a so-called laser ablation process, even for high melting point materials that could not be vaporized by conventional thermal heating techniques. In the PLD, the gaseous precursors are then deposited on a substrate to fabricate crystalline films [29]. In the PLD-VLS, in which PLD is used to supply gaseous precursors, as already pointed out, any chemical decomposition processes of source gases are not involved, different from the CVD-VLS process. Furthermore, the diffusion process of the precursors in a liquid layer, as discussed in the previous section, is not influential on the crystal growth as long as the supply rate of the gaseous precursors is not so high. Hence, the PLD-VLS approach enables the crystal growth to be focused on at the liquid-substrate interface and thereby potential catalytic-like effects of additives in liquid flux on the thin film growth can be more efficiently investigated.

   Figure 4(a) is a schematic of the high-pressure type PLD-VLS chamber system that we originally developed, together with a photograph of the appearance of the chamber [30]. A single crystal substrate is mounted on a sample holder using carbon paste. There are two types of sample holders: one is made of carbon with a carbon heater, the other is of stainless steel from the backside of which a Nd:YAG laser is irradiated for sample heating. Si, Al, and Pt in an arbitrary bulk amount are placed as flux on the substrate in advance. The growth temperature can reach as high as 1800°C by the carbon heater, while the maximum temperature is 1300°C by the Nd:YAG laser heating. The UV laser used for laser ablation is a KrF excimer laser (LPX200), which is introduced from the side wall of the chamber through a quartz viewing port and irradiated on a SiC target placed inside the chamber, as shown in Figure 4(a), to deposit SiC on the heated substrate. During deposition, the ablation laser is two-dimensionally scanned over the SiC target so that the target is homogeneously ablated. A 0.1 Torr H_2 gas, and ca. a 10 Torr Ar gas if needed, is (are) introduced into the chamber. The H_2 gas is used as a reducing reagent, while the Ar gas is used to suppress a possible evaporation of the flux liquid at high temperature. Figure 4(b) displays the deposition-time dependence of the deposited SiC film thickness for both the cases without and with a Si-Pt

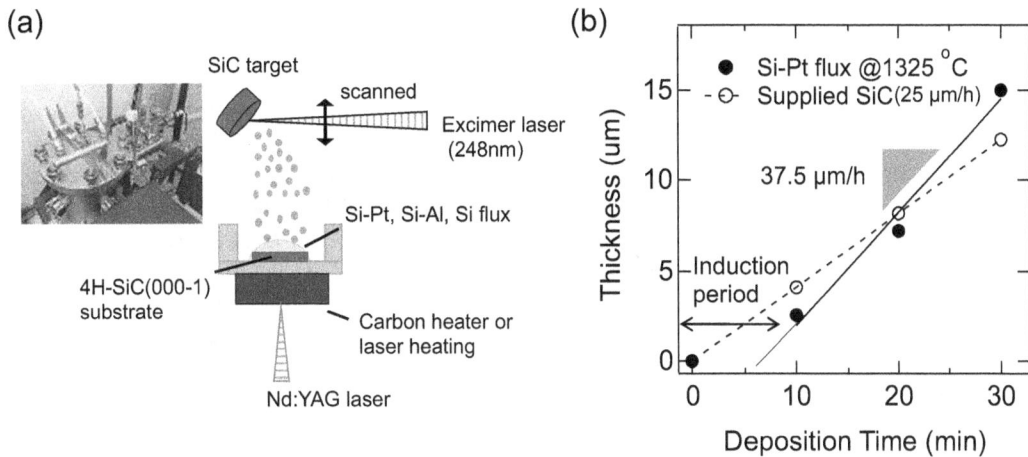

FIGURE 4
(a) Schematic of a high-pressure type PLD-VLS chamber system, together with a photograph of the appearance of the chamber. Reproduced with permission from [30]. Copyright 2017, The Royal Society of Chemistry. (b) The deposited SiC film thickness plotted against the deposition-time for the both cases without and with a Si-Pt flux, respectively. Reproduced with permission from [43]. Copyright 2020, Elsevier.

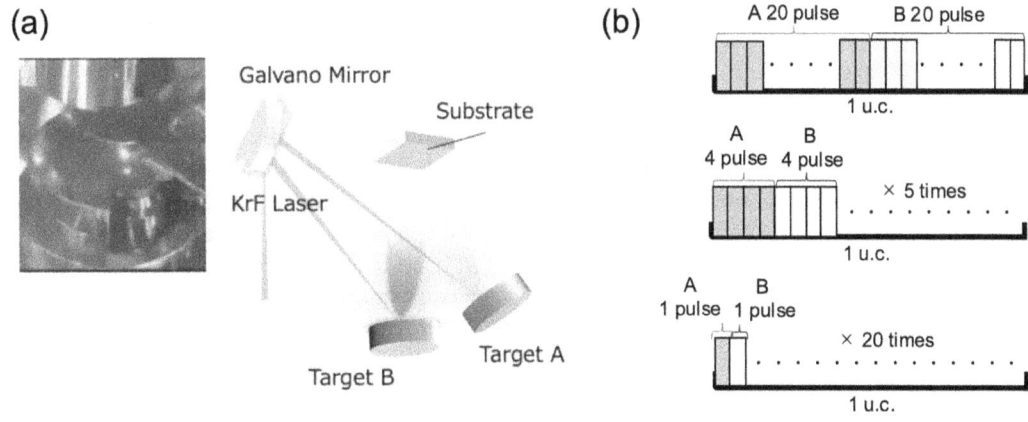

FIGURE 5
(a) Schematic of a high-vacuum type PLD-VLS chamber system, (b) Alternate deposition sequence, where the one cycle consisting of depositions with 20 pulses for target materials of A and B, respectively, i.e., the deposition with 40 pulses in total corresponds to the growth of one-unit cell layer. For example, three different model deposition sequences are shown within the one cycle with different degrees of mixing the A and B target materials. Reproduced with permission from [31]. Copyright 2019, AIP publishing.

flux, respectively. In this case, as the experimental results will be later discussed in more detail in section 4.2, the supply rate of gaseous precursors is 25 μm/h, which is estimated from the slope of the plots (○), while SiC films grow via the Si-Pt flux almost in accordance with the supply rate of the gaseous precursors (see the plots (●)). It should be pointed out that there seems an induction period for the first 10 min, which is attributed to the time required for reaching the steady-state condition as already mentioned before, with a temporary growth rate of 37.5 μm/h just after the induction period.

A high-vacuum type PLD-VLS chamber system [31] is schematically shown in Figure 5, which is, as will be later explained in section 5, used for demonstrating the VLS-like growth process that we have first proposed. In this PLD-VLS system, the same excimer laser as is used in the high-pressure type PLD-VLS system is also used for target ablation, while the substrate heating is performed by using a high-power semiconductor laser, reaching the maximum temperature of 1050°C. As shown in Figure 5(a), the utilization of a Galvano-mirror placed on the optical path of the excimer laser enables one to alternately ablate two different targets with a time interval as short as on the order of several ms. As a consequence, different from conventional VLS processes where a certain amount of a flux material is placed on a substrate before starting SiC deposition, the Galvano-mirror PLD-VLS system allows one to alternately deposit SiC and flux liquid sources on a nanometer scale, achieving the VLS-like growth of high-quality SiC films. Figure 5(b) shows examples of such an alternate deposition sequence, where the one cycle consisting of depositions with 20 pulses for target materials of A and B, respectively, i.e., the deposition with 40 pulses in total corresponds to the growth of one-unit cell layer. In the one cycle, there are, for example, three different model deposition sequences: a sequential deposition of A and B with 20 pulses for each, a five times repetition of a sequential deposition with four pulses for each, and a 20 times repetition of a sequential deposition with one pulse for each. As a result, the alternate deposition process can be digitally controlled to different degrees of mixing the two target material sources.

3.2 Confocal Laser Scanning Microscope (CLSM) at Solution Growth Interfaces

Crystal growth at substrate-liquid interfaces in VLS is essentially equivalent to that at solution-growth interfaces. Therefore, *in situ* observation of the solution-growth interface is useful also to understand the VLS growth mechanism, especially investigating potential catalytic-like effects of additives in flux. There are, on the one hand, many reports on *in situ* nanoscale observation of solution-growth interfaces using a variety of microscopy techniques, but the observation temperature is limited to up to 300°C in most cases [32–35]. On the other hand, *in situ* nanoscale observation at high-temperature such as over 1000°C, which is required for the solution growth of SiC crystals, has been so far technically difficult. In fact, optical microscopy techniques are often used for such high-temperature *in situ* observation, but its spatial resolution is in the range of several hundred micrometers [36, 37].

In order to overcome the problems pointed out above, we have developed a new *in situ* observation system for solution-growth interfaces, in which the combination of confocal laser scanning microscopy (CLSM) and vacuum chamber technologies enables the *in situ* nanoscale observation at high temperature over 1000°C and its quantitatively data analyses. In the CLSM, the microscope image is constructed by mapping out the reflectivity intensity of its own laser light, which is two-dimensionally scanned over an object material surface. Thanks to the use of its own laser light, instead of the white light in the environment like in the conventional optical microscopes, the CLSM is powerful to directly observe material surfaces even when they emit thermal radiation at high temperature without any filters to cut the radiation. Furthermore, the CLSM has a long focus working distance, as will be explained later, allowing observation of a distant object placed in a vacuum chamber by introducing the laser light through a viewing port of the vacuum chamber. The most useful feature of the CLSM is that it is able to clearly observe a buried substrate-liquid interface by adjusting its focus point on a nanoscale [38].

A schematic of the CLSM system, which was originally developed in our group, is shown in Figure 6(a) [39]. As already mentioned above, the CLSM is set up over the outside of a vacuum chamber, whose base pressure is less than 1×10^{-6} Torr. The probe laser is introduced into the inside of the chamber through a viewing port and scanned over a sample surface placed on a XYZ stage just under the viewing port to detect its reflected laser from the sample surface. The sample temperature is controlled by heating up to max. 1800°C with Nd:YAG laser irradiation under monitoring the temperature with a pyrometer from the backside of the sample holder. In this system, since a rubber O-ring gasket is used for vacuum sealing of the viewing port, the viewing port and its surrounding are cooled by flowing water inside of the top flange plate to avoid a possible heat damage on the viewing port and the O-ring by thermal radiation from the high temperature sample. Figure 6(b) illustrates a schematic of the configuration of the sample holder. The sample consisting of a pair of SiC single crystal plates, between which a Si-based powder is sandwiched as a flux, is mounted on a carbon plate with carbon paste. The Si-based powder flux melts by elevating the sample temperature over its melting or liquidus point, while the heating from the backside of the carbon plate causes a not-negligible temperature gradient from the bottom to top of the SiC plates. Hence, the bottom SiC plate works as a source of SiC, partially dissolving into the flux liquid, and the dissolved SiC travels to and crystallizes on the top SiC plate, a seed substrate. The probe laser introduced into the chamber is focused on the substrate-liquid interface from the backside of the seed substrate to directly observe the crystal growth behavior.

Figure 6(c) illustrates a good example of *in situ* observation of the solution growth of SiC with a Si-Ni flux at around 1600°C, displaying from the left to the right representative time course images of its step advancing [39]. The step lines, each of which is observed as a black line with a different contrast, are seen to move and change their line shapes with time. Although the lateral resolution is not less than 1 μm, the observable step heights

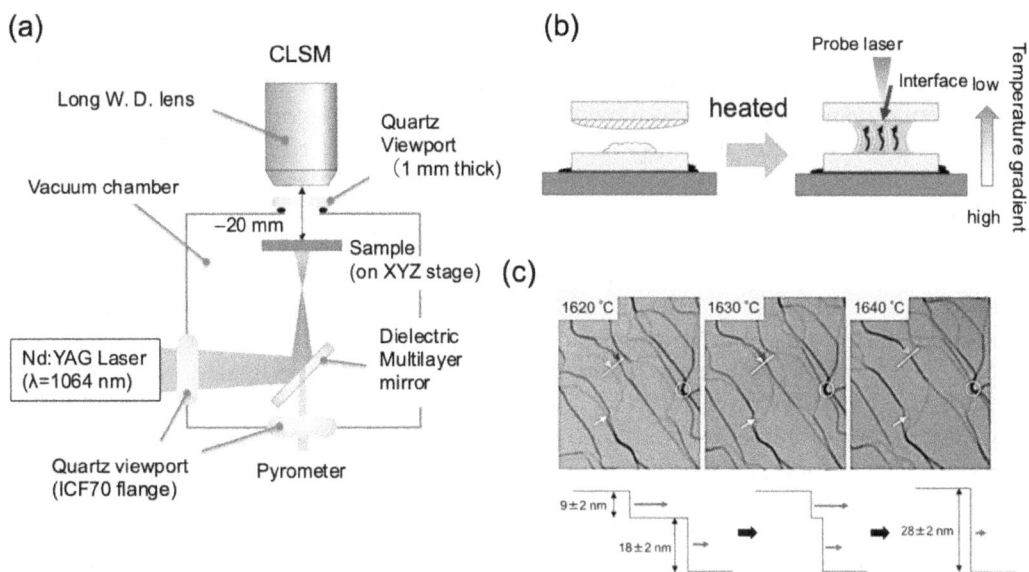

FIGURE 6
(a) Schematic of the CLSM system, which was originally developed in our group. (b) Configuration of the sample holder, (c) *in situ* observation result of the solution growth of SiC with a Si-Ni flux at around 1600°C, displaying from the left to the right representative time course images of its step advancing. Reproduced with permission from [39]. Copyright 2017, American Chemical Society.

are on a nanometer scale with the minimum value reaching down to ~5 nm. A more interesting finding is that there is a good linear relationship between the observed step line contrasts and their real height values estimated by atomic force microscopy, making it possible to evaluate each step height from its contrast. The evaluated step height values are quantitative to as much degree as they satisfy the sum rule in the step-bunching process. In fact, as shown in Figure 6(c), it can be seen that a bunched step with a step height of 9±2 nm indicated by the white arrow advances faster and merges with the anterior bunched step with a step height of 18±2 nm indicated by the black arrow, forming a new bunched step with a step height of 28±2 nm. Furthermore, more recently *in situ* observation of SiC growth in a Si-Cr solvent was attempted at over 1800°C with the CLSM technique [40]. This has revealed that the reversibility, i.e., not only the crystallization process, but also the dissolution one during the solution growth becomes more important to obtain high-quality SiC crystals with higher growth rate.

4 VLS Growth of SiC Films

4.1 Origin of the Flattening Effect of Al Addition on the VLS Growth of SiC [30]

Among various metal elements as additives that have been attempted in the solution growth of SiC, the additive effects of Al have recently drawn much attention. Its potential effect on the crystal growth behavior of SiC can, at least, date back to a report in 1970 by Mitomo et al [41]. Although it was not in the solution-growth process, it has been already found that the most stable phase was the 4H polytype in SiC ceramics sintered with the addition of Al. In the solution growth of SiC bulk crystals, a flattening effect on the growing surface has been reported in addition to a similar stabilizing effect on the 4H- polytype. Komatsu and coworkers discussed a possible reason for the flattening effect based on their experimental results and thermodynamic calculations [42]. According to their proposed mechanism, the addition of Al increases the liquid/solid interfacial energy, and consequently the two-dimensional nucleation energy increases, lowering the frequency of the two-dimensional nucleation on the growing surface. Furthermore, G. Ferro and coworkers also reported the stabilization of the 4H-SiC phase by adding Al even in the VLS mechanism [22]. From these previous results, a hypothesis may arise: there should be a certain common mechanism to explain such Al additive effects on the growth of SiC crystals independent of the crystal growth processes.

Motivated by the hypothesis mentioned above, we further investigated the Al additive effects on the growth of SiC films in the PLD-VLS process. Figure 7 displays a set of differential interference contrast microscope (DIC) and atomic force microscope (AFM) images of 4-μm-thick SiC films grown with $Si_{100-x}Al_x$ fluxes (x=0, 25, 50, 75) on 4° off 4H-SiC (000-1) substrates, respectively, using the high-pressure type PLD-VLS chamber explained above. The growth temperature was, except for the growth with the pure Si flux (x=0: 1530°C), set to 1250°C, a temperature much lower than those in the conventional solution-growth and CVD processes. This is because the addition of Al lowers the temperature at which Si-Al alloys become liquid [43]. For the SiC film grown with the pure Si flux, two kinds of step lines with step edge angles of 120° and 60° coexist on the surface, and they are attributable to 4H(6H)-SiC and 3C-SiC polytypes, respectively. In contrast, for the SiC films grown with the $Si_{100-x}Al_x$ (x=25, 50) fluxes, a uniform steps-and-terraces structure with a step edge angle of 120° is observed all over the surface. Further increase of Al addition up to

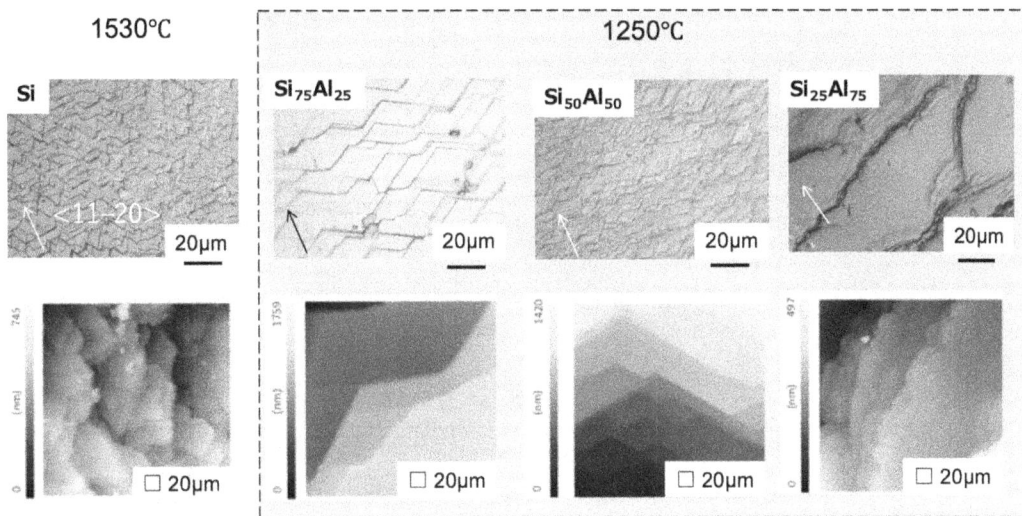

FIGURE 7

Set of differential interference contrast microscope (DIC) and atomic force microscope (AFM) images of 4-μm-thick SiC films grown with $Si_{100-x}Al_x$ fluxes (x=0, 25, 50, 75) on 4° off 4H-SiC (000-1) substrates, respectively, using the high-pressure type PLD-VLS chamber. Reproduced with permission from [30]. Copyright 2017, The Royal Society of Chemistry.

x=75 results in larger terrace widths with round-shaped steps. In the corresponding AFM images, the difference in the surface morphology is more clearly found for different Al addition: the pure Si flux-grown SiC film shows a significant rough surface, while the $Si_{100-x}Al_x$ flux-grown films (x=25, 50) exhibit a regular steps-and-terraces structure. In addition, the round-shaped steps on the $Si_{25}Al_{75}$ flux-grown film surface, macroscopically observed by DIC, are confirmed to consist of many small steps with a step edge angle of 120°.

Raman spectroscopy measurements were performed in order to identify the polytypes and get some insight into the Al doping in the SiC films, and the results of the measurements are summarized in Figure 8. As shown in the Raman spectra scanned in a wide range of 100 to 1200 cm^{-1}, there is found only the 4H FTO peak (776 cm^{-1}) in any SiC films grown with the $Si_{100-x}Al_x$ flux, while the 3C FTO peak (796 cm^{-1}) with a non-negligible intensity additionally appears in the pure Si flux-grown SiC film. Focusing on the 4H-FTO and 3C-FTO peaks, a comparison of the micro-Raman mappings of their intensity ratio $I_{3C\text{-}FTO}/I_{4H\text{-}FTO}$ clearly confirms that a large portion of 3C-SiC grains are included in the pure Si flux-grown SiC film and the amount of 3C-SiC tends to decrease with an increase of Al added in Si flux. In fact, when using the $Si_{25}Al_{75}$ flux, almost perfect single phase 4H-SiC could be obtained. The result is not inconsistent to the different existence ratio of the two kinds of step lines with step edge angles of 120° and 60° on the SiC film surfaces depending on the Al content in the flux used in the VLS growth, as was observed by DIC and AFM. On the other hand, the information on the Al doping level in the SiC films grown with the $Si_{100-x}Al_x$ flux could be inferred from the spectral change of the 4H-FTA(+) peak at around 200 cm^{-1}, as shown in the expanded Raman spectra in the range of 100 to 220 cm^{-1}. The peak position is found to shift to the lower wavenumber side, accompanied by a gradual increase of the background, with an increase of Al added in the flux; this effect is known as a Fano interference phenomenon for heavily doped SiC crystals [22]. Taking into account a gradual broadening of the A1 LOPC peak around 1000 cm^{-1} with the Al content in the

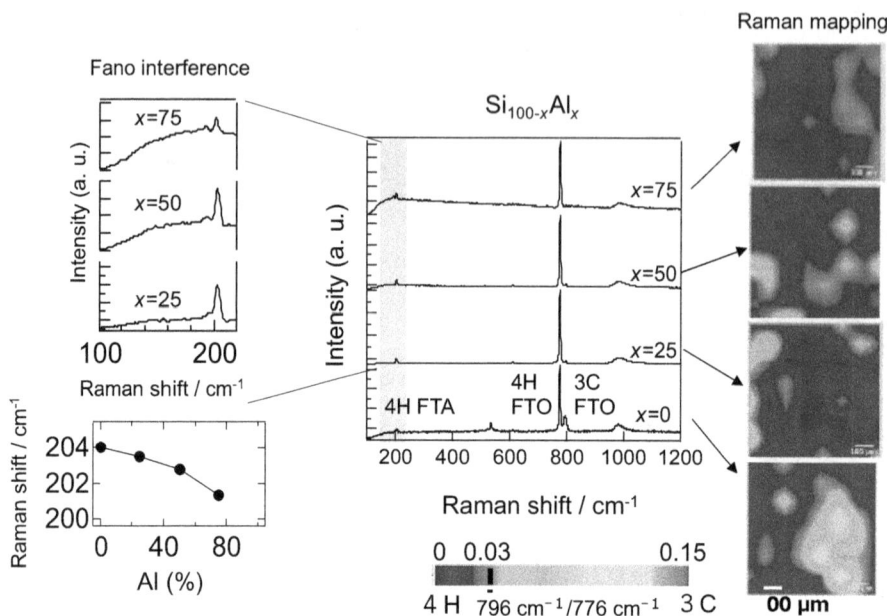

FIGURE 8

Raman spectroscopy measurements for the SiC films grown with the $Si_{100-x}Al_x$ flux. Wide-scanned spectra in a range of 100 to 1200 cm^{-1} is centered in the figure; and the micro-Raman mappings of the peak intensity ratio *I*3C-FTO/*I*4H-FTO and the spectral change of the 4H-FTA(+) peak at around 200 cm^{-1} are displayed on the right- and left-hand sides of the wide-scanned spectra, respectively. Reproduced with permission from [30]. Copyright 2017, The Royal Society of Chemistry.

flux, a sizable amount of Al should be incorporated as a dopant into the SiC films grown with the $Si_{100-x}Al_x$ flux.

To verify the high-level doping of Al in the SiC films speculated above, we took EDX elemental mappings of Al for the SiC films for different Al contents in the flux, as shown in Figure 9(a) for the SiC films grown with the $Si_{25}Al_{75}$ flux, along with the corresponding SEM surface image. The Al signal could be detected all over the SiC film area, indicative of Al being homogeneously doped in the film. If the VLS process proceeds under the conditions almost close to the thermodynamic equilibrium, the Al contents in the films for different Al contents in the flux should be uniquely determined by the partition equilibrium. In fact, as shown in Figure 9(b), the Al doping level reaches the order of 10^{20} cm^{-3}; and there is a good linear relationship between the Al contents in the film and in the flux, giving an equilibrium constant K_d of $(1.55\pm0.2) \times 10^{-2}$ at 1250°C; this result is evidence for the present VLS process proceeding under the conditions almost close to the thermodynamic equilibrium.

To investigate a possible origin of the dominant growth of the 4H-SiC polytype in a solution with the $Si_{100-x}Al_x$ flux, the initial stage of the SiC growth at the solution-growth interface was observed with the CLSM technique introduced in the previous section. Figure 10 displays the observation results for the pure Si and $Si_{90}Al_{10}$ fluxes, respectively, on on-axis 4H-SiC(000-1) substrates at the initial stage of starting the temperature increasing. The pure Si flux became liquid completely at around its melting point of 1414°C, and was found to etch back the original seed substrate, especially along its polishing marks, at 1450°C. As the temperature further increased up to 1500°C, step-and-terrace structures appeared to

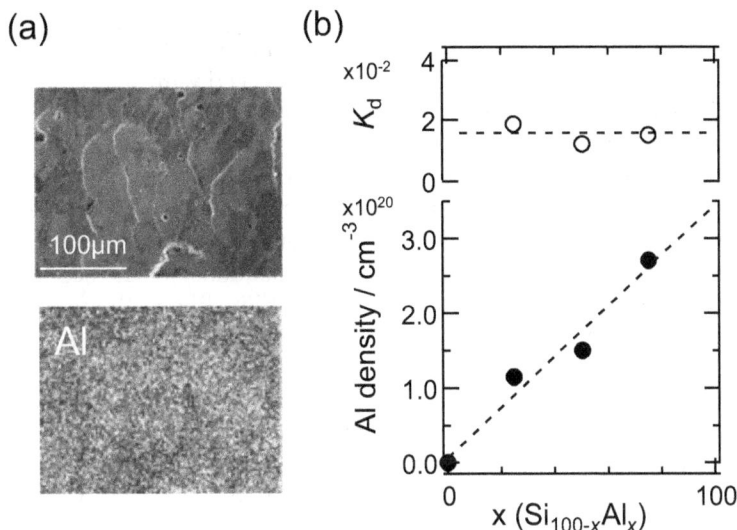

FIGURE 9
(a) EDX elemental mapping of Al for the SiC film grown with the $Si_{25}Al_{75}$ flux, along with the corresponding SEM surface image. (b) Al density (cm^{-3}), evaluated for each SiC film by EDX, plotted as a function of the content of Al in the flux. Reproduced with permission from [30]. Copyright 2017, The Royal Society of Chemistry.

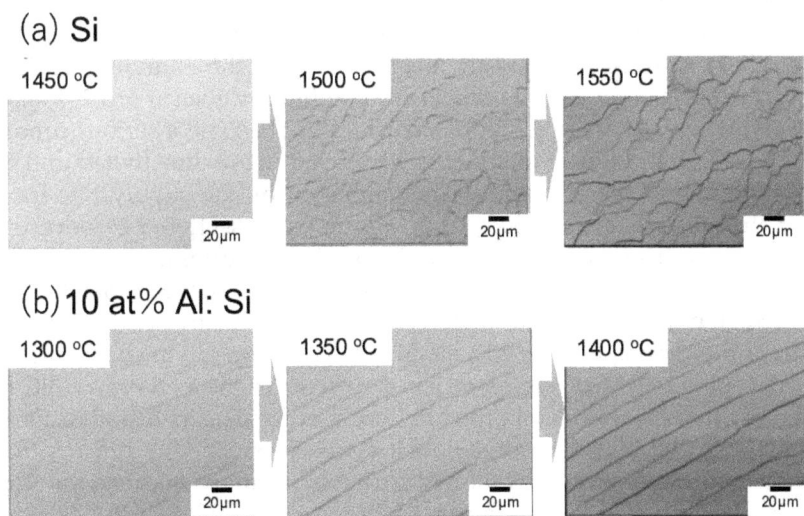

FIGURE 10
Comparison of sequential CLSM images for (a) the pure Si flux and (b) the $Si_{90}Al_{10}$ flux on on-axis 4H-SiC(000-1) substrates at the initial stage of starting the temperature increasing. Reproduced with permission from [30]. Copyright 2017, The Royal Society of Chemistry.

initiate the step advancing, accompanying the growth of SiC, but the meandering step lines were found to be peculiar to the case of the pure Si flux. The step edge angles of 60° as well as 120° indicate inclusion of the 3C- polytype even from the initial stage of the growth (Figure 10(a)). On the other hand, as shown in Figure 10(b), when a 10 at% Al was added in the Si flux, the flux was ready to become liquid at around 1300°C, which is close to its liquidus temperature according to the binary phase diagram of Si-Al [43]. The etching back process was first observed at the temperature, not long after which it turned to the step flow growth, i.e., the step-advance was observed with step lines almost parallel to each other. The step heights were rather small as long as the temperature was relatively low, as compared with the growth of SiC with the pure Si flux, while above 1600°C, the terrace width became large, accompanied by step-bunching. Although these results were obtained for the on-axis 4H-SiC(00-1) substrate, such a well-regulated step-flow growth around the growth temperature of 1300°C, which has never experimentally been reported, should be favored in succeeding to the 4H-stacking sequence of the seed substrate when the 4° off substrate is used for the growth of the SiC films like in the present experiment. This might be one of the possible reasons for the dominant growth of 4H-polytype SiC in the thin-film VLS process as well as in the solution process of bulk single crystals. It should be pointed out that there has never been direct observation of the step-flow growth of SiC at such a low temperature of 1300°C, and the low temperature step-flow process, which is realized by adding Al in the Si flux, will result in a non-negligible contribution to the reduction of electric power consumption for the growth of high-quality SiC films.

4.2 Pt Additive Effect on the Step-Bunching in the Growth of SiC on Vicinal Substrates [44]

In the solution growth of SiC, Fe [37], Ni [45], Cr [46], Al [22,42], Ge [47], and Ti [48] etc. have been intensively tested as additives in Si melt; these results can provide more than a little insight into the VLS process of SiC films as well. Among them, Cr is added as a representative metal to increase the carbon solubility in the Si melt, thereby increasing the growth rate, otherwise the crystal growth rate is too low for the solution process to become commercially available because of its small carbon solubility even at process temperatures over 2000°C. However, the carbon solubility would not become a serious problem in the VLS process, in which the source materials such as carbon are directly fed into the Si melt from the gas phase and the growth rate is dominated by the supply rate from the gas phase under the steady-state conditions, as discussed in section 2.2. Nevertheless, the additive effects, as in the case of Al, are, on the one hand, still important and potentially expected to improve the surface morphology and control the polytype of SiC films even in the VLS process. On the other hand, there is a possible inclusion of such additive metals as impurities in the crystal, leading to a significant deterioration in characteristics of SiC. In fact, the Al doping level reaches the order of 10^{20} cm^{-3} in the VLS grown SiC films with the Al-added Si flux as mentioned in the previous section. The inclusion of such additive impurity metals can become a more serious drawback of the VLS process than that of the solution-growth process toward the industrial applications because the VLS process is assumed as an alternative to the existing CVD process for high-speed growth of high-purity epitaxial SiC layers on a SiC wafer. Accordingly, it would be, at least, desirable to use metal elements that have atomic and/or ionic radii relatively larger than carbon and Si atoms, to avoid their substitution in the SiC crystal.

With this background, we introduce the VLS growth of SiC films with a Pt-added Si flux in this section. There are mainly three reasons for the use of Pt as an additive in the VLS growth. The first one is in terms of the chemical stability of Si-Pt alloy as liquid. Figure 11 is a schematic of the phase diagram of a Si-Pt binary system, in which the liquidus temperature is lowered by adding even a small amount of Pt, e.g., the temperature goes down to ~1250°C by adding 20 at% Pt. Furthermore, no alloy compounds are formed at such a low temperature, including any compounds with carbon (Pt carbides), as long as the

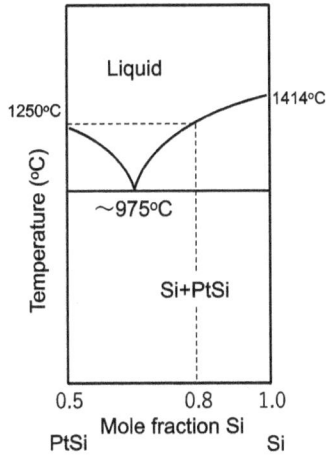

FIGURE 11
Schematic of the binary phase diagram of Pt-Si.

TABLE 2

Comparison of atomic radii of C, Al, Si, and Pt. The values in parentheses are those calculated [50]

	atomic radius [pm]
C	70 (67)
Al	125 (118)
Si	110 (111)
Pt	135 (177)

Pt composition less than 50 at% [49]. The second one is in terms of its atomic radius, as discussed above. As compared in Table 2, the atomic radius of Pt is, experimentally as well as theoretically [50], larger than those of Si and C atoms, and thus its almost no substitution in SiC is expected and even if Pt atoms are substituted in the SiC, it may work as neither donor nor acceptor dopants. The third one is a possible catalytic-like effect of Pt in the growth of SiC crystals and films. According to H. Hara et al, Pt is found to work as a catalyst in a chemical polishing process for a step-and-terrace structured SiC surface in HF aqueous solution, which is known as CARE treatment [51]. Hence, Pt is expected to play a similar role on a growing SiC surface even in the VLS process, improving surface smoothness and flatness, though its conditions are very far from those of CARE treatment.

Figure 12 displays a series of DIC (upper row) and AFM images (lower row) of VLS-grown SiC films on 4° off 4H-SiC (000-1) substrates via a pure Si, 5 at% Pt-added, and 10 at% Pt and 18 at% Al co-added Si fluxes, respectively. The growth temperature is 1500 °C for the pure Si and the Pt-added fluxes, while it is 1325°C for the Pt and Al co-added flux. The white arrows on the DIC images indicate the off-cut direction of the substrate. The SiC film grown with the pure Si flux exhibits a step-and-terrace structure with step edge angles of 120° and 60°, and bunched and wide terraces are observed as more clearly in the AFM image. In contrast, just a 5 at% addition of Pt in the flux causes a drastic surface morphology change, the step lines regularly running along the direction perpendicular to the

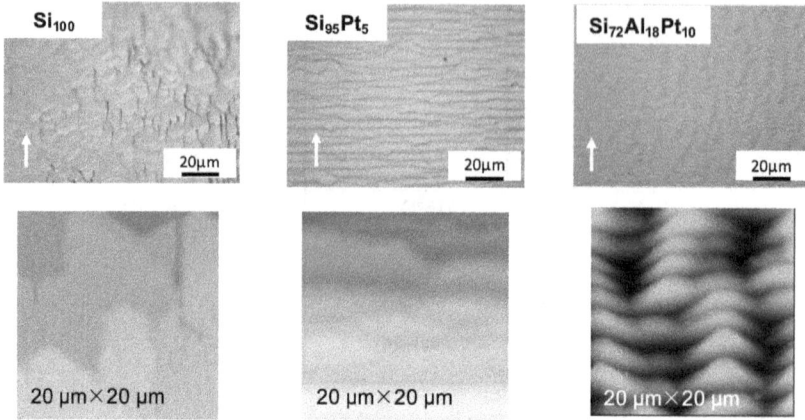

FIGURE 12

Series of DIC (upper row) and AFM images (lower row) of VLS-grown SiC films on 4° off 4H-SiC (000-1) substrates via a pure Si, 5 at% Pt-added, and 10 at% Pt and 18 at% Al co-added Si fluxes, respectively. Reproduced with permission from [44]. Copyright 2020, Elsevier.

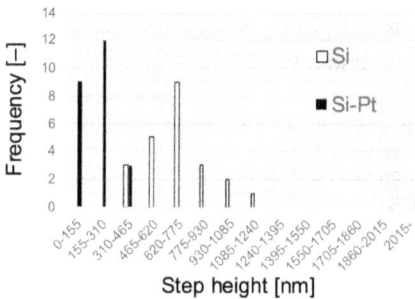

Flux	Si	Si-Pt
Sample N	24	24
Ave. [nm]	661	196
SD	238	90

FIGURE 13

Step-height distributions for SiC films with Si and Si-Pt fluxes, respectively, along with the list of their average step-height and standard deviation in the statistics analysis. Reproduced with permission from [44]. Copyright 2020, Elsevier.

off-cut direction and the terraces between neighboring steps having rather narrow widths. That is, the step-bunching is significantly suppressed. In fact, as shown in Figure 13, from a statistical point of view, the step height distributions of the two SiC film surfaces are clearly different, resulting in the average step height values of 196 nm and 661 nm for the SiC films grown with the Pt-Si flux and the pure Si flux, respectively; the former is almost one-third smaller than the latter. An additional feature that can be found by AFM observation is that the step lines are even microscopically rather straight, different from those of SiC films grown with the pure Si as well as Al-added fluxes, in which the step edge angles are 120° and/or 60°, as discussed above. It should be pointed out that a further addition of Al to a Pt-Si flux induces, as is more clearly found in the AFM image, the step morphology with a unique step edge angle of 120°, while the straight steps and narrow terraces characteristic of the Pt-Si flux grown SiC film surface are still observed macroscopically in the DIC image. This result indicates that Al and Pt as a flux additive have their own roles independently to each other, i.e., the former is to form the step lines with their edge angle

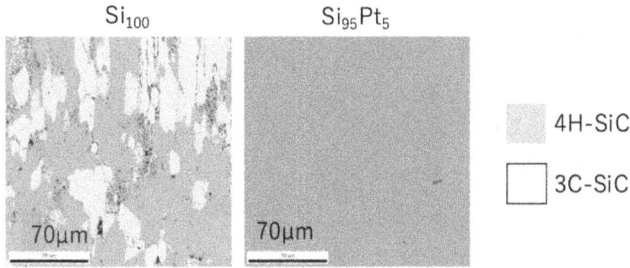

FIGURE 14

Polytype mappings (200 μm × 200 μm) of SiC films grown with the pure Si and 5 at% Pt added-Si fluxes, identified by EBSD, respectively. Reproduced with permission from [44]. Copyright 2020, Elsevier.

of 120° and the latter is to suppress the step-bunching, respectively. As a consequence, Pt is confirmed to have a catalytic-like surface flattening effect.

Regarding a possible additive effect of Pt on the polytype selectivity of 3C- or 4H-SiC, it was investigated by using an electron back scattered diffraction (EBSD) pattern method, which is a more surface sensitive technique than the Raman spectroscopy one in the identification of SiC polytypes; and the result is shown in Figure 14. A sizable portion of 3C-SiC polytype domains can be found in addition to 4H-SiC polytype domains on the SiC film grown with the pure Si flux. In contrast, when only a small amount of 5 at% Pt was added in the flux, 4H-polytype domains were found to cover all over the SiC surface, at least in a 200 μm × 200 μm square area, which leads to a conclusion of Pt being a more effective additive to stabilize the 4H polytype than Al in the flux under the present VLS growth conditions.

As for the effect of Pt suppressing the step-bunching on SiC crystal surfaces, there may arise one question about whether the straight step-and-narrow terrace surfaces are formed during or after the VLS growth of SiC films with the Pt-Si flux. A possibility that such a surface structure characteristic of the Pt-Si flux-grown SiC film is formed during the chemical etching with a mixture solution of HF and HNO_3 to remove the remaining flux cannot be ruled out. This is because Pt is originally known to work as a catalyst under the presence of a HF solution in CARE treatment to reconstruct the step-and-terrace surface structures of SiC bulk single crystals. In order to clarify at which step in the process the surface flattening occurs when Pt is added, the initial stage of the SiC growth at the solution-growth interface was again observed with the CLSM technique. Figure 15 displays the observation results for the pure Si flux and the $Si_{95}Pt_5$ flux, respectively, on 4° off-axis 4H-SiC(000-1) substrates at the initial stage of starting the temperature increasing. In the case of no adding Pt, round-shaped surface structures first appeared at random, following which step-and-terrace-like structures gradually developed (Figure 10(a)). In contrast, in the case when Pt was added, Figure 15(b) shows that the etching back process occurred upon the Si-Pt flux melted, thereby unveiling many polishing marks on the SiC seed crystal surface; and it was then followed by the appearance of similar straight step-and-narrow terrace structures as those found on the surfaces of VLS-grown SiC films with the Si-Pt flux; this process is schematically illustrated together in Figure 15(b). This observation is direct evidence for the straight step-and-narrow terrace structures being formed during the film growth, but not after the post chemical etching process. In addition, the resultant stable proceeding of step-flow growth, as is found in the case of adding Al, allows the growing SiC layer to more

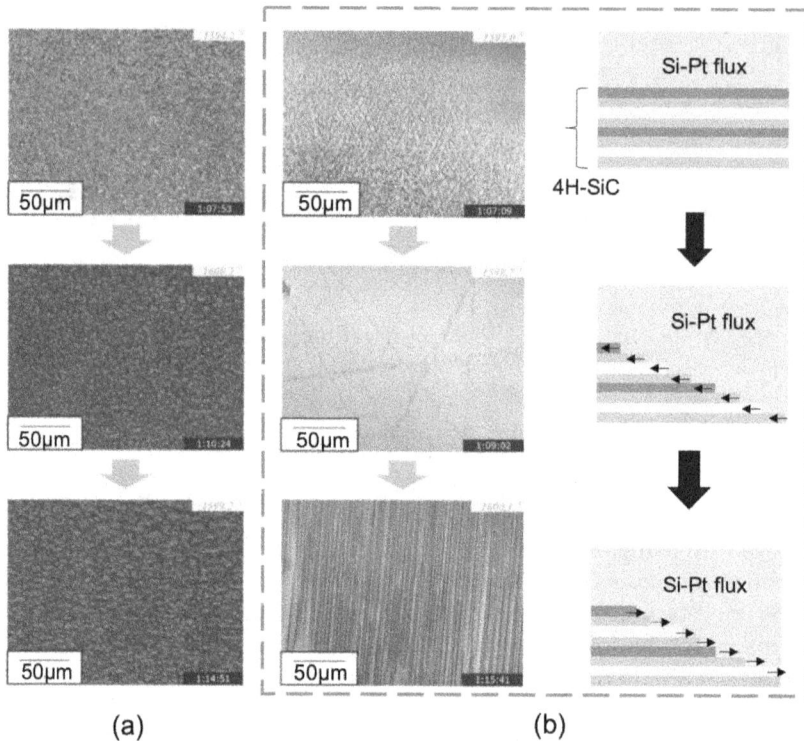

FIGURE 15
Comparison of sequential CLSM images for (a) the pure Si flux and (b) the $Si_{95}Pt_5$ flux on 4° off-axis 4H-SiC(000-1) substrates at the initial stage of starting the temperature increasing around 1600°C, along with a schematic model for the initial etching back process and its subsequent growth of SiC in the Si-Pt flux. Reproduced with permission from [44]. Copyright 2020, Elsevier.

effectively succeed to the 4H-stacking sequence of the 4° off seed substrate, leading to the dominant growth of the 4H-polytype SiC.

As shown above, it was found that the effects that Pt adding has on the VLS growth of SiC films include the suppression of the step-bunching and thereby the preferential growth of the 4H-SiC polytype. However, there is a concern about a possible incorporation of impurity elements such as Pt and Al in the flux into the grown SiC films, as similarly discussed in the solution growth of SiC bulk single crystals, which often causes a critical controversy as to the possibility of the VLS process being practically used for the epitaxial layer and wafer production processes in the SiC industry. Hence, the amount of impurity there was for a 400-nm-thick 4H-SiC film, which was VLS-grown with a $Pt_{0.2}Si_{0.8}$ flux at 1250°C by use of SIMS, was examined. The signal intensities of 28Si, 13C, and 195Pt are plotted against the sputtering time as shown in Figure 16(a). For the first 250 sec in the beginning of the sputtering, i.e., within ~50 nm in depth from the surface, which is estimated from the present sputtering rate of ~0.2 nm/sec, a non-negligible intensity of the Pt signal was detected. However, the prompt decrease of the Pt signal down to the detection limit suggests that the signal, which was detected at the very beginning of the sputtering, should come from the remaining flux even after its removal by the chemical etching, leading to a conclusion of almost no inclusion of Pt in the bulk. This result is, as

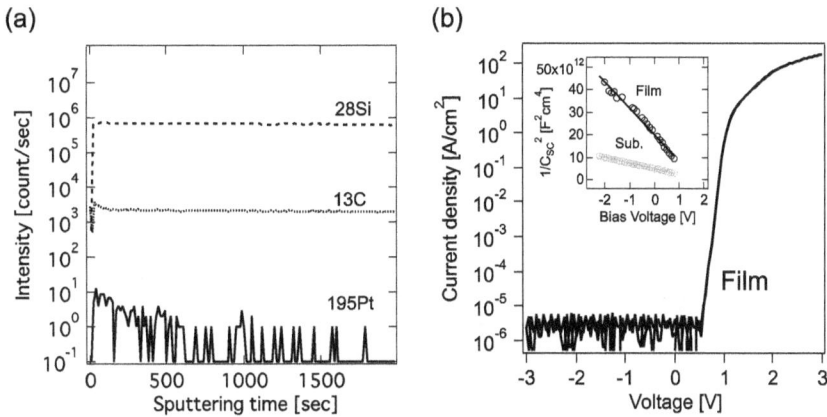

FIGURE 16

(a) SIMS signal intensities of 28Si, 13C, and 195Pt plotted against the sputtering time for a VLS-grown 4H-SiC film with a $Pt_{0.2}Si_{0.8}$ flux at 1250°C. The sputtering rate is 0.2 nm/sec. (b) Typical current-voltage curve for the Schottky junctions of the VLS-grown 4H-SiC film (a), along with the Mott-Schottky plots for the film and a bulk single crystal substrate, respectively (insets). Reproduced with permission from [44]. Copyright 2020, Elsevier.

the possibility was already pointed out, partly owing to the larger atomic radius of Pt than those of Si and C. Schottky junctions of the SiC film were then fabricated by evaporating Ni on the film surface through a hole mask for forming the top electrode pads (ϕ = 200 μm), while the ohmic contact was made on the bottom substrate with In-Ga alloy. Figure 16(b) shows a typical current-voltage curve for the Schottky junction, exhibiting a rectifying characteristic of the *n*-type semiconductor Schottky junction. The avalanche voltages (VR) are larger than -2 V, which is comparable value to that of an *n*-type 4H-SiC bulk single crystal used as a substrate. The donor density, which was estimated from the Mott-Schottky plot (inset), was 2.1×10^{18} cm^{-3}, which is on the same order as that of the substrate (5.5×10^{18} cm^{-3}). These results demonstrate that a proper choice among impurity flux elements such as Pt allows one to avoid unfavorable incorporation of the impurity element into the grown SiC crystals and, thereby, to obtain even in the VLS process high-quality 4H-SiC single crystal films whose electric properties are comparable to those of 4H-SiC bulk single crystals.

5 VLS-Like Growth of SiC Films

5.1 Basic Concept of VLS-Like Growth

In the VLS process, there are, at least two good reasons why the amount of flux used in the process should be as small as possible, from the engineering point of view. One is for increasing the maximum growth rate, which is expressed by eqn (4) as derived in section 2.2 for the VLS process. According to eqn (4), it is self-evident that the thinner the liquid layer is, the larger the achievable growth rate is. Second is for avoiding a possible dissolution of the substrate used in the case of heteroepitaxial growth of SiC films. In the VLS process, but it is not limited to the present SiC case, not a small amount of flux is prepared

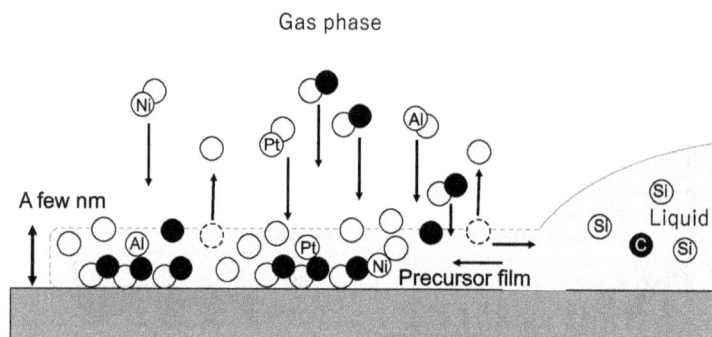

FIGURE 17

Schematic of a Si-based flux liquid droplet and its precursor film extending from the liquid droplet on a substrate. In the VLS process at high temperature, Si atoms and clusters are ready to re-evaporate from the precursor film as well as the liquid droplet, while the source of Si and C atoms are supplied from the gas phase for crystal growth.

on the substrate and melted by heating the substrate up to a temperature exceeding the melting or liquidous temperature of the flux prior to starting film deposition. In the meantime, the flux may dissolve part of the substrate to possibly cause unexpected damage of the substrate surface that would not be favored for the subsequent film growth.

On the other hand, from a basic scientific point of view, an interesting question would arise about whether the dissolution and crystallization in bulky liquid flux is essential or not in the VLS process. If more emphasis is put on possible catalytic-like effects of flux, such as the additive effects of Pt and Al as the impurity flux, the molecular-level interactions between individual atoms/molecules composing the flux and the precursors or surface atoms of the crystal would be rather important, in terms of directly activating the bond breaking and formation involved in the crystal growth process. The bulky liquid flux is, as shown in Figure 17, just a source of such atoms and molecules, and their spreading area around the liquid flux might be identical to what has been so far known as "precursor film" with a nanoscale thickness. In fact, there is a good reason to expect such a precursor film to play a similar role to the bulky liquid flux in the conventional VLS process. In our previous CLSM observation of micro-liquid droplets of NiSi$_2$ on a 4H-SiC single crystal surface, we found that the surface reforming occurred not only on the areas underneath the liquid droplets, but also on their surrounding areas in spite of no visible existence of liquid droplets [38]. This result strongly suggests that precursor films spread around the liquid droplets and work in the same way as the liquid flux, leading us to the following working hypothesis: if only such a flux precursor film is formed to homogeneously spread on a substrate and the subsequent growing film surface, the VLS or VLS-like growth should be realized without using the bulky liquid flux. However, the re-evaporation is not negligible for most kinds of flux at a temperature high enough for the VLS growth, and thus its molecular-level precursor film may not be stable alone. In order to compensate for the re-evaporation, one possible solution is that the flux source is continuously or sequentially supplied throughout the process such that, as schematically shown in Figure 17, the supply rate and the re-evaporation rate become balanced, enabling formation of the precursor or precursor-like film with a constant thickness on a nanoscale.

For this purpose, the high-vacuum type PLD-VLS chamber system that was introduced in section 3.1 is employed, in which SiC and Si-based flux sources can be alternately supplied many times throughout the deposition process. The degree of mixing in alternating the supply of the flux and SiC is controlled on a nanoscale by varying their deposition amounts in one cycle, the total of which is usually, e.g., comparable to or less than the unit cell length of SiC. In fact, the new approach has turned out to be successful to show a similar improvement in crystal quality of SiC films. Furthermore, the amount of the flux could be reduced down to 5 vol.% of the SiC thickness, e.g., a 10-nm-thick flux is enough for the growth of a 200-nm-thick SiC film. Based on this result, the new process has been named "VLS-like growth", the details of which will be described in the following sections.

5.2 Homoepitaxial Growth of SiC Films [52]

To demonstrate VLS-like growth by the alternating deposition approach, the seven SiC film samples were prepared under different deposition conditions, as listed in Table 3, with a growth temperature of 1000°C using a $NiSi_2$ flux. Samples #1~#4 are SiC films VLS-grown at a laser frequency of 5 Hz, with different flux amounts of 0%, 3%, 5%, and 15% in volume ratio to the 200-nm-thick SiC films with the other conditions being the same. On the other hand, sample #5~#7 are SiC films grown at a laser frequency of 10 Hz, w/ and w/o a $SiNi_2$ flux, respectively, and that only w/ a Si flux, to elucidate the effect of adding Ni in the Si-based flux.

Figure 18(a) is a comparison of the SEM images of SiC film surfaces for sample #1~#4, clearly showing that the grain size increases with an increase of the amount of SiN_2 flux supplied by the one-by-one cycle in the alternating deposition sequence. However, supplying an excess amount of the flux, which is 15% in volume ratio in the present case, just after deposition some droplet-like precipitates can be found, which are the $NiSi_2$ flux and the related compounds, on the SiC surface, indicating that the amount of flux should be optimized. A comparison of the Raman spectra for sample #1 and #3 is shown in Figure 18(b). In the SiC film, which was grown by just PLD without adding the $NiSi_2$ flux (sample #1), the D- and G-bands of graphite carbon are significantly observed, which suggests that the graphite carbon was formed from the excess carbon atoms resulting from a not-small amount of Si atoms re-evaporating even at the growth temperature of 1000°C. In contrast, the D- and G-bands completely disappeared, leaving only the SiC peak observed in this range of wavenumbers in the SiC film, which was grown with adding the

TABLE 3

Experimental alternating deposition conditions

Sample	Deposition thickness of $NiSi_2$ and SiC at 1 cycle	Number of cycle	Excimer laser frequency for SiC deposition (deposition time)
(#1)	Only SiC, 200m	-	5 (52 min)
(#2)	$NiSi_2$ 0.003 nm, SiC 0.1 nm	2000	5 (52 min)
(#3)	$NiSi_2$ 0.005 nm, SiC 0.1 nm	2000	5 (52 min)
(#4)	$NiSi_2$ 0.015 nm, SiC 0.1 nm	2000	5 (52 min)
(#5)	Only SiC, 200m	-	10 (27 min)
(#6)	$NiSi_2$ 0.005 nm, SiC 0.1 nm	2000	10 (27 min)
(#7)	Si 0.005 nm, SiC 0.1 nm	2000	10 (27 min)

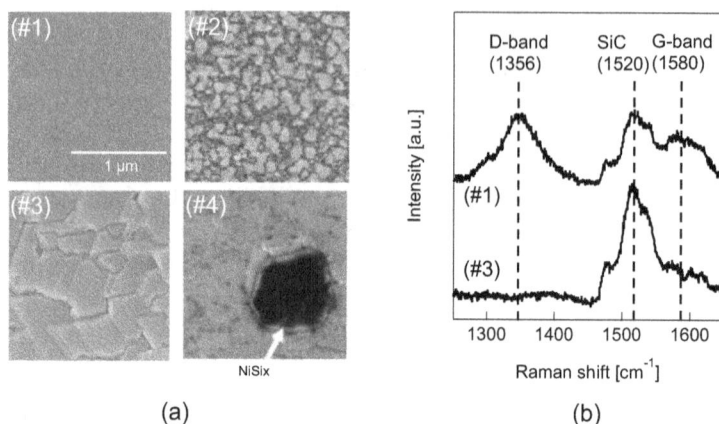

FIGURE 18
(a) Comparison of the SEM images of SiC film surfaces for sample #1~#4. (b) The Raman spectra for sample #1 and #3. Reproduced with permission from [52]. Copyright 2020, Elsevier.

optimal amount of the $NiSi_2$ flux (sample #3). It is thus concluded that one of the effects of the flux is to compensate for the amount of Si atoms that have re-evaporated in a vacuum during deposition at high temperature.

On the other hand, it should be further clarified if the compensation of the Si deficit alone is essential for the increase in the grain size, e.g., by the suppression of carbon aggregates, which would prevent the grains from being well developed in the crystal growth. Next, the results of sample #5~#7 are then compared, as shown in Figure 19. Even though the laser frequency is different from that of sample #1~#4, a carbon aggregation occurs when a SiC film is grown without adding any flux and its grain size is small, as indicated by the Raman spectrum and SEM image of sample #5. In contrast, the addition of only Si as the flux is effective to suppress such a carbon aggregation in SiC, as indicated by a complete disappearance of the D- and G-bands of graphite carbon in the Raman spectrum of sample #7. However, its SEM image shows that the grain size remains as small as those of the SiC film grown without adding any flux, allowing one to conclude that the Ni addition is indispensable for the increase of the grain size. Considering that Si melt alone is known to work as the flux in solution-growth processes for SiC single crystals, the difference could be whether the flux additive is able to become liquid during the VLS process. That is, since the present growth temperature of 1000°C is higher than the liquidous temperature of $NiSi_2$, but much lower than the melting point of Si (1414°C), Si alone could not behave as liquid, thereby being ineffective at increasing the grain size. If this is the case, being liquid flux, even though it is on the nanoscale, would be indispensable for achieving the VLS-like growth, the liquid state being favored for the flux to diffuse enough over the surface. Incidentally, in the SiC films of sample #3 and #6 both of which showed the increase of the grain size as well as the suppression of carbon aggregation, the EBSD pattern method confirmed that they were in the single phase of the 3C polytype with double domains relative to each other around the [111] axis by 60 degrees. The 3C polytype is known as being thermodynamically the most stable phase at a low growth temperature such as 1000°C, which suggests that the process proceeds under the conditions close to the thermodynamic equilibrium even in the VLS-like growth.

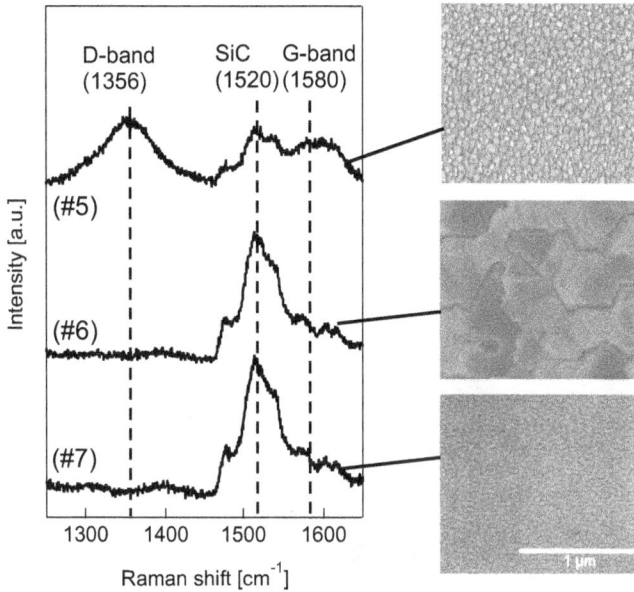

FIGURE 19

Comparison of the Raman spectra and SEM images for sample #5~#7. Reproduced with permission from [52]. Copyright 2020, Elsevier.

5.3 Heteroepitaxial Growth of SiC Films [53]

The VLS-like growth method, whose capability has been verified above, would become more useful in the heteroepitaxial growth of SiC. In the conventional VLS processes, where a bulky flux is prepared on a substrate prior to starting deposition, the bulky flux can erode the substrate surface, which may not have become a serious problem in the homoepitaxial growth with SiC single crystals as substrates, and, to make matters worse, some of the constituent atoms of the substrate may be incorporated as impurities into the grown SiC layer. Figure 20 shows a contrastive result of the experiments designed to mimic the VLS and VLS-like processes (Figure 20(a)) to illustrates how a NiSi$_2$ flux possibly erodes the surface of an α-Al$_2$O$_3$(0001) substrate. Even in the VLS-mimic experiment, the amount of the NiSi$_2$ flux was just 10 nm in thickness, which was first deposited on an α-Al$_2$O$_3$(0001) substrate, and it was then followed by annealing at 1000°C in a vacuum for a time corresponding to e.g., a deposition time in the VLS-like process described in the previous section. While in the VLS-like mimic experiment, the NiSi$_2$ flux was subsequently deposited on an α-Al$_2$O$_3$(0001) substrate at 1000°C, but without deposition of SiC, until the flux thickness reached 10 nm in total taking the same time as in the VLS-mimic experiment. From an AFM image of the substrate surface in the VLS-mimic experiment as shown in Figure 20(b), the substrate surface became very rough, where the RMS value significantly increased up to 3.93 nm from the original value of 0.24 nm before the treatment, and the etched pits could be as much as 15 nm in depth in the cross-sectional surface profile denoted by "VLS-treatment" in Figure 20(c). In contrast, in the VLS-like mimic experiment, no significant surface etching of the substrate was found, the RMS value remaining almost the same: less than 1 nm (0.82 nm). This result allows us to expect that the VLS-like process will be validated even in the heteroepitaxial growth of SiC films, without any more substrate etching.

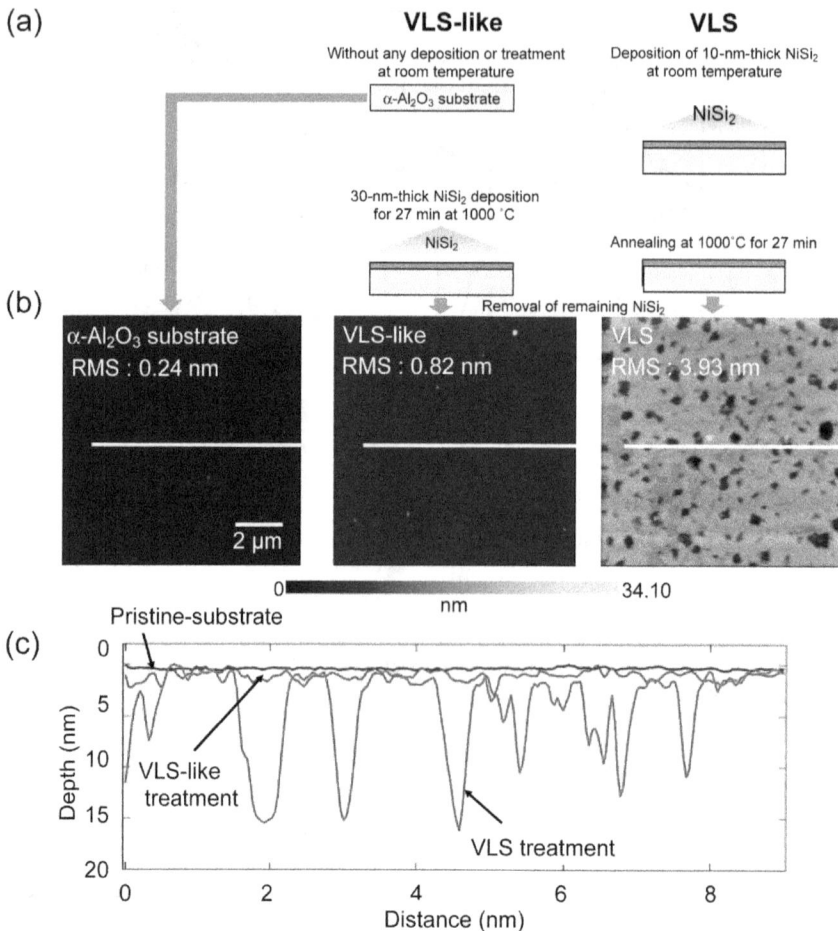

FIGURE 20
(a) Experiments designed to mimic the VLS and VLS-like processes to illustrates how a NiSi$_2$ flux possibly erodes the surface of an α-Al$_2$O$_3$(0001) substrate. (b) AFM images of α-Al$_2$O$_3$(0001) substrates after different flux treatments, respectively, along with a non-treated α-Al$_2$O$_3$(0001) substrate for reference. (c) Comparison of the cross-sectional surface profiles measured along the white lines on the AFM images of (b), respectively. Reproduced with permission from [53]. Copyright 2017, The Royal Society of Chemistry.

In order to demonstrate the VLS-like mechanism in the heteroepitaxial growth of SiC films, the six 3C-SiC film samples, whose polytype and double domain structure similarly found in the homoepitaxial case had been confirmed by XRD analysis prior to this experiment, were prepared by depositing SiC on α-Al$_2$O$_3$(0001) substrates at a growth temperature of 1000°C and at a laser frequency of 10 Hz using a NiSi$_2$ flux. As listed in Table 4, the total thickness of the flux was varied from 0 up to 60 nm, i.e., 30% in volume ratio to the 200-nm-thick 3C-SiC films, keeping the total process time constant (27 min), and the dependence of the flux amount on the VLS-like growth of SiC films was investigated as shown in Figure 21. In Figure 21(a), the intensity of the 111 reflection of 3C-SiC and its rocking curve FWHM value are plotted as a function of the flux amount (in thickness), where the intensity for each sample is normalized by that of the 0006 reflection of α-Al$_2$O$_3$. The 111 peak intensity gradually increases with the flux amount and saturates for the flux

TABLE 4

Experimental alternating deposition conditions

Sample	Deposition thickness of NiSi$_2$ and SiC at 1 cycle	Number of cycle	Excimer laser frequency for SiC deposition (deposition time)	Total flux amount
(#1)	Only SiC, 200m	-	10 (27 min)	0 nm
(#2)	NiSi$_2$ 0.005 nm, SiC 0.1 nm	2000	10 (27 min)	10 nm
(#3)	NiSi$_2$ 0.010 nm, SiC 0.1 nm	2000	10 (27 min)	20 nm
(#4)	NiSi$_2$ 0.015 nm, SiC 0.1 nm	2000	10 (27 min)	30 nm
(#5)	NiSi$_2$ 0.020 nm, SiC 0.1 nm	2000	10 (27 min)	40 nm
(#6)	NiSi$_2$ 0.030 nm, SiC 0.1 nm	2000	10 (27 min)	60 nm

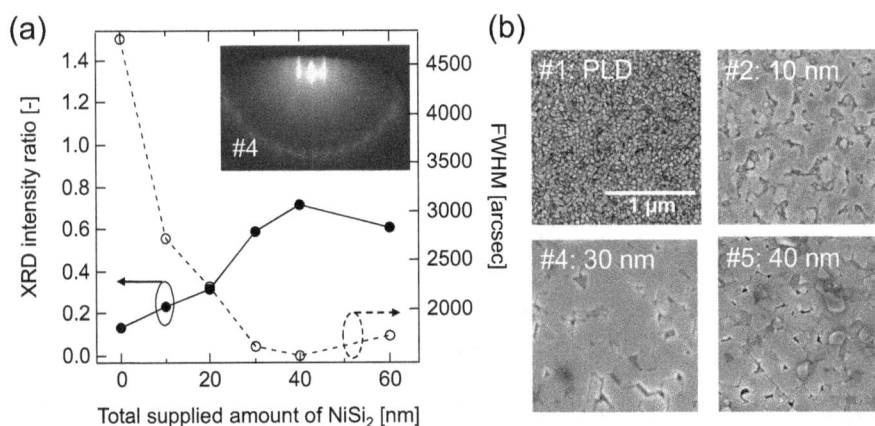

FIGURE 21

(a) The intensity of the 111 reflection of 3C-SiC and its rocking curve FWHM value plotted as a function of the flux amount (in thickness). (b) Typical SEM images of the representative SiC film surfaces of sample #1, #2, #4, and #5. Reproduced with permission from [53]. Copyright 2017, The Royal Society of Chemistry.

amounts over 30 nm. In contrast, the rocking curve FWHM value drastically decreases even by adding the flux of only 10 nm in thickness, and similarly saturated to be around 1500 arcsec for the flux amounts over 30 nm. In the VLS-like growth at such a heterointerface, the amount of the flux needed to improve the quality of SiC films to a degree of single crystalline is clearly larger than the optimal flux amount of 10 nm in the VLS-like growth of homoepitaxial films. As shown in Figure 21(b), comparing the surface morphology of the SiC films observed by SEM, it is also for around the flux amount of 30 nm that the increase in the grain size is most significant, and thus the optimal flux amount is inferred to be 30 nm of sample #4, exhibiting a sharp reflection high-energy electron diffraction (RHEED) (see the inset of Figure 21(a)), under the present conditions.

5.4 Visible Light Photocurrent Response

In the VLS-like growth, it has become possible to grow high-quality 3C-SiC films on not only 4H-SiC single crystal substrates, but also α-Al$_2$O$_3$(0001) substrates at a growth temperature of 1000°C. Finally, the optical property and photocatalytic ability of these 3C-SiC films are discussed as one of their physicochemical properties in this section. Figures 22(a)

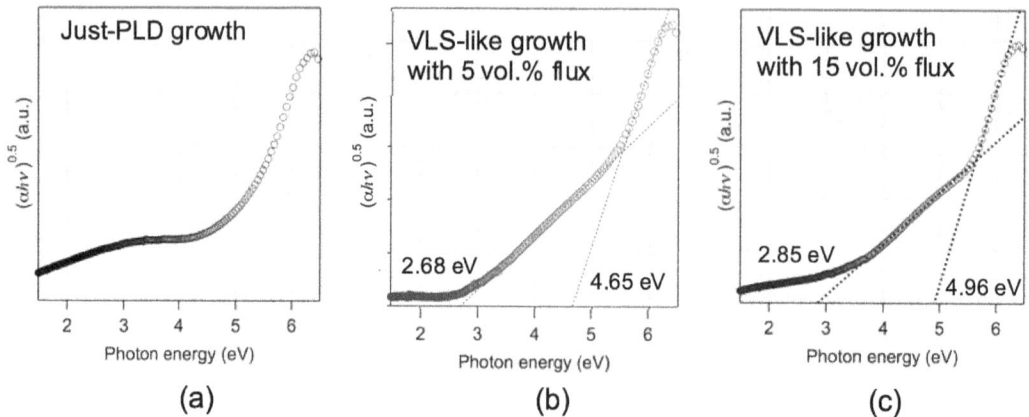

FIGURE 22

(a) Tauc plots for indirect allowed transitions of the SiC films fabricated by (a) PLD and VLS like growth with (b) 5 vol.% and (c) 15 vol.% flux, respectively. Reproduced with permission from [53]. Copyright 2017, The Royal Society of Chemistry.

and **(b)** are the Tauc plots of 50-nm-thick just-PLD and VLS-like (5 vol.% flux) grown SiC films on α-Al$_2$O$_3$(0001) substrates, which were constructed from their UV-visible absorption spectra respectively, on the basis of 3C-SiC reportedly having an indirect band gap of ~2.4 eV ($E_{c1}(\mathbf{X})$- $E_{v1}(\mathbf{\Gamma})$) [54, 55]. In the VLS-like grown film even with a 5 vol.% flux, the absorption edge was clearly observed at a photon energy between 2 and 3 eV, and then its indirect band gap was estimated to be 2.68 eV, though the value was somewhat larger than those previously reported, whose reasons will be later discussed in more detail. On the other hand, in the just-PLD grown film, the indirect band gap was difficult to estimate because the absorption edge was not so clear. This is due to additional absorptions near the band edge originating from defects possibly related to Si deficiency in the PLD film. In fact, from the micro-Raman mappings of the intensity ratio of the G-band of carbon (1600 cm^{-1}) and the A1g peak of sapphire (418 cm^{-1}) [56] over the film region (200 μm × 200 μm) of the SiC films, carbon aggregates are uniformly present in the entire film for the just-PLD grown sample, while they are almost not for the VLS-like grown sample. Furthermore, the second narrowest indirect transition was also observed in both SiC films, but more clearly in the VLS-like grown film, giving a value of 4.65 eV as the second indirect band gap for the VLS-like grown film. Some of the indirect transitions were also experimentally reported for 3C-SiC [57, 58], the most probable transition corresponds to the energy difference of $E_{c1}(\mathbf{L})$-$E_{v1}(\mathbf{\Gamma})$ of 4.2 eV as the second narrowest indirect transition in 3C-SiC. It should be pointed out that both the band gap and the second narrowest indirect transition energy are larger than those previously reported. A widening of the band gap in 3C-SiC is theoretically predicted to occur when Si is deficient in the SiC [59], but it would not be plausible in the VLS-like grown SiC films. In contrast, this is because they tend to shift toward higher energy with an increase of the total amount of NiSi$_2$ flux used in the VLS-like growth, e.g., shifting up to 2.85 eV and 4.96 eV, respectively, at 15 vol.% flux, as shown in Figure 22(c). Even if Ni were incorporated into the SiC films as an impurity, it would induce some in-gap states [60], the absorption edge rather shifting to lower photon energy. The origin of the band gap widening and its higher energy shift has not been clear at the moment, which should be clarified in our future work. As a consequence, not only the crystallinity and surface morphology, but also the optical property was much improved even in PLD-grown 3C-SiC films by employing the VLS-like growth approach.

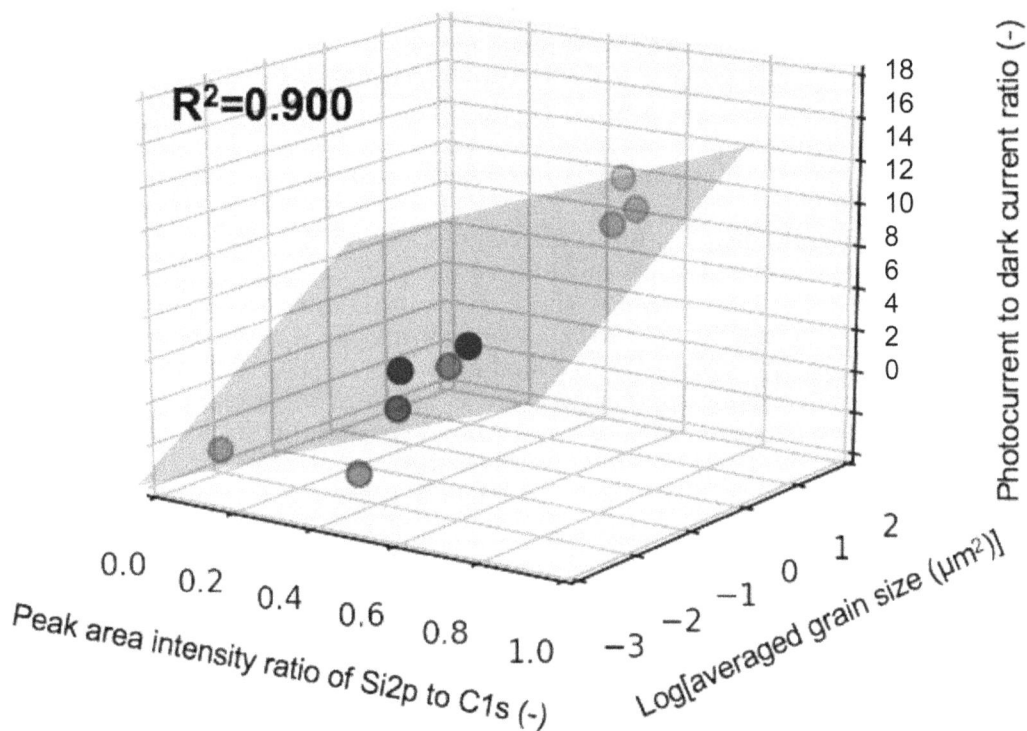

FIGURE 23

Three-dimensional plots of the time-averaged photocurrent to dark current ratio as a function of the peak area intensity ratio of Si2p to C1s in XPS and the averaged grain size estimated from the SEM images, for each sample, together with the best fitted linear regression plane. Reproduced with permission from [52]. Copyright 2020, Elsevier.

In measuring the visible light photocurrent response, the just-PLD and VLS-like grown 3C-SiC films on *n*-type C-face 4H-SiC substrates were used as a photoanode. The photo-irradiation with a high-pressure mercury lamp was carried out at an electrode potential of +0.5 V vs. Ag in a 0.3 M K_2SO_4 solution, where the light wavelengths shorter than 432 nm were cut by a dielectric multilayer filter. Figure 23 is a three-dimensional plot of the photo-current to dark current ratio against the peak area intensity ratio of Si2p to C1s in XPS data and the averaged grain size (in logscale) in SEM images, along with the best fitted linear regression plane; the correlation coefficient was 0.9. The photocurrent relative to the dark current is found to increase with decreasing the Si deficiency and increasing the grain size. This tendency is reasonably understood by the theoretically expected impact of the Si defi-ciency and grain size on the electric properties of SiC. That is, when Si is less deficient, SiC is more insulating, having a smaller dark current; when the averaged grain size is larger, the number of the recombination site is smaller, producing a larger photocurrent.

6 Conclusions and Future Prospects

In conclusion, this chapter has given a brief review of our recent progresses in developing an innovative, VLS mechanism-based thin film process for fabrication of SiC films with single crystal quality. As already pointed out in the Introduction, it has been almost two decades since the first challenge of VLS growth for SiC single crystal films, during which there has been substantial progress in the VLS process and related techniques, with a gradually growing interest in the possibility of industrial applications of VLS to the SiC film process. However, we would, unfortunately, have to say that we need more efforts to further boost the potentiality of the VLS process being commercially adopted in the mass-production of SiC films.

In newly establishing the VLS growth process for a material film, like conventional bulk solution-growth processes, the exploration of new additive materials to the flux and optimization of the conditions for their use are generally carried out in a try-and-error manner, thereby taking much time and effort. This situation would be, I believe, a practical reason why such additives can be regarded as a kind of catalyst, besides a scientific one of the additives chemically playing important roles in crystal growth like catalysts in chemical reactions. However, most of the researchers who are engaged in crystal growth of SiC and the related technical issues seem hesitant to embark on intensive development of the VLS process for SiC single crystal films. This might be because the current technical situation has not been matured enough to allow them to have any successful vision that will counterbalance their time and effort. As already mentioned before, one of the most significant technical drawbacks that has been pointed out in developing the VLS process for semiconductor films is the undeniable possibility of flux additives being incorporated into the semiconductor film; thus, the VLS process is concluded to be not appropriate as a semiconductor process technique.

However, it has been found that VLS-grown 4H-SiC films with the Pt-Si alloy as a new flux include almost no Pt atoms detectable by SIMS, exhibiting excellent Schottky junction properties comparable to those of a 4H-SiC bulk single crystal. In fact, our recent attempt of the VLS growth of 4H-SiC films with the Pt-Si alloy flux by CVD has demonstrated that the residual donor density in the resultant films can be reduced to as low as the order of 10^{15} cm^{-3} [61]. This result could do away with the long existing concern about possible high residual donor densities in the VLS-grown SiC films. Furthermore, the x-ray topography experiment also has revealed that the propagation behaviors of dislocations in the VLS process are rather similar to those in the solution-growth process than in the conventional CVD process, illustrating its potential of effectively reducing the total dislocation density in the resultant SiC thick films [61]. All these recent results, I personally hope, could convince the researchers that it is worth their time and effort to develop the VLS process for SiC films, as in the development of new catalysts. The VLS-like growth, which was introduced in the latter part of this chapter, is also a next-generation approach to high-quality SiC films. Although the crystal quality is still inferior to those by the original VLS growth, it can improve the film quality more significantly than other vapor deposition processes and has enabled the heteroepitaxial growth of SiC films, which has been difficult in the conventional VLS process. In addition, the amount of flux can be much decreased in the VLS-like growth process, thereby further reducing the possibility of the flux additives being incorporated into the film as impurities.

Finally, I will close this chapter with expecting that much more researchers will be engaged in development of the VLS and/or VLS-like growth process(es) and that the VLS and the related process will become commercially adopted in mass-production of SiC films in the near future.

Acknowledgements

The author would like to give his sincere thanks to all the collaborators and students: Assoc. Prof. S. Maruyama, Prof. M. Kubo, Assist. Prof. K. Kaminaga, Dr. A. Onuma, Dr. N. Sannodo, Mr. R. Yamaguchi, Mr. K. Nakano, and Ms. A. Osumi in Tohoku University; Dr. H. Okumura, Dr. T. Kato, Dr. T. Mitani, Dr. N. Komatsu, Dr. Y. Yonezawa and Dr. K. Kojima in National Institute of Advanced Industrial Science and Technology (AIST). This work was partially supported by Collaborative Research Matching Grant Program between Tohoku University and AIST. This work was also supported by the Novel Semiconductor Power Electronics Project Realizing Low Carbon Emission Society under the New Energy and Industrial Technology Development Organization (NEDO), and by the Advanced Low Carbon Technology Research and Development Program (ALCA) of the Japan Science and Technology Agency (JST).

References

1. L. Patrick, W.J. Choyke, and D.R. Hamilton, "Luminescence of 4H-SiC, and Location of Conduction-Band Minima in SiC Polytypes", Phys. Rev. 137, A1515-A1520 (1965).
2. O. Konstantinov, Q. Wahab, N. Nordell, and U. Lindefelt, "Study of Avalanche Breakdown and Impact Ionization in 4H Silicon Carbide", J. Electron mater., 27, 335–341 (1998).
3. R. Mickevicius and J.H. Zhao, "Monte Carlo Study of Electron Transport in SiC", J. Appl. Phys., 83, 3161–3167 (1998).
4. G. Muller, G. Krotz, and E. Niemann, "SiC for Sensors and High-Temperature Electronics", Sens. Actuat. A: Phys., 43, 259–268 (1994).
5. D.M. Brown, E. Downey, M. Grezzo, J. Kretchmer, V. Krishnamethy, W. Hennessy and G. Michon, "Silicon Carbide MOSFET Technology", Solid State Electron, 39, 1531–1542 (1996).
6. S. Mounce, B. McPherson, R. Schupbach, and A.B. Lostetter, "Ultra-Lightweight, High … for Extreme Environments", Proc. IEEE Aerosp. Conf., 1–19 (2006).
7. R. Oka, K. Yamamoto, H. Akamine, D. Wang, H. Nakashima, S. Hishiki, and K. Kawamura, "High Interfacial Quality Metal-Oxide-Semiconductor Capacitor on (111) Oriented 3C-SiC with Al_2O_3 Interlayer and its Internal Charge Analysis", Jpn. J. Appl. Phys., 59, SGGD17 (2020).
8. R. Anzalone, S. Privitera, M. Camarda, A. Alberti, G. Mannino, P. Fiorenza, S. Di Franco, and F. La Via, Materials Science and Engineering B, 198, 14–19 (2015).
9. M. Kobayashi, H. Uchida, A. Minami, T. Sakata, R. Esteve, and A. Schoner, "3C-SiC MOSFET with High Channel Mobility and CVD Gate Oxide", Mater. Sci. Forum, 679–680, 645–648 (2011).
10. Y. Tokuda, E. Makino, N. Sugiyama, I. Kamata, N. Hoshino, J. Kojima, K. Hara, and H. Tsuchida, "Stable and High-Speed SiC Bulk Growth Without Dendrites by the HTCVD Method", J. Cryst. Growth, 448, 29–35 (2016).
11. M.A. Fanton, B.E. Weiland, D.W. Snyder, and J.M. Redwing, "Thermodynamic Equilibrium Limitations on the Growth of SiC by Halide Chemical Vapor Deposition", J. Appl. Phys. 101 014903-1–5 (2007).
12. N. Hoshino, I. Kamata, Y. Tokuda, E. Makino, T. Kanda, N. Sugiyama, H. Kuno, J. Kojima, and H. Tsuchida, "Fast Growth of n-Type 4H-SiC Bulk Crystal by Gas-Source Method", J. Cryst. Growth 478, 9–16 (2017).

13. R.S. Wagner and W.C. Ellis, "Vapor-Liquid-Solid Mechanism of Single Crystal Growth" Appl. Phys. Lett. 4, 89–90 (1964).

14. D.E. Bugaris and Hans-Conrad zur Loye, "Materials Discovery by Flux Crystal Growth: Quaternary and Higher Order Oxides", Angew. Chem. Int. Ed. 51, 3780–3811 (2012).

15. E. Kuphal, "Liquid Phase Epitaxy", Appl. Phys. A., 52, 380–409 (1991).

16. T.F. Kuech (ed.), Thin Films and Epitaxy, Handbook of Crystal Growth, Chapter 6 "Liquid-Phase Epitaxy", Michael G. Mauk, (2nd Ed.), (2015).

17. Y. Matsumoto, R. Takahashi, and H. Koinuma, "Flux-Mediated Epitaxy: A General Application in Vapor Phase Epitaxy to Single Crystal Quality of Complex Oxide Films", J. Cryst. Growth 275, 325–330 (2005).

18. K.S. Yun, B.D. Choi, Y. Matsumoto, J.H. Song, N. Kanda, T. Itoh, M. Kawasaki, T. Chikyow, P. Ahmet, and H. Koinuma, "Vapor-Liquid-Solid Tri-Phase Pulsed-Laser Epitaxy of $RBa_2Cu_3O_{7-y}$ Single-Crystal Films", Appl. Phys. Lett. 80, 61–63 (2002).

19. R. Takahashi, Y. Yonezawa, M. Ohtani, M. Kawasaki, K. Nakajima, T. Chikyow, H. Koinuma, and Y. Matsumoto, "Perfect $Bi_4Ti_3O_{12}$ Single-Crystal Films via Flux-Mediated Epitaxy", Adv. Funct. Mater. 16, 485–491 (2006).

20. G. Ferro, D. Chaussende, F. Cauwet and Y. Monteil, "Effect of the Si Droplet Size on the VLS Growth Mechanism of SiC Homoepitaxial Layers", Mater. Sci. Forum, 389–393, 287–290 (2002).

21. J. Lorenzzi, G. Zoulis, M. Marinova, O. Kim-Hak, J.W. Sun, N. Jegenyes, H. Peyre, F. Cauwet, P. Chaudouet, M. Soueidan, D. Carole, J. Camassel, E.K. Polychroniadis, and G. Ferro, "Incorporation of Group III, IV and V Elements in 3C–SiC(1 1 1) Layers Grown by the Vapor–Liquid–Solid Mechanism", Journal of Crystal Growth, 312, 3443–3450 (2010).

22. G. Ferro and C. Jacquier, "Growth by a Vapor–Liquid–Solid Mechanism: A New Approach for Silicon Carbide Epitaxy", New. J. Chem., 28, 889–896 (2004).

23. Yu-Ling Liang, Shih-Zong Lu, Hsin-Ying Lee, Xiaoding Qi, and Jow-Lay Huang, "Growth of 3C–SiC Films on Si Substrates by Vapor–Liquid–Solid Tri-Phase Epitaxy", Ceramics International 41, 7640–7644 (2015).

24. Y. Takeyama, S. Maruyama, and Y. Matsumoto, "Growth of Single-Crystal Phase Pentacene in Ionic Liquids by Vacuum Deposition", Cryst. Growth Des., 11, 2273–2278 (2011).

25. Y. Takeyama, S. Maruyama, H. Taniguchi, M. Itoh, K. Ueno, and Y. Matsumoto, "Ionic Liquid-Mediated Epitaxy of High-Quality C_{60} Crystallites in a Vacuum", CrystEngComm, 14, 4939–4945 (2012).

26. Kohei Kuroishi, Shingo Maruyama, Noboru Ohashi, Mio Watanabe, Kenta Naito, and Yuji Matsumoto, "Ionic Liquid-Assisted Growth of DBTTF-TCNQ Complex Organic Crystals by Vacuum Co-deposition", J. Cryst. Growth, 453, 34–39 (2016).

27. T. Torimoto, K. Okazaki, T. Kiyama, K. Hirahara, N. Tanaka, S. Kuwabata, "Sputter Deposition onto Ionic Liquids: Simple and Clean Synthesis of Highly Dispersed Ultrafine Metal Nanoparticles", Appl. Phys. Lett. 89, 243117 (2006).

28. Keisuke Okawara, Tomonobu Nishimura, Shingo Maruyama, Masaki Kubo and Yuji Matsumoto, "In-Vacuum Electropolymerization of Vapor-Deposited Source Molecules into Polymer Films in Ionic Liquid", React. Chem. Eng., 5, 33–38 (2020).

29. Mihai Stafe, Aurelian Marcu, Niculae N. Puscas, "Pulsed Laser Ablation of Solids, Basic, Theory and Applications", SSSUR, vol. 53, Springer (2014).

30. R. Yamaguchi, A. Osumi, A. Onuma, K. Nakano, S. Maruyama, Y. Mitani, T. Kato, H. Okumura and Y. Matsumoto, "Effects of Al Addition to Si-Based Flux on the Growth of 4H-Sic Films by Vapor-Liquid-Solid Pulsed Laser Deposition", CrystEngComm, 19, 5188–5193 (2017).

31. S. Maruyama, N. Sannodo, R. Harada, Y. Anada, R. Takahashi, M. Lippmaa and Y. Matsumoto, "Pulsed Laser Deposition with Rapid Beam Deflection by a Galvanometer Mirror Scanner", Rev. Sci. Instrum. 90, 093901 (2019).

32. P. Dold, E. Ono, K. Tsukamoto, G. Sazaki, "Step Velocity in Tetragonal Lysozyme Growth as a Function of Impurity Concentration and Mass Transport Conditions", J. Cryst. Growth 293, 102–109 (2006).

33. M. Azhagurajan, T. Kajita, T. Itoh, Y.-G. Kim, K. Itaya, "In Situ Visualization of Lithium Ion Intercalation into MoS_2 Single Crystals Using Differential Optical Microscopy with Atomic Layer Resolution", J. Am. Chem. Soc. 138, 3355–3361 (2016).
34. K. Tsukamoto, P. Dold, "Interferometric Techniques for Investigating Growth and Dissolution of Crystals in Solutions", AIP Conf. Proc. 916, 329–341 (2007).
35. G. D. Saldi, G. Jordan, J. Schott, E. H. Oelkers, "Magnesite Growth Rates as a Function of Temperature and Saturation State. Geochim. Cosmochim. Acta, 73, 5646–5657 (2009).
36. K. Fujiwara, Y. Obinata, T. Ujihara, N. Usami, G. Sazaki, K. Nakajima, "In-Situ Observations of Melt Growth Behavior of Polycrystalline Silicon", J. Cryst. Growth, 262, 124–129 (2004).
37. S. Kawanishi, M. Kamiko, T. Yoshikawa, Y. Mitsuda, K. Morita, "Analysis of the Spiral Step Structure and the Initial Solution Growth Behavior of SiC by Real-Time Observation of the Growth Interface", Cryst. Growth Des. 16, 4822–4830 (2016).
38. Aomi Onuma, Shingo Maruyama, Takeshi Mitani, Tomohisa Kato, Hajime Okumura and Yuji Matsumoto, "Uniform Growth of SiC Single Crystal Thin Films via a Metal-Si Alloy Flux by Vapour-Liquid-Solid Pulsed Laser Deposition: The Possible Existence of a Precursor Liquid Flux Film", CrystEngComm, 18, 143–148 (2016).
39. A. Onuma, S. Maruyama, N. Komatsu, T. Mitani, T. Kato, H. Okumura and Y. Matsumoto, "Quantitative Analysis of Nanoscale Step Dynamics in High-Temperature Solution-Grown Single Crystal 4H-SiC via In Situ Confocal Laser Scanning Microscope", Cryst. Growth Des. 17, 2844–2851 (2017).
40. K. Nakano, S. Maruyama, T. Kato, Y. Yonezawa, H. Okumura and Y. Matsumoto, "Direct Visualization of Kinetic Reversibility of Crystallization and Dissolution Behavior at Solution Growth Interface of SiC in Si-Cr Solvent", Surf. Interfaces, 28, 101664 (2022).
41. M. Mitomo, Y. Inomata and M. Kumanomido, "The Effect of Doped Aluminium on Thermal Stability of 4H- and 6H-SiC", Yogyo Kyokaishi, 78, 224–228 (1970).
42. N. Komatsu, T. Mitani, Y. Hayashi, T. Kato, S. Harada, T. Ujihara and H. Okumura, "Modification of the Surface Morphology of 4H-SiC by Addition of Sn and Al in Solution Growth with SiCr Solvents", J. Cryst. Growth, 458, 37–43 (2017).
43. Binary Alloy Phase Diagrams, ed. T. B. Massalski, et, al., ASM International, Materials Park, Ohio, 2nd edn, 1998, vol. 1.
44. A. Osumi, K. Nakano, N. Sannodo, S. Maruyama, Y. Matsumoto, T. Mitani, T. Kato, Y. Yonezawa, H. Okumura, "Platinum Additive Impacts on Vapor-Liquid-Solid Growth Chemical Interface for High-Quality SiC Single Crystal Films", Materials Today Chemistry, 16, 100266-1–8 (2020).
45. Y. Yonezawa, M. Ryo, A. Takigawa, and Y. Matsumoto, "Screening of Metal Flux for SiC Solution Growth by a Thin-Film Combinatorial Method", Sci. Technol. Adv. Mater. 12, 054209-1–4 (2011).
46. K. Danno, H. Saitoh, A. Seki, H. Daikoku, Y. Fujiwara, I. Ishii, H. Sakamoto, and Y. Kawai, "High-Speed Growth of High-Quality 4H-SiC Bulk by Solution Growth Using Si-Cr Based Melt", Mater. Sci. Forum 645–648, 13–16 (2010).
47. G.O. Filip, B. Epelbaum, M. Bickermann, and A. Winnacker, "Micropipe Healing in SiC Wafers by Liquid-Phase Epitaxy in Si–Ge Melts", J. Cryst. Growth 271, 142–150 (2004).
48. K. Kamei, K. Kusunoki, N. Yashiro, N. Okada, K. Moriguchi, H. Daikoku, M. Kado, H. Suzuki, H. Sakamoto, and T. Bessho, "Crystallinity Evaluation of 4H-SiC Single Crystal Grown by Solution Growth Technique using Si-Ti-C Solution", Mater. Sci. Forum 717, 45–48 (2012).
49. R.H. Davies, A.T. Dinsdale, J.A. Gisby, J.A.J. Robinson, and S.M. Martin, Calphad, "MTDATA – Thermodynamic and Phase Equilibrium Software from the National Physical Laboratory", 26, 229–271 (2002).
50. G. Aylward and T. Findlay's, SI Chemical Data, fifth ed., John Wiley & Sons, Queensland, Australia, 2008.
51. H. Hara, Y. Sano, H. Murata, K. Arima, A. Kubota, K. Yagi, J. Murata, and K. Yamauchi, "Novel Abrasive-Free Planarization of 4H-SiC (0001) using Catalyst", J. Electron. Mater., 35, L11-L14 (2006).

52. N. Sannodo, A. Osumi, S. Maruyama and Y. Matsumoto, "Vapor-Liquid-Solid-Like Growth of Thin Film SiC by Nanoscale Alternating Deposition of SiC and $NiSi_2$", Appl. Surf. Sci., 530, 147153 (2020).

53. N. Sannodo, A. Osumi, K. Kaminaga, S. Maruyama and Y. Matsumoto, "Vapour-Liquid-Solid-Like Growth of High-Quality and Uniform 3C-SiC Heteroepitaxial Films on α-Al_2O_3(0001) Substrates", CrystEngComm, 23, 1709–1717 (2021).

54. J.B. Casady and R.W. Johnson, "Status of Silicon Carbide (SiC) as a Wide-Bandgap Semiconductor for High-Temperature Applications: A Review, Solid-State Electronics, 39, 1409–1422 (1996).

55. G.L. Zhao and D. Bagayoko, "Electronic Structure and Charge Transfer in 3C- and 4H-SiC", New Journal of Physics 2, 16.1–16.12 (2000).

56. S.P.S. Porto and R.S. Krishnan, "Raman Effect of Corundum", J. Chem. Phys., 47, 1009–1012 (1967).

57. C. Persson and U. Lindefelt, "Relativistic Band Structure Calculation of Cubic and Hexagonal SiC Polytypes", J. Appl. Phys., 82, 5496–5508 (1997).

58. Physics of Group IV Elements and III–V Compounds, edited by O. Made-lung et al., Landolt-Börnstein, New Series, Group III, Vol. 17a Springer, Berlin, 1982.

59. A. Trejo, M. Calvino and M. Cruz-irisson, "Chemical Surface Passivation of 3C-SiC Nanocrystals: A First-Principle Study", International Journal of Quantum Chemistry, 110, 2455–2462 (2010).

60. Yankun Dou, Hai-bo Jin, Maosheng Cao, Xiaoyong Fang, Zhiling Hou, Dan Li, and S. Agathopoulos, "Structural Stability, Electronic and Optical Properties of Ni-doped 3C–SiC by First Principles Calculation", J. of Alloy and Compounds, 509, 6117–6122 (2011).

61. Naoki Sanoodo, Tomohisa Kato, Yoshiyuki Yonezawa, Kazutoshi Kojima, and Yuji Matsumoto, "Vapor-Liquid-Solid Growth of 4H-SiC Single Crystal Films with Extremely Low Carrier Densities in Chemical Vapor Deposition with Pt-Si Alloy Flux and X-Ray Topography Analysis of its Dislocation Propagation Behaviors", Cryst.EngComm., 23, 5039–5044 (2021).

3

Spectroscopic Investigations for the Dynamical Properties of Defects in Bulk and Epitaxially Grown 3C-SiC/Si (100)

Devki N. Talwar

Department of Physics, University of North Florida, Jacksonville, Florida, USA
Department of Physics, Indiana University of Pennsylvania, Pennsylvania, USA

1 Background

Since 1957, power electronics has been dominated by the well-established silicon (Si) technology. With the inception of thyristors, many Si-based switching devices were commercially launched to meet several industrial applications and performance requirements [1-17]. In the past 65 years, the working operation of Si-based devices has now been approaching intrinsic limitations due to its basic characteristics in terms of the blocking voltage, operational temperature, conduction and switching traits, etc. Several novel wide bandgap (WBG) semiconductor switches have emerged in the last few years as the favorite with substantial advancements in the power electronic converters. The major devices to become accessible with astounding commercial values are being engineered by using either silicon carbide (SiC) or gallium nitride (GaN) materials. While the SiC-based devices and circuits are aimed at high-temperature, high-power, and high-radiation conditions, the GaN materials are intended for lower voltage and low power usage. Our focus here is to concentrate on comprehending experimentally and theoretically the basic characteristics of SiC.

2 Silicon Carbide: A Wide Bandgap Material for Power and Microelectronics

WBG semiconductors such as SiC, GaN, and diamond (C) offer the potential to overcome the problems of temperature and high voltage blocking that bedevil Si. With excellent mechanical and electrical properties, SiC has proven to be the most attractive alternative to offer significant advantages at high temperature and high voltage limits in developing electronic devices and sensors for modern technology. Despite the exceptional traits of SiC

DOI: 10.1201/9780429198540-4

over Si for designing the next-generation power electronics, the issues related to higher wafer cost, material processing, controlling defect structures, anisotropies, mechanical and thermal properties have restricted its limited use. None of these problems are too critical. Extensive research efforts over the past decades around the World have been quite dramatic for the development of SiC technology, with major improvements in material processing and wafer growth. The successful development of SiC-based devices will promise a significant reduction in the switching losses – permitting their usage not only at higher frequencies, high temperature, high power but also in hostile-environment conditions. The ability of SiC to function under such extreme situations is anticipated improvements for supporting significant global expansions to far-ranging diversity of industrial applications. These undertakings are envisioned valuable from a substantially improved high-voltage switching for energy savings in the local electric power distribution and electric motor drives to expand powerful microwave electronics for radars, communications to sensors and controlling the cleaner-burning for fuel-efficient jet aircraft and automobile engines. For implementing SiC as a power device material, it is essential to adopt the device engineering concepts and processing skills of the matured Si technology while administering different reliability tests required of the novel SiC system. The ultimate success of SiC as an electronic technology will depend on the close interplay of research in fundamental material science and progress in the design of electronic devices and packaging. In the next section we have outlined the main characteristics of SiC with emphasis on its crystalline structure and electronic properties. These characteristics are ultimately responsible for the highest interest SiC has in power- and micro-electronics [1-17].

2.1 Crystalline Structure

SiC exhibits in a crystal structure like bulk silicon and diamond materials – except its base is made up of two different types of atoms Si and C, which are tetrahedrally coordinated by means of sp^3 hybridized bonds. The nearest-neighbor bond length of SiC is around 1.89 Å – longer than diamond but shorter than silicon with a bond angle of 109° identical to other materials having the zincblende (zb) structure.

2.1.1 Polytypism

The SiC exhibits in several hundred polymorphs and their structures can be visualized (cf. Figure 1) from Si-C subunits if organized in a diverse stacking arrangement [18].

The simplest structure is a cubic (or zb): 3C-SiC or β-SiC, while the non-cubic structures are either the hexagonal (wurtzite (wz)) or rhombic (referred to as α-SiC). For each Si-C bilayer (A), there are two possible sites (B, C) used for identifying various polytypes by stacking them along the c-axis of a hexagonal close-packed system. The structure of 3C-SiC displayed in Figure 1 (left) reveals bilayers oriented by three possible positions of ABC with respect to the crystal lattice for maintaining a tetrahedral bonding and periodicity. The 4H-SiC polytype (cf. Figure 1 middle) consists of an equal number of cubic (k) and hexagonal (h) bonds with a stacking sequence of ABCB while the 6H-SiC polytype (cf. Figure 1 right) is composed of two-thirds of cubic and one-third of hexagonal bonds with a stacking sequence of ABCACB. The overall symmetries of α-SiC polytypes are hexagonal despite the presence of cubic bonds in different structures. The stacking sequence with major differences among the five commonly established SiC polytypes are summarized in Table 1.

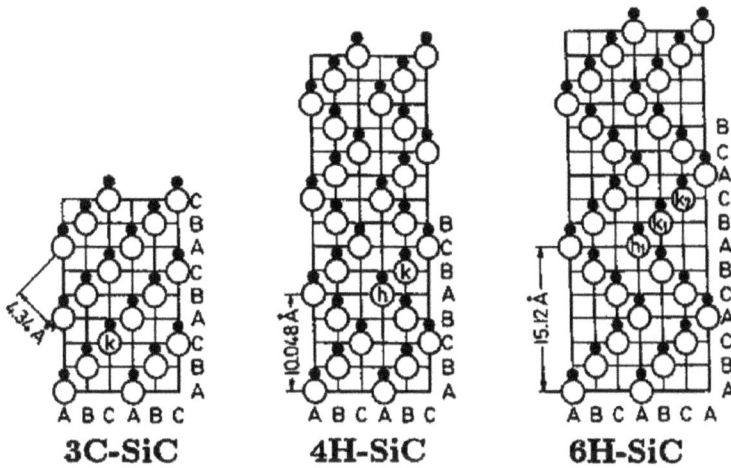

FIGURE 1
Schematic models of the three most common, also commercially available, SiC polytypes. H and kC denote the hexagonal and cubic sites in the lattice, respectively. Open and full circles stand for the silicon and carbon atoms, respectively. From [18] (Figure 1), with permission for reproduction from Elsevier.

TABLE 1

Stacking sequences in the *c*-axis direction for three major SiC polytypes [18]

	Stacking sequence	No. hexagonal (*h*)	No. cubic (*k*)
3C	ABC	0	1
4H	ABCB	1	1
6H	ABCACB	1	2

2.1.2 Growth of SiC

Different techniques are employed in preparing SiC materials [19-36]. For the bulk SiC growth, a seeded sublimation or "modified-Lely method" has been commonly used [26-31]. Due to the phase equilibrium in Si and C, the method is based on a physical vapor transportation (PVT) where the source (SiC) in the form of a powder is sublimed at ~ 2300–2500 °C in a closed crucible and deposited on a seed crystal maintained at lower temperature [26]. Breakthrough in the modified-Lely method has enabled the growth of large diameter single high-quality SiC crystals, which resulted in the rapid progress for the epitaxial growth of SiC thin films [19–25].

Another alternative technique is the high-temperature chemical vapor deposition (HT-CVD) method where transport of growth species to the seed crystal is directly offered by the high purity gas precursors containing Si- and C-species. The thermal environment and growth rates achievable in this technique are to a larger extent close to the PVT method [28-34]. It is to be noted that the sublimation epitaxy is a proven suitable approach for the growth of nearly ~100 μm thick layers with smooth as-grown surfaces. Reproducible quality of these surfaces is obtained with the growth rates between 2 to 100 μm/h in the temperature range from 1600 to 1800 °C. Structural quality of the samples are improved compared to the substrate and the surface roughness diminished in the sublimation epilayer.

Another simple and elegant method is the liquid phase epitaxy (LPE) with several advantages viz., the low process temperature, high growth rate, easy for technical implementations in different geometries including doped layers and multilayer structures. The main advantage of LPE [35–36] is that the process is carried out at a relatively low temperature and close to thermodynamic equilibrium conditions, which presumes low concentration of point defects in epitaxial layers. Consequently, the quality of grown materials is mainly limited by morphological features.

2.1.3 Emerging Interest in Cubic SiC

Among the growth of SiC, major research efforts for device engineering are focused only on three important structures (i.e., 3C-SiC, 6H-SiC, and 4H-SiC) having diverse physical properties (see Table 2). For many years 6H-SiC was preferred in power electronics – it is only recently that 4H-SiC has gained the supremacy. Major pitfalls in the epitaxial growth of 4H- and 6H-SiC are related, however, to the high cost of good quality substrates. To realize cost-effective electronic devices, an alternative route adopted in recent years is to prepare 3C-SiC on Si substrate [37–47].

From Table 2, one may note that 3C-SiC has the smallest bandgap and higher electron mobility. From an engineering standpoint, the cubic phase exhibits an improved critical electric field over Si. For device fabrications, the use of 3C-SiC needs no expensive and challenging "super-junction" processes and require only practices employed in Si-based manufacturing. Although, 3C-SiC/Si has been effectively grown in research environments, the material is not currently available commercially. The crystal growers are, however, persistently exploring many other opportunities of using cubic SiC beyond power electronics. The prospects include investigating its employment in micro-electro-mechanical systems (MEMS), light-emitting diodes (LEDs), bio-medical devices as well as substrate/buffer layers for growing III-nitrides and graphene [37–47].

2.1.4 3C-SiC/Si Processing Issues

Despite several advantages of 3C-SiC/Si there still exist some serious material processing issues. While both 3C-SiC and Si exhibit the same structure and stacking sequence, the two materials have a large (~19.8%) lattice mismatch. This disparity in lattice constants implies

TABLE 2

Comparison of the basic characteristics for major SiC polytypes with Si and diamond

Property Crystal structure	Si diamond	4H-SiC wurtzite	6H-SiC wurtzite	3C-SiC cubic	Diamond diamond
Lattice constant (Å)	5.431	3.07;10.05	3.08;15.12	4.3596	3.57
Bandgap E_g (eV)	~1.1	~3.23	~3.05	~2.36	~5.5
Density (g/cm³)	2.329	3.21	3.21	3.213	3.52
Bulk modulus (GPa)	97.8	220	220	250	442
Melting temperature (K)	1690	3003	3003	2746	5000
Electron mobility (cm² V⁻¹ s⁻¹)	1400	900	400	800	2200
Hole mobility (cm² V⁻¹ s⁻¹)	600	120	90	320	1600
Saturation electron velocity (10⁷ cm/s)	2.5	2.7	2.7	2.5	2.5
Thermal conduction (W/cm K)	1.5	3.7	4.9	3.63	20
Breakdown field (10⁶ V/cm)	0.25	3.5	3.5	3.0	10

four layers of Si equal to five layers of SiC, which create faults after every five layers to cause stacking disorders and crystalline defects at the 3C-SiC/Si interface. Another issue has been the large (~ 8%) difference in their thermal expansion coefficients, which induces stress at the interface of 3C-SiC and Si to instigate cracks and/or macroscopic bending [37-47]. While thin epitaxial films of 6H- (4H- SiC) are grown at higher temperature (> 1450 °C) on commercially available substrates, the preparation of 3C-SiC/Si requiring lower (~1350 °C) temperature below the melting point of Si is a challenge. However, many improvements have been made in the qualities of 3C-SiC/Si epifilms since its inception in the late 1980s. The prospective of low production cost and easy integration with Si technology, have rekindled new awareness in 3C-SiC as an attractive alternative to 4H- and 6H-SiC.

To continue the development of SiC-based technology and sustain improvements in the efficiency and performance of devices, research efforts need to be continued at the level of comprehending material's physical characteristics.

2.1.5 Basic Properties of SiC

Despite significant success in the growth front and employing SiC in emerging technologies, only a limited number of investigations are carried out for comprehending its basic properties – especially the physics behind those aspects that determine the importance of SiC polytypes at a practical level. Assessing the local arrangement of constituent atoms in SiC and epifilms has been and still is a challenge. Appraising the quality of materials, evaluating stress, identifying defects, understanding crystalline disorder, surface/interface morphology, film thickness etc. are all imperative for improving the materials' electrical, optical, and device characteristics. From the existing studies on physical properties [1-5] – a major portion of investigations concentrates on the electronic structure and optical properties of undoped/doped SiC materials. The results of such studies are summarized in a review article by Choyke and Devaty [48].

Besides studying the electronic traits, the exploration of phonon characteristics of SiC has been equally crucial for comprehending not only the lattice dynamics and polytypism of SiC but also assessing the site selectivity of defects by studying their vibrational properties. Except for limited results of phonon dispersions by Raman scattering spectroscopy (RSS) on SiC polytypes [49–62], inelastic x-ray scattering (IXS) for 3C-SiC [63] and inelastic neutron-scattering (INS) on 6H-SiC and 4H-SiC [64, 65], spectroscopic ellipsometry (SE) [66–72] and other [73] methods – thorough investigations on the phonon characteristics are still not available. While the experimental measurements by RSS are only confined to the Γ - A direction in the Brillouin zone (BZ), complete phonon dispersions in the low symmetry directions are simulated by various theoretical methods [74–78]. Based on these efforts, it is strongly felt that there is a compelling reason for systematic investigations on the structural, lattice dynamical, thermodynamic, and defect properties of SiC.

The purpose of this chapter is to provide an overview of both the experimental and theoretical (cf. Sec. 3) methods of phonons for comprehending the structural and dynamical characteristics of bulk SiC materials and their heterostructures. As the structure of various polytypes are linked to one another through the stacking of Si-C layers along the c-axis with distinctive unit cells [18], atypical energies of phonon modes are expected in both α- and β-SiC. The study of lattice dynamics of SiC (cf. Sec. 3) not only reflects on the structural behavior but also adds insight into its polytypism. As the electronic contributions to the free energy of different polytypes vary only by exceedingly small amounts, the role of phonons in stabilizing certain phases is a matter of careful theoretical investigations.

Group theoretical classifications of phonons in SiC polytypes are given in Sec. 3.1. From experimental standpoints, the spectroscopic methods (cf. Sec. 3.2) including Fourier transform infrared (FTIR) spectroscopy has a potential of providing valuable supplementary information (cf. Sec. 3.2.1–3.2.4) to the RSS results on SiC polytypes. Careful analyses of IR reflectivity and transmission spectra of 3C-SiC/Si (100) epifilms based on theoretical models (cf. Sec. 4) within an effective medium approximation is equally valuable. From the Kramers-Krönig analysis on the IR data of bulk SiC we have accurately determined not only the phonon peak positions but also the widths or damping constants using the Drude-Lorentz method (cf. Sec. 4.1, 4.1.1–4.1.2). In the ideal configurations the reflectance spectra of hetero-epitaxial materials are studied (cf. Sec. 4.2) by adopting a three-phase model (air/epifilm(s)/substrate) with dielectric functions $\tilde{\varepsilon}_1 = 1$ (air), $\tilde{\varepsilon}_2 = \tilde{\varepsilon}_{tf}$ (thin-film), and $\tilde{\varepsilon}_3 = \tilde{\varepsilon}_s$ (substrate). The effects of surface roughness and "conducting" interfacial transition layer are appropriately included (cf. Sec. 4.3) in modifying the classical theory to elucidate the observed damping in the Fabry-Perot (FP) interference fringes at high frequency in some of the 3C-SiC/Si (100) samples. Polarization-dependent infrared spectroscopy (transmission or reflection) at oblique incidence in 3C-SiC/Si epilayers (cf. Sec. 4.4) and superlattices (cf. Sec. 4.5) is equally valuable for an unambiguous assessment of their optical phonons and also determining the LO-plasmon coupling modes for estimating the charge carrier concentration in doped materials. Bruggeman's "two-component" effective medium theory has provided not only a very good account for the atypical structures seen at the top of the Reststrahlen band but also offered valuable insights into the microstructural geometry and topology for the heterogeneous mixture of individual media structures. Comprehensive simulations of the infrared spectra of V-CVD 3C-SiC/Si epilayers is reported in Sec. 5 using a classical model within the effective-medium approximation (EMA). The effects of surface roughness and "conducting" interfacial transition layer are appropriately included in modifying the classical theory to elucidate observed damping FP interference fringes at high frequency in some of the V-CVD grown 3C-SiC/Si (100) samples (cf. Secs. 5.1, 5.1.1–5.1.7, 5.2, 5.2.1, 5.2.2). Theoretical results of reflectivity/transmission spectra are appraised with the experimental data for a diverse group of 3C-SiC/Si (100) samples grown under various conditions having different thicknesses. The "Si-C" and "Si-Si" bond lengths are evaluated by a careful analysis of the recent synchrotron radiation extended x-ray absorption fine structure (SR-EXAFS) data (cf. Secs. 6, 6.1). In Sec. 7, we have reported the results of impurity vibrational modes in 3C-SiC by incorporating phonons from a realistic rigid-ion model (RIM) fitted to the IXS [63] data (cf. Sec. 7.1) and expending apposite group-theoretic selection rules. The relaxation around Si/C atoms is accomplished for isolated defects by a first-principles bond-orbital model (BOM) [76]. This study has assisted evaluating the necessary force constant variations to simulate the local vibrational modes (LVMs) of various defect-centers using Green's function (GF) methodology (cf. Secs.7.2, 7.2.1–7.2.4) [76]. The simulated results of the vibrational properties for both perfect and imperfect 3C-SiC are reported and compared with the existing experimental/theoretical [76–77] data. The summary of our findings with concluding remarks are presented in Sec. 8.

3 Phonons

In crystalline solids, the phonons are the quanta of atomic vibrations. If the crystal unit cell contains N atoms, then 3N degrees of freedom result in 3 acoustic phonons and 3N-3 optical modes. These phonons can propagate in the lattice of a single crystal as a wave and exhibit dispersions depending on their wavelength or equivalently their wavevector in the Brillouin zone (BZ). The atomic vibrational frequencies in the crystalline solids range from zero to about ~ 100 THz. A more common unit for describing the mode frequencies is wavenumber (cm^{-1}) obtained by dividing the actual frequency by the velocity of light or by inverting the wavelength. The acoustic phonons have frequencies from zero to about a few hundred wavenumbers while the optical phonons have higher frequencies. Consequently, methods exploited to study the phonons included the use of FTIR, RSS, SE, INS, and IXS, etc. Besides these experimental techniques, there exist extensive numerical simulations [63–65] for investigating the influence of doping on the optical properties of both β- and α-SiC.

Theoretically, the major spectral features in SiC materials, resulting from the cation-anion charge transfer and/or electronic bandgap transitions in the ultraviolet region, are studied from their lattice vibrations. While the dynamical motions involving charges in the dipole moment are IR active, the nearest-neighbor interactions between Si and C atoms, however, produce a strong band in the infrared region regardless of the polytypes. Several experimental measurements are carried out in the past for extracting the vibrational, microstructures, elastic, thermal, mechanical, doping, and defect characteristics of SiC polytypes. Here, we have summarized the phonon, structural and defect properties of SiC polytypes from the standpoint of both experimental [62] and numerical simulations [63–65].

3.1 Group Theoretical Classification of Phonons

From the perspective of lattice dynamics, the cubic 3C-SiC is the simplest structure (space group T_d^2) having two atoms per unit cell with six phonon dispersions. In contrast, the hexagonal (4H- and 6H-) SiC polytypes with 8 and 12 atoms per unit cell (space group C_{6v}^4) pose critical problems as the existing resolution of INS and IXS is not sufficient for resolving the 24 and 36 closely lying phonon dispersions falling within the same energy range as the six phonon branches of 3C-SiC.

This problem can be eluded by using RSS, which provides a much better energy resolution but yields only frequencies of the optical modes at the center (i.e., Γ point or $\bar{q} \sim 0$) of the BZ. One must note that both 4H- and 6H-SiC hexagonal polytypes contain altogether 12 operations, which satisfy the symmetry conditions [79]. For 4H-SiC the total number of phonons at Γ point for the C_{6v} group is given by

$$\Gamma_{C_{6v}} = 4A_1 + 4B_1 + 4E_1 + 4E_2, \tag{1}$$

with 24 distinct phonon branches. One of the four A_1 and E_1 modes are acoustic (three in total, as E modes are doubly degenerate) while $3A_1, 4B_1, 3E_1$, and $4E_2$ are optical. Among these phonons, the A_1, E_1, and E_2 modes are Raman active while A_1 and E_1 are IR active. In each polytype while the long-wavelength modes are Raman active, they also strung out along the axial dispersion curves in the large-zone picture.

3.2 Spectroscopic Methods

Structural differences in the SiC polytypes affect their interaction with light in two different ways: First of all, if the light passes through β-SiC it lacks a preferred direction, however, in the layers of α-SiC it causes interaction with Si-C dipoles to differ for the electric field vector (E) traveling perpendicular to the crystal's *c*-axis (ordinary ray) or parallel to the *c*-axis (extraordinary ray). Secondly, one also expects different numbers of vibrational modes for the SiC polytypes based on their symmetry [62]. Therefore, in SiC polytypes we infer the number, type, and frequency of vibrational modes from the increase in size of the unit cells over β-SiC, i.e., from the folding of phonons in the first Brillouin zone [62]. Here, we are considering high-resolution IR absorption, infrared spectroscopic ellipsometry (IR-SE), RSS, and SR-EXAFS) to study the vibrational modes and the structural characteristics of SiC polytypes.

3.2.1 Infrared Spectroscopy

Infrared spectroscopy provides valuable information for the identification of phonon characteristics of the bulk and epitaxially grown materials – complementary to RSS. The analysis of IR spectra in SiC samples has been commonly used for determining their crystal structures (see Sec. 4) as well as identifying the site selectivity of different defect centers (see Sec. 7).

3.2.2 Infrared Spectroscopic Ellipsometry

The study of IR-SE has recently gained sufficient interest in the quantitative determination of both the optical properties associated with the free charge carriers and the lattice vibrations [66–72] of SiC polytypes. Quantitative measurement for the complex dielectric response of the nitrogen-doped 4H and 6H silicon carbide crystals in the midinfrared ~700 to 4000 cm^{-1} range has also been available for comprehending the free-carrier absorption, as well as assessing the effective mass anisotropy (cf. Sec. 4).

3.2.3 Raman Scattering Spectroscopy

Raman scattering spectroscopy is one of the fastest, non-destructive, easily operable, practical, and quite efficient methods for probing the phonon features in both the perfect and doped SiC wafers and epifilms (cf. Sec. 4).

3.2.4 X-ray Absorption Fine Structure Spectroscopy

X-ray absorption fine structure (XAFS) is extensively used for investigating the atomic-scale local structures and localized chemical states around specific elements in different materials [80]. The SR-EXAFS region of XAFS provides information on the chemical species, bond lengths, and coordination numbers of the neighbors of absorbing atoms in SiC polytypes (cf. Sec. 4).

4 Optical Response Theory

In this section we have briefly outlined the significance of light-phonon interactions, provided basic equations to help comprehend the infrared spectral traits as well as the optical characteristics of 3C-SiC, which can be measured experimentally. Here we start with an optical response of the bulk materials (cf. Sec. 4.1) to the IR radiation by using a classical methodology. The theory will be extended in Secs. 4.2–4.3 to the heteroepitaxial 3C-SiC/Si epilayers.

4.1 Drude-Lorentz Model of IR Spectroscopy on Bulk 3C-SiC

In the far IR region, the physical processes involved in crystal lattices can be described by the interactions between radiation and matter. This interaction is articulated by the wave-vector and frequency-dependent dielectric response function $\tilde{\varepsilon}(\omega,\vec{q})$. In zb type polar semiconductor materials there are two main processes that contribute to $\tilde{\varepsilon}(\omega,\vec{q})$: (a) the free-carrier effect $[\tilde{\varepsilon}_e(\omega,\vec{q})]$ from the electrons in the conduction band or from the holes in the valence band, and (b) the lattice effect $[\tilde{\varepsilon}_i(\omega,\vec{q})]$ from the optical phonons [81]. In the limiting case, where the wave-vector $\vec{q}$ approaches zero, the general form of the dielectric function in the classical "Drude-Lorentz" model takes the form

$$\tilde{\varepsilon}(\omega) = \tilde{\varepsilon}_e(\omega) + \tilde{\varepsilon}_i(\omega). \tag{2}$$

In this approach, the phonons in bulk 3C-SiC are treated like the charged harmonic oscillators, which can be set in motion by the light alternating electric field. The process described by the interaction of electromagnetic radiation with solids is expressed in terms of the complex refractive index $\tilde{n}(\omega)$ via the dielectric function $\tilde{\varepsilon}(\omega)[\tilde{n}(\omega)=\sqrt{\tilde{\varepsilon}(\omega)}]$. In zb type 3C-SiC one can consider two main practices, which contribute to $\tilde{\varepsilon}(\omega)$: (a) the free-charge carrier effect in either n- or p-doped materials caused by electrons or holes, and (b) lattice effect from the optical phonons. Therefore, a classical form of $\tilde{\varepsilon}(\omega)$ for doped zb binary materials developed by Drude-Lorentz, and others can be adopted using

$$\tilde{\varepsilon}(\omega) = \tilde{\varepsilon}_e(\omega) + \tilde{\varepsilon}_i(\omega)$$

$$\tilde{\varepsilon}(\omega) = \varepsilon_\infty\left[1-\frac{\omega_p^2}{\omega^2+i\omega\gamma}\right]+\frac{(\varepsilon_0-\varepsilon_\infty)\omega_{TO}^2}{\omega_{TO}^2-\omega^2-i\omega\Gamma},$$

$$= \varepsilon_\infty\left[1-\frac{\omega_p^2}{\omega^2+i\omega\gamma}\right]+\frac{S\omega_{TO}^2}{\omega_{TO}^2-\omega^2-i\omega\Gamma}. \tag{3}$$

Here, the first term in Eq. (3) describes the interaction between charge carriers and electromagnetic wave of frequency ω; $\gamma(=\frac{1}{\tau})$ is the free carrier damping constant with a relaxation time of τ; $\omega_p\left\{\equiv\left[\frac{4\pi n e^2}{m^*\varepsilon_\infty}\right]^{1/2}\right\}$ is the plasma frequency of free carrier concentration Γ with the effective mass m^*, and e is the magnitude of electron charge. The second term in Eq. (3) describes the interaction of electromagnetic wave with the optical phonons. Here ε_0 and ε_∞ are, respectively, the static and high-frequency dielectric constants; ω_{TO} is the TO-phonon, $S(\equiv(\varepsilon_0-\varepsilon_\infty)$ or $\equiv\varepsilon_\infty(\frac{\omega_{LO}^2}{\omega_{TO}^2}-1))$ an oscillator strength and Γ is the phonon damping. The ω_{LO} phonon frequency is related to the well-known Lyddane-Sachs-Teller (LST) relationship, i.e., $\omega_{LO}\left[\equiv\omega_{TO}\left(\frac{\varepsilon_0}{\varepsilon_\infty}\right)^{1/2}\right]$.

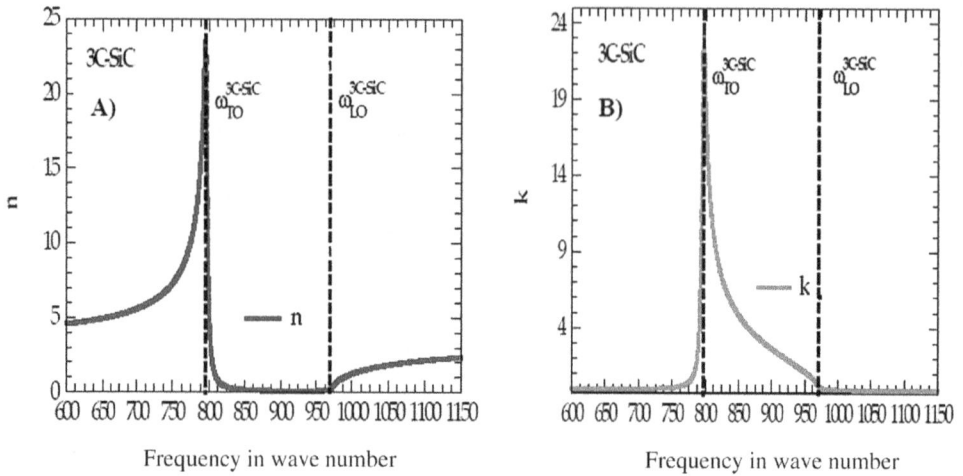

FIGURE 2
Calculated n (left A) and κ (right B) of 3C-SiC based on Eq. (3).

4.1.1 The Refractive Index

In the FIR spectroscopy, two directly observed quantities viz., the refractive index n, and absorption coefficient α are encapsulated via the complex refractive index $\tilde{n}(\omega)(= n + i\kappa)$ and the extinction coefficient $\kappa = \alpha c/2\omega$. A monochromatic light $[E(z,t) = E_0 e^{i(\tilde{\omega} \tilde{n} z/c - \omega t)}]$ of frequency ω propagating in a material of refractive index $\tilde{n}$ exhibits reduction (Beer's law) in both the electric field amplitude and the intensity by $e^{-z\alpha/2}$ and $e^{-z\alpha}$, which reflects the foundation of absorption coefficient. The real part of $\tilde{n}$ with a constant value at low frequency exhibits a sharp (see Figure 2) maximum at ω_{TO}. This reflects the fact that close to resonance the high amplitude of phonon oscillations makes the material optically dense. In fact, the speed of light in a material (proportional to c/n) is significantly reduced close to ω_{TO}. The small value of refractive index n would be zero (without damping Γ) for frequencies between ω_{TO} and ω_{LO} is another manifestation of the fact that in such a region the light cannot propagate in the material. The behavior of κ is somehow complementary, being zero for frequencies outside the ω_{TO}-ω_{LO} region. In such a region, κ has a maximum value close to ω_{TO}, where the absorption of light is ascribed to the high amplitude of the phonon coordinate, leading to energy dissipation towards other phonons through lattice anharmonicities. The measured complex refractive index can be compared to the dielectric constant $\tilde{\varepsilon}$ (cf. Eq. (3)) arising from the Drude-Lorentz model via the relations

$$\epsilon_1 = n^2 - \kappa^2; \epsilon_2 = 2n\kappa, \tag{4 a}$$

where

$$n(\omega) = \left[\frac{1}{2}\left\{\left(\varepsilon_1^2 + \varepsilon_2^2\right)^{1/2} + \epsilon_2\right\}\right]^{1/2} ; \kappa(\omega) = \left[\frac{1}{2}\left\{\left(\varepsilon_1^2 + \varepsilon_2^2\right)^{1/2} + \epsilon_2\right\}\right]^{1/2}. \tag{4 b}$$

4.1.2 Reflectivity

Another relevant optical property of infrared active phonons is the reflectivity, which can also be measured experimentally. The reflectivity R is related to the refractive index by the relationship

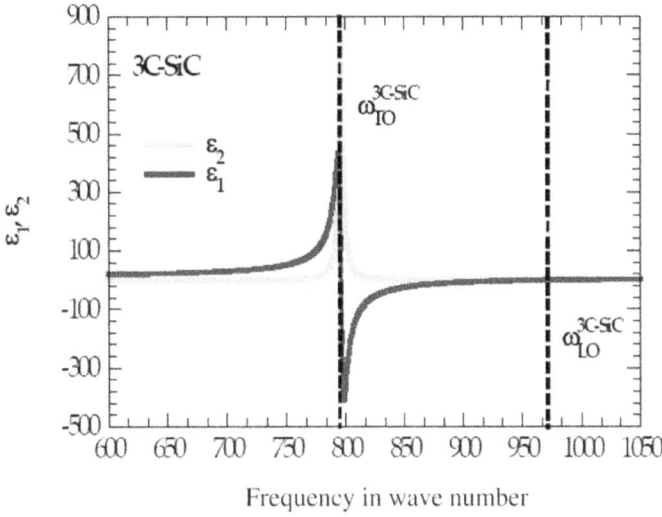

FIGURE 3
Calculated and ϵ_1 and ϵ_2 of 3C-SiC based on Eq. (4a).

$$R = \left|\frac{1-\tilde{n}}{1+\tilde{n}}\right|^2 = \left|\frac{1-\sqrt{\tilde{\varepsilon}}}{1+\sqrt{\tilde{\varepsilon}}}\right|^2 . \tag{5}$$

From Eq. (5) one can notice that unlike $\tilde{n}$ and $\sqrt{\tilde{\varepsilon}}$, which are complex, R is a real quantity and therefore does not completely describe the interaction of light with the phonon. The complete information of phonons is, however, contained in the reflection coefficient $\tilde{r}$, which is the ratio between the incoming and reflected electric field. These two quantities are related by

$$R = |\tilde{r}|^2 , \tag{6}$$

which shows that the phase information is lost when going from $\tilde{r}$ to R. The peculiar reflectivity of an infrared active phonon exhibits a constant value at low frequency, is highly enhanced towards ω_{TO}, where it reaches values close to one, and remains very high (see Figure 4) up to $\omega \approx \omega_{LO}$. This high reflectivity region is called the "Reststrahlen band" because the light that cannot enter the material (as already mentioned when discussing $\tilde{n}$) gets mostly reflected. At a frequency slightly higher than ω_{LO} where R drops to zero. This happens as ϵ_1 approaches value ~ 1, implying that the light can propagate in the material like it is in air, and therefore no light gets reflected at the sample surface.

The reflectivity, albeit containing only part of the material properties, is easy to measure, since in most cases the detectors used in the experiments can only measure the light intensity, and not the electric field profile. The full material properties can still be retrieved, in the case of infrared active phonons, by fitting R with the model presented here. All the optical constants ε_0, ω_{TO}, and Γ of silicon carbide used throughout this chapter are obtained in this manner, with ε_∞ taken from the literature.

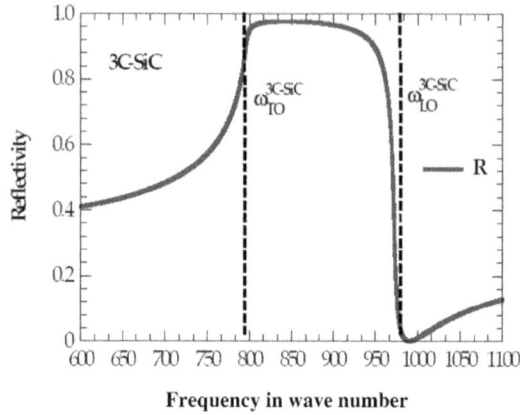

FIGURE 4
Calculated reflectivity of 3C-SiC based on Eq. (6).

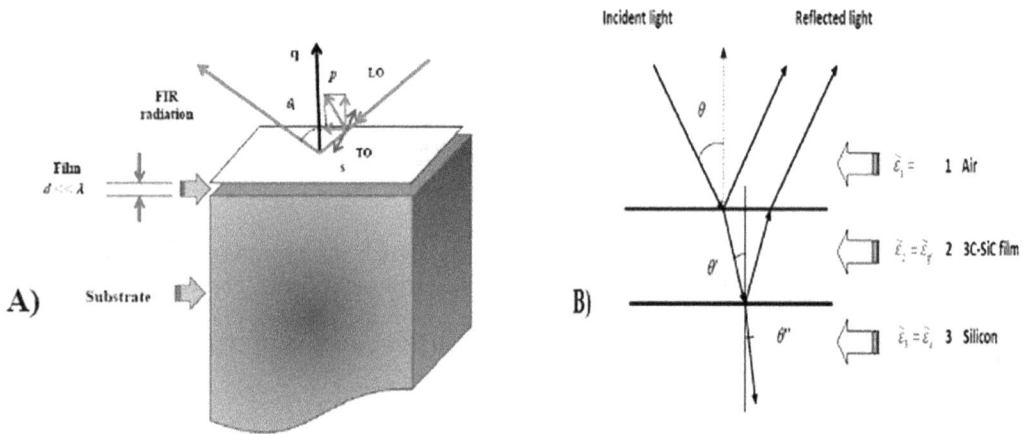

FIGURE 5
(A) 3C-SiC polar semiconductor thin film grown on a thick Si substrate. The directions of the s (perpendicular, $\perp$) and p (parallel, $||$) components of the far-infrared radiation incident are at an oblique angle to the surface of a thin film (perpendicular to the phonon wave-vector $\tilde{q}$) of thickness $d \ll \lambda$ grown on a thick substrate. (B) Sketch of a three-phase ideal model (air/epifilm(s)/substrate) with dielectric functions $\tilde{\varepsilon}_1 = 1$ (air), $\tilde{\varepsilon}_2 = \tilde{\varepsilon}_{tf}$ (thin-film), and $\tilde{\varepsilon}_3 = \tilde{\varepsilon}_s$ (substrate) for studying the reflectance and transmission spectra of thin films grown on a substrate.

4.2 Infrared Reflectivity of 3C-SiC/Si Epilayers: Ideal Configuration

In the ideal configurations (see Figure 5) the reflectance spectra of hetero-epitaxial materials can be studied by adopting a three-phase model (air/epifilm(s)/substrate) with dielectric functions $\tilde{\varepsilon}_1 = 1$ (air), $\tilde{\varepsilon}_2 = \tilde{\varepsilon}_{tf}$ (thin-film), and $\tilde{\varepsilon}_3 = \tilde{\varepsilon}_s$ (substrate). In this scheme, the coherent light reflected by the epifilm/substrate and air/epifilm interfaces can interfere to give rise to FP fringes above the Reststrahlen band. Following Cadman and Sadowski [82], the reflection $\tilde{r}$ and transmission $\tilde{t}$ coefficient, at near normal incidence for such an epifilm of thickness d can be expressed as

$$\tilde{r}_{123} = \frac{\tilde{r}_{12} + \tilde{r}_{23}\exp[i2\beta]}{1 + \tilde{r}_{12}\tilde{r}_{23}\exp[i2\beta]}, \tag{7}$$

$$\text{and } \tilde{t}_{123} = \frac{(1+\tilde{r}_{12})(1+\tilde{r}_{23})\exp[i2\beta]}{1 - \tilde{r}_{12}\tilde{r}_{23}\exp[i2\beta]}, \tag{8}$$

in terms of the Fresnel coefficients $\tilde{r}_{ij} = \dfrac{\tilde{n}_i - \tilde{n}_j}{\tilde{n}_i + \tilde{n}_j}$ and phase multiplier $\beta = 2\pi d\omega\sqrt{\tilde{\varepsilon}_2}$ for calcu-

lating the power reflection $R(\omega) = \left|\tilde{r}_{123}\right|^2$ and transmission $T(\omega) = \left|\tilde{t}_{123}\right|^2$. The above approach for simulating reflectivity spectra of hetero-epitaxial thin films at near normal incidence can be extended to oblique incidence (cf. Sec. 4.4) as well as to superlattice structures (cf. Sec. 4.5).

4.3 Infrared Reflectivity of 3C-SiC/Si Materials: Modified Model

In the modified model we have assumed a thin interfacial transition layer (TL) of thickness d_2 near the 3C-SiC/Si interface and meticulously included surface roughness δ at 3C-SiC/air, and δ_2 at 3C-SiC/TL interfaces, respectively. The notion of surface roughness and "conducting" TL has been insinuated in many recent surface characterization experiments [e.g., x-ray photoelectron spectroscopy (XPS), inverse photoelectron spectroscopy (IPES), and scanning tunneling microscopy (STM)]. From Shokhovets *et al.* the scattering factors χ, χ_2 of the ripples at interfaces follow the Gaussian distributions [81]:

$$\chi = \exp\{-16(\pi\delta/\lambda)^2\} \text{ and } \chi_2 = \exp\{-16(\pi\delta_2/\lambda)^2\} \tag{9}$$

which depend upon interfacial conditions (δ, δ_2 – rms surface roughness in µm) and the wavelength λ of incident photon. By using Eq. (7), the total reflection coefficient $\tilde{r}$ and the power reflection $[R(\omega) = \left|\tilde{r}_{123}\right|^2]$ at near normal incidence in the modified model can be evaluated by using

$$\tilde{r} = \frac{\chi \cdot \tilde{r}_{12} + \chi_2 \cdot \tilde{r}'\exp[i2\beta]}{1 + \chi \cdot \chi_2 \cdot \tilde{r}_{12}\tilde{r}'\exp[i2\beta]} \quad \text{with} \quad \tilde{r}' = \frac{\tilde{r}_{22'} + \tilde{r}_{2'3}\exp[i2\beta']}{1 + \tilde{r}_{22'}\tilde{r}_{2'3}\exp[i2\beta']} \tag{10}$$

and appropriately including the Fresnel coefficients at the 3C-SiC/TL interface with phase multiplier $\beta' = 2\pi d_2\omega\sqrt{\tilde{\varepsilon}_{2'}}$.

4.4 Infrared Reflectivity of 3C-SiC/Si Epifilms at Oblique Incidence

In the conventional FTIR spectroscopy, the measurements of reflection and transmission at near normal incidence are used to extract optical constants in compound semiconductors including 3C-SiC (cf. Sec 4.2). Due to the transverse characteristics of the electromagnetic radiations one expects to observe in polar materials only the zone center ω_{TO} phonons at normal incidence. In thin films, however, the interaction of electromagnetic field with the material can be enhanced by an oblique incidence. Using infrared spectroscopy (transmission or reflection) at an oblique incidence it was Berreman [83] who first predicted theoretically and demonstrated experimentally the observation of ω_{TO} and ω_{LO} phonons in thin LiF films grown on a collodion film. For *p*-polarized light the ω_{LO} structure is generated by the surface charges due to the normal component of the electric field. Many transmission measurements at oblique incidence have now been reported in the litera-ture on a diverse group of semiconductors of free-standing thin films, epilayers, and SLs [84-85] offering a direct ratification of the Berreman effect [83]. More recently, the detec-tion of the *p*-polarization minima associated with the ω_{TO} phonon and the high-frequency ω_{LO}-phonon plasmon coupled modes ω_{LPP}^+ has made the far infrared transmission spec-troscopy a highly sensitive tool for estimating the free charge carrier density in doped semiconductor thin films [84].

In the framework of a multi-layer optics and using the effective medium dielectric tensor (cf. Sec. 4.2) one can simulate the transmission and reflection spectra at oblique incidence for free-standing thin films, epilayers, and SLs. As mentioned earlier, the polar semiconductors exhibit a distinct response to the FIR radiation between *s*- and *p*-polarization when the light is incident obliquely to the surface of the material. Here the terms *s* and *p* imply the directions of the linear polarization, which are, respectively, perpendicular, and parallel to the plane of incidence. Berreman's argument consisted in the demonstration that for thin films ($d \ll \lambda$) the long wavelength phonons have the wave vector $\vec{q}$ perpendicular to the film surface, such that the normal incidence radiation can interact only with the ω_{TO} modes (parallel to the surface). On the other hand, the *p*-polarized component of the oblique incident radiation, as shown in Figure 4.4, (cf. Figure 9) has subcomponents parallel and perpendicular to the film surface, which can excite both the ω_{TO} and ω_{LO} phonons, respectively.

4.5 Infrared Reflectivity of Superlattice Structures

Following Piro [86] the analytical expressions for $R(\omega)$ and $T(\omega)$ at an arbitrary angle of incidence θ_i can be derived for an isotropic layer of thickness d grown on a thick iso-tropic substrate. The dielectric tensor $\tilde{\varepsilon}_{SL}$ of the SL consisting of two optically isotropic layers *AC*, *BC* grown alternately on a substrate has the form of a uniaxial crystal of D_{2d} symmetry:

$$\tilde{\varepsilon}_{SL}(\omega) = \begin{bmatrix} \tilde{\varepsilon}_\perp & 0 & 0 \\ 0 & \tilde{\varepsilon}_\perp & 0 \\ 0 & 0 & \tilde{\varepsilon}_{||} \end{bmatrix}. \tag{11}$$

Here, the *z*-axis is set along the optical axis perpendicular to the plane of the SL. If the wavelength of the radiation is large compared to the period d ($=d_{AC} + d_{BC}$) of the SL, the electromagnetic boundary conditions give the values of $\tilde{\varepsilon}_\perp$ and $\tilde{\varepsilon}_{||}$:

$$\tilde{\varepsilon}_\perp(\omega) = \frac{\tilde{\varepsilon}_{AC}(\omega)d_{AC} + \tilde{\varepsilon}_{BC}(\omega)d_{BC}}{d_{AC} + d_{BC}} \tag{12a}$$

$$\text{and} \quad \tilde{\varepsilon}_{||}(\omega) = \frac{(d_{AC} + d_{BC})\tilde{\varepsilon}_{AC}(\omega)\tilde{\varepsilon}_{BC}(\omega)}{d_{AC}\tilde{\varepsilon}_{BC}(\omega) + d_{BC}\tilde{\varepsilon}_{AC}(\omega)}. \tag{12b}$$

For a SL grown on a thick substrate, the amplitude reflectance $\tilde{r}_{s(p)}$ (or amplitude trans-

mittance $\tilde{t}_{s(p)}$) and hence the power reflectance $R_{s(p)}(\omega) = \left|\tilde{r}_{s(p)}\right|^2$ (or the power transmittance

$T_{s(p)}(\omega) = \left|\tilde{t}_{s(p)}\right|^2$) can be obtained using

$$\tilde{r}_{s(p)} = \frac{(1-P)\cos Q - i(R-S)\sin Q}{(1+P)\cos Q - i(R+S)\sin Q}, \tag{13a}$$

$$\tilde{t}_{s(p)} = \left(\cos\xi - \frac{i}{2}\zeta\sin\xi\right)^{-1}, \tag{13b}$$

where the terms P, Q, R, S, and ξ, ζ in the $s(p)$-polarization are given elsewhere [81].

5 Spectroscopic Analysis of Infrared Spectra

5.1 Ideal 3C-SiC/Si Films

We have summarized in Table 3 the values of optical phonons and dielectric constants for 3C-SiC and Si reported by various research groups [80–81]. To study the reflectance and transmission spectra of the bulk 3C-SiC we used values of ε_∞ (= 6.7), Γ (= 5.5 cm^{-1}), the TO (= 794 cm^{-1}), LO (= 973 cm^{-1}) modes and m_e^* ($\equiv 0.313 m_e$) – the effective electron mass [80-81]. As 3C-SiC exhibits a dipole moment having distinct ω_{TO}, ω_{LO} phonon values, and the material can be doped, all three terms in Eq. (3) are significant. On the other hand, since Si (substrate) is not a polar material as its phonon dispersion near $\vec{q} = 0$ does not present a splitting between LO and TO phonons ($\omega_{TO} = \omega_{LO} = 520$ cm^{-1}), only the free charge carriers can have interaction with the infrared radiations and therefore, the Drude-Lorentz term in Eq. (3) may be neglected. Thus, we modeled the undoped Si substrate by selecting the value of $\varepsilon_\infty = 11.7$, with $n_s = 3.43$ and $\kappa_s = 0$, respectively.

TABLE 3

Optical constants for 3C-SiC and Si. The carrier concentration η (cm^{-3}), ω_{TO} (TO), ω_{LO} (LO) phonon modes, Γ phonon damping constant (in cm^{-1}); ε_∞ for 3C-SiC and Si used in Eq. (3) to calculate the infrared reflectivity/transmission of V-CVD grown 3C-SiC/Si samples.

Optical constants for 3C-SiC and Si						
Material	ε_∞	η	ω_{TO}	ω_{LO}	Γ	Ref. [81]
3C-SiC (film)	6.7	1.2×10^{17}	794	973	5.5	Our
Si (substrate)	11.7		520	520		Our

5.1.1 Effects of Film Thickness

By using the set of parameters from Table 3, we have displayed in Figure 6(A) the simulated results of the reflectance spectra at normal incidence for bulk 3C-SiC [Figure 6(A)] and 3C-SiC/Si epifilms [Figure 6(A) curves of varied thickness]. In Figure 6(B), we show the calculated reflectivity results of both undoped and doped 5 µm thick free-standing 3C-SiC film [see Figure 6(B) curve a], and a 5 µm thick 3C-SiC/Si epilayer [see Figure 6(B) curve b].

In these figures [cf. Figures 6 (A–B)] there are three frequency regions of interest, these are $\omega < \omega_{TO}$; ω lying between ω_{TO} and ω_{LO}; and $\omega > \omega_{LO}$. For the bulk 3C-SiC [see Figure 6(A) curve a], a high reflectivity (R) in the Reststrahlen region (i.e., between ω_{TO} and ω_{LO}) with a maximum value (~95%) occurs at ω_{TO}- due to the resonance between incident photons and lattice vibrations. The reflectivity exhibits minimum near ~1000 cm⁻¹, just above ω_{LO}, and reaches ~20% for $\omega > 2000$ cm⁻¹. The inspection of Figure 6(A) reveals that the reflectance spectrum of an ultrathin 3C-SiC film ($d = 0.04$ µm) has a sharp and narrow peak near ω_{TO} with no interference fringes. This is simply because the optical path difference of the film fails to meet the required interference condition. As the film thickness d increases to several micrometer [e.g., 4–8 µm, see Figure 6(A)], the reflectance shows a well-developed spectral feature attaining similarity to the bulk-like 3C-SiC [see Figure 6(A)] with interference fringes appearing on both sides of the Reststrahlen band. It is to be noted that in the frequency range $\omega \gg \omega_{LO}$ the contrast in the interference fringes [81] varies with the refractive indices of the epifilms (n_f) and substrates (n_s) while the film thickness d depends upon the fringe spacing ($\Delta\omega$) and the refractive index n_f. For 3C-SiC/Si, with $\kappa_f = 0$ and $n_f < n_s$, the reflectance values at the fringe maxima and minima can be calculated by using [81]

$$R\left[\frac{\left(n_s-1\right)}{\left(n_s+1\right)}\right]^2_{max} \quad \text{and} \quad R\left[\frac{\left(n_s-n_f^2\right)}{\left(n_s+n_f^2\right)}\right]^2_{min}. \tag{14}$$

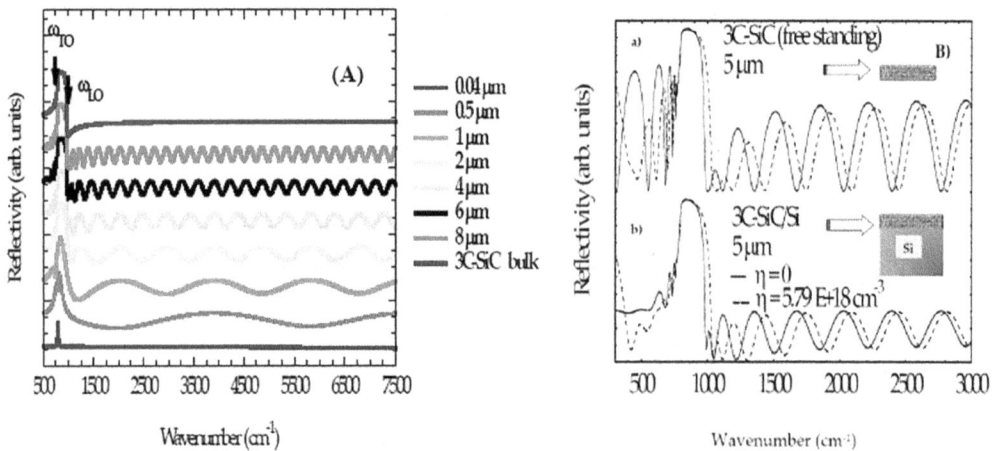

FIGURE 6

(A) Comparison of the calculated infrared reflectance spectrum of 3C-SiC/Si of different film thickness with the bulk 3C-SiC: 8 µm, 6 µm, 4 µm, 2 µm, 1 µm, 0.5 µm, and 0.04 µm [see Figure 6(A)]. Figure 6(B) we have displayed the calculated infrared reflectance spectrum [undoped $\eta = 0$, full line and doped $\eta = 5.79 \times 10^{18}$ cm⁻³, dotted line] of a 5 µm thick free-standing 3C-SiC film (curve a) and 3C-SiC/Si epilayer (curve b).

It is possible to calculate approximately the film thickness in the spectral region ($\omega > 2000$ cm^{-1}) by using $d \approx (2n_f \Delta\omega)^{-1}$ where n_f is nearly constant. Considering $n_f \approx 2.5$ for 3C-SiC we find $\Delta\omega \approx 50{,}000$ cm^{-1} for 0.04 µm thick epilayer, which falls well beyond the simulated spectral range of 500–7500 cm^{-1}.

As Si substrate can be removed from the thicker 3C-SiC films by a KOH etching solution, we have displayed in Figure 6(B) our simulated reflectance spectra of undoped $\Gamma = 0$ (full line) and doped $\Gamma = 5.79 \times 10^{18}$ cm^{-3} (dotted line) 5 µm thick free-standing 3C-SiC film (curve a) and contrasted them with the results obtained for the 3C-SiC/Si epilayer (curve b). As compared to the full-line curves (undoped) there are two obvious differences in the dotted-line (doped) spectra, these are (*a*) above ω_{LO} the first minimum shifts to the higher frequency, and (*b*) below ω_{TO} the interference fringes are either severely distorted or essentially disappeared for higher Γ (not shown here). Again, for the free-standing 3C-SiC film and 3C-SiC/Si epilayer, the reflectance [using Eq. (14)] at the fringe maxima and minima for $\omega > 2000$ cm^{-1} is found to be $R_{max} = 54.8\%$, $R_{min} = 0$ and $R_{max} = 30.1\%$, $R_{min} = 8.5\%$, respectively.

5.1.2 Reflectivity and Transmission of 3C-SiC/Si Epifilms

The calculated results for the reflectivity and transmission spectra near normal incidence ($\theta_i \approx 0°$) FTIR data on the V-CVD grown 3C-SiC/Si (100) films of diverse thicknesses are displayed in Figures 7–8. They are matched well as compared with the experimental data [81]. For a very thin film with growth time t of only 2 min, the transmission spectrum exhibits a sharp dip $\sim \omega_{TO}$ indicating the initial stage of the formation of SiC but not yet fully developed epifilm of 3C-SiC.

As t increases to 30 min, a deep and flat band appears between 770 and 920 cm^{-1} [see Figure 8(A) curve b] with both edges falling between TO and LO phonon frequencies. With the increase of film thickness, the peak in the reflectance (see Figure 7) and dip

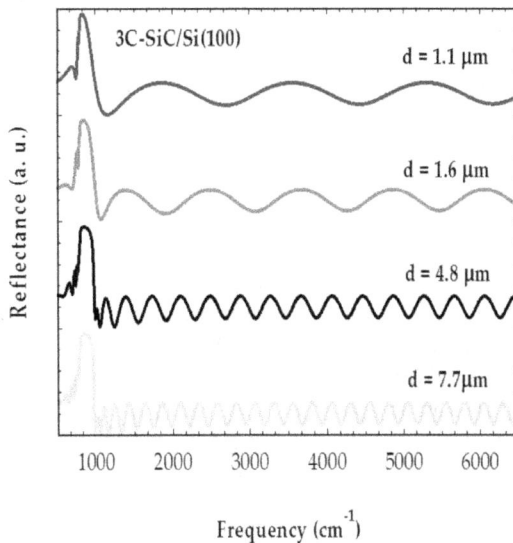

FIGURE 7

Simulated room temperature infrared reflectance spectra from four different 3C-SiC/Si (100) epilayers of thickness 1.1, 1.6, 4.8, and 7.7 µm, respectively, which are matched well with the experimental results [81].

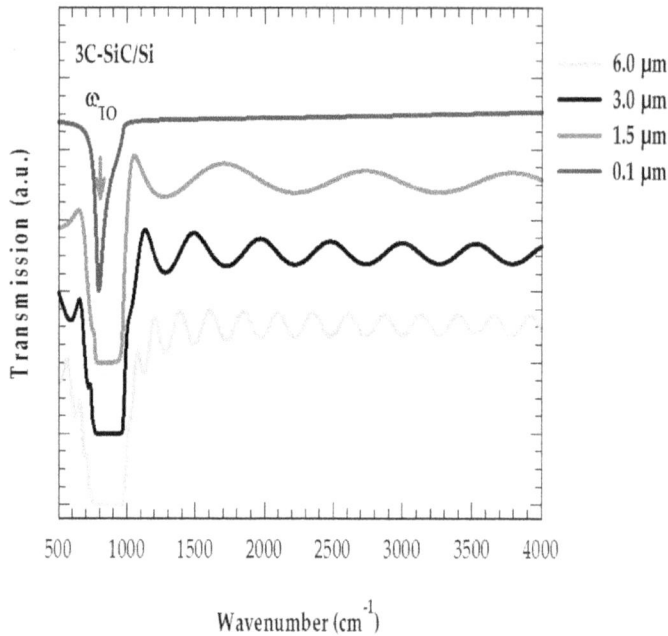

FIGURE 8
Simulated room-temperature infrared transmission spectra from four 3C-SiC/Si (100) epilayers of thickness 0.1, 1.5, 3, and 6 μm, respectively, which are matched well with the experimental results [81].

in the transmission (cf. Figure 8) spectra broadens (flattens) achieving 3C-SiC bulk-like Reststrahlen band for $d \geq 4.8$ μm, demonstrating the improvement to the cubic crystalline perfection of 3C-SiC epilayers grown on Si substrate by V-CVD. Except for a few noticeable dips observed in the experimental transmission spectra [see Figure 8 between 1300–1650 cm^{-1} (possibly due to the multi-phonon [81] absorption), an overall agreement between the simulated and experimental results for the reflectivity (see Figure 7) and transmission (see Figure 8) spectra is very good. Additional ratifications for the high crystalline quality of these samples have also been provided by Rutherford backscattering, Raman spectroscopy, and XRD studies.

5.1.3 Impact of Oblique Incidence on Transmission/Reflection: Berreman Effect

The impact of incident angle on the polarization dependent (s- and p-type) infrared reflectance/transmission spectra is calculated for a 0.5 μm thick 3C-SiC epilayer on Si substrate. The simulated results are displayed in Figures 9(a–b) and in Figures 9(c–d) for the frequency range ~500 to 1100 cm^{-1} and ~500–6500 cm^{-1}, respectively, confirming the Berreman effect [83]. In the transmission spectra both ω_{TO} and ω_{LO} modes appear as dips in the p-polarization while only the dip at the ω_{TO} mode is noticed in the s-polarization. In the reflectance spectra, however, a peak is observed at ω_{TO} in the s-polarization while a peak at ω_{TO} and a dip at ω_{LO} is noticed in the p-polarization.

In Figure 10 we have displayed the calculated results of the transmission spectra of a 1 μm thick 3C-SiC/Si film at oblique ($\theta_i = 45°$) incidence, which are matched well as compared with the existing experimental data of Dean Sciacca *et al.* [85]. Clearly it offered a strong

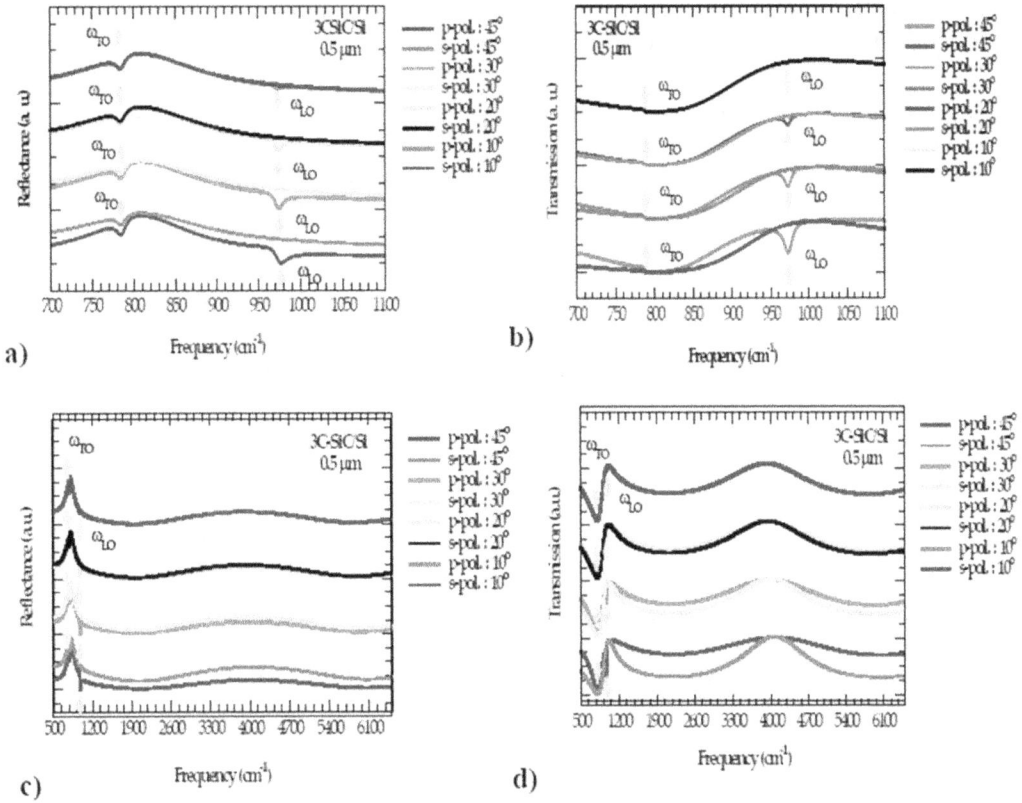

FIGURE 9
Impact of incident angle on the reflectance (a) and (c); transmission (b) and (d) spectra of 0.5 μm 3C-SiC/Si epifilm in the frequency range of 500 to 1100 cm^{-1} (500–6500 cm^{-1}).

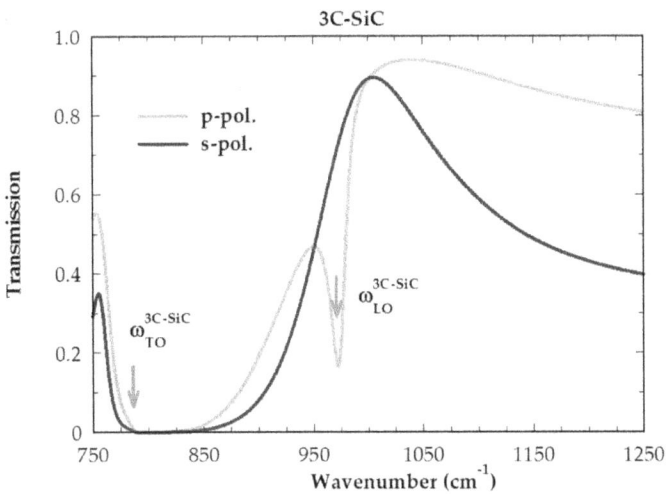

FIGURE 10
Theoretical simulated transmission spectra of the 3C-SiC/Si 1 μm thick film as a function of frequency (cm^{-1}) for $\theta_i = 45°$ p-polarization, which are matched well with the existing experimental data from Dean Sciacca et al. [85].

corroboration to Berreman's [83] effect revealing a sharp minimum at ω_{TO} for $\theta_i = 45°$ with an additional dip at ω_{LO} in the p-polarization.

5.1.4 Transmission at Oblique Incidence: Impact of Film Thickness of 3C-SiC/Si

Following the method outlined in Sec. 4.3 and using the optical constants of 3C-SiC/ Si (Table 3) we have displayed in Figure 11 the calculated results of the transmission spectra at oblique incidence $\theta_i = 45°$ in both s- and p-polarization for 3C-SiC epifilms with varying thickness. The results have clearly revealed transmission dips near $\omega_{TO} = 794$ cm^{-1} and $\omega_{LO} = 973$ cm^{-1} frequencies in very good agreement with the existing inelastic x-ray scattering and Raman spectroscopy data. The study has clearly demonstrated the progression from the transmission dip at ω_{TO} for ultrathin films to a broad region of transmission covering the entire Reststrahlen band as the film thickness increases. With the increase of 3C-SiC film thickness, the results show some interference fringes emerging below the TO frequency region for thick films – caused primarily by the rapid increase in the dielectric function $\tilde{\varepsilon}(\omega)$ for frequencies lying close to the pole at ω_{TO}.

5.1.5 LO-plasmon Coupling in n-doped 3C-SiC

In doped semiconductors, one would expect strong coupling between the ω_{LO} (Γ) phonons and collective oscillations of the free charge carriers (plasmons) ω_p. The coupling is maximum when ω_{LO} (Γ) and ω_p have comparable energies – as ω_p depends upon carrier concentration η via $\omega_p \left\{ \equiv \left[\dfrac{4\pi\eta e^2}{m^* \varepsilon_\infty} \right]^{1/2} \right\}$.

According to the theory of coupled ω_{LO}-plasmon modes, the frequencies of the two ω_{LO}-plasmon coupled $\omega_{LPP}^{\pm}$ modes in n-doped 3C-SiC can be estimated from the singularity of dielectric function [cf. Eq. (3)] in the long wavelength (i.e., $\vec{q} \to 0$) limit with no phonon or plasmon damping:

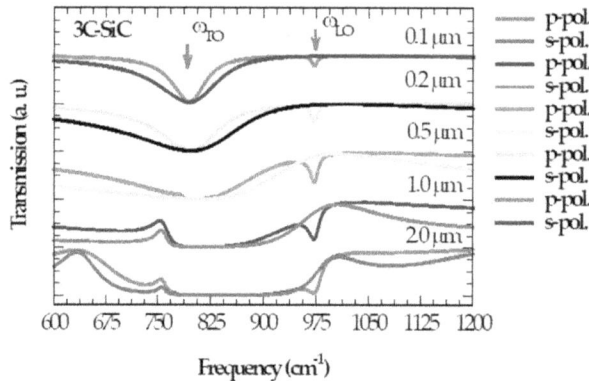

FIGURE 11

Calculated transmission spectra of 3C-SiC epifilms with increasing thicknesses in s- and p-polarization at oblique incidence $\theta_i = 45°$ as a function of frequency (cm^{-1}). Spectra of different films are vertically displaced for clarity.

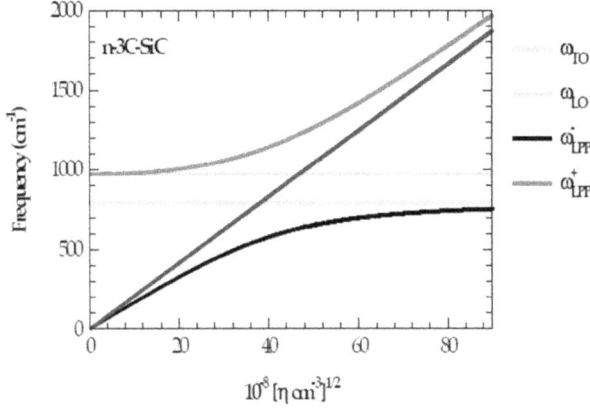

FIGURE 12
In *n*-doped 3C-SiC, the calculated $\omega^{\pm}_{LPP}$ modes using Eq. (15) are reported as a function of charge carrier concentration η.

$$\omega^{\pm}_{LPP} = \frac{1}{2}\left\{\left(\omega^2_{LO} + \omega^2_{P}\right) \pm \left[\left(\omega^2_{LO} + \omega^2_{P}\right)^2 - 4\omega^2_{P}\omega^2_{LO}\right]^{1/2}\right\}. \tag{15}$$

In Figure 12 we have displayed the calculated $\omega^{\pm}_{LPP}$ modes as a function of carrier concentration η. It is to be noted that ω^{+}_{LPP} is larger than ω_{LO} while ω^{-}_{LPP} is smaller than ω_{TO} modes. Again, ω^{-}_{LPP} mode exhibits plasmon-like behavior for smaller charge carrier concentration, and it becomes phonon-like for the larger values of Γ. On the other hand, for the lower value of charge carrier concentration Γ the mode ω^{+}_{LPP} lies close to ω_{LO} displaying phonon-like behavior and turns into a plasmon-like mode for the higher value of charge carrier concentration.

5.1.6 Effects of Plasma Damping

The plasma damping coefficient related to mobility $\gamma(\eta) = \dfrac{e}{m^{*}\mu(\eta)}$ in doped polar materials also depends upon the charge carrier concentration η. By using an empirical equation of Caughey and Thomas [87] we have calculated the effects of η on plasma frequency ω_{P} [Figure 13(a)], electrical mobility $\mu(\eta)$ and plasma damping coefficient $\gamma(\eta)$ [Figure 13(b)]. Consistent with Hall studies [81] our results displayed in Figure 13(b) have revealed reduction in μ with the increase of η. The parameter values extracted from Figure 13(b), are integrated into Eq. (3) for monitoring the effects of η and γ on the reflectance and transmission spectra of 3C-SiC/Si epilayers.

5.1.7 Transmission Spectra at Oblique Incidence in Doped 3C-SiC/Si Epilayers

In Figure 12 we have reported results of the coupled $\omega^{\pm}_{LPP}$ modes in 3C-SiC as a function of η. The simulated transmission $[T_{s}(\omega)/T_{p}(\omega)]$ spectra of a 0.5 μm thick *n*-type 3C-SiC/Si epilayer are displayed in Figure 14 at oblique incident angle ($\theta_{i} = 45°$) for different doping levels. The minima (maxima) in $T_{s}(\omega)$ (see Figure 14) linked to $\omega^{3C\text{-}SiC}_{TO}$ modes are

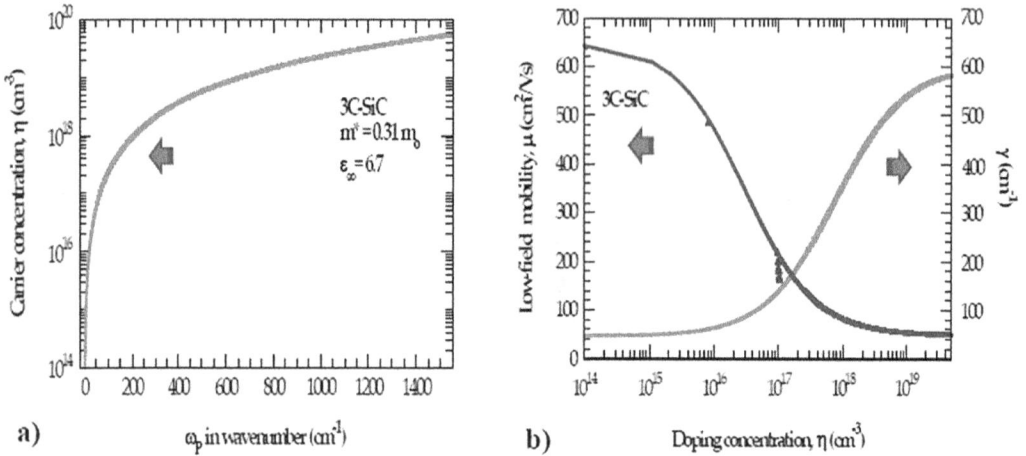

FIGURE 13

(a) Calculated plasma frequency ω_P in (cm^{-1}) versus charge carrier concentration η (cm^{-3}) in n-type 3C-SiC. (b) Calculated low field mobility μ in (cm^2/Vs) (left), plasmon coupling coefficient γ in (cm^{-1}) versus charge carrier concentration η (cm^{-3}) in n-type 3C-SiC (see text).

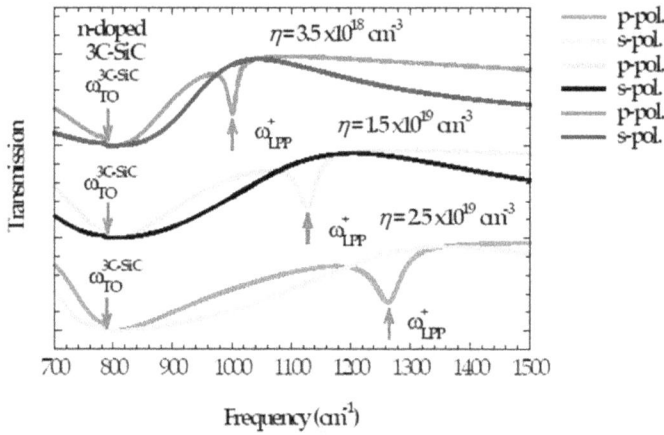

FIGURE 14

Polarization-dependent transmission spectra of 0. 5 μm thick n-3C-SiC/Si epifilms. The first minima shown by vertical gray color arrow in the s-polarization represents $\omega_{TO}^{3C\text{-}SiC}$ mode while in the p-polarization besides perceiving first minima representing $\omega_{TO}^{3C\text{-}SiC}$ mode the second dip represents ω_{LOPL}^{+} mode, which shifts to higher frequency with the increase of η.

asymmetrically broadened – possibly due to the free-carrier concentration. The other features noticed in the $T_p(\omega)$ spectra are the dips at higher frequencies. These traits are indicated by the gray color vertical arrows, which shift to higher frequency with the increase of Γ. One must note that the transmission minima near ~1001 cm^{-1}, ~1128 cm^{-1}, and ~1263 cm^{-1} in Figure 14 are not perceived in the s-polarization $T_s(\omega)$ and we assign these structures as ω_{LOPL}^{+} modes based on their dependence on carrier concentration Γ, and their appearance only in the p-polarization spectra. This assignment is further supported by the calculations carried out for ω_{LOPL}^{+} mode frequencies derived from the zeros of the real part of $\tilde{\varepsilon}(\omega)$[Eq. (15)] with $\gamma = 0$, $\Gamma = 0$. Again, we are unable to identify the features related to the ω^-_{LOPL} modes. This result is not surprising as ω^-_{LOPL} mode being plasmon-like at lower Γ has a relatively large damping constant and weaker strength while at higher Γ it is expected to merge with a stronger and a broader ω^{3C-SiC}_{TO} phonon minimum. On the other hand, with a larger plasmon contribution at higher Γ, the ω^+_{LOPL} mode in n-doped 3C-SiC though broad is distinctly noticeable (cf. Figure 14).

5.2 Typical Reflectance Spectra of 3C-SiC/Si (100) Epilayers

In Figure 15, we have displayed the experimental reflectivity results for five of our V- CVD grown 3C-SiC/Si (100) samples exhibiting atypical spectral features. The selection of these samples is made to exemplify the unusual characteristics from the IR data recorded on several nearly typical cases. At first sight, one may be tempted to attribute the observed fringes in samples #113, #121, #125, #409, and #713 (see Figure 15) as typical multiple interferences between 3C-SiC epilayers and Si substrate.

After close inspection of the reflectance spectra above the Reststrahlen band region, one may realize that the fringe contrasts [81] are in fact decreasing with the increase of frequency. To a varying degree, the diminishing fringe disparities are transpired in all the samples with an average reflectance ($\bar{R}$) falling between ~11 to 19% at $\omega \approx 4000$ cm^{-1}. Clearly, this range of estimated values for the reflectivity is lower than the simulated reflectance for the bulk 3C-SiC (R~20%) or ideal (3C-SiC/Si) epilayers ($\bar{R}$~20%). The declining fringe contrast can be attributed to the sample's surface or interface roughness – simply because the light from a rough surface is expected to disperse diffusely rather than reflect specularly causing the scattering losses to amplify with increasing frequency. As compared to Figure 15, the measured infrared reflectivity in samples #113, #409, and #713 (see Figure 15) are revealed to exhibit aberrant spectral characteristics. While a dip or divot is observed within the Reststrahlen region, the damping in interference fringes has also occurred away from the zone-center lattice phonons. The film thicknesses of five 3C-SiC/Si samples are estimated by fitting the infrared spectra (cf. Sec. 5) and using a modified EMA methodology as 9.89, 2.10, 7.69, 3.37, and 5.86 μm, respectively, in excellent agreement with XRD measurements.

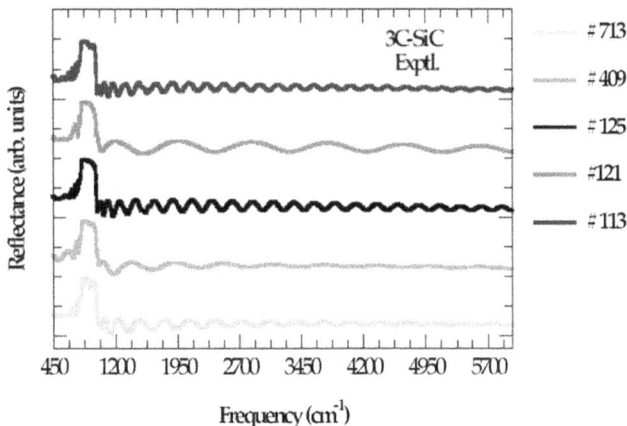

FIGURE 15
Atypical experimental near normal incidence reflectance spectra for five of the 3C-SiC/Si (100) V-CVD grown samples #113, #121, #125, #409, and #713, respectively.

Clearly, the unusual features observed in some of the V-CVD grown samples (cf. Figure 15) cannot be explicated in terms of the ideal EMA methodology (cf. Secs. 4.1–4.2). In the next section we have appraised a few cases of the atypical reflectance spectra with simulated results based on the modified model (cf. Sec. 4.3).

5.2.1 Effect of Transition Layer and Surface Roughness

In Figure 16a we have compared our measured reflectivity spectrum (blue color full-line) of a 7.68 μm thick 3C-SiC/Si sample (#125) with theoretical results based on the modified model having film thickness with surface roughness $\delta = 0.07$ μm. Although the ideal model has a well-described reflectance pattern it failed, however, to elucidate the observed damping behavior in interference fringes away from the Reststrahlen band. The effect of $\delta = 0.07$ μm at higher frequencies exhibited reduction in the average reflectivity by providing the correct trend of interference-fringe contrasts. The inset exhibits values

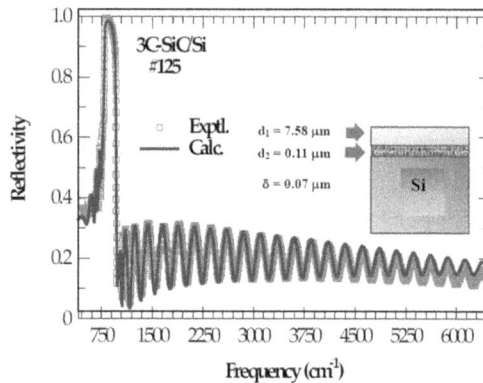

FIGURE 16A

Comparison of experimental (red color square) infrared reflectance spectra for V-CVD grown 3C-SiC/Si (100) sample #125 with the simulated results (blue color line). The calculated reflectance spectra is based on the modified EMA methodology using $d_1 = 7.58$ μm, $d_2 = 0.11$ μm surface roughness, i.e., $\delta = 0.07$ μm.

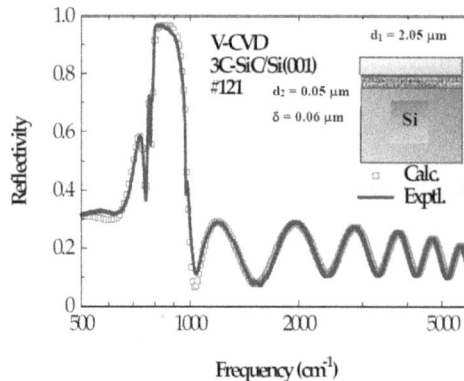

FIGURE 16B

Comparison of experimental (red color square) infrared reflectance spectra for V-CVD grown 3C-SiC/Si (100) sample #121 with the simulated results (blue color line). The calculated reflectance spectra is based on the modified EMA methodology using $d_1 = 2.05$ μm, $d_2 = 0.05$ μm surface roughness, i.e., $\delta = 0.06$ μm.

used in our simulation for epilayer d_1 (= 7.58 μm), TL d_2 (= 0.11 μm), and surface roughness (δ = 0.07 μm, δ_2 = 0). This provided strong corroboration to the fact that the total thickness [$d = (d_1 + d_2)$ 7.69 μm] remains unchanged.

Similar results (see Figure 16b) in another V-CVD sample (#121) offered excellent agreement between measured and simulated spectra – indicating that reflectivity analysis by modified theory presents a simple and accurate method for assessing the thicknesses of epilayer and TL, as well as surface roughness.

5.2.2 Two-Component Bruggeman's Model

In Figures 17(a–b) we have displayed atypical results of IR reflectance measurements recorded on 3C-SiC/Si samples #409 and #713. In addition to the decreasing fringe contrasts at higher frequency, the uncommon features observed in the Reststrahlen band are the dips or divots near ~895 cm^{-1} [Figure 17(a)] and ~894 cm^{-1} [Figure 17(b)], respectively. Although, similar characteristics were seen earlier [81] near ~900 cm^{-1} in the CVD grown 3C-SiC/Si samples having pits or bumps on the surface of epilayers – yet the exposition of their origin remained rather sparse to being qualitative. In the reflectivity study of surface treated α-SiC, a dip perceived near ~885 cm^{-1} was ascribed [88] to a second oscillator having mode strength ~0.3% compared to that of the fundamental resonance. The change in relative height of the main peak and dip seems to exclude this possibility. It is likely that the crystalline or topological defects in heterogeneous 3C-SiC/Si materials may cause the $\vec{q} \rightarrow 0$ conservation rule to relax for the IR spectroscopy. Thus, relating the dip near ~900 cm^{-1} to defect-activated mode cannot be completely ruled out – its justification would require, however, cumbersome computations.

In perpetuation with our current methodology, we adopted a modified "two-component" Bruggeman's effective medium method [81] for generating the effective dielectric function $\tilde{\varepsilon}_{ff}(\omega)$ to simulate and fit atypical features observed in the IR reflectance spectra of samples #409 and #713. In the quasi-static approximation, the dielectric function $\tilde{\varepsilon}_{ff}(\omega)$

$$\sum_{i=a,b} \frac{f_i \left[\tilde{\varepsilon}_i(\omega) - \tilde{\varepsilon}_{ff}(\omega) \right]}{\tilde{\varepsilon}_i(\omega) + 2\tilde{\varepsilon}_{ff}(\omega)} = 0, \tag{16}$$

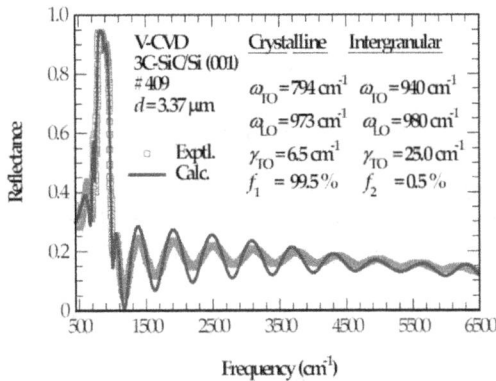

FIGURE 17A

Comparison of the experimental (full line) infrared reflectance spectra for V-CVD grown 3C-SiC/Si (100) sample #409 with the simulated results based on Bruggeman's "two-component" effective medium approach.

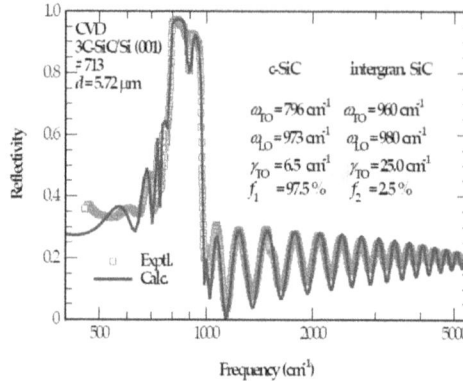

FIGURE 17B

Comparison of the experimental (full line) infrared reflectance spectra for V-CVD grown 3C-SiC/Si (100) sample #713 with the simulated results based on Bruggeman's "two-component" effective medium approach.

of the heterogeneous 3C-SiC material system is implicitly modeled by assuming the co-existence of both crystalline $\tilde{\varepsilon}_a(\omega)$ and intergranular materials $\tilde{\varepsilon}_b(\omega)$ with volume fractions f_a and f_b, respectively. Here, the choice of crystalline grains is obvious as we are closely scrutinizing the Reststrahlen band region resulting from the coupling of optical phonons with IR photons. The selection of intergranular material and its role, at first in Bruggeman's model, was less apparent.

However, from the appearance of atypical mode (divot) within the Reststrahlen band and by comparison of the simulation with experimental data has certainly fortified its importance. The optical parameters required for evaluating the dielectric functions [$\tilde{\varepsilon}_a(\omega)$, $\tilde{\varepsilon}_b(\omega)$] of the two media phases with volume fractions (f_a, f_b) are reported in Ref. [81]. The calculated results of reflectivity for samples #409 (see Figure 17a) and #713 (see Figure 17b) are appraised to offer very good agreement with the experimental data. While higher TO_b frequencies (close to ~940 cm^{-1}) are deduced for intergranular materials in both of the samples, the variation of volume fraction f_b from 0.5% (sample #409) to 2.5% (sample #713) is required, however, to attain the correct and unique features (i.e., the depth of dips ~895 cm^{-1}) observed in the Reststrahlen band region of the reflectivity spectra.

6 Structural Characteristics of V-CVD Grown 3C-SiC

The extended x-ray absorption fine-structure (EXAFS) is a powerful technique, which is commonly exploited for probing the physical and chemical structure of matter at an atomic scale without requiring the special needs for material preparations.

6.1 Synchrotron Radiation X-ray Absorption Spectroscopy (SR-XAFS)

We have carried out SR-XAFS measurements on some of the V-CVD grown 3C-SiC samples (#409, #413, #429A, and #440) in the x-ray fluorescence yield mode by expending a double crystal monochromatic beam line 16 A at the National Synchrotron Radiation Research Center (NSRRC), Hsinchu, Taiwan. In these experiments, while the applied incident photon flux covered the energy range between 1750–2700 eV including the K absorption edge (1839 eV) of the Si atom (cf. Figure 18a) – its intensity was monitored carefully by N_2 filled ionization chamber and the fluorescence emitted from the samples assessed by an argon-filled Stern-Heald-Lytle-type detector. A Si (111) double-crystal monochromator was used with a 0.5 mm entrance slit. A filter between the sample and detector window was inserted for reducing the noise from scattering and thus improving the spectrum quality. All the RT spectral measurements performed here were normalized to the intensity of the incident I_o photon flux.

The x-ray absorption coefficient μ as a function of E shown in Figure 18a revealed a sharp rise in intensity at Si K-edge (~1839 eV) – initiating almost a step-like function with weak oscillatory wiggles beyond several hundred eV above the absorption edge. In each sample, the region closer to the edge is dominated by the strong scattering processes including

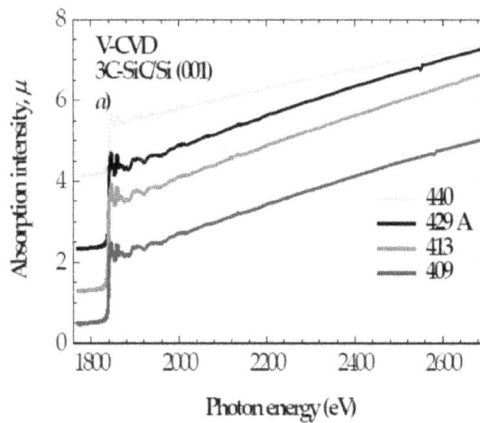

FIGURE 18A

Si K edge EXAFS absorption coefficient m versus x-ray photon energy E for four V-CVD grown 3C-SiC/Si (001) samples (#409, #413, #429A, and #440). http://dx.doi.org/10.1016/j.apsusc.2017.07.266.

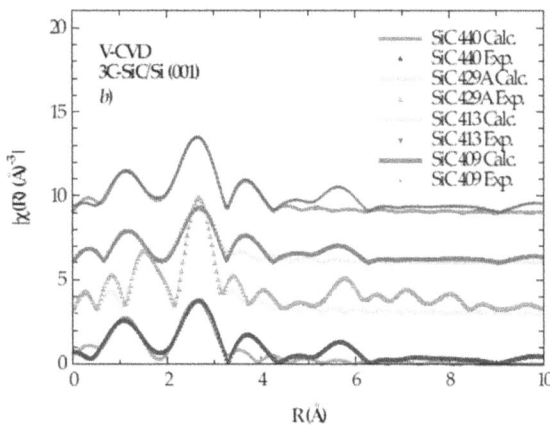

FIGURE 18B

Fourier transformed EXAFS spectra: χ (R) (Å)$^{-3}$ versus R between 0 and 10 Å for different V-CVD grown 3C-SiC samples. http://dx.doi.org/10.1016/j.apsusc.2017.07.266.

TABLE 4

Fitting results from the EXAFS data of four
different thickness 3C-SiC/Si (001) samples for
extracting local atomic structures [i.e., nearest-
neighbor (R_1) and next-nearest-neighbor (R_2)
bond lengths in Å]

Sample	# 409	# 413	# 429 A	# 440
R_1 (Å)	1.88	1.85	1.86	1.91
R_2 (Å)	3.09	3.07	3.08	3.06

local atomic resonances. The x-ray absorption near edge structures (XANES) are custom-
arily excluded from EXAFS. Here, our interest has only been in the fine structures that
contained precise information of the local structures around the atom that absorbed x-rays.
The ATHENA package codes are exploited for removing the background contributions
using AUTOBK for extracting EXAFS oscillations from the k-space signals. The information
of local structure around Si in 3C-SiC is obtained by utilizing FEFF 8.2 code implemented
in the ARTEMIS software. Figure 18b shows the magnitudes of the Fourier transformed
EXAFS results for four of the samples with major peaks implying the reflective waves from
the atoms in the vicinity of the absorbed Si atom. By using the simulations and fitting the
EXAFS data we have obtained (see Table 4) the nearest neighbor Si-C and next-nearest
neighbor Si-Si bond lengths. The changes in Si-C and Si-Si bond lengths as compared to the
bulk 3C-SiC values may be related to the possibility of nitrogen (N) substituting for the C
atoms and/or stress-induced distortions near the absorbed atom.

7 Lattice Dynamics of Defects in 3C-SiC/Si

Due to significant differences in the lattice constants (19.8%) and coefficients of thermal
expansion (8%) between 3C-SiC and Si, a large number of crystalline defects are formed
in 3C-SiC/Si. To assess the local arrangement of constituent atoms in 3C-SiC/Si is still a
challenge – especially for understanding the atomic structures of the "intrinsic-defects".
This is notably true for the low-temperature photoluminescence (PL) D_I center, which has
been known for decades [76]. While the D_I center in 3C-SiC exhibited two closely spaced
gap-modes (661.3 and 668.7 cm^{-1}), its atomic structure impacting the vibrational features
has not been completely understood. Given its relevance, it is important to investigate the
characteristics of impurity vibrations in 3C-SiC.

Earlier, we have performed comprehensive FIR, RSS, and SR-EXAFS studies to extract
valuable lattice dynamical parameters related to the structural traits [76] of 3C-SiC/Si (001).
In the first-order RSS, besides observing the conventional long-wavelength optical modes
(~794 cm^{-1}, 973 cm^{-1}) we have also recognized two additional features near ~ 625 cm^{-1} and
670 cm^{-1} in 3C-SiC. It is quite likely that these defect modes may fall within the forbidden
gap of the acoustic and optical branches of 3C-SiC (cf. Sec. 7.1). As a relatively broad
unresolved Raman band near 670 cm^{-1} lies close to the doublet (661.3 cm^{-1} and 668.7 cm^{-1}
ascribed to the D_I center) it is important to study the vibrational modes of a close-by Si_C –
C_{Si} anti-site pair defect.

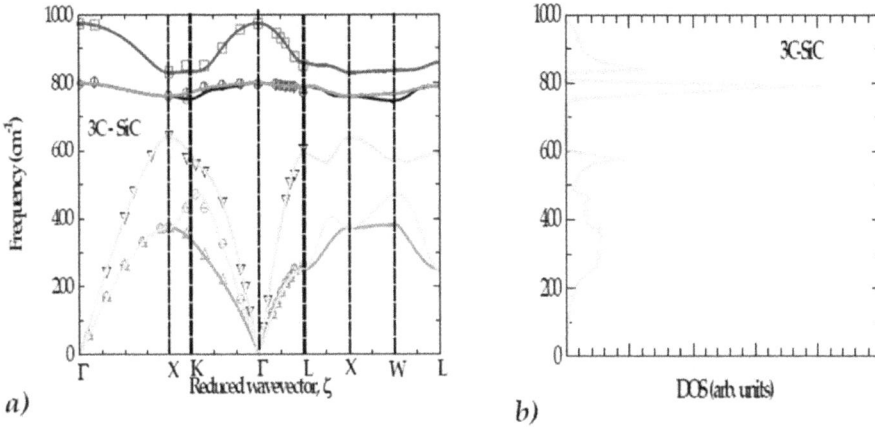

FIGURE 19
(a) Experimental (Δ,□,∇,⊕,∇) [63] and calculated solid lines (-) phonon dispersion curves along high-symmetry directions based on rigid-ion model [76], and (b) one-phonon density of states $g(\omega)$. DOI: 10.1016/j.mseb.2017.08.018.

7.1 Phonon Characteristics of 3C-SiC

To simulate the dynamical behavior of the atomic structure for the D_I center, it is essential to use a credible theoretical method of the host lattice phonons with appropriate perturbation. Hence, we have adopted a realistic RIM of 3C-SiC and extracted phonons by accurately fitting the IXS [63] data with the simulations [76] [see Figure 19(a) of phonon dispersions and Figure 19(b) one-phonon density of states].

7.2 Green's Function Theory

7.2.1 The Perfect Green's Function Matrix $G_o(\omega)$

To determine the dynamical behavior of defects by Green's function (GF) method, a detailed knowledge of the host lattice phonons is a prerequisite [76]. As compared to *ab initio* calculations [63], the advantage of using GF theory is that it yields appropriate spectral functions and impurity traits for correlating them with optical experiments. In *zb* materials, a complete description of the GF methodology is reported in Ref. [76] for simulating impurity-induced first-order RSS profiles. Here, we summarized a pertinent portion of the GF theory required for elucidating the lattice dynamics of a nearest-neighbor (NN) anti-site pair 'Si$_C$-C$_{Si}$' defect (C_{3v}-symmetry) that best explained the gap modes and phonon broadening of an unresolved ~670 cm^{-1} mode observed in RSS experiments. For simulating the impurity vibrational modes of an anti-site pair defect in 3C-SiC, we have incorporated the host lattice phonons from a RIM [76] fitted to the IXS [63] data (cf. Sec. 7.1) and calculated the necessary GF matrix elements $G_o(\omega)$.

7.2.2 The Perturbation Matrix $P(\omega)$

In the tetrahedral 3C-SiC [see Figure 20(a)], the NN pair-defect involves two impurity atoms [see Figure 20(b)] occupying lattice sites 1 and 2, respectively, causing the changes in masses ε_1 [$\equiv(M_1 - M_1')/M_1$] and ε_2 [$\equiv(M_2 - M_2')/M_2$], at the impurity sites and variations

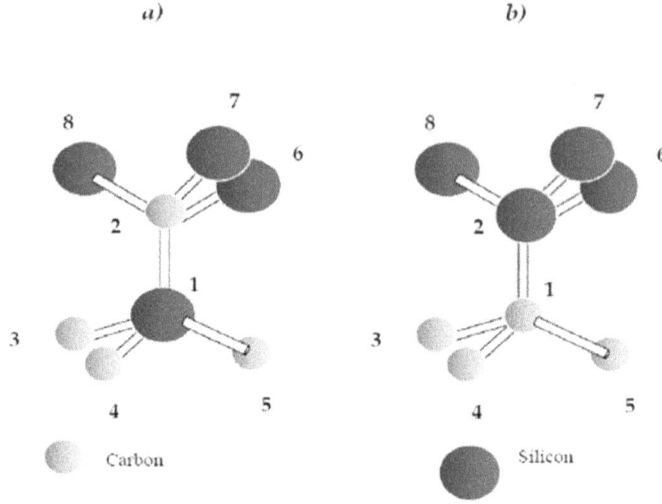

FIGURE 20

(a) Perfect zb 3C-SiC of T_d symmetry with Si atom at site 1 surrounded by C atoms at sites 2, 3, 4, and 5 C atoms and C atom at site 2 surrounded by Si atoms at sites 1, 6, 7, and 8, respectively. (b) The nearest-neighbor (NN) anti-site pair Si_C-C_{Si} defect (C_{3v}) symmetry in the zb 3C-SiC.

in force constants t (1–2, 1–3, 1–4, 1–5) and u (2–1, 2–6, 2–7, 2–8) between the impurity and its nearest-neighbor (NN) bonds, respectively. An effective force constant between the pair impurities F_{12} ($\equiv u + t - ut + \Gamma_{12}$) is also included (see Ref. [76], using Γ_{12}) to account for the changes in u and t of the isolated impurities involved in the formation of the anti-site pair-defect. The term $F_{12} < 0$ (or > 0) signifies stiffening (or softening) between the impurity-pair-bond. The point group symmetry of the NN "pair-defect-center" is C_{3v} with axis along the pair-bond involving eight atoms, causing the size of the perturbation matrix **P** to increase to the size 24×24.

7.2.3 Group-Theoretic Analysis of Impurity Vibrational Modes

The total representation of $\Gamma_{C_{3v}}$ [76] in the 24-dimensional space is used to block-diagonalize the $\mathbf{G_o}$ and **P** matrices with each block along the diagonal belonging to the following irreducible representations:

$$\Gamma_{C_{3v}} = 6a_1 \otimes 2a_2 \otimes 8e \tag{7.1}$$

From group-theoretical analysis it has been perceived that as the impurity atoms in the "pair-defect" remain stationary in the a_2 representation – only the a_1 and e type of modes are optically (IR and Raman) active. As the degeneracies are lifted at the defect sites (#1 and #2), one would expect observing four LVMs for the pair-defect with very light impurity atoms: *two* non-degenerate modes due to the movement of impurity atoms along the bond [i.e., ω_1 ($a_1^+ \leftarrow \rightarrow$) and ω_4 ($a_1^- \rightarrow \leftarrow$)] and *two* doubly degenerate modes as a result of their vibration perpendicular to it [i.e., ω_2 ($e^+_{\uparrow\downarrow}$) and ω_3 ($e^-_{\uparrow\uparrow}$)] generally with $\omega_1 > \omega_2 > \omega_3 > \omega_4$. However, the number of vibrational modes can differ if one of the impurity masses is heavier than the host atom and/or having strongly modified interactions.

By deliberating appropriate perturbation matrices of the pair defects, one can obtain the frequencies of their local and/or gap modes in different irreducible representations by solving the real part of the determinant:

$$\prod_{\mu\Gamma} \det |I - G_o^{\mu\Gamma}(\omega) P^{\mu\Gamma}(\omega)| = 0. \tag{7.2}$$

Here, $G_o^{\mu\Gamma}$ is Green's function of a perfect crystal projected on to the defect space and $P^{\mu\Gamma}(\omega)$ is the perturbation matrix in a given irreducible representation.

7.2.4 Impurity Vibrational Modes of NN Anti-site Pairs

In 3C-SiC, while the formation of "complex-defect-centers" involving intrinsic and doped defects are speculated, the identification of such defect-centers has not been firmly established. Moreover, there exists a limited knowledge about the realization of "intrinsic-defect complexes". From the first-order RSS study in V-CVD grown 3C-SiC samples we have identified two weak modes near ~ 625 and 670 cm^{-1}. The former mode (at 625 cm^{-1}) falls close to the edge of the acoustic phonon band continuum and was assigned as a triply degenerate F_2 gap mode of C_{Si}. The latter unresolved Raman band at ~670 cm^{-1} exhibited a relatively larger (~8 cm^{-1}) full width at half maximum (FWHM) and was designated as the F_2 gap mode of Si_C. The temperature-dependent profile of the band at ~670 cm^{-1} indicated disordering caused either by adjacent defects and/or stress that made the phonon lifetime shorter that triggered the mode broadening.

To assess the vibrational frequencies of a NN Si_C-C_{Si} pair, we retained the same force constant variations (u, t) estimated for isolated anti-site defects Si_C, C_{Si} and included interaction between them i.e., F_{12} $(\equiv u + t - ut + \Gamma_{12})$ by setting $\Gamma_{12} = 0$. As expected, our calculations have revealed splitting of the triply degenerate F_2 gap-mode of an isolated Si_C defect into a singlet (a_1) ~664.8 cm^{-1} (or 83.1 meV) and a doublet (e) ~660.6 cm^{-1} (or 82.6 meV) – the other two modes, however, merge into the 3C-SiC phonon-continuum band. One must also note that the doublet gap mode frequencies of ASP pair defect fall well within the FWHM (~8 cm^{-1}) of the broad band observed near ~670 cm^{-1} (or 83.8 meV) in RSS experiments. By exploiting the high-resolution IR and Raman scattering spectroscopy we are in the process of studying the local/gap vibrational modes in co-implanted 3C-SiC materials to identify the NN (B_{Si}-N_C) and/or NNN (B_C-N_C; Al_{Si}-P_{Si}) donor-acceptor-pairs (DAP). The simulations by GF theory reported here can certainly help us to evaluate the impurity mode fingerprints to classify the DAP structures.

---

8 Summary

In summary, we have performed room-temperature infrared reflectance and transmission measurements at near-normal incidence to characterize 3C-SiC epitaxial layers grown on (100) Si by V-CVD using different Si/C ratio and growth time ranging from 2 min to 4 hr. In many cases we perceived nearly typical infrared spectral results with customary characteristics, however, atypical infrared reflectance features are also observed in some of the V-CVD grown samples. By using classical effective medium theory in the ideal con-figuration, we performed systematic simulations and reported results for the reflectance and transmission spectra of various 3C-SiC bulk, free-standing film, and 3C-SiC/Si epilayers. The effects of film thickness, charge carrier concentration, angle of incidence, phonon (plasma) damping, LO-phonon plasmon coupled modes on the infrared reflectance profiles are evaluated by comparing the simulated spectra with the measured data from different samples. To account for the observed deviations from the ideal reflectivity spectra in a few V-CVD samples, we adopted a modified methodology by meticulously including the "conducting" transition layer near the substrate and surface roughness at the 3C-SiC/air and 3C-SiC/TL interfaces. By analyzing the reflectivity spectra our estimated values of surface roughness and TL thickness are very well correlated with the recent measurements of optical interference and scanning probe microscopy. The best match to uncommon features observed in the Reststrahlen band near 895 cm^{-1} were elucidated by the modified "two-component" Bruggeman model assuming the coexistence of both crystalline grains and intergranular media forming the heterogeneous 3C-SiC material system. These results suggest that careful analyses of the spectral shape of infrared reflection in the optical phonon resonance region can offer a versatile and microscopic means to characterize, diagnose, and improve the quality of V-CVD grown 3C-SiC/Si and other materials of technological importance. Accurate assessments of the lattice dynamical and defect properties are achieved by exploiting phonons from a RIM fitted to the IXS data and expending apposite group-theoretical selection rules. Lattice relaxations around Si/C atoms attained by first-principles bond-orbital model for isolated defects has helped us evaluating the necessary force constant variations to construct perturbation matrices of "complex-defect-centers". For the isolated intrinsic C_{Si} and Si_C defects (T_d-symmetry), our methodical Greens function (GF) theory has predicted triply degenerate F_2 gap modes near ~630 cm^{-1} and ~660 cm^{-1}, respectively. The GF simulations of impurity vibrations for a neutral "anti-site" C_{Si}–Si_C pair (C_{3v}-symmetry) provided gap-modes to appear within the broad ~670 cm^{-1} band at 664.8 cm^{-1} (a_1) and 660.6 cm^{-1} (e). The results are compared and discussed with phonon features observed in RS experiments as well as with density function theory calculations of localized vibrational modes for an anti-site pair defect.

Acknowledgements

The author sincerely acknowledges Profs. W.J. Choyke and H.H. Lin for fruitful conversation on the subject matter, C.C. Tin for providing the V-CVD grown samples used in this chapter and Editor/Prof. Zhe Chuan Feng for formatting figures and getting permissions.

References

1. N. Jiya, and R. Gouws, "Overview of Power Electronic Switches: A Summary of the Past, State-of-the-Art and Illumination of the Future" Micromachines 2020, 11, 1116; doi:10.3390/mi11121116.
2. M. Yamamoto, T. Kakisaka, and J. Imaoka, "Technical Trend of Power Electronics Systems for Automotive Applications" Jpn. J. Appl. Phys. 2020, 59 SG0805.
3. Y. Ji, W. He, S. Cheng, J. Kurths, and M. Zhan, "Dynamic Network Characteristics of Power-Electronics-Based Power Systems" Scientific Reports, 2020, 10, 9946.

4. X. Guo, Q. Xun, Z. Li, and S. Du, "Silicon Carbide Converters and MEMS Devices for High-temperature Power Electronics: A Critical Review" Micromachines 2019, 10, 406; doi:10.3390/mi10060406.

5. D. Baierhofer, "Current SiC Power Device Development, Material Defect Measurements and Characterization at Bosch" 978-1-7281-1539-9/19/2019 IEEE.

6. H. Amano, et al. "The 2018 GaN Power Electronics Roadmap" J. Phys. D: Appl. Phys. 2018, 51, 163001.

7. I. C. Kizilyalli, E. P. Carlson, D. W. Cunningham, J. S. Manser, Y. A. Xu and A. Y. Liu, "Wide Band-Gap Semiconductor Based Power Electronics for Energy Efficiency" https://arpae.energy.gov/sites/default/files/documents/files/ARPAE_Power_Electronics_Paper-April2018.pdf.

8. P. J. Wellmann, "Power Electronic Semiconductor Materials for Automotive and Energy Saving Applications – SiC, GaN, Ga_2O_3, and Diamond" Z. Anorg. Allg. Chem. 2017, 643, 1312–1322.

9. J. Hayes, K. George, P. Killeen, B. McPherson, K. J. Olejniczak, and T. R. McNutt, "Bidirectional, SiC Module-based Solid-State Circuit Breakers for 270 VDC MEA/AEA Systems," in IEEE Workshop on Wide Bandgap Power Devices and Applications, 2016.

10. J. Hayes, K. George, P. Killeen, B. McPherson, K. J. Olejniczak, and T. R. McNutt, "Bidirectional, SiC Module-based Solid-State Circuit Breakers for 270 VDC MEA/AEA Systems," in IEEE Workshop on Wide Bandgap Power Devices and Applications, 2016.

11. T. Daranagama, F. Udrea, T. Logan, and R. McMahon, "A Performance Comparison of SiC and Si Devices in a Bi-Directional Converter for Distributed Energy Storage Systems," in 2016 IEEE 7th International Symposium on Power Electronics for Distributed Generation Systems (PEDG), 2016, pp. 1–8.

12. F. Wang and Z. Zhang, "Overview of Silicon Carbide Technology: Device, Converter, System, and Application" CPSS Trans. Power Electr. and Appl., 1, 2016, pp. 13–32.

13. F. Wang, Z. Zhang, T. Ericsen, R. Raju, R. Burgos, and D. Boroyevich, "Advances in Power Conversion and Drives for Shipboard Systems," Proceedings of the IEEE, 2015, 103, 2285–2311.

14. T. Lagier and P. Ladoux, "A Comparison of Insulated DC-DC Converters for HVDC Off-Shore Wind Farms," in 2015 International Conference on Clean Electrical Power (ICCEP), 2015, pp. 33–39.

15. J. Millan, P. Godignon, X. Perpina, A. P. Tomas, and J. Rebollo, "A Survey of Wide Bandgap Power Semiconductor Devices," IEEE Transactions on Power Electronics, 2014, 29, 2155–2163.

16. P. Wolfs, Y. Fuwen, and H. Q. Long, "Distribution Level SiC FACTS Devices with Reduced DC Bus Capacitance for Improved Load Capability and Solar Integration," in 2014 IEEE 23rd International Symposium on Industrial Electronics (ISIE), 2014, pp. 1353–1358.

17. Y. Liu et al., "A Silicon Carbide Fault Current Limiter for Distribution Systems," in 2014 IEEE Energy Conversion Congress and Exposition (ECCE), 2014, pp. 4972–4977.

18. A. Fissel, "Artificially Layered Heteropolytypic Structures Based on SiC Polytypes: Molecular Beam Epitaxy, Characterization, and Properties", Physics Reports, 2003, 379, pp.149–255; DOI: 10.1016/S0370-1573(02)00632-4.

19. P. Schuh, F. L. Via, M. Mauceri, M. Zielinski, and P. J. Wellmann," Growth of Large-Area, Stress-Free, and Bulk-Like 3C-SiC (100) Using 3C-SiC-on-Si in Vapor Phase Growth" Materials, 2019,2179; doi:10.3390/ma12132179.

20. O. M. Ellefsen, M. Arzig, J. Steiner, P. Wellmann and P. Runde, "Optimization of the SiC Powder Source Material for Improved Process Conditions During PVT Growth of SiC Boules" Materials 2019, 12, 3272; doi:10.3390/ma12193272.

21. A.A. Lebedev, S.P. Lebedev, V.Yu. Davydov, S.N. Novikov, Yu.N. Makarov, "Growth and Investigation SiC Based Heterostructures" International Baltic Electronics Conference, BEC, 2016, doi:10.1109/BEC.2016.7743717 https://ieeexplore.ieee.org/xpl/conhome/7738033/proceeding.

22. D. Nakamura, "Simple and Quick Enhancement of SiC Bulk Crystal Growth using a Newly Developed Crucible Material" Appl. Phys. Express 2016, 9, 055507.

23. T. Kimoto, "Bulk and Epitaxial Growth of Silicon Carbide" Progress in Crystal Growth and Characterization of Materials" 2016, 62, 329–351.

24. M. Asghar, M. Y. Shahid, F. Iqbal, K. Fatima, M. A. Nawaz, H. M. Arbi, and R. Tsu, "Simple Method for the Growth of 4H Silicon Carbide on Silicon Substrate" AIP Advances 2016, 6, 035201; https://doi.org/10.1063/1.4943399.

25. I. Chowdhury, M. V. S. Chandrasekhar, P. B. Klein, J. D. Caldwell and T. Sudarshan, "High Growth Rate 4H-SiC Epitaxial Growth using Dichlorosilane in a Hotwall CVD Reactor" https://arxiv.org/ftp/arxiv/papers/1011/1011.1039.pdf.

26. K. Grasza, R. Diduszko, R. Bożek and M. Gała, "Initial Stages of SiC Crystal Growth by PVT Method" Crystal 2007, https://doi.org/10.1002/crat.200711011.

27. K. Semmelroth, et al. "Growth of 3C-SiC Bulk Material by the Modified Lely Method" Materials Science Forum May 2004, 457–460, 151–156 DOI: 10.4028/www.scientific.net/MSF.457-460.151.

28. R. Yakimova and E. Janzén, "Current Status and Advances in the Growth of SiC" Diamond and Related Materials, 2000, 9, 432–438.

29. V. D. Heydemann, N. Schulze, D. L. Barrett, and G. Pensl "Growth of 6H and 4H Silicon Carbide Single Crystals by the Modified Lely Process Utilizing a Dual-Seed Crystal Method" Appl. Phys. Lett. 1996, 69, 3728.

30. C. Meyer and P. Philip, "Optimizing the Temperature Profile During Sublimation Growth of SiC Single Crystals: Control of Heating Power, Frequency, and Coil Position" Crystal Growth & Design, 2005, 5, 1145–1156.

31. P. G. Neudeck, A. J. Trunek, D. J. Spry and J. A. Powell, "CVD Growth of 3C-SiC on 4H/6H Mesas" http://citeseerx.ist.psu.edu/viewdoc/download?doi=10.1.1.560.5607&rep=rep1&type=pdf.

32. A. Ellison, B. Magnusson, B. Sundqvist, G. Pozina, J. P. Bergman, E. Janzén and A. Vehanen, "SiC Crystal Growth by HTCVD" Materials Science Forum, 2004, 457–460, 9–14.

33. O. Kordina, "High Temperature Chemical Vapor Deposition of SiC" Appl. Phys. Lett. 1996, 69, 1456; https://doi.org/10.1063/1.117613.

34. D. H. Nam, B. G. Kim, J. -Y. Yoon, M. -H. Lee, W.-S. Seo, S.-M. Jeong, et al., "High-Temperature Chemical Vapor Deposition for SiC Single Crystal Bulk Growth Using Tetramethylsilane as a Precursor" Cryst. Growth Des. 2014, 14, 5569.

35. S. -C. Wang, P. K Nayak, Y. -L. Chen, J. C. Sung and J. -L. Huang, "Growth of Single Crystal Silicon Carbide by Liquid Phase Epitaxy using Samarium/Cobalt as Unique" Proc I Mech E Part N: J Nanoengineering and Nanosystems 2012, 1 1–5.

36. R. Hattori, et al. "LPE Growth of Low Doped n-Type 4H-SiC Layer on On-Axis Substrate for Power Device Application" www.scientific.net/author-papers/ryo-hattori.

37. A. Pradeepkumar, D. K. Gaskill, and F. Iacopi, "Electrical Challenges of Heteroepitaxial 3C-SiC on Silicon" Appl. Sci. 2020, 10, 4350; doi:10.3390/app10124350.

38. M. Fraga and R. Pessoa, "Progresses in Synthesis and Application of SiC Films: From CVD to ALD and from MEMS to NEMS" Micromachines 2020, 11, 799; doi:10.3390/mi11090799.

39. G. Fisicaro, C. Bongiorno, I. Deretzis, F. Giannazzo, F. L. Via, F. Roccaforte, M. Zielinski, M. Zimbone, and A. L. Magna, "Genesis and Evolution of Extended Defects: The Role of Evolving Interface Instabilities in Cubic SiC" Appl. Phys. Review 2020, DOI: 10.1063/1.5132300.

40. F. L. Via, A. Severino, R. Anzalone, C. Bongiorno, G. Litrico, M. Mauceri, M. Schoeler, P. Schuh, and P. Wellmann, "From Thin Film to Bulk 3C-SiC Growth: Understanding the Mechanism of Defects Reduction" Materials Science in Semiconductor Processing, 2018, 78, 57–68; ibid Materials Science Forum 2017-09-04 ISSN: 1662-9752, Vol. 924, pp. 913–918.

41. M. Zimbonea, M. Mauceri, G. Litrico, E. G. Barbagiovanni, C. Bongiorno, F. L. Via, "Protrusions Reduction in 3C-SiC Thin Film on Si" Journal of Crystal Growth, 2018, 498 248–257.

42. Y. Yang, B. Duan, S. Yuan and H. Jia, "Novel Developments and Challenges for the SiC Power Devices" Intech 2015, Ch. 6, pp. 175–195.

43. H. -P. Phan, D. V. Dao, K. Nakamura, S. Dimitrijev, and N.-T. Nguyen, "The Piezoresistive Effect of SiC for MEMS Sensors at High Temperatures: A Review" J. Microelectromech. Syst., 2015, 24, 1663.

44. A. Henry, X. Li, H. Jacobson, S. Andersson, A. Boulle, D. Chaussende and E. Janzén, "3C-SiC Heteroepitaxy on Hexagonal SiC Substrates", Mat. Sci. Forum, 2013, 740–742, 257–262. http://dx.doi.org/10.4028/www.scientific.net/MSF.740-742.257.

45. M. A. Fraga, M. Bosi and M. Negri, "Silicon Carbide in Microsystem Technology Thin Film Versus Bulk Material" Intech 2015, Ch. 1, pp. 3–31.

46. C.L Frewin, M. Reyes, J. Register, S. W. Thomas and S. E. Saddow," 3C-SiC on Si: A Versatile Material for Electronic, Biomedical and Clean Energy Applications" Mater. Res. Soc. Symp. Proc. Vol. 1693 © 2014 Materials Research Society DOI: 10.1557/opl.2014.567.

47. X.M.H. Huang, K.L. Ekinci, Y.T. Yang, C.A. Zorman, M. Mehregany and M.L. Roukes, "Nanoelectromechanical Silicon Carbide Resonators for Ultrahigh Frequency Applications" Solid-State Sensor, Actuator and Microsystems Workshop Hilton Head Island, South Carolina, June 2–6, 2002.

48. W.J. Choyke and R.P. Devaty, "Progress in the Study of Optical and Related Properties of SiC Since 1992" Diamond and Related Materials, 1997, 6, 1243–1248; ibid "Optical Properties of SiC: 1997–2002 Silicon Carbide" Eds. W. J. Choyke, H. Matsunami, G. Pensl © Springer-Verlag Berlin Heidelberg 2004; W. M. Klahold, W. J. Choyke, and R. P. Devaty, "Band structure properties, phonons, and exciton fine structure in 4H-SiC measured by wavelength-modulated absorption and low-temperature photoluminescence", Phys. Rev. B 2020,102, 205203.

49. Z. Xu, Z. He, Y. Song, X. Fu, M. Rommel, X. Luo, et al., "Topic Review: Application of Raman Spectroscopy Characterization in Micro/Nano-machining" Micromachines 2018, 9, 361; doi:10.3390/mi9070361.

50. R. Sugie and T. Uchida, "Determination of Stress Components in 4H-SiC Power Devices via Raman Spectroscopy" J. Appl. Phys. 2017, 122, 195703. https://doi.org/10.1063/1.5003613.

51. X. Feng and Y. Zang, "Raman Scattering Properties of Structural Defects in SiC" 3rd International Conference on Mechatronics and Information Technology (ICMIT 2016).

52. G. Chikvaidze, N. M.-Ulmane, A. Plaude, and O. Sergeev, "Investigation of Silicon Carbide Polytypes by Raman Spectroscopy" Lat. J. Phys and Tech. Sci. 2014, doi: 10.2478/lpts-2014-0019.

53. S. Lin, Z. Chen, L. Li, and C. Yang, "Effect of Impurities on the Raman Scattering of 6H-SiC Crystals" Materials Research, 2012, 15, 833–836; DOI: 10.1590/S1516-1439201200 5000108.

54. J. Wasyluk, T. S. Perova, S. A. Kukushkin, A. V. Osipov, N. A. Feoktistov, and S.A. Grudinkin, "Raman Investigation of Different Polytypes in SiC Thin Films Grown by Solid-Gas Phase Epitaxy on Si (111) and 6H-SiC Substrates" Materials Science Forum 2010, 645–648, 359–362.

55. T. Tomita, S. Saito, M. Baba, M. Hundhausen, T. Suemoto, and S. Nakashima, "Selective Resonance Effect of the Folded Longitudinal Phonon Modes in the Raman Spectra of SiC" Phys. Rev. 2000, B 62, 12896; Shin-ichi Nakashima, Makoto Higashihira, Kouji Maeda and Hidehiko Tanaka, "Raman Scattering Characterization of Polytype in Silicon Carbide Ceramics: Comparison with X-ray Diffraction", J. Amer. Ceramic Soc. 2004, 20, 823–829. https://doi.org/10.1111/j.1151-2916.2003.tb03382.x.

56. P. Borowicz, T. Gutt, and T. Malachowski, "Structural Investigation of Silicon Carbide with Micro-Raman Spectroscopy" MIXDES 2009, 16th International Conference "Mixed Design of Integrated Circuits and Systems", June 25–27, 2009, Lódz, Poland.

57. H. Harima, "Raman Scattering Characterization on SiC" Microelectronic Engineering, 2006 https://doi.org/10.1016/j.mee.2005.10.037.

58. J. C. Burton, L. Sun, M. Pophristic, S. J. Lukas, F. H. Long, Z. C. Feng, and I. T. Ferguson, "Spatial Characterization of Doped SiC Wafers by Raman Spectroscopy", J. Appl. Phys. 1998, 84, 6268.

59. H. Okumura, E. Sakuma, J. H. Lee, H. Mukaida, S. Misawa, K. Endo and S. Yoshida, "Raman Scattering of SiC: Application to the Identification of Heteroepitaxy of SiC Polytypes" J. Appl. Phys. 1987, 61, 1134; https://doi.org/10.1063/1.338157.

60. Z. C. Feng and A. J. Mascarenhas and W. J. Choyke, "Raman Scattering Studies of Chemical-Vapor-Deposited Cubic SiC Films of (100) Si" Journal of Applied Physics 1988, 64, 3176; https://doi.org/10.1063/1.341533.

61. Z. C. Feng and W. J. Choyke, "Raman Determination of Layer Stresses and Strains for Heterostructures and its Application to the Cubic SiC/Si System", J. Appl. Phys. 1988, 64, 6827; https://doi.org/10.1063/1.341997.

62. D. W. Feldman, J. H. Parker, Jr., W. J. Choyke, and Lyle Patrick, "Phonon Dispersion Curves by Raman Scattering in SiC Polytypes 3C, 4H, 6H, 15 R and 21 R", Phys. Rev. 1968, 173, 787; ibid "Raman Scattering in 6H SiC", Phys. Rev. 1968, 170, 698; Lyle Patrick, Phys. Rev. 1968, 167, 809.

63. J. Serrano, J. Stempfer, M. Cardona, M. S. -Böhning, H. Requardt, M. Lorenzen, B. Stojetz, P. Pavone and W. J. Choyke, "Determination of the Phonon Dispersion of Zinc-Blende (3C) Silicon Carbide by Inelastic X-Ray Scattering" Appl. Phys. Lett. 2002, 80, 4360.

64. B. Dorner, H. Schober, A. Wonhas, M. Schmitt, and D. Strauch, "The Phonon Dispersion in 6H-SiC Investigated by Inelastic Neutron Scattering", Eur. Phys. J B5, 1998, 839–846; G. Lorenz, H. Boysen, F. Frey, H. Jagodzinski and G. Eckold, "Akustische Phononen in SiC, Neutronenspektrometer" UNIDAS, Ergebnisbericht 81–86, Juel-Spez 1987, 410, 39.

65. D. Strauch, B. Dorner, A. Ivanov, M. Krisch, J. Serrano, A. Bosak, W. J. Choyke, B. Stojetz, and M. Malorny, "Phonons in SiC from INS, IXS, and Ab-Initio Calculations" Materials Science Forum Vols. 2006, 527–529, 689–694.

66. T. E. Tiwald, J. A. Woollam, S. Zollner, J. Christiansen, R. B. Gregory, T. Wetteroth, S. R. Wilson, and Adrian R. Powell, "Carrier Concentration and Lattice Absorption in Bulk and Epitaxial Silicon Carbide Determined using Infrared Ellipsometry", Phys. Rev. 1999, 60, 11464.

67. S. Zollner, J. G. Chen, E. Duda T. Wetteroth, S. R. Wilson, and J. N. Hilfiker, "Dielectric Functions of Bulk 4H and 6H SiC and Spectroscopic Ellipsometry Studies of Thin SiC Films on Si", J. Appl. Phys. 1999, 85, 8353.

68. S.-G. Lim, T. N. Jackson, W. C. Mitchel, R. Bertke and J. L. Freeouf, "Optical Characterization of 4H-SiC by Far Ultraviolet Spectroscopic Ellipsometry", Appl. Phys. Lett. 2001, 79, 162.

69. S. A. Kukushkin, and A. V. Osipov, Determining Polytype Composition of Silicon Carbide Films by UV Ellipsometry, Technical Physics Letters, 2016, 42, 175–178.

70. C. Cobet, K. Wilmers, T. Wethkamp, N. V. Edwards, N. Esser and W. Richter, "Optical Properties of SiC Investigated by Spectroscopic Ellipsometry from 3.5 to 10 eV" Thin Sol. Films, 2000, 364, 111–113.

71. A. Boosalis, T. Hofmann, V. Darakchieva, R. Yakimova and M. Schubert, "Visible to Vacuum Ultraviolet Dielectric Functions of Epitaxial Graphene on 3C and 4H SiC Polytypes Determined by Spectroscopic Ellipsometry" Appl. Phys. Lett. 2012, 101, 011912.

72. R. Ossikovski, M. Kildemo, M. Stchakovsky, and M. Mooney, "Anisotropic Incoherent Reflection Model for Spectroscopic Ellipsometry of a Thick Semitransparent Anisotropic Substrate" Appl. Opt., 2000 39, 2071.

73. A. T. Tarekegne, B, Zhou, K. Kaltenecker, K. Iwaszczuk, S. Clark, and P. U. Jepsen, "Terahertz Time-Domain Spectroscopy of Zonefolded Acoustic Phonons in 4H and 6H Silicon Carbide" Opt. Exp. 2019, 27, 3618.

74. Z. Tong, L. Liu, L. Li and H. Bao, "Temperature-Dependent Infrared Optical Properties of 3C-, 4H- and 6H-SiC" Physica B: Cond. Matter, 2018, 537, 194–201

75. N. H. Protik, A. Katre, L. Lindsay, J. Carrete, N. Mingo and D. Broido, "Phonon Thermal Transport in 2H, 4H and 6H Silicon Carbide from First Principles" Materials Today Physics, 2017, 1, 31–38.

76. Devki N. Talwar, "On the Pressure-Dependent Phonon Characteristics and Anomalous Thermal Expansion Coefficient of 3C-SiC" Mat. Sci. and Eng. B, 2017, 226, 1–9, and references cited therein.

77. K. Karch, P. Pavone, W. Windl, D. Strauch and F. Bechstedt, "Ab initio Calculation of Structural, Lattice Dynamical, and Thermal Properties of Cubic Silicon Carbide", International Journal of Quantum Chemistry, 1995, 56, 801–817.
78. C. H. Hodges, "Theory of Phonon Dispersion Curves in SiC Polytypes", Phys. Rev. 1969, 187, 994.
79. W. Hayes and R. Loudon, "Scattering of light by crystals" (Dover Publications, 1978); A. A. Maradudin, "Group-theoretic analysis of long-wavelength vibrations of polar crystals" (Department of Physics, University of California, 1975); J. F. Cornwell "Group theory in Physics I & II" (Academic Press 1984); Michael Tinkham "Group theory and Quantum Mechanics" (McGraw Hill 1964).
80. Devki N. Talwar, Linyu Wan, Chin-Che Tin, Hao-Hsiung Lin, and Zhe Chuan Feng, Spectroscopic Phonon and Extended X-Ray Absorption Fine Structure Measurements on 3C-SiC/Si (001) Epifilms, Applied Surface Science, 2018, 427, 302–310.
81. Devki N Talwar, Z. C. Feng, C. W. Liu, and C. -C. Tin, "Influence of Surface Roughness and Interfacial Layer on the Infrared Spectra of V-CVD Grown 3C-SiC/Si (1 0 0) Epilayers" Semicon. Sci. Technol. 2012, 27, 115019, and references cited therein.
82. T. W. Cadman and D. Sadowski, "Generalized Equations for the Calculation of Absorptance, Reflectance, and Transmittance of a Number of Parallel Surfaces" Appl. Opt. 1978, 17, 531
83. D.W. Berreman, "Infrared Absorption at Longitudinal Optic Frequency in Cubic Crystal Films" Phys. Rev., 1963, 130, 2193.
84. Devki N. Talwar, "Direct Evidence of LO Phonon-Plasmon Coupled Modes in n-GaN" Appl. Phys. Lett. 2010, 97, 051902; https://doi.org/10.1063/1.3473826.
85. M. Dean Sciacca, A. J. Mayur, E. Oh, A. K. Ramdas, S. Rodriguez, J. K. Furdyna M. R. Melloch, C. P. Beetz, and W. S. Yoo, "Infrared Observation of Transverse and Longitudinal Polar Optical Modes of Semiconductor Films: Normal and Oblique Incidence" Phys. Rev., 1995, B 51. 7744.
86. O. E. Piro, "Optical Properties, Reflectance, and Transmittance of Anisotropic Absorbing Crystal Plates" Phys. Rev., 1987, B 36, 3427.
87. D. M. Caughey and R. E. Thomas, "Carrier Mobilities in Silicon Empirically Related to Doping and Field" Proc. IEEE 1967, 55, 2192.
88. H. Y. Sun, S.-C. Lien, Z. R. Qiu, H. C. Wang, T. Mei, C. W. Liu, and Z. C. Feng, "Temperature Dependence of Raman Scattering in Bulk 4H-SiC with Different Carrier Concentration", Opt. Exp. 2013, 21, 26475.

4

SiC Materials, Devices, and Applications: A Review of Developments and Challenges in the 21st Century

Min Lu

Compound Semiconductor China (CSC)
Changzhou Perfect Crystal Semiconductor Co. Ltd

Chengling Lu

Beijing University of Posts and Telecommunications, China
Suzhou Institute of Nano-Tech and Nano-Bionics (SINANO), CAS

1 Introduction

As we all know, because the main driver of Tesla Model 3 adopts all-silicon carbide modules, it opens the way for the mass application of silicon carbide. So far, model 3 has sold nearly 1 million vehicles worldwide, and it has been crowned the world's sales champion model for three consecutive years, which is enough to prove that the application of silicon carbide power devices in electric vehicles has no problems in technology, industry, and market. Compared with silicon devices, the superior performance of silicon carbide modules can increase the car's cruising range by 5–10%, speed up faster, charge faster with more power saving, but the price is also very expensive. It can be said that the bottleneck of the silicon carbide industry is that the price of silicon carbide substrates is too high. Although in recent years, with the substantial increase in market demand and the development of silicon carbide substrate manufacturing technology to promote the improvement of yield, the price of silicon carbide substrates has been declining year by year, but is still very high, the price is about 60 times more than the same size silicon substrate, the silicon carbide substrate accounts for up to 50% of the device cost, which ultimately leads to the high price of silicon carbide devices. Because the cost of device manufacturing has slightly increased with the increase in material size, large-size materials are the first choice to reduce the cost of devices. Nowadays, 4–6 inches of silicon carbide is the mainstream of the market, and it is estimated that 8 inches will also enter the market in 3 years. Therefore, the SEMI M55-0817 standard "Specification for Polished Mono-crystalline Silicon Carbide Wafers" is under revision. The core content of the revision is to increase the specification requirements for 8-inch silicon carbide to prepare for the industrialization of 8-inch silicon carbide. In addition, it is compatible with the SEMI M55 epitaxial wafer specification standard "Specification for silicon carbide homogeneous epitaxial wafer" and three test method standards "Test Method for Micropipe Density of Silicon Carbide Wafer by Laser Reflection", "Test Method for Flatness of Silicon Carbide Wafers by Optical Interference", and "Test Method for Residual Stress of Silicon Carbide Wafers by Photo-"elastic" are both being created. These four standards also cover 8 inches in material size.

DOI: 10.1201/9780429198540-5

Silicon carbide "getting on the car" is not an exclusive technology and application, because there is already a mature technology on the car – the "silicon", so the silicon carbide "getting on the car" will inevitably lead to a strong interception of "silicon". The "territories" of the competition include main inverters, OBC, and DC/DC. Of course, the other chips on the car, such as MCU, AI, and CMOS image sensors, information, intelligence, and control chips are still in the bag of "silicon", which no one can match. Although silicon carbide has the above-mentioned superior properties, whether these are hard demands of consumers or pain points of use, which determines whether people are willing to pay for these benefits. This is actually a very sensitive market issue. Let the market solve it. In addition, the automobile industry is a traditional industry, and automobile safety is related to the issue of "life and death". Therefore, the supply chain system has always been rigorous and conservative. Therefore, it is very difficult to enter the qualified supplier list of OEM manufacturers. The threshold of technology, reliability, and supply capacity of the company is very high, and the certification cycle is also very long. Although electric vehicles have the characteristics of their emerging business formats, because they are born out of the traditional automobile industry, the above-mentioned high threshold and long certification cycle characteristics are the same as before. We often hear that the car regulations AEC-Q101, AQG-324 product certification, and IATF16949 system certification are just the stepping stone to the application of car regulations, that is, the necessary minimum requirements, and the more stringent is customer certification. These certification requirements for devices/modules will inevitably be passed on to the certification and requirements for silicon carbide materials. Although there are no relevant standards and certifications for car-regulated silicon carbide materials, I think it is necessary to carry out research and creation in this area. Because the quality of silicon carbide materials is far from the level of silicon materials, it is necessary to give a standard to select high-reliability materials that meet automotive regulations, so as to promote the application of silicon carbide materials in electric vehicles.

Nowadays, in addition to a Tesla model that uses silicon carbide, BYD also uses a full silicon carbide module main inverter on BYD Han models. There are probably dozens of car companies that use silicon carbide technology on OBC. However, the penetration rate of silicon carbide in the automotive power device market is only about 5%. Of course, many device manufacturers, such as Wolf-speed, Infineon, Rohm, ON Semiconductor, etc., are cooperating with Tier1 or automakers., Making full preparations for silicon carbide to "get on the car", because after all, this is a future "big cake" of tens of billions of dollars a year, and no one wants to miss it.

Based on the above reasons, the road of silicon carbide "getting on the car" will be long-lasting, and its market penetration rate depends on the rate of decline in the price of silicon carbide (the core is the decline in the price of silicon carbide substrates), and it is estimated to be around 2030 or 2035. The penetration rate of silicon carbide in the automotive power device market will reach about 50%. Therefore, silicon carbide and silicon will coexist in the car for a long time. After silicon carbide's price falling to a certain level, it will enter the middle and low-end car market with the largest car market capacity. A long penetration cycle is actually a good thing. This can give the EV industry more time to grow, and at the same time give the "latecomers" more time and opportunities to catch up with the "first movers", which will facilitate the formation of a relatively healthy and benign market competition.

2 A Review of Developments and Challenges on SiC Substrate

Nowadays, the mainstream manufacturing technology of silicon carbide substrates are PVT growth, multi-wire cutting, grinding, mechanical polishing, and CMP. The mainstream sizes are 6 inches (conductive) and 4 inches (semi-insulating), 6-inch semi-insulating substrates have begun to enter the market, and 8-inch conductive substrates have been successfully developed and are expected to enter the market in 2–3 years. The quality of the 6-inch conductive substrate can reach a micropipe density of less than $0.1/cm^2$ and a dislocation density of less than $1000/cm^2$, the resistivity of the 6-inch semi-insulating substrate can reach more than $10^9 \Omega \cdot cm$. However, the growth rate is slow, with an average of 200 um/h, the maximum thickness of the crystal is about 2–5cm, and the crystal processing efficiency is relatively low. In addition, the overall yield is also low. Therefore, the price of silicon carbide substrate remains high. At present, the 6-inch conductivity is about 7000 RMB/piece, and the 6-inch semi insulation is about 20000 RMB/piece, which is about 60–100 times the price of silicon wafer of the same size. Considering the devices, the substrate accounts for about 50% of the device cost, which is the biggest bottleneck in the promotion of the silicon carbide industry. Therefore, the main research at this stage focuses on how to further improve the quality, reliability, and manufacturing efficiency of materials to reduce the cost and price of silicon carbide.

It still has many disadvantages although PVT silicon carbide technology is the most mature industrialization technology. Such as the growth chamber of the black box cannot monitor the internal growth parameters of the crucible in real time, slow growth rate and unbalanced growth, etc. Therefore, simulation is a very effective means, which can help us develop and research more efficiently, and provide qualitative and directional guidance in solving some key problems.

The Alexander-Haasen (AH) model has been applied to analyze the plastic deformation and dislocation generation during the crystal growth process of 4H-SiC (silicon carbide). Plastic parameters are obtained by fitting the predicted curves to the experimental data on the plastic deformation of α-SiC crystals under uniaxial compression. The relationship between the activity energy (Q) and stress exponent (n) is considered when using the AH model. This relationship explicitly represents two deformation mechanisms around the critical temperature. The ratio of the activity energy and stress exponent, Q/n, equals 0.3 eV when the temperature is below the transition temperature, and 1.3 eV when the temperature is above the transition temperature. Then, the model is used to predict the dislocation density and thermal stresses in the crystals. The largest dislocation density is found to occur near the graphite/SiC interface, and the dislocation density gradually decreases with the thickness of the ingot. The plastic model can be used to improve the understanding of plastic deformation of SiC crystals and to improve the SiC crystal growth process.

When single crystal silicon carbide is used in epitaxial film growth, it needs to show a smooth, defect free, and damage free surface. However, because silicon carbide has the characteristics of high hardness, high brittleness, and good chemical stability, it is difficult to flatten silicon carbide efficiently and ultra accurately. Chemical mechanical polishing (CMP) is a commonly used method for overall planarization polishing of semiconductor wafers.

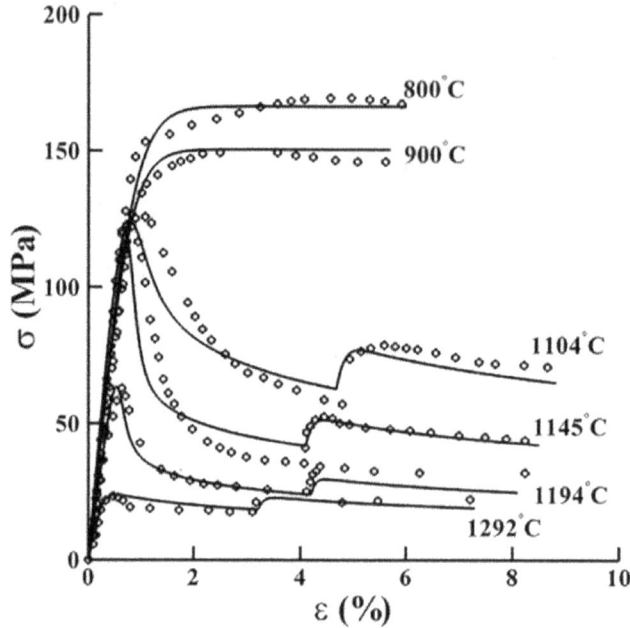

FIGURE 1
Simulated stress(MPa)-versus-strain(%) curves for 4H-SiC at temperatures between 800 and 1300°C. The circles represent experimental data from [3]. The initial strain rate is 1.5×10^{-5} s^{-1} for all tests. For tests at temperatures above 1000°C, the strain rate jumps to 3.0×10^{-5} s^{-1} from the initail strain rate when the strain is close to 4%. (After [3] Figure 3, reproduced with permission of Elsevier.)

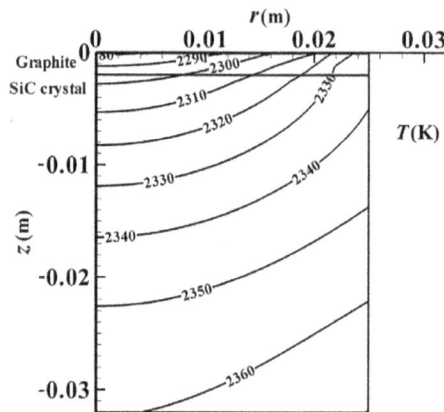

FIGURE 2
Temperature distribution in the graphite and SiC crystal layers. (After [3] Figure 4, reproduced with permission of Elsevier.)

The environment of photo catalytic reaction exerted no significant influence on the stability of particle size of colloidal silica abrasives while it reduced the zeta potential of colloidal silica abrasives, therefore they can be used as abrasives for photo catalysis-assisted polishing.

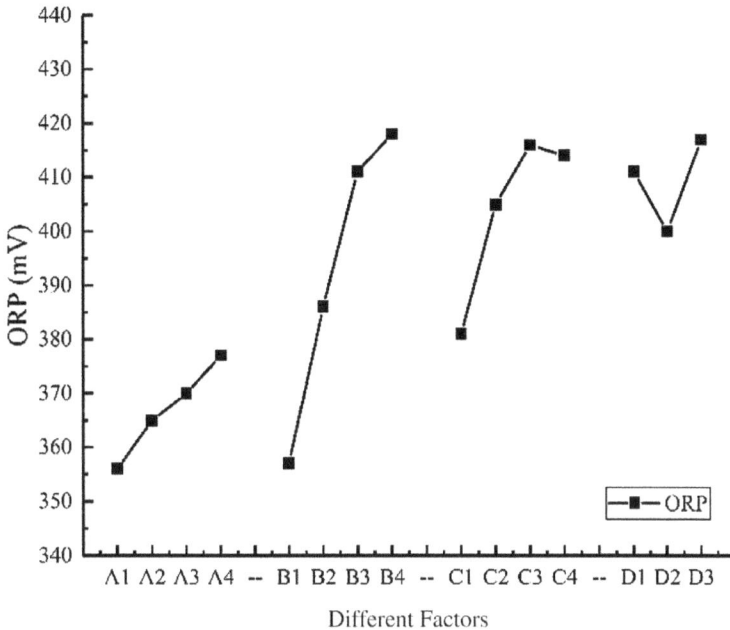

FIGURE 3
Rates of photo catalytic reaction under different conditions. (After [8] Figure 10, reproduced with permission of Elsevier.)

Different factors exhibited diverse levels of influence on the rate of this photo catalytic reaction. The H_2O_2 concentration exerted the most significant influence on the rate of photo catalytic reaction. In an alkaline environment, OH- ions in solutions can function as the electron capture agent, which can promote the rate of photo catalytic reaction.

The greater the rate of the UV photo catalytic reaction, the greater the MRR of 6H-SiC wafer and the lower the final surface roughness Ra. Under optimized experimental conditions: a light intensity of 1000 mW/cm^2, an H_2O_2 concentration of 4.5 vol%, a TiO$_2$ concentration of 3 g/L, and a pH of 11, the 6H-SiC surface with the surface roughness of Ra 0.423 nm can be obtained and the MRR was 107 nm/h.

3 A Review of Developments and Challenges on SiC Epitaxy

Different from traditional silicon power devices, if silicon carbide power devices want to give full play to the characteristics of this material, they cannot be directly fabricated on silicon carbide single crystal substrates, but grow high-quality epitaxial materials on substrate, and manufacture various devices on epitaxial layers. The research and development of silicon carbide epitaxial materials began in the mid-1990s. At present, great progress has been made in material growth technology and device research. 6-inch SiC epitaxial wafer has been successfully developed and commercialized. It can meet the development of SiC power electronic devices with a voltage level of 3.3 kV and below. However, it cannot meet the needs of developing 10 kV and above, and bipolar devices. The high-quality

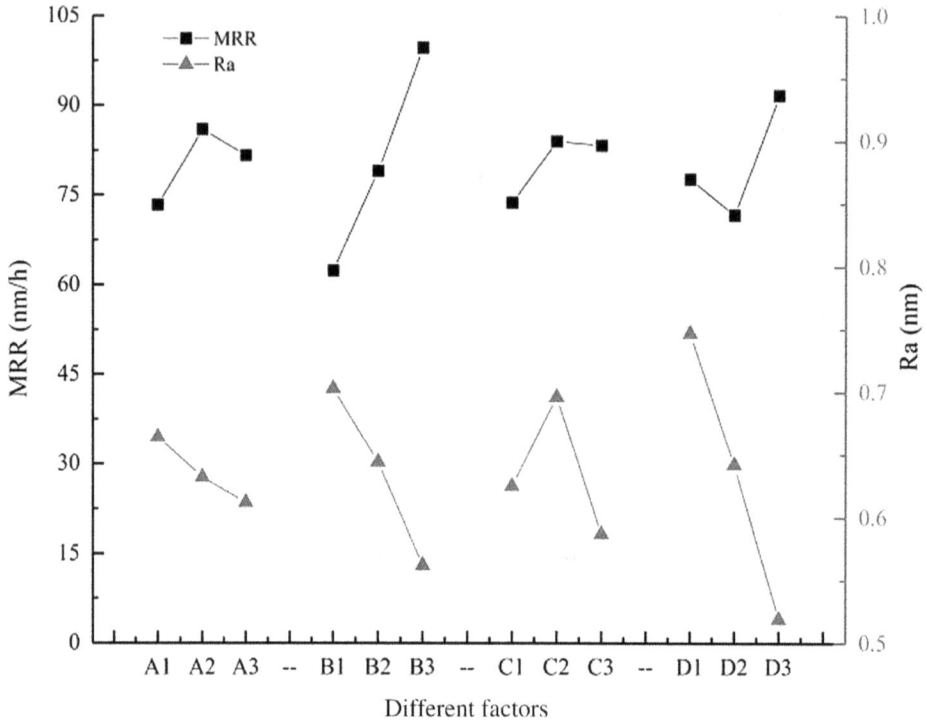

FIGURE 4

The influence of various factors on the MRR of polishing materials and surface roughness. (After [8] Figure 11, reproduced with permission of Elsevier.)

production of epitaxial wafers can be guaranteed in the medium and low voltage field, but there are still many defects and difficulties to be overcome in the high voltage field.

4H-SiC epilayers have been grown on on-axis and 4° off-axis substrates by our home-made horizontal LPCVD with different C/Si ratio. The crystallinity was characterized by XRD and SEM. SEM, AFM, and Nomarski microscope were used to investigate the effect of C/Si ratio on the growth of 4H-SiC thin film. With the increase of C/Si ratio, the growth rate shows an upward trend, but the growth rate reaches saturation with C/Si ratio up to 1.0, which is caused by the balance of supply between C and Si sources. Furthermore, smooth surface can be obtained with the C/Si ratio between 0.6 and 1.2 for growth on 4° off-axis substrate. However, C/Si ratio window of on-axis epitaxial growth is very narrow. We have considered the C/Si ratio as a major parameter in 4H-SiC homoepitaxial growth on on-axis and 4° off-axis substrates.

Silicon carbide has a variety of crystal forms, among which commercial 4H and 6H SiC wafers can be prepared and have been frequently used in the manufacture of electronic devices. At the same time, 3C silicon carbide is also promising because it has the largest electron mobility ($\sim$800 cm^{2}V^{-1}s^{-1}) and saturation drift velocity ($\sim$2.5 $\times$10^{7}cm/s) among all polytypes. Last but not least, 3C silicon carbide can be heteroepitaxially grown on silicon, which not only provides the advantages of lower production cost and larger silicon carbide wafer size, but also allows the process integration of silicon carbide based devices to be compatible with silicon technology.

A flow-modulated carbonization process is applied to grow a 3C-SiC thin film of high crystal quality on the Si (100) substrate using low pressure chemical vapor deposition.

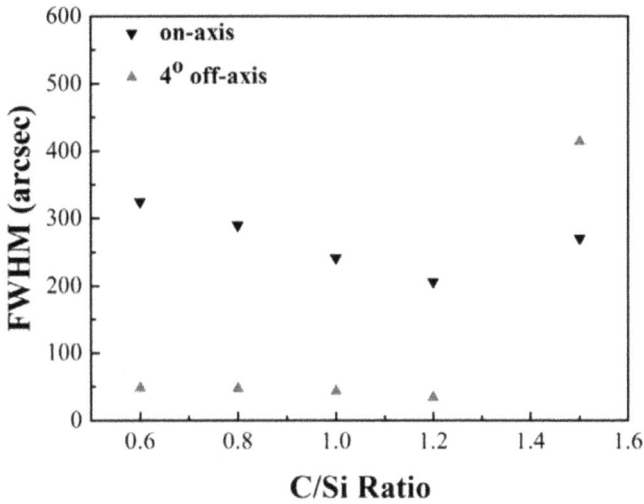

FIGURE 5
The dependence of FWHM on C/Si ratio of 4H-SiC epitaxial layers grown on on-axis and 4° off-axis substrates. (After [14] Figure 1, reproduced with permission of Elsevier.)

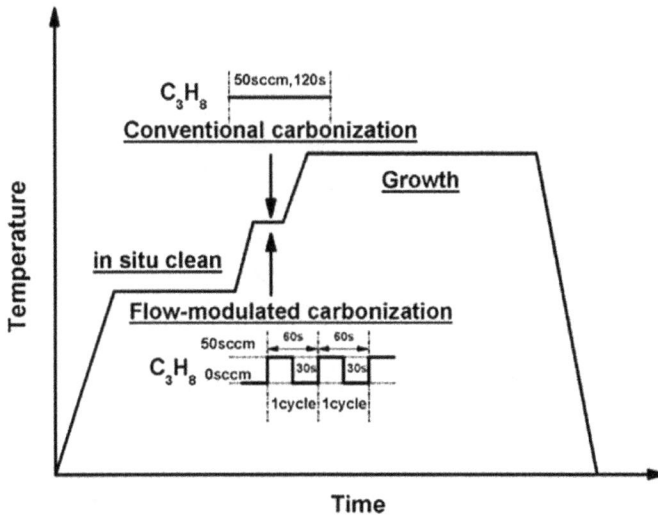

FIGURE 6
Schematic growth program and sequence of the carbon-based precursor introduced into the reactor for the conventional carbonization process and the flow-modulated carbonization process. (After [22] Figure 1, reproduced with permission of Elsevier.)

The flow-modulated carbonization was performed by flowing intermittent carbon-based precursor. The crystal quality of the so-obtained 3C-SiC is compared with that fabricated via the conventional carbonization process, using x-ray diffractometry and Raman spectra data, indicates a better crystal quality using this flow-modulated carbonization process. Moreover, Si outdiffusion from Si substrates is suppressed in the flow-modulated carbonization process, resulting in a reduced density of voids. Consequently, the flow-modulated

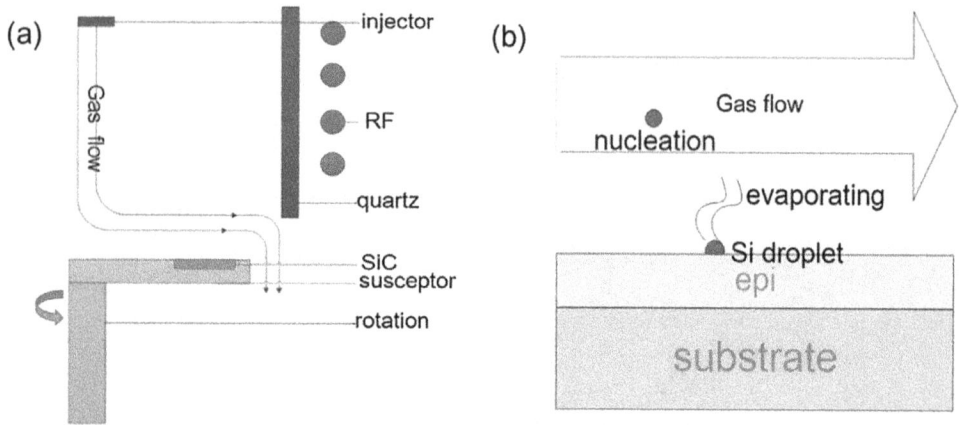

FIGURE 7
(a) Schematic diagram of the home-built vertical hot-wall CVD reactor; (b) model of silicon droplet evolution during the epitaxial layer growth. (After [26] Figure 3, reproduced with permission of Elsevier.)

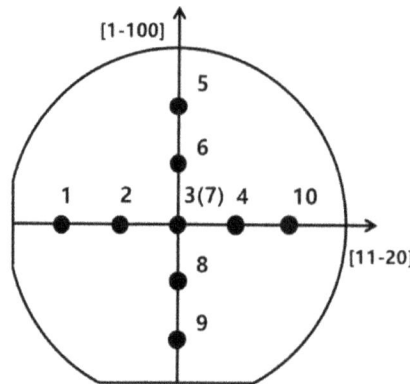

FIGURE 8
The locations of nine sample points on the wafer. (After [30] Figure 1, reproduced with permission of Elsevier.)

carbonization process plays an active role in enhancing the crystal quality and reducing the void formation. This is the first report of the heteroepitaxial growth of 3C-SiC layers using the flow-modulated carbonization process.

The homoepitaxial growth of 4H-SiC epilayers was conducted by hot-wall vertical chemical vapor deposition (CVD). The dependence of silicon droplets on the growth temperature on 4°off-axis 4H-SiC substrates and its mechanism have been investigated, which were characterized by a Nomarski optical microscope, scanning electronic microscope (SEM), and micro-Raman spectrometer. The results indicated that the silicon droplets were highly crystalline. It was also found that the silicon droplet generation could be suppressed by increasing the growth temperature, though the growth rate declined slightly.

To promote the wide application of silicon carbide in practical applications, there are still many problems to be solved. For example, intrinsic deep level defects usually act as traps or recombination centers in wafers, which will reduce carrier lifetime and affect the

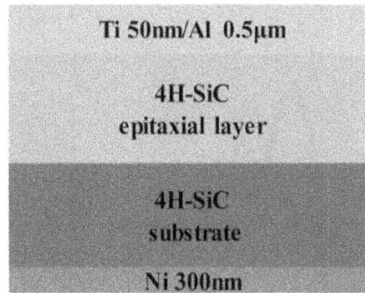

FIGURE 9
Schematic diagram of Schottky structure of the samples. (After [30] Figure 2, reproduced with permission of Elsevier.)

performance of bipolar 4H-SiC devices. Therefore, the study of deep level defects is one of the main research fields of silicon carbide.

Schottky structures were fabricated on the epitaxial wafers and the deep levels of the wafers were measured by deep level transient spectroscopy (DLTS). The thickness of the commercial n-type 4°off-axis 4H-SiC (0001) epitaxial wafers was 100 μm, with a doping concentration of 2×10^{14} cm^{-3}. Nine sample points were selected along with the direction of [1–100] and [11–20] on the wafer. Through the peak fitting analysis of the test results, the detailed information of deep level defects in the epitaxial layer was obtained. The typical deep level defect $Z_{1/2}$ was found, but the Ti center did not appear, which indicates that the Ti center was not introduced in our technological process. Defect concentration uniformity of the 4-inch 4H-SiC epitaxial wafer is 5.74%. The results demonstrate that although the concentration of $Z_{1/2}$ defect differs in different locations, but the difference is so small that $Z_{1/2}$ defect concentration can be considered to be evenly distributed across the wafer.

4 A Review of Developments and Challenges on SiC Devices

Compared with traditional silicon devices, silicon carbide power devices have the advantages of high temperature, high voltage, high frequency, and high efficiency, which greatly improves the performance of semiconductor power devices and promotes the development of power electronics. According to the working form of devices, silicon carbide power devices are mainly divided into power diodes and power switches. The power diode includes a junction barrier Schottky (JBS) diode, pin diode, and super junction diode; power switches mainly include metal oxide semiconductor field effect switch (MOSFET), junction field effect switch (JFET), bipolar switch (BJT), insulated gate bipolar transistor (IGBT), gate turn off thyristor (GTO), and emitter turn off thyristor (ETO). However, there are still some problems in silicon carbide devices, such as low output, high price, few types of commercial devices, and lack of high-temperature packaging. Major manufacturers and research institutions are still making in-depth research. These problems will also be solved step by step to expand the application of silicon carbide in the field of power electronics.

Gate oxide film on silicon carbide (SiC) severely affects the performance of SiC metal-oxide-semiconductor field effect transistor (MOSFET). The authors investigated the

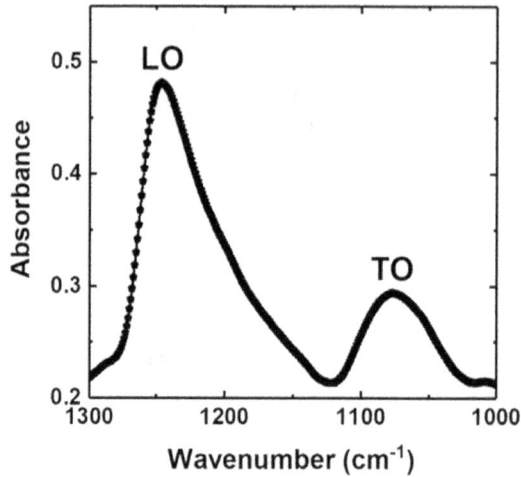

FIGURE 10

Infrared spectra of thermally grown SiO_2. (After [37] Figure 5, reproduced with permission of Elsevier.)

influence of curvature induced stress/strain to flatband voltage (V_{fb}) and interface density (D_{it}) on SiO_2/SiC by capacitance-voltage (*C-V*) measurement. The curvature of epitaxy wafers has been identified by the Thin Film Stress Measurement system. The compress/tensile curvature led to increase of the positive V_{fb} shift (negative fixed charge) of SiO_2 and the interface density of SiO_2/SiC during dry thermal oxidation process. In addition, the transverse optical (TO) phonon wavenumber of the samples related with the curvature of the film, indicating that stress mainly affected the interface of SiO_2/SiC. According to the experimental result, the authors suggested that a "free" stress oxide film might be a best choice for the application of SiC-MOSFET.

High performance 4H-SiC TJBS diodes have been successfully designed and fabricated. For making a larger design window that enables good reverse blocking and forward conduction capabilities at the same time in the TJBS diode, trench profile, especially bevel angle (θ) of a tapered trench sidewall, is further studied. The optimal design is verified by the experimental results measured from the fabricated 4H-SiC PJBS and TJBS diodes. Experimental results show that the reverse leakage current (I_r) of the TJBS is only 1.62×10^{-7} A/cm² at V_r of 1200V, which is about one order of magnitude less than that of the PJBS. The FOM of the TJBS is improved by 56.3% when compared with that of the PJBS.

The effect of P+ region design on the simulation and fabrication of 6500V 4H-SiC JBS diodes is studied. The P+ region spacings of the active area in the 6500V SiC diodes decrease during the fabrication because of the implantation spreading and the mask etching. The simulation of the P+ region design has a deviation of about 0.9μm from the actual device for the 6500V SiC diodes. Based on the above analysis, relatively large area 6500V/40A 4H-SiC JBS diodes have been fabricated. The device has a leakage current of 17uA at 6.5kV and typical turn-on voltage is about 0.8V. The forward conduction current is about 40A when the forward voltage drops to 4V. The simulation fits well with the measured results considering the broadening of 0.9μm of the P+ region.

Due to the high density of near interface electron traps and low gate oxide integrity, silicon carbide MOSFET has the problems of low channel mobility and gate oxide reliability. In the past few decades, some post oxidation annealing (POA) treatments, such

TABLE 1

Device parameters used in the simulations. (After [46] Table I, reproduced with permission of Elsevier.)

Parameters	PJBS	TJBS
Trench depth, t	/	0.5 μm
Trench bevel angle, θ	/	90°, 96°, 102°, 108°, 114°
P+ junction depth, xj	0.9 μm	0.9 μm
P+ junction width, W	3 μm	3 μm
Junction Spacing, S	4 μm	4,5,6 μm
N-epilayer thickness, d	12 μm	12 μm
N-epilayer doping, N_D	6.8×10^{15} cm^{-3}	6.8×10^{15} cm^{-3}
N+ substrate doping, N_{sub}	5×10^{18} cm^{-3}	5×10^{18} cm^{-3}

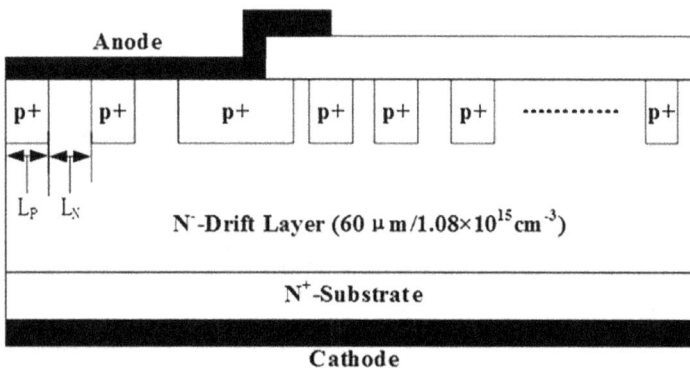

FIGURE 11

The cross-sectional view of the 4H-SiC JBS diode. (After [48] Figure 1, reproduced with permission of Elsevier.)

as high temperature annealing in nitrogen or phosphorus rich environment, have been proved to improve the electrical properties of silicon carbide MOSFET.

A novel nitrogen-phosphorus (N&P) hybrid passivation technique by NO-containing dry oxidation of phosphorus-implanted 4H-SiC epilayer has been proposed to grow high-property gate oxide film. For demonstration and contrast, *n*-type 4H-SiC MOS capacitors with N&P hybrid and only N passivation have been fabricated and tested. The results from secondary ion mass spectroscopy (SIMS) reveal that about 60% of the implanted phosphorus atoms remain in the SiO$_2$ film after sacrificial oxidation of the implantation layer. And the distribution of phosphorus is approximately uniform throughout the grown SiO$_2$ film with a concentration about 4×10^{19} cm^{-3}, avoiding the SiO$_2$ film converting to phosphosilicate glass (PSG). *I-V* and x-ray photoelectron spectroscopy (XPS) data show that the grown SiO$_2$ film with appropriate phosphorus passivation can effectively improve the quality of oxide film, leading to the gate oxide band gap having improved integrity and the leakage current being reduced by above two orders of magnitude. The bias stress measurements with ultraviolet light (UVL) irradiation also show that the N&P hybrid passivation treatment can completely restrain the near-interface electron traps (NIETs) in 4H-SiC MOS devices, indicating that incorporation of an appropriate amount of phosphorus atoms in the grown SiO$_2$ film can also make the electron trap density reduce effectively.

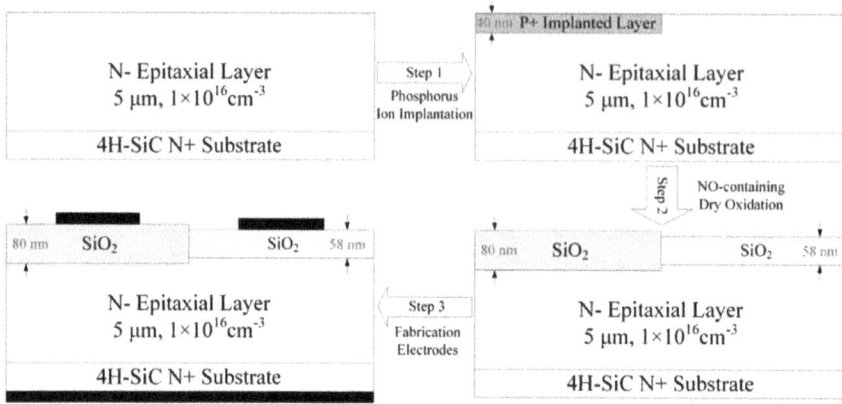

FIGURE 12

The process flow for the samples with N&P and N passivation. (After [56] Figure 1, reproduced with permission of Elsevier.)

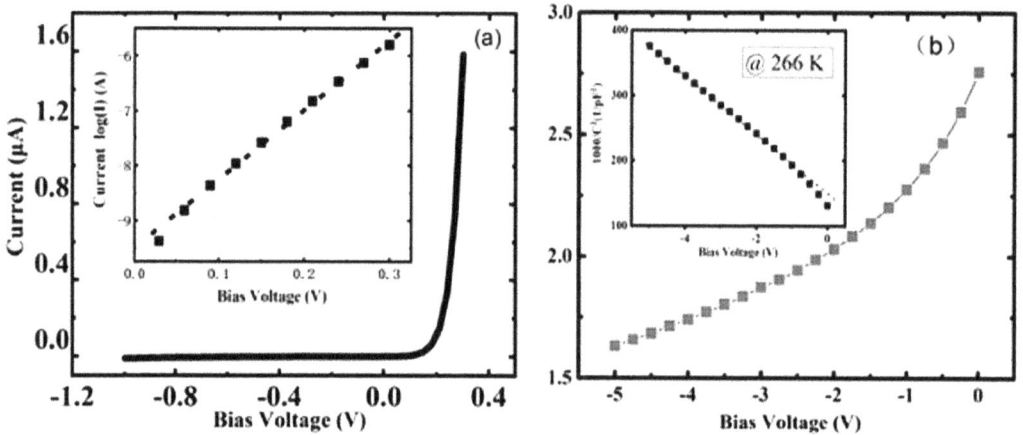

FIGURE 13

(Left) Current density and (right) capacitance versus voltage characteristics of Ge-doped 6H-SiC at room temperature, inset of (right) shows $1/C^2$ versus V plot. (After [57] Figure 3, reproduced with permission of Elsevier.)

In recent years, after solving the *n*-type and *p*-type in silicon carbide, some researchers have focused on doping group IV elements into silicon carbide to change conductivity and promote better contact characteristics.

Crystal structure and deep levels of Ge doped 6H-SiC were systematically investigated. Raman shows the Ge doped 6H-SiC has a good crystal structure. Then, the RBS/c verifies the doped Ge atoms should be 190 substitutional impurity. The DLTS test and Arrhenius fitting show the induced defect energy levels are below the conductive band bottom 0.298 eV and 0.323 eV, respectively. The first-principles calculations show that only the Ge replacing C atoms will induce deep energy below the middle position of the band gap. However, this configuration needs large form energy. Thus, of the measured energy levels, 195 are primarily caused by native defects arising in the PVD process. Thus, two effects lead to the improvement of ohmic characteristics. First, the Ge-C bond is easier to break to generate electrons under the influence of an applied field than the Si-C bond in the same situation. Second, the generated electrons can jump into the conductive band bottom (CBD) through energy levels generated by native defects. In total, 200 small cross sections near to the CBD position make the energy levels trap and emit electrons to the CBD. In the future, high temperature annealing and other methods for defect characterization should be preformed to further investigate the function of native defect.

5 A Review of Developments and Challenges on the SiC Package and Module

In recent years, the performance requirements of semiconductor power devices in the field of power electronics are higher and higher. The third generation semiconductor materials represented by silicon carbide are developing rapidly and commercialized gradually. At the same time, the packaging technology of silicon carbide devices lags, which makes the development of device performance into a bottleneck, especially the requirements for heat dissipation and high-frequency performance in the field of high frequency and high power need to be solved urgently.

Compared with traditional silicon semiconductor materials, silicon carbide has the advantages of higher operating temperature and lower resistance due to its excellent material properties. With the increase of the working frequency of the power module, the stray inductance of the traditional packaging structure limits the development of the power module to high frequency applications. Stray inductance has become a challenge in power electronics applications, especially in high frequency and high power applications.

The simulation, fabrication, and measurement of the high-voltage H-bridge SiC diode module is reported. The SiC module consists of eight self-designed 3.3 kV 30A SiC Schottky diodes (SBDs), in which each bridge arm is connected by double SBD chips to achieve 60A current. Q3D software is used to establish the simulation model of 3.3 kV 60A high-voltage H-bridge SiC diode module. The parasitic parameters are extracted from the current circuits. By establishing the geometric model and finite-element model, finite-element analysis software ANSYS is used to calculate steady-state thermal conduction, and the temperature gradient distribution shows that the distribution of the temperature field is reasonable. Under the condition of the test at room temperature and static, the module voltage drop is about 2.1 V, the leakage current is less than 5 uA, and the break-down voltage is more than 3700 V. The fabricated 3300 V devices exhibit a large safety margin. The insulation voltage exceeds 7000 V, thus ensuring the safety of the system. The thermal resistance of the chip is about 0.21 K/W, which is basically consistent with the simulation results.

The brief status review of the SiC power module market and packaging is addressed, with the new concept of structure for efficient cooling and high reliable packaging technologies. It is reported that the thermal resistance of the SiC module is a bit higher than the SiC module because of high current density, so the power cycling capability will be not as good as the Si module if the SiC module is packaged by conventional solutions. However, the power dissipation of the SiC module is much less than the Si module, especially at high temperature and high switching areas as the power loss of SiC is less sensitive to junction temperature, and its switching energies are very low compared to the Si device.

With the increasing usage of silicon carbide (SiC) devices for renewable energy and electrified transportation, a customized and reliable package for SiC power modules is urgently needed. To elevate the advanced performances of SiC devices and optimize the package of power modules, the parasitic sensitivity and temperature dependency of the SiC power module should be carefully addressed, compared with the traditional Si power module by using the specific package. In this paper, under the same wire-bonding package, a full SiC power module is proposed, designed, and evaluated, compared with the full Si and hybrid power modules. In an inductive-clamped double-pulse test rig, these power modules are comparatively investigated in conditions of different load current and junction temperatures. Experimental results demonstrate the SiC power module achieves less switching loss and is less sensitive to junction temperature; however, it is more sensitive to package parasitics. Comprehensively experimental results also reveal the limitations of the traditional package for the SiC power module. Furthermore, based on abundant failure samples, several typical destructive mechanisms of power modules are illustrated, which can improve the package of SiC power modules.

A novel fan-out panel-level PCB embedded package for the phase-leg SiC MOSFET power module was presented. Electro-thermo-mechanical co-design was conducted and found that the maximum package parasitic inductance was about 1.24 nH at 100 kHz. Compared with wire-bonded packages, the parasitic inductances of the PCB embedded package decreased by 87.6% at least. Compared with blind via structure, the thermal resistance of the proposed blind block structure reduced by about 26% at most, and the stress of the SiC MOSFETs decreased by about 45.2%. Then, a novel PCB embedded packaging process was presented and three key packaging processes were analyzed. Furthermore, the effect of the PCB embedded package on static characterization of SiC MOSFET was analyzed and the following was found: (1) output current of the PCB embedded package was decreased under a certain gate-source voltage compared to SiC die. (2) Miller capacitance of SiC MOSFET was increased thanks to parasitic capacitance induced by package. (3) Compared with SiC die, non-flat miller plateau of the PCB embedded package extends, and as drain-source voltage increases, the non-flat miller plateau extends. Lastly, switching characteristics of the PCB embedded package and TO-247 package were compared. The results show the PCB embedded package has smaller parasitic inductances.

Ceramic substrates and thick-film metallization based packaging systems including chip-level packages and compatible printed circuit board were demonstrated with high temperature durable die-attach and Au wire-bonding technologies for 500°C SiC electronics. Packaged SiC JFET circuits have been successfully tested for over 10,000 hours at 500°C. A spark-plug type sensor package with low parasitic effects for high temperature capacitive pressure sensors has been characterized and tested with SiC sensors at temperatures up to 500°C. This sensor package applies to high temperature and high differential pressure environments.

SiC devices are promising for outperforming Si counterparts in high frequency applications due to their superior material properties. A conventional wire-bonded packaging scheme has been one of the most preferred package structures for power modules. However, the technique limits performance of a SiC power module due to parasitic inductance and heat dissipation issues that are inherent with aluminum wires. Low parasitic inductance and high-efficient cooling interconnection techniques for Si power modules, which are the foundation of packaging methods of SiC ones, are reviewed firstly. Then, attempts on developing packaging techniques for SiC power modules are thoroughly overviewed. Lastly, scientific challenges in packaging of the SiC power module are summarized.

6 A Review of Developments and Challenges on SiC Applications

In recent years, the third generation semiconductor materials represented by silicon carbide have become one of the most promising semiconductor materials because of their significant advantages in band-gap width, breakdown field strength, electron drift rate, thermal conductivity, and radiation resistance, which further meet the requirements of high power, high temperature, high voltage, and high frequency in power electronics and other fields. It has attracted more research interests. Semiconductor power devices based on silicon carbide have also attracted much attention and expectation. Compared with gallium nitride, the application of silicon carbide wafer as substrate material has gradually matured and entered the stage of industrialization. Chemical vapor deposition (CVD) is usually used to grow an epitaxial layer on the silicon carbide single crystal surface. Epitaxial wafer is mainly used to manufacture power devices, RF devices, and other discrete devices. It can be widely used in modern industrial fields such as new energy vehicles, 5G communication, photovoltaic power generation, rail transit, smart grid, aerospace, and so on.

Compared with the most advanced silicon devices, silicon carbide can work at higher temperatures and has greatly improved power consumption and power processing capacity. However, due to the lack of packaging technology, the rated operating temperature of commercial silicon carbide devices and modules is usually much lower than the inherent capacity of silicon carbide devices.

Si-based power devices have been widely used in electric locomotive and EMU, however, there are urgent requirements from the industry to have power converters with smaller size and higher performance. To meet such demands, wide band-gap (WBG) devices such as SiC power chips and modules are developed for the traction systems, and a 1.7kV hybrid SiC power module has been used in the Metro system and all SiC 3.3kV power module has been developed. In this chapter, the development of hybrid SiC and all SiC power modules are introduced from the perspectives of devices, modules, and applications. SiC SBD and MOSFET chips and modules are manufactured and tested, both static and dynamic testing results are evaluated and showed that SiC devices present the better performance and higher efficiency than the traditional Si IGBT modules, especially in the conditions of high temperature and high operation frequency. Based on the developed SiC modules, the urgent demand of traction converters with smaller size and lighter weight can be achieved while present advantages such as higher efficiency, power density, operation temperatures and frequency. The current challenges of SiC devices and future applications are discussed for the next generation of rail transportation.

It is generally believed that carbon atoms will remain at the silicon carbide/silica interface at low oxidation temperature, and oxygen vacancies will dominate at high temperature. Therefore, it is possible to reliably form gate oxides by carefully controlling the oxidation temperature to effectively remove carbon atoms and inhibit the decomposition of silica only by thermal oxidation.

Research confirmed the SiC/SiO_2 interface state density obtained from the ultrahigh-temperature dry oxidation process on 4H-SiC Si-face (0001) at up to 1550°C without any other passivating techniques. Our results were consistent with those of previous reports. Furthermore, we also considered the reliability of SiO_2, which is important for its practical application, by time-dependent dielectric breakdown measurements (TDDB). The optimal interface state density was obtained for the gate oxide formed at 1450°C at EC-E= 0.2~0.6eV, whereas the gate oxide was relatively the most reliable for the oxidation at

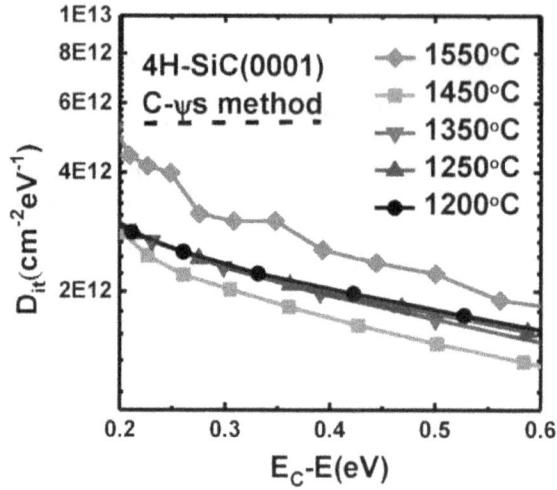

FIGURE 14

Interface state density (D_{it}) of SiO$_2$/SiC at room temperature as a function of energy level below the conduction band of SiC estimated by the C-ψs method. Five samples oxidized at temperatures from 1200 to 1550°C are compared. (After [75] Figure 2, reproduced with permission of Elsevier.)

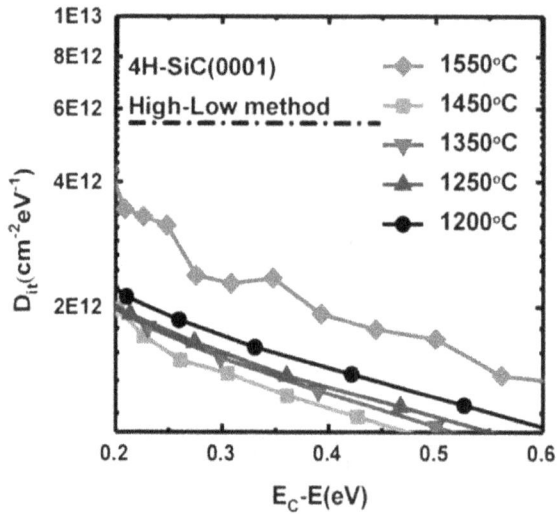

FIGURE 15

Interface state density (D_{it}) of SiO$_2$/SiC interface at room temperature as a function of energy level below the conduction band of SiC, estimated by high-low method. Five samples oxidized at temperatures from 1200 to 1550°C are compared. (After [75] Figure 3, reproduced with permission of Elsevier.)

1250°C. It suggests that the effects of oxidation temperature of 4H-SiC (0001) had a trade-off between gate oxide reliability and SiC/SiO$_2$ interface properties.

7 A Review of Developments and Challenges on SiC Technical Standardization

Standards are documents that provide rules, guidelines, or characteristics for various activities or their results for common use and reuse through standardization activities and formulated by consensus according to specified procedures. Standardization is an activity to obtain the best order within a given scope, promote common benefits, and formulate, publish, and implement standards for practical or potential problems. Generally speaking, standards are rule documents and the result of standardization activities. Standards are democratic, authoritative, systematic, and scientific. Standards include management standards and technical standards, and technical standards include product standards and method standards. At present, there is a popular saying in the industry: "third rate enterprises sell coolies, second rate enterprises sell products, first-class enterprises sell technology, and super first-class enterprises sell standards", which reflects the key supporting role of standards for the industry. In addition, technology patenting, patent standardization, and standard branding are also the development path of standards in the industrial ecology. It can be said that the foundation of high-quality development is high standards.

At present, the international silicon carbide standardization organization has SEMI. SEMI is the international semiconductor industry association, which mainly provides a set of practical environmental protection, safety and health standards for semiconductor process equipment, which is applicable to all equipment used for chip manufacturing, measurement, assembly, and testing. SEMI (International Semiconductor Industry Association) is a global industry association dedicated to promoting the overall development of the supply chain of microelectronics, flat panel displays, solar photovoltaic, and other industries. Members include manufacturing, equipment, materials and service companies in the above industrial supply chain, which is the core driving force to improve the quality of human life. Since 1970, SEMI has been continuously committed to helping member companies quickly obtain market information, improve profitability, create new markets, and overcome technical challenges. Using the proceeds from the exhibition, SEMI founders reinvested their funds into their industry and published SEMI standard plans. Since the earliest successful standardization of silicon wafers and commitment to factory automation and software, SEMI international standard program has helped protect the free market and reduce conductor manufacturing costs. SEMI standard program has been for more than 30 years. SEMI standard also promotes the development of new industries, including today's wide band-gap conductor, white light lighting and photovoltaic industries.

In conclusion, the technical standardization of silicon carbide semiconductor has begun to layout and has a certain foundation. However, compared with the development trend of silicon carbide semiconductor industry, there are still some lags and deficiencies, especially in terms of vehicle regulation standards, we are eager for the introduction of corresponding standards and certification, because the application of vehicle regulations is, after all, the largest market of silicon carbide semiconductor and is growing rapidly.

8 Conclusion

There is no perfect material, and there is no perfect device. Tailor-made is the way of devices. For SiC power electronic devices, the epitaxial thickness determines the voltage, and 1um achieves about 100V, so the 650V device is generally about 6um; the chip area determines the current, because the current density is an inherent characteristic of the material, and the silicon carbide is about several hundred Acm^{-2}. Compared with silicon devices, the process of silicon carbide devices is similar. The difference lies in several high-temperature processes, such as high-temperature ion implantation, high-temperature oxidation, and high-temperature annealing. Device characteristics include performance and reliability. Performance refers to the current test indicators under normal conditions, including static characteristics and dynamic characteristics. Static characteristics are the characteristics of the non-switching state, that is, the characteristics in the two states of on and off, such as on-resistance, Leakage current, threshold voltage, withstand voltage, etc.; dynamic characteristics are switching characteristics, such as turn-off tail current, switching loss, switching frequency, etc.; reliability is the test index change after long-term use under various harsh environmental conditions. Performance is like the explosive power of an athlete, and reliability is like endurance. It needs to be tailored according to different application scenarios of the device. Different designs, different materials, and different cutting processes can be used to create the most suitable sprint, middle and long distance or marathon devices. Therefore, there is no best device, only the most suitable device.

Silicon carbide semiconductor is very useful, but at least it has not been used in integrated circuits yet, and it is unlikely in the future. However, integrated circuits are the largest application market for semiconductor devices (the world of silicon), accounting for more than 80% (currently about US$400 billion). The current power electronic device market is about 40 billion U.S. dollars (stock market), which is shared by all semiconductors, of which silicon carbide semiconductors now only share about 1.5 billion U.S. dollars. Of course, the killer applications of silicon carbide semiconductors are new energy vehicles and 5G base stations, which belong to the incremental market. In the future, the largest market for new energy vehicles will be tens of millions, with a device volume of about tens of billions of dollars; the largest market for 5G base stations in the future will be tens of millions. It can be seen that silicon carbide semiconductors in the entire semiconductor world, in terms of market volume, are at most named "little brothers", and the over-lord big brother is still "silicon". Therefore, although it is on the cusp of the storm and is attracting attention, do not let the silicon carbide semiconductor float and forget your "position"!

In summary, silicon carbide semiconductor is a very personal "expensive, rich, handsome" material, with a noble background (expensive), capital sought (rich), and out-standing advantages (handsome), but in fact it is only a "hairy guy". There is the need to adjust (key technology research and development) and grow (industry cultivation), and as long as everyone in place cares (policy, capital, market, talent, etc.), it will become great. Therefore, the silicon carbide semiconductor industry has a lot to do, but it must be remembered that the prerequisite for doing so is to abandon what you cannot do, that is, to admit your own limitations; the more important quality of doing so is to be fearful, that is, to fear nature, fear opponents, fear that there is a sky outside the sky, and a mountain outside the mountain.

References

1. Tingxiang Xu, Xuechao Liu, Shiyi Zhuo, Wei Huang, Pan Gao, Jun Xin, Erwei Shi, "Effect of thermal annealing on the defects and electrical properties of semi-insulating 6H-SiC", Journal of Crystal Growth, 531, 125399(2020). https://doi.org/10.1016/j.jcrysgro.2019.125 399.
2. Fu Fen, Ying Min Wang, Ru Sheng Wei, "Investigation of the 6-Folded Pattern in the Facetted Region of 4° Off-Axis 4H-SiC", Materials Science Forum, 954, 9–13(2019). https://10.4028/www.scientific.net/msf.954.9.
3. Qi-Sheng Chen, Peng Zhu, Meng He, "Simulations of dislocation density in silicon carbide crystals grown by the PVT-method", Journal of Crystal Growth, 531, 125380(2020). https://doi.org/10.1016/j.jcrysgro.2019.125380.
4. Y.B. Yang, J. Wang, Y.M. Wang, "Thermal stress simulation of optimized SiC single crystal growth crucible structure, Journal of Crystal Growth", 504, 31–36(2018). https://doi.org/10.1016/j.jcrysgro.2018.09.021.
5. Fei Qu, He Zhang, Meng Han, Hui Li, Fa Zhu Ding, Hong Wei Gu, "Study of the Growth Temperature Measurement and Control for Silicon Carbide Sublimation", Materials Science Forum, 954, 65–71(2019). https://doi.org/10.4028/www.scientific.net/msf.954.65.
6. Zeng Ze Wang, Zhou Li Wu, Ming Ming Ge, Hui Qiang Bao, Zhi Fang Ma, Jun Wu, "Study on Carbon Particle Inclusions during 4H-SiC Growth by Using Physical Vapor Transport System", Materials Science Forum, 954, 46–50(2019). https://10.4028/www.scientific.net/msf.954.46.
7. Fang Jiao, Zhou Li Wu, Dian Peng Cui, Mu Long Yang, Bo Yu Dong, "Study on the Synthesis of SiC Powder Material by Using Induction Heating System", Materials Science Forum, 1014, 38–42(2020). https://10.4028/www.scientific.net/msf.1014.38.
8. Qiusheng Yan, Xin Wang, Qiang Xiong, Jiabin Lu, Botao Liao, "The influences of technological parameters on the ultraviolet photocatalytic reaction rate and photocatalysis-assisted polishing effect for SiC", Journal of Crystal Growth, 531, 125379(2020), https://doi.org/10.1016/j.jcrysgro.2019.125379.
9. Lixia Zhao, Huiwang Wu, "A correlation study of substrate and epitaxial wafer with 4H-N type silicon carbide", Journal of Crystal Growth, 507, 109–112(2019). https://doi.org/10.1016/j.jcrysgro.2018.10.030.
10. G.G. Yan, X.F. Liu, Z.W. Shen, Z.X. Wen, J. Chen, W.S. Zhao, L. Wang, F. Zhang, X.H. Zhang, X.G. Li, G.S. Sun, Y.P. Zeng, Z.G. Wang, "Improvement of fast homoepitaxial growth and defect reduction techniques of thick 4H-SiC epilayers", Journal of Crystal Growth, 505, 1–4(2019). https://doi.org/10.1016/j.jcrysgro.2018.09.023.
11. X.F. Liu, G.G. Yan, L. Sang, Y.X. Niu, Y.W. He, Z.W. Shen, Z.X. Wen, J. Chen, W.S. Zhao, L. Wang, M. Guan, F. Zhang, G.S. Sun, Y.P. Zeng, "Defect appearance on 4H-SiC homoepitaxial layers via molten KOH etching", Journal of Crystal Growth, 531, 125359(2020). https://doi.org/10.1016/j.jcrysgro.2019.125359.
12. Yingxi Niu, Xiaoyan Tang, Pengfei Wu, Lingyi Kong, Yun Li, Jinghua Xia, Honglin Tian, Liang Tian, Lixin Tian, Wenting Zhang, Renxu Jia, Fei Yang, Junmin Wu, Yan Pan, Yuming Zhang, "Effect of growth rate on morphology evolution of 4H-SiC thick homoepitaxial layers", Journal of Crystal Growth, 507, 143–145(2019). https://doi.org/10.1016/j.jcrysgro.2018.10.040.
13. Ying Xi Niu, Xiao Yan Tang, Li Xin Tian, Liu Zheng, Wen Ting Zhang, Ji Chao Hu, Ling Yi Kong, Xin He Zhang, Ren Xu Jia, Fei Yang, Yu Ming Zhang, "Low Defect Thick Homoepitaxial Layers Grown on 4H-SiC Wafers for 6500 V JBS Devices", Materials Science Forum, 954, 114–120(2019). https://10.4028/www.scientific.net/msf.954.114.
14. G.G. Yan, Y.W. He, Z.W. Shen, Y.X. Cui, J.T. Li, W.S. Zhao, L. Wang, X.F. Liu, F. Zhang, G.S. Sun, Y.P. Zeng, "Effect of C/Si ratio on growth of 4H-SiC epitaxial layers on on-axis and 4°

off-axis substrates",Journal of Crystal Growth, 531,125362(2020). https://doi.org/10.1016/j.jcrysgro.2019.125362.

15. L.X. Zhao, L. Yang, H.W. Wu, "High quality 4H-SiC homo-epitaxial wafer using the optimal C/Si ratio", Journal of Crystal Growth, 530,125302(2020). https://doi.org/10.1016/j.jcrysgro.2019.125302.

16. L. Yang, L.X. Zhao, H.W. Wu, "Effect of temperature on conversion of basal plane dislocations to treading edge dislocations during 4H-SiC homoepitaxiy", Journal of Crystal Growth, 531, 125360(2020). https://doi.org/10.1016/j.jcrysgro.2019.125360.

17. Zhe Li, Xuan Zhang, Ze Hong Zhang, Li Guo Zhang, Tao Ju, Bao Shun Zhang, "Microstructure of Interfacial Basal Plane Dislocations in 4H-SiC Epilayers", Materials Science Forum, 954, 77–81(2019). https://10.4028/www.scientific.net/msf.954.77.

18. Chuan Gang Li, Tao Ju, Li Guo Zhang, Xiang Kan, Xuan Zhang*, Juan Qin, Bao Shun Zhang, Ze Hong Zhang, "Elimination of Silicon Droplets Formation during 4H-SiC Epitaxial Growth by Chloride-Based CVD in a Vertical Hot-Wall Reactor", Materials Science Forum, 1014, 3–7(2020). https://10.4028/www.scientific.net/msf.1014.3.

19. Zhifei Zhao, Yun Li, Xianjun Xia, Yi Wang, Ping Zhou, Zhonghui Li, "Growth of high-quality 4H-SiC epitaxial layers on 4° off-axis C-face 4H-SiC substrates", Journal of Crystal Growth, 531, 125355(2020). https://doi.org/10.1016/j.jcrysgro.2019.125355.

20. Yun Li, Zhifei Zhao, Le Yu, Yi Wang, Ping Zhou, Yingxi Niu, Zhonghui Li, Yunfeng Chen, Ping Han, "Reduction of morphological defects in 4H-SiC epitaxial layers", Journal of Crystal Growth, 506, 108–113(2019). https://doi.org/10.1016/j.jcrysgro.2018.10.023.

21. Jichao Hu, Renxu Jia, Yingxi Niu, Yuan Zang, Hongbin Pu, "Study of a new type nominal 'washboard-like' triangular defects in 4H-SiC 4° off-axis (0 0 0 1) Si-face homoepitaxial layers", Journal of Crystal Growth, 506, 14–18(2019) . https://doi.org/10.1016/j.jcrysgro.2018.10.026.

22. Yun Li, Zhifei Zhao, Le Yu, Yi Wang, Zhijun Yin, Zhonghui Li, Ping Han, "Heteroepitaxial 3C-SiC on Si (1 0 0) with flow-modulated carbonization process conditions", Journal of Crystal Growth, 506, 114–116(2019). https://doi.org/10.1016/j.jcrysgro.2018.09.037.

23. X.F. Liu, G.G. Yan, Z.W. Shen, Z.X. Wen, J. Chen, Y.W. He, W.S. Zhao, L. Wang, M. Guan, F. Zhang, G.S. Sun, Y.P. Zeng, "Homoepitaxial growth of multiple 4H-SiC wafers assembled in a simple holder via conventional chemical vapor deposition", Journal of Crystal Growth, 507, 283–287(2019). https://doi.org/10.1016/j.jcrysgro.2018.10.055.

24. Guo Guo Yan, Xing Fang Liu, Feng Zhang, Zhan Wei Shen, Wan Shun Zhao, Lei Wang, Ying Xin Cui, Jun Tao Li, Guo Sheng Sun, "Homoepitaxial Growth on Si-Face (0001) On-Axis 4H-SiC Substrates", Materials Science Forum, 954, 31–34(2019). https://10.4028/www.scientific.net/msf.954.31.

25. G.G. Yan, X.F. Liu, Z.W. Shen, W.S. Zhao, L. Wang, Y.X. Cui, J.T. Li, F. Zhang, G.S. Sun, Y.P. Zeng, "The influence of growth temperature on 4H-SiC epilayers grown on different off-angle (0001) Si-face substrates", Journal of Crystal Growth, 507, 175–179(2019). https://doi.org/10.1016/j.jcrysgro.2018.10.041.

26. Yingxi Niu, Xiaoyan Tang, Ling Sang, Yun Li, Lingyi Kong, Liang Tian, Honglin Tian, Pengfei Wu, Renxu Jia, Fei Yang, Junmin Wu, Yan Pan, Yuming Zhang, "The influence of temperature on the silicon droplet evolution in the homoepitaxial growth of 4H-SiC", Journal of Crystal Growth, 504, 37–40(2018). https://doi.org/10.1016/j.jcrysgro.2018.09.022.

27. Qiankun Yang, Dongguo Zhang, Zhonghui Li, Weike Luo, Lei Pan, "Influence of surface step width of 4H-SiC substrates on the GaN crystal quality", Journal of Crystal Growth, 504,41–43(2018). https://doi.org/10.1016/j.jcrysgro.2018.09.028.

28. Zhi Fei Zhao, Yun Li, Yi Wang, Ping Zhou, Wu Yun, Zhong Hui Li, "Influence of the Etching Process on the Surface Morphology of 4H-SiC Substrate Used in the Epitaxial Graphene", Materials Science Forum, 954, 21–25(2019). https://10.4028/www.scientific.net/msf.954.21.

29. Yi Wang, Pengfei Gu, Zhifei Zhao, Ping Zhou, Zhonghui Li, Yun Li, "Synthesis of graphene-like carbon film on SiC substrate", Journal of Crystal Growth, 531, 125356(2020). https://doi.org/10.1016/j.jcrysgro.2019.125356.

30. Yawei He, Guoguo Yan, Zhanwei Shen, Wanshun Zhao, Lei Wang, Xingfang Liu, Guosheng Sun, Feng Zhang, Yiping Zeng, "Investigation of the distribution of deep levels in 4H-SiC epitaxial wafer by DLTS with the method of decussate sampling", Journal of Crystal Growth, 531, 125352(2020). https://doi.org/10.1016/j.jcrysgro.2019.125352.

31. X.F. Liu, G.G. Yan, B. Liu, Z.W. Shen, Z.X. Wen, J. Chen, W.S. Zhao, L. Wang, F. Zhang, G.S. Sun, Y.P. Zeng, "Process optimization for homoepitaxial growth of thick 4H-SiC films via hydrogen chloride chemical vapor deposition", Journal of Crystal Growth, 504,7–12(2018). https://doi.org/10.1016/j.jcrysgro.2018.09.030.

32. Lingqin Huang, Mali Xia, Xiaogang Gu, "A critical review of theory and progress in Ohmic contacts to p-type SiC", Journal of Crystal Growth, 531, 125353(2020). https://doi.org/10.1016/j.jcrysgro.2019.125353.

33. Rui Liu, Hao Wu, Hongdan Zhang, Chen Li, Liang Tian, Ling Li, Jialin Li, Junmin Wu, Yan Pan, "A dry etching method for 4H-SiC via using photoresist mask", Journal of Crystal Growth, 531, 125351(2020). https://doi.org/10.1016/j.jcrysgro.2019.125351.

34. Yi Wen, Xiao Jie Xu, Meng Ling Tao, Xiao Chuan Deng, Xuan Li, Zhi Qiang Li, Bo Zhang, "A Study of the High-K Enhanced Depletion-JTE for Ultra-High Voltage SiC Power Device with Improved JTE-Dose Window", Materials Science Forum, 1014, 109–114(2020). https://10.4028/www.scientific.net/msf.1014.109.

35. Jinghua Xia, Shihai Wang, Lixin Tian, Wenting Zhang, Hengyu Xu, Jun Wan, Caiping Wan, Yan Pan, Fei Yang, "Al(ON) gate dielectrics for 4H-SiC MOS devices", Journal of Crystal Growth, 532, 125434(2020). https://doi.org/10.1016/j.jcrysgro.2019.125434.

36. Heng Yu Xu, Cai Ping Wan, Jin Ping Ao, "Improved Electrical Properties of 4H-SiC MOS Devices with High Temperature Thermal Oxidation", Materials Science Forum, 954, 99–103(2019). https://10.4028/www.scientific.net/msf.954.99.

37. Hengyu Xu, Caiping Wan, Ling Sang, Jin-Ping Ao, "Influence on curvature induced stress to the flatband voltage and interface density of 4H-SiC MOS structure", Journal of Crystal Growth, 505, 59–61(2019). https://doi.org/10.1016/j.jcrysgro.2018.09.024.

38. Yasuto Hijikata, "Macroscopic simulations of the SiC thermal oxidation process based on the Si and C emission model", Diamond and Related Materials, 92, 253–258(2019). https://doi.org/10.1016/j.diamond.2019.01.012.

39. Heng Yu Xu, Cai Ping Wan, Jin Ping Ao, "Reliability of 4H-SiC (0001) MOS Gate Oxide by NO Post-Oxide-Annealing", Materials Science Forum, 954, 109–113(2019). https://10.4028/www.scientific.net/msf.954.109.

40. Kai Tian, Jing Hua Xia, Jin Wei Qi, Shen Hui Ma, Fei Yang, Anping Zhang, "An Improved 4H-SiC Trench Gate MOSFETs Structure with Low On-Resistance and Reduced Gate Charge", Materials Science Forum, 954, 151–156(2019). https://10.4028/www.scientific.net/msf.954.151.

41. Wei Jiang Ni, Xiao Liang Wang, Chun Feng, Hong Ling Xiao, Li Juan Jiang, Wei Li, Quan Wang, Ming Shan Li, Holger Schlichting, Tobias Erlbacher, "Design and Fabrication of 4H-Sic Mosfets with Optimized JFET and p-Body Design", Materials Science Forum, 1014, 93–101(2020). https://10.4028/www.scientific.net/msf.1014.93.

42. Yi Wen, Hao Zhu, Wenchi Yang, Xiaochuan Deng, Xuan Li, Wanjun Chen, Bo Zhang, "Design and Simulation on Improving the Reliability of Gate Oxide in SiC CDMOSFET", Diamond and Related Materials", 91, 213–218(2019). https://doi.org/10.1016/j.diamond.2018.11.020.

43. Chenxi Fei, Song Bai, Qian Wang, Runhua Huang, Zhiqiang He, Hao Liu, Qiang Liu, "Influences of Pre-Oxidation Nitrogen Implantation and Post-Oxidation Annealing on Channel Mobility of 4H-SiC MOSFETs", Journal of Crystal Growth, 531, 125338(2020). https://doi.org/10.1016/j.jcrysgro.2019.125338.

44. Noriyuki Iwamuro, "Recent Progress of SiC MOSFET Devices", Materials Science Forum, 954, 90–98(2019). https://10.4028/www.scientific.net/msf.954.90.

45. Chunyan Jiang, Hao Wu, Liang Tian, Jialin Li, Rui Liu, Tao Zhu, Ling Li, Hongdan Zhang, Qianqian Jiao, Bin Wu, Xiang Qi, Junmin Wu, Yan Pan, "Angular Rotation Ion Implantation

Technology in SiC for 4H-SiC Junction Barrier Schottky Rectifiers", Journal of Crystal Growth, 531, 125354(2020). https://doi.org/10.1016/j.jcrysgro.2019.125354.

46. Wentao Dou, Qingwen Song, Hao Yuan, Xiaoyan Tang, Yuming Zhang, Yimen Zhang, Li Xiao, Liangyong Wang, "Design and Fabrication of High Performance 4H-SiC TJBS Diodes", Journal of Crystal Growth, 533, 125421(2020). https://doi.org/10.1016/j.jcrysgro.2019.125 421.

47. Ling Sang, Li Xin Tian, Fei Yang, Jing Hua Xia, Rui Jin, Yi Ying Zha, Liang Tian, Xi Ping Niu, Jun Min Wu, "Design and Fabrication of Non-Uniform Spacing Multiple Floating Field Limiting Rings for 6500V 4H-SiC JBS Diodes", Materials Science Forum, 1014, 120–125(2020). https://10.4028/www.scientific.net/msf.1014.120.

48. Ling Sang, Lixin Tian, Jialin Li, Yingxi Niu, Rui Jin, "Effect of P+ Region Design on the Fabrication of 6500 V 4H-SiC JBS Diodes", Journal of Crystal Growth, 530, 125317(2020). https://doi.org/10.1016/j.jcrysgro.2019.125317.

49. Ling Sang, Jing Hua Xia, Liang Tian, Fei Yang, Rui Jin, Jun Min Wu, "Effect of the Field Oxidation Process on the Electrical Characteristics of 6500V 4H-SiC JBS Diodes", Materials Science Forum, 1014, 144–148(2020). https://10.4028/www.scientific.net/msf.1014.144.

50. Xi Wang, Hong Bin Pu, Ji Chao Hu, Bing Liu, "SiC Trenched Schottky Diode with Step-Shaped Junction Barrier for Superior Static Performance and Large Design Window", Materials Science Forum, 1014, 62–67(2020). https://10.4028/www.scientific.net/msf.1014.62.

51. Hang Gu, Yi Dan Tang, Lan Ge, Yun Bai, You Run Zhang, Ya Fei Luo, Xin Yu Liu, Guan Song, Ben Tan, "Simulation of Electrothermal Characteristics of 1200V/75A 4H-SiC JBS", Materials Science Forum, 954, 139–143(2019). https://10.4028/www.scientific.net/msf.954.139.

52. Meng Ling Tao, Xiao Chuan Deng, Hao Wu, Yi Wen, "Design, Fabrication and Characterization of 10kV/100A 4H-SiC PiN Power Rectifier", Materials Science Forum, 1014, 115–119(2020). https://10.4028/www.scientific.net/msf.1014.115.

53. Yue Wei Liu, Rui Xia Yang, Xiao Deng, "Design, Fabrication and Characterization of a 4.5kV / 50A 4H-SiC PiN Rectifiers", Materials Science Forum, 954, 85–89(2019). https://.4028/www.scientific.net/msf.954.85.

54. Wen Ting Zhang, Yun Lai An, Yi Ying Zha, Ling Sang, Jing Hua Xia, Fei Yang, "Enhancement of Minority Carrier Lifetime in Ultra-High Voltage 4H-SiC PiN Diodes by Carbon-Film Annealing", Materials Science Forum, 1014, 137–143(2020). https://10.4028/www.scientific. net/msf.1014.137.

55. Shi Hai Wang, Cai Ping Wan, Heng Yu Xu, Jin Ping Ao, "Effect of Grinding-Induced Stress on Interface State Density of SiC/SiO$_2$", Materials Science Forum, 954, 121–125(2019). https:// 10.4028/www.scientific.net/msf.954.121.

56. Yifan Jia, Hongliang Lv, Xiaoyan Tang, Chao Han, Qingwen Song, Yimen Zhang, Yuming Zhang, Sima Dimitrijev, Jisheng Han, "Growth and Characterization of Nitrogen-Phosphorus Hybrid Passivated Gate Oxide Film on N-type 4H-SiC Epilayer", Journal of Crystal Growth, 507, 98–102(2019). https://doi.org/10.1016/j.jcrysgro.2018.09.029.

57. Yutian Wang, Zuoyi Zhang, Ke Zhou, Zeyu Guo, Ming Lei, Ye Tian, Hui Guo, Chen Xiufang, "Ohmic Contact Formation Mechanism of Ge-doped 6H-SiC", Journal of Crystal Growth, 534, 125363(2020). https://doi.org/10.1016/j.jcrysgro.2019.125363.

58. Jianing Su, Ying Yang, Xuhui Zhang, Hao Wang, Longxiang Zhu, "Sulfur Passivation of 3C-SiC Thin Film", Journal of Crystal Growth, 505, 15–18(2019). https://doi.org/10.1016/j.jcrys gro.2018.09.025.

59. Heng Yu Xu, Cai Ping Wan, Jin Ping Ao, "The Correlation between the Reduction of Interface State Density at the SiO$_2$/SiC Interface and the NO Post-Oxide-Annealing Conditions", Materials Science Forum, 954, 104–108(2019). https://10.4028/www.scientific.net/ msf.954.104.

60. Ying Xi Niu, Dong Bo Song, Ling Sang, "The Influence of Tri-Defects of Epitaxial Layers on the Performance of 4H-Sic Diodes", Materials Science Forum, 1014, 33–37(2020). https:// 10.4028/www.scientific.net/msf.1014.33.

61. Feng Bin Hao, Xiao Xing Jin, Ao Liu, Shi Yan Li, Song Bai, Gang Chen, "Research of 3.3kV, 60A H-Bridge High-Voltage SiC Diode Modules", Materials Science Forum, 1014, 163–170(2020). https://10.4028/www.scientific.net/msf.1014.163.

62. Yangang Wang, Xiaoping Dai, Guoyou Liu, Yibo Wu; Daohui Li, Steve Jones, "Status and Trend of SiC Power Semiconductor Packaging", IEEE, 396–402(2015). https://10.1109/ICEPT.2015.7236613.

63. Liang Yu Chen, R. Wayne Johnson, Philip G. Neudeck, Glenn M. Beheim, David J. Spry, Roger D. Meredith, Gary W. Hunter, "Packaging Technologies for 500°C SiC Electronics and Sensors", Materials Science Forum, 717–720, 1033–1036(2012). https://10.4028/www.scientific.net/msf.717-720.1033.

64. Fengze Hou, Wenbo Wang, Liqiang Cao, Jun Li, Meiying Su, Tingyu Lin, Guoqi Zhang, Braham Ferreira, "Review of Packaging Schemes for Power Module", IEEE, 8, 223–238(2020). https://10.1109/JESTPE.2019.2947645.

65. Wei Ping Zhang, Ya Dong Duan, Mao Peng, Zhang Liang, "Characteristics and Drive Design Analysis of SiC", Materials Science Forum, 1014, 55–61(2020). https://10.4028/www.scientific.net/msf.1014.55.

66. Yu Jie Du, Jin Yuan Li, Peng Wang, Mei Ting Cui, "Comparison of High Voltage SiC MOSFET and Si IGBT Power Module Thermal Performance", Materials Science Forum, 954,194–201(2019). https://10.4028/www.scientific.net/msf.954.194.

67. Guoyou Liu, Yibo Wu, Kongjing Li, Yangang Wang, ChengZhan Li, "Development of High Power SiC Devices for Rail Traction Power Systems", Journal of Crystal Growth, 507, 442–452(2019). https://doi.org/10.1016/j.jcrysgro.2018.10.037.

68. Ming Chang He, Li Xia Hu, Jun Ding Zheng, Wen Sheng Wei, Hai Lin Xiao, Jian Zhu Ye, Guan Jun Qiao, "Effect of Tunneling on Small Signal Characteristics of IMPATT Diodes with SiC Heteropolytype Structures", Materials Science Forum, 954, 176–181(2019). https://10.4028/www.scientific.net/msf.954.176.

69. Wei Jie Li, Dong Yi Meng, Yu Jie Chang, Chun Yang, Yi Lv, Li Jun Diao, "Research on Performance Contrast between SiC MOSFET and Si IGBT Based on the Converter of Urban Rail Vehicles", Materials Science Forum, 954, 188–193(2019). https://10.4028/www.scientific.net/msf.954.188.

70. Yu Hao Lee, Mu Zhang, Ping Juan Niu, Ping Fan Ning, Lei Liu, Shu Shu Lee, "Simplified Silicon Carbide MOSFET Model Based on Neural Network", Materials Science Forum, 954, 163–169(2019). https://10.4028/www.scientific.net/msf.954.163.

71. Ao Liu, Song Bai, Run Hua Huang, Tong Tong Yang, Hao Liu, "Research on Threshold Voltage Instability in SiC MOSFET Devices with Precision Measurement", Materials Science Forum, 954, 133–138(2019). https://10.4028/www.scientific.net/msf.954.133.

72. Mei Ting Cui, Jin Yuan Li, Xiao Liang Yang, Yu Jie Du, "The Effect of Circuit Parameters on Reverse Biased Safe Operating Area of SiC MOSFET", Materials Science Forum, 954, 170–175(2019). https://10.4028/www.scientific.net/msf.954.170.

73. Zhi Peng Luo, Cai Ping Wan, Jing Hua Xia, Zhi Jin, Heng Yu Xu, "The Effect on the Interface and Reliability of SiC MOS by Ar/O_2 Annealing", Materials Science Forum, 1014, 102–108(2020). https://10.4028/www.scientific.net/msf.1014.102.

74. Zhi Qiang Bai, Xiao Yan Tang, Chao Han, Yan Jing He, Qing Wen Song, Yi Fan Jia, Yi Men Zhang, Yu Ming Zhang, "The Influence of Temperature Storage on Threshold Voltage Stability for SiC VDMOSFET", Materials Science Forum, 954, 144–150(2019). https://10.4028/www.scientific.net/msf.954.144.

75. Caiping Wan, Hengyu Xu, Jinghua Xia, Jin-Ping Ao, "Ultrahigh-Temperature Oxidation of 4H-SiC (0 0 0 1) and Gate Oxide Reliability Dependence on Oxidation Temperature", Journal of Crystal Growth, 530, 125250(2020). https://doi.org/10.1016/j.jcrysgro.2019.125250.

Part II

SiC Materials Growth and Processing

5

CVD of SiC Epilayers – Basic Principles and Techniques

Chin-Che Tin and Rongxiang Hu
Department of Physics, Auburn University, Alabama, USA.

Roman Drachev
ON Semiconductor, Hudson, New Hampshire, USA.

Alireza Yaghoubi
Center for High Impact Research, University of Malaya, Kuala Lumpur, Malaysia.

1 Introduction

In any electronic materials development program, epitaxial growth is a key enabling technology that affects the development and progress of materials processing and device fabrication technologies. Although various studies of SiC crystal growth have been conducted as far back as the early 19th century [1–7], albeit as small samples, a resurgence of interest only started in the early 1980s catalyzed by the successful demonstration of large area heteroepitaxial growth of 3C-SiC on silicon [8]. However, 3C-SiC-based device technology development was impeded by the large density of defects originating at the 3C-SiC/Si interface. The introduction of the modified Lely technique [9], also known as the physical vapor transport technique, to grow large boules of both 6H- and 4H-SiC resulted in the commercial availability of 6H- and 4H-SiC substrate materials in the early 1990s [10–12]. The availability of these substrate materials made homoepitaxial growth of 6H- and 4H-SiC polytypes possible, resulting in growth of high-quality device-grade 6H- and 4H-SiC epilayers, which then led to a rapid progress in SiC material and device technology.

Several techniques such as liquid phase epitaxy [13], molecular beam epitaxy [14–16], and sublimation epitaxy [14,17] have been applied for epitaxial growth of different polytypes of SiC; but by far the most popular technique that has consistently provided high quality device-grade epilayers is the chemical vapor deposition (CVD) technique [14,18–21].

Variations of the CVD technique have been used but basically they can be differentiated into cold-wall [18,19,22–26] and hot-wall [27–29] techniques employing either vertical [19,23,25-27,29] or horizontal [18,22,24,25,28] reactor tube configuration. The hot-wall CVD technique was developed to produce faster growth rate allowing for more efficient growth of thicker epilayers needed in high-power devices. Later development of high-temperature CVD (HTCVD) allows substantial increase in growth rate [27,29,30]. Although

DOI: 10.1201/9780429198540-7

the horizontal configuration has and is still being used quite widely, the vertical config-uration is also popular due to its sample rotation capability. In the CVD technique, the growth pressure can be either low pressure (LPCVD) [25,26,28] or atmospheric pressure (APCVD) [22,23,27,29].

Both vertical and horizontal CVD systems are available commercially but their high acquisition costs make them out of reach for many researchers interested in SiC CVD. This chapter serves to provide those interested in SiC CVD with the basic principles and procedures covering the areas of system design, material characteristics, sample prepar-ation, epitaxial growth, and doping, which are essential to obtain high quality device-grade SiC epilayers. The main focus is on an affordable SiC CVD system that graduate students and faculty members can build in-house. Such a SiC CVD system can provide researchers in academia with a powerful tool to participate effectively in the research and development of wide bandgap materials and devices. These different aspects of SiC CVD are discussed next.

2 SiC CVD System

2.1 Overview

The basic system design is shown in Figure 1 where the shaded items are optional depending on the complexity and budget. Figure 2 shows an example of an in-house built SiC CVD system.

The system can be as simple or as complicated as one would like. A simple system would consist of the gas delivery lines, which include mass flow controllers, a pressure control system consisting of a pressure sensor and either a downstream throttle valve or an upstream mass flow controller, a vacuum pumping system, a heat source, a cooling system, and a gas detection system to detect process gas leaks. A more complicated system would include various safety features such as double-wall stainless steel tubing for gas delivery lines, excess flow valves, flame-retardant valves at the gas cylinders, run-and-vent fea-ture, a high-vacuum grade gas plumbing lines, partial pressure transducer and a residual gas analyzer (RGA) system, high vacuum capability, and motorized sample rotation with ferrofluid feedthrough to prevent gas leakage. A run-and-vent feature produces abrupt interface without the compositional distortion caused by flow instability that is caused when mass flow controllers attempt to regulate gas flows when process gases are first introduced directly into the growth chamber. An exhaust gas scrubber is also used to treat the exhaust gas before venting into the atmosphere. The choice of exhaust gas scrubber basically narrows down to either dry or wet scrubbing. A dry scrubber can be combustion type where the exhaust gas is heated to a high temperature to decompose the gas, or an absorbent type where exhaust gas is passed through a cylinder of an absorbent element to remove the toxic or acidic components. A wet scrubber is one where the exhaust gas is passed through a series of water jets to remove the acidic components. The latter type of scrubber is not highly recommended for a system where hydrogen chloride is used in pre-growth treatment of the sample. This is because the wet scrubber would be a source of moisture that may backstream into the exhaust line. The presence of the moisture makes the hydrogen chloride gas more corrosive to the valves, connectors, and tubing in the exhaust line. Backstreaming of the moisture into the system will also have a detrimental effect on the moisture content in the growth chamber.

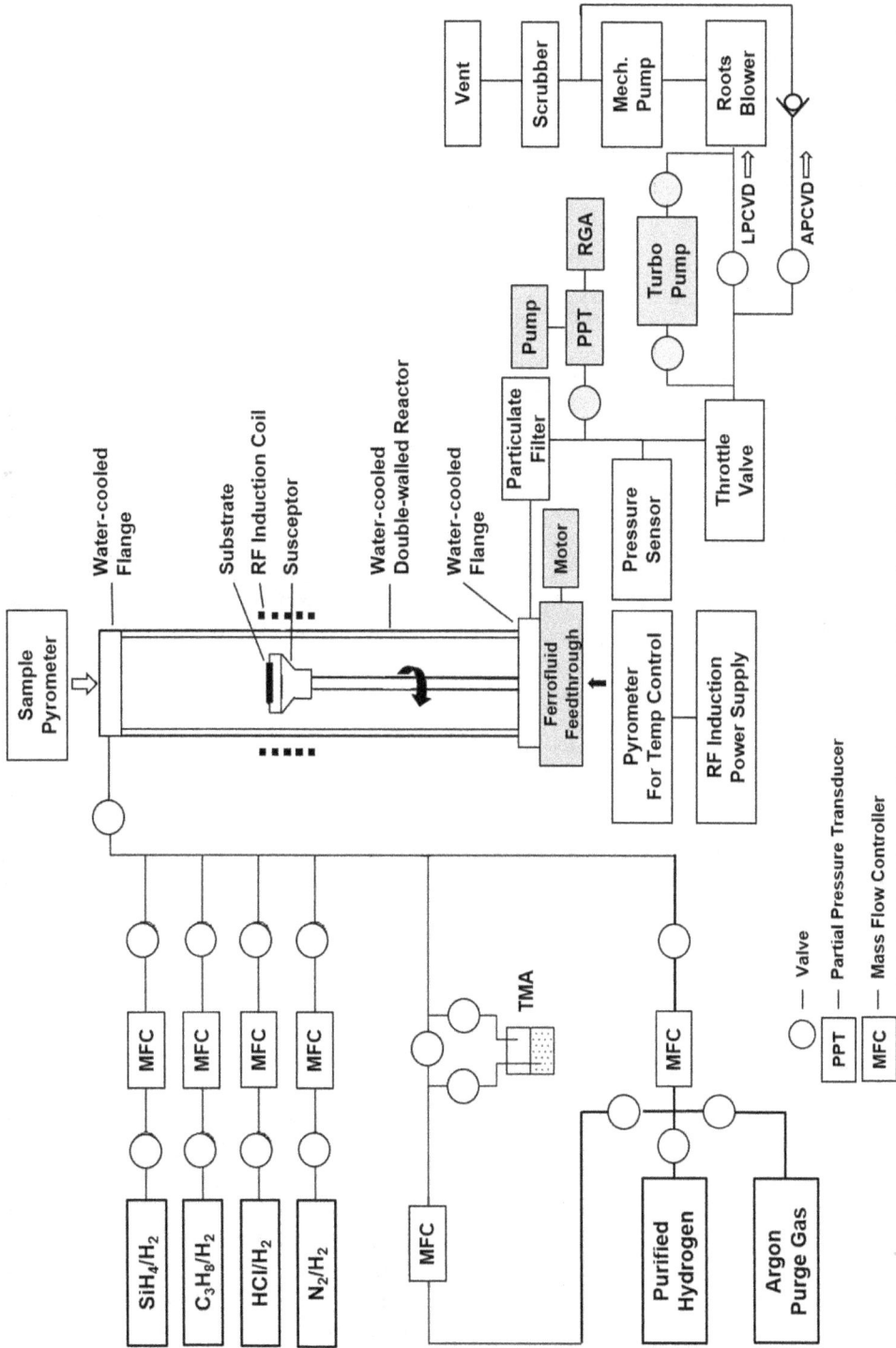

FIGURE 1

Schematics of a SiC CVD system. Modified from Figure 1 of [31], with permission of Elsevier.

FIGURE 2
An in-house built SiC CVD system.

2.2 Heating Technique

A common laboratory heating method is a furnace using resistive heating. The maximum temperature of such a furnace depends on the heating element used. A furnace using a nichrome 80/20 (80% nickel, 20% chromium) heating element can provide a maximum operating temperature of up to 1200°C. However, a furnace with MoSi$_2$ heating element can reach 1850°C, whereas a SiC heating element can provide temperatures as high as 1625°C. Furnace heating does have the capability to provide the required high temperature for SiC CVD. But furnace heating is not usually preferred in SiC CVD because of the limitation imposed by the reactor tube. In furnace heating, the whole reactor tube is heated up to the growth temperature. Commonly available high-temperature materials for reactor tubes are fused quartz and alumina. The recommended maximum operating temperature for fused quartz is 1200°C whereas the temperature for 3C-SiC epitaxial growth is about 1350°C and that for 4H- and 6H-polytypes is about 1450–1500°C. The maximum operating temperature limit of 1200°C for fused quartz is insufficient for SiC CVD and thus precludes its use as tubing material in a conventional furnace. On the other hand, the maximum operating temperature for alumina is about 1700°C and high purity alumina tubings are commonly used in high-temperature furnaces. Furnace heating using alumina tube can theoretically be used for SiC CVD of 4H- and 6H-polytypes but is not suitable for 3C-polytype. This is because of the rapid temperature ramp required for initial carbonization in 3C-SiC CVD. Such a rapid temperature rise would be difficult in furnace heating but is easily produced by induction heating. Another point to consider is that, generally, in crystal growth the fixtures inside the reactor as well as the wall of the reactor itself are known to release impurities when heated at high temperatures resulting in unwanted autodoping of the epilayer. The common impurities in fused quartz and alumina tubings are mostly alkali and transition metal elements.

Due to the high growth temperature in SiC CVD, the preferred heating method is induction heating. Induction heating is both efficient and fast, and is especially useful for the cold-wall CVD configuration. In induction heating, the alternating current in the induction coil generates an alternating magnetic field, which induces an opposing alternating magnetic field in the workpiece, which is the sample holder or susceptor in this discussion. The opposing alternating magnetic field in the susceptor then produces an eddy current, which generates heat by Joule heating. If the susceptor material were to be magnetic, then heating by hysteresis can also occur. Even then, Joule heating by eddy current is still the predominant heating mechanism. Since heat is generated by Joule heating, only low resistivity or electrically conductive materials within the induction coil will generate heat through the induction process. Nonconductive materials such as quartz will not generate heat from the applied radio-frequency (RF) power due to lack of eddy current. Except for heat due to radiation and heat conduction through the gaseous medium from the heated sample holder or susceptor, the quartz reactor wall will not heat up from the input RF power. In cold-wall CVD, the reactor tube is double-walled with cooling water flowing in the space between the outer and inner tubings. The water jacket helps limit the temperature rise of the quartz tubing. When the heated susceptor is at 1500°C, the wall of the reactor tubing, at a distance of about 1 cm away from the edge of the sample holder, could be as high as 350°C. Within the water jacket, the water would be boiling if the rate of water flow is insufficient. To conserve water a high capacity water chiller is needed to drive the cooling system with the RF power supply, the power transfer tubing, the induction coil, and the reactor tube all connected sequentially or in series. In other words, the reactor tube cooling water is the same cooling water the goes through the RF power supply and the induction coil. Since the water exiting from the reactor tube would be hot, the reactor tube should be the last item in the series of components cooled by the cooling water.

In induction heating terminology, coupling is the transfer of electrical energy between the induction coil and the susceptor. Coupling efficiency, or coil efficiency, defines the fraction of the power supplied to the coil that is converted to heat inside the susceptor. Coil efficiency can never achieve 100% due to energy loss from several factors such as stray inductive elements, susceptor dimension and poor coil design. A system with poor coil efficiency would entail requiring much higher power input than is necessary to reach a required temperature. If the power supply cannot provide the required power, then the required temperature can never be reached. Induction power supply is the single most expensive equipment essential to the SiC CVD system. A wrong estimate in the power requirement can result in either purchasing an overly expensive high power unit that far exceeds the need of the project or an under-powered unit that does not have the capacity to generate the heat required to attain the growth temperature. Online calculators are now available at vendors' websites to estimate the power requirement of an induction heating system based on a user's input of system parameters.

The eddy current that generates heat travels within a certain depth from the surface, known as the skin depth. Heat is generated at this near-surface region and flows to the interior of the material by conduction. Skin depth affects coil efficiency. Skin depth is dependent on power supply frequency and resistivity of the material according to the following equation:

$$\delta = \sqrt{\frac{2\rho}{\omega\mu_r\mu_o}} \tag{1}$$

where ρ is the resistivity of the material, ω is the angular frequency of the RF power supply, μ_r is relative permeability of the material and μ_o is permeability of free space. The skin depth is shallower at high frequency and deeper at low frequency. Table 1 shows the skip depth values at different frequencies for copper and graphite since the induction coil is commonly made of copper and the susceptor is usually graphite.

The coil efficiency, η, in the case of a copper induction coil around a graphite susceptor, is approximately given by [34]

$$\eta = \frac{1}{1 + \dfrac{D + \delta_c}{d - \delta_g}\sqrt{\dfrac{\rho_c}{\rho_g\mu_r}}} \tag{2}$$

where D = internal diameter of the copper coil, d is the diameter of the graphite susceptor, δ_c is the skin depth of copper, δ_g is the skin depth of graphite, ρ_c is the resistivity of copper, ρ_g is the resistivity of graphite, and μ_r is the relative permeability of graphite. Figure 3 show the coil efficiencies for different ratios of D/d with and without inclusion of skin depth at 10 kHz. As can be seen from Figure 3, the skin depth makes a significant difference to the heating efficiency. Heating efficiency of a large susceptor may degrade at high frequency because of surface heating predominantly arising from a reduced skin depth.

Coil efficiency is also affected by the fill factor or the volume within the coil occupied by the susceptor. To maximize coil efficiency, the susceptor must maximize the space that it occupies within the induction coil. Since the magnetic field is stronger nearer to the induction coil, the susceptor must be as close as possible to the induction coil and this is accounted for by the ratio D/d in Eq. (2) above. When the susceptor occupies most of the space within the coil then the coupling is said to be tight (Figure 4a). A loose coupling is one where a susceptor is placed far from the coil leaving much of the volume within the coil vacant (Figure 4b). A more powerful power supply will be required if the coil diameter is much larger than that of the susceptor. However, in a water-cooled double-walled reactor tube there is a limit to how close one can get to the coil because of the space occupied by the walls of the reactor tube and its water jacket, as shown in Figure 4c. Consequently, a significant amount of coil efficiency is lost using a water-cooled double-walled reactor tube.

TABLE 1

Skin depths of copper (δ_c) and graphite (δ_g) at different frequencies

Resistivity: copper $\rho_c = 1.678 \times 10^{-8}$ Ωm [32];
graphite $\rho_g = 1.375 \times 10^{-5}$ Ωm [33]

Frequency (kHz)	δ_c (mm)	δ_g (mm)
10	0.65	18.7
25	0.41	11.8
50	0.29	8.35
75	0.24	6.82
100	0.21	5.90

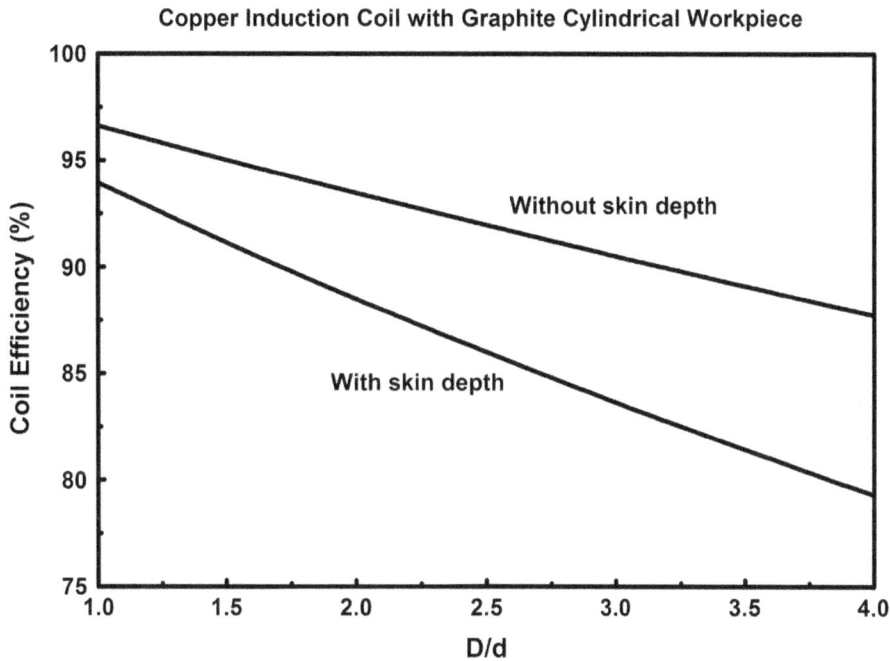

FIGURE 3
Dependence of coil efficiency on the ratio *D/d* and skin depth. Value of *d* is fixed at 40 mm.
Frequency = 10 kHz.

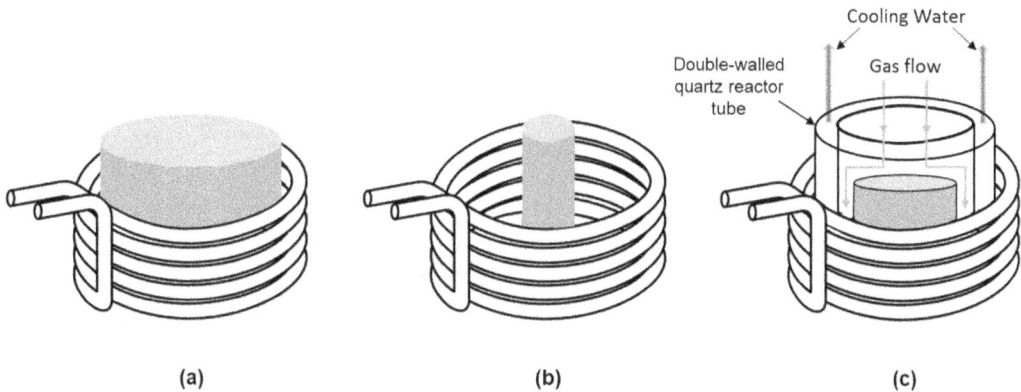

FIGURE 4
Fill factor. (a) Tight coupling, (b) loose coupling, and (c) physical limitation imposed by double-walled reactor.

Such is the case of the SiC CVD system using a water-cooled reactor tube where tight coupling is impossible because the susceptor cannot be placed much closer to the induction coil.

In the design of the induction coil, consideration should also be given to the fact that the coil efficiency is also proportional to the ratio l/D where l is the length and D is the inner diameter of the coil. Coil efficiency is degraded when the length of the coil is excessively greater than its diameter.

To avoid preferential heating, or excessive heating, of the end surface of the susceptor, known as the edge effect, the end of the coil should not extend further than the end of the susceptor, but should be even or less. When the length of the coil is longer than the length of the susceptor, the unused portion of the coil would be wasted as lead loss.

Long power conducting leads, with their attendant large lead inductance, can reduce coil efficiency. In most induction power supplies, the induction coil around the reactor tube forms a tuned circuit with a bank of capacitors inside the power supply. The capacitance in this bank of capacitors can be adjusted to achieve resonance condition and improve coil efficiency.

Assuming an induction heating system consisting of a tuned circuit represented by a simplified equivalent circuit of Figure 5a, where L_1 and L_2 are the inductances of the two power conducting leads between the RF power supply and the induction coil, L_{coil} is the inductance of the induction coil and C is the tank capacitor. Since the leads and the induction coil are in series, the lead-coil combination can be represented as two inductances in series. The total voltage generated by the RF power supply, V_T, is the sum of the total voltage drop across the leads, V_{lead}, and that across the coil, V_{coil}, as given by

$$V_T = V_{lead} + V_{coil} \tag{3}$$

To reduce power loss at the leads, the voltage drop V_{lead} must be reduced and this can be done by reducing the lead impedance, Z_{lead}.

To determine the lead impedance, Z_{lead}, we can consider the power conducting leads from the RF power supply to the induction coil as a pair of parallel conductors with current flowing in opposite direction. Such parallel conductors can be modeled using the equivalent circuit of Figure 5b, where an inductor and capacitor are connected in parallel, which is exactly that of a lossless parallel transmission line. A parallel transmission line model has been widely used not only in a power utilities' system but also at circuit board level and device design. Using the equation for the characteristic impedance of a long parallel transmission lines, we can approximate the impedance of the leads to be

$$Z_{lead} = \sqrt{\frac{L_{lead}}{C_{lead}}} \tag{4}$$

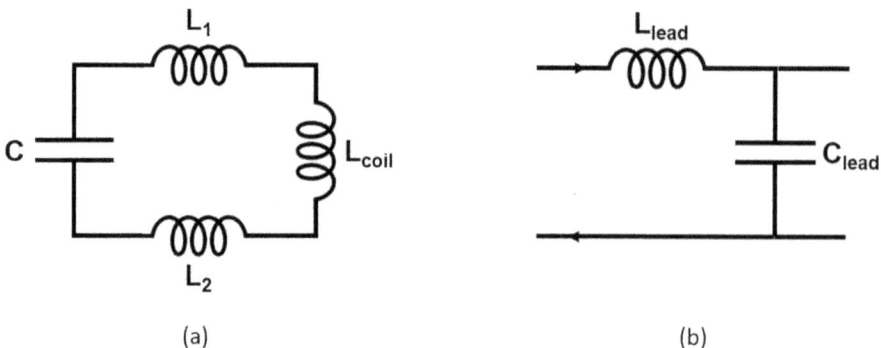

(a)   (b)

FIGURE 5

Equivalent circuits. (a) Simplified tuned circuit of induction heating coil system with leads in series with the induction coil and (b) leads modeled as an inductor and capacitor connected in parallel.

It can be seen that Z_{lead} can be reduced by reducing the lead inductance and increasing the lead capacitance. The lead inductance can be reduced by keeping the two leads short and close to each other. The close proximity of the leads allows the mutual inductance M to cancel the corresponding amount of self-inductance L to ultimately reduce the effective inductance L_{lead} of the two leads according to the equation $L_{lead} = L - M$ [35]. The power conducting leads from the induction power supply to the CVD growth chamber are inevitably long because of the need for working space around the reactor tube. One method to reduce the inductance of the long leads and increase the capacitance of the leads is to use a bus bar as shown in Figure 6.

A bus bar can provide both low inductance and a high capacitance as required by the transmission line impedance equation. The bus bar consists of a thin sheet of a dielectric material sandwiched between two rectangular copper plates. The copper power conduction tubings are brazed to the copper plates. The broad copper plates now assume the task of electric power conduction while the role of the copper tubing is now reduced to water cooling the bus bar. The bus bar allows the two conductors to be as close as possible to produce a low effective inductance. The sandwich structure increases the capacitance of the two leads. This bus bar replaces the round copper tubings or cables that connect between the power supply and the induction coil in the growth chamber. The rectangular bus bar has significantly lower self-inductance compared to a round conductor of identical length and cross sectional area.

The most common material used for the induction coil is the widely available annealed high conductivity thick-walled round copper tubing of 6.35 mm (or ¼ inch) outer diameter (OD) and wall thickness of about 0.89 mm (or 0.035 inch). A large wall thickness is preferred because current travels through the tubing within the skin depth of each of the outer and inner surfaces of the tubing as previously mentioned. This means the total conductive

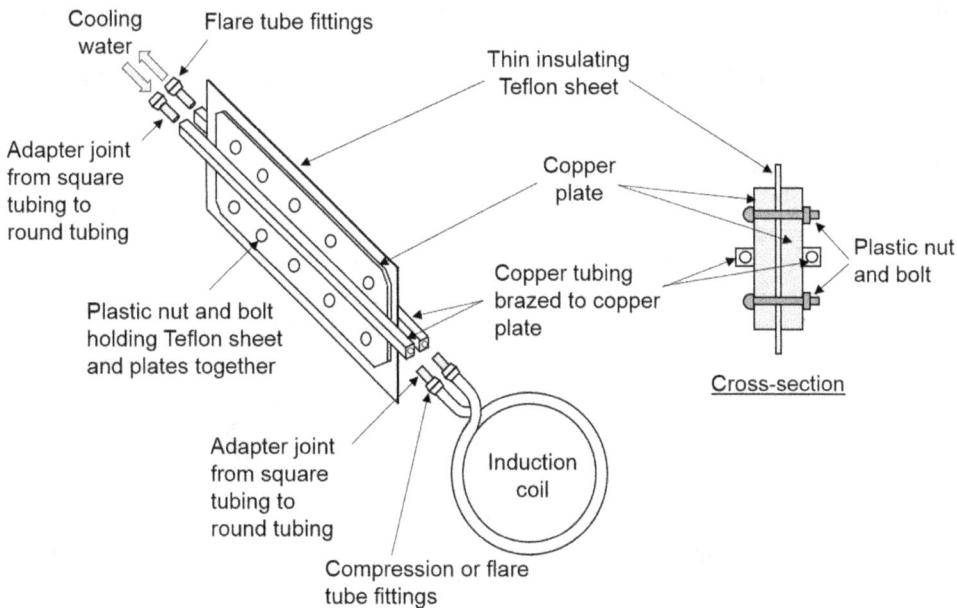

FIGURE 6
A bus bar to replace the long leads.

region is twice the skin depth. According to Table 1, at 10 kHz the skin depth of copper is 0.65 mm and this requires a tubing wall thickness of 1.30 mm, which exceeds the 0.89 mm wall thickness of commercially available 6.35 mm OD copper tubing. At 50 kHz, the skin depth of copper is 0.29 mm, which requires a tubing wall thickness of 0.58 mm, and this is adequately accommodated by commercial 6.35 mm OD copper tubing. Although hollow round copper tubing is efficient at higher frequency, the overall coil efficiency is reduced at high frequency due to a thinner skin depth of the graphite susceptor. However, even if the available wall thickness of commercial copper tubing is less than the required wall thickness, the heating efficiency is not greatly degraded because other factors have greater influence.

A better choice of round copper tubing is the plastic insulated copper tubing, which is also commercially available. This plastic coated copper tubing allows the tubes to be placed close together to further improve coil efficiency. Use of plastic coated copper tubing for induction coil is highly recommended for safety reasons because a bare induction coil carrying high tension electricity can be an electrical hazard to the operators. Square copper tubing is also commercially available albeit expensive. Due to its flat surface facing the susceptor, square tubing provides a more uniform magnetic field pattern than round tubing. Square tubing coil is also more rigid but needs workshop expertise to perform brazing work to make corners and adaptors to connect to round tube fittings. A flat piece of copper, known as a liner, spot brazed to the inside surface of a round copper coil can also give the same effect as a square copper tubing. In this case the round copper tubing plays the role of water cooling the coil. The power output connectors of the induction power supply are usually flare fittings. So the ends of the copper tubings of the bus bar connected to the power supply must also be flare fittings. But the other ends of the copper tubings of the bus bar connected to the induction coil are usually compression fittings. These compression fittings will deteriorate over time due to frequent dismantling of the induction coil. Consequently, the compression fittings need to be replaced and this also results in replacement of the induction coil. A commercially made induction coil can be expensive. The design of the induction coil must therefore be simple and be easily fabricated in the laboratory. An induction coil can be easily fabricated by winding a round copper tubing around a mandrel.

Common susceptor material for SiC CVD is silicon carbide-coated graphite. SiC-coated graphite is high purity graphite machined into required shape then coated with a thin layer of 3C-SiC. SiC-coated graphite susceptor can get corroded by hydrogen chloride (HCl) gas and degrades after repeated use. As discussed later, a degraded susceptor affects the quality of the epilayer. Since the susceptor is a custom-made item purchased commercially, one has to consider simplicity in design to reduce cost.

It must be noted and emphasized that epitaxial growth of SiC is an expensive venture that requires a large operating budget. The exception is CVD of 3C-SiC, which is cheap because Si wafers are cheap. However, CVD of 4H- and 6H-SiC is expensive because 4H- and 6H-SiC wafers are expensive. As in any experimental work that requires repeated runs and iterations, the rate of consumption of wafer materials is high. One has to consider the objective of having a SiC CVD system. In most R&D work, in an academic research environment, the sample size that is adequate for both materials characterizations and device fabrication is 5-mm × 5-mm square. An example of a SiC-coated susceptor that handles multiple 5-mm × 5-mm samples is shown in Figure 7.

In subsequent work, such as photolithography or reactive ion etching, that requires whole wafers, then these 5-mm × 5-mm samples can be mounted onto a Si wafer of required size such as the common 100-mm wafer. In the SiC program at Auburn University, initial R&D

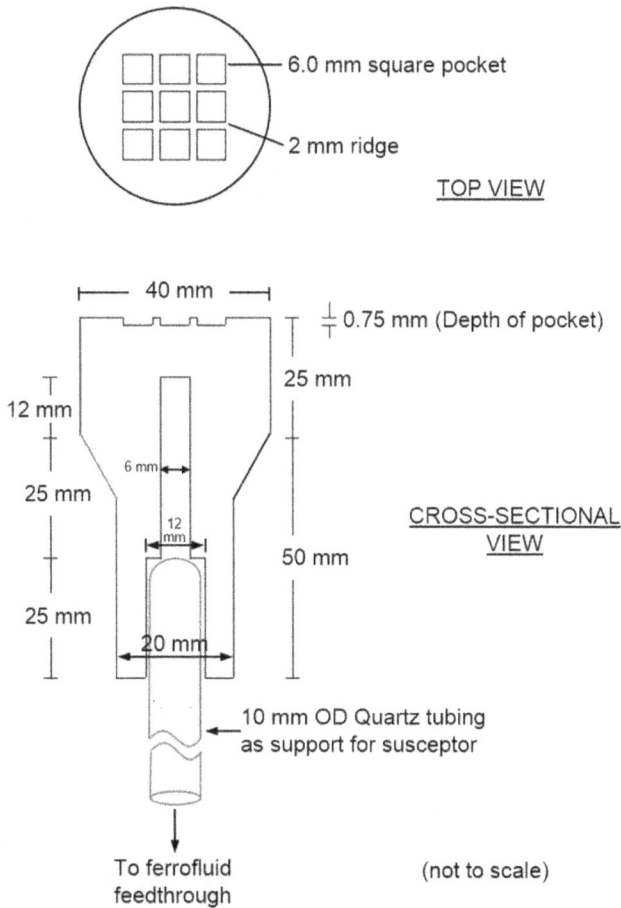

FIGURE 7
A SiC-coated susceptor for a vertical reactor SiC CVD system.

work was carried out on such 5-mm × 5-mm samples. Full wafer size was only utilized when the technology developed was ready for process verification at the production stage in a commercial setting.

Due to the fact that induction heating requires a conductive material, sintered silicon carbide, which is highly resistive, should not be used as the susceptor material in SiC CVD because its conductivity is insufficient for eddy current heating and thus will not be heated up by induction heating.

2.3 Temperature Control

The sample temperature is measured using a pyrometer. Pyrometer operates either as a single-wavelength or a dual-wavelength ratiometric version. In a single-wavelength pyrometer, the accuracy of the temperature reading depends on knowing the absolute spectral emissivity of the target. The emissivity of the target is affected by surface condition and is temperature dependent. The intensity of the emitted radiation detected by the sensor is affected by various factors such as the optical absorption of the gas molecules in the reactor, the transparency of the optical window, the light collection efficiency, and the spectral response of the sensor. The pyrometer is calibrated directly in the CVD system using the melting point of silicon as a calibration point. Such direct calibration will take into account the optical transmission and detection characteristics of the system. In the dual-wavelength version, the ratio of the intensities at two different wavelengths is computed to remove the spectral emissivity from the equation. The assumptions made in this case are that the emissivity of the target does not vary between the two wavelengths, that the spectral response of the sensor does not vary much (or at most varies linearly) at the two wavelengths, and that the optical characteristics of the light collection system does not vary at the two wavelengths. In line with those assumptions, it is often thought that *in situ* calibration is not necessary. In practice, however, a dual-wavelength pyrometer still needs to be calibrated using the melting point of silicon. In view of the many uncertainties affecting the response of a pyrometer, the pyrometer reading from the sample is not usually used for temperature control. Instead a second pyrometer looking at the inside of the susceptor is used for temperature control. The susceptor is supported by a clear quartz tubing that is sealed at one end, as shown in Figure 7. Besides acting as support, this clear quartz tubing also allows the temperature-control pyrometer to detect the radiation from inside the susceptor. The radiation from inside the susceptor is a true black-body radiation with unambiguous emissivity and is very consistent. The temperature reading of this temperature-control pyrometer may not be the true temperature of the wafer; but this is not critical as long as this 'pseudo-temperature' reading varies directly with, or provides a measure of, the temperature of the wafer. The consistency of this second temperature reading provides a more reliable control signal for the RF induction power supply.

2.4 Pressure, Reactor Configuration, and Susceptor Design

The commonly used pressure control system consists of a pressure sensor, throttle valve, pressure readout, and a throttle valve controller. The throttle valve controls the reactor pressure by varying the conductance of the tubing. If reactor pressure of 100 Torr or higher is desired in a CVD system that has a large mechanical pump, an adjustable valve of smaller conductance (e.g. metering valve) may be necessary to reduce the pumping speed further. An upstream pressure control technique, which uses a downstream pressure sensor and an upstream mass flow controller is not a popular method because this technique varies the gas flow rate to maintain a certain pressure. A constant gas flow rate into the growth chamber is preferred to maintain the concentration ratio of the gaseous precursors. Hence, a downstream pressure control is a better approach to control the reactor pressure. APCVD is the easiest mode to implement. In the simplest implementation of the APCVD technique, the pump can be bypassed and the reactor tube's exhaust is connected through a check valve to the gas scrubber or vent. Both low-pressure and atmospheric pressure processes have their own advantages and disadvantages.

The background doping concentration of the epilayer depends on reactor pressure [36]. Reactor pressure also affects the gas flow pattern and thermal distribution inside the reactor. Low pressure is the preferred mode of operation where purity and interfacial abruptness are important. The ability to obtain a sharp or abrupt interface between layers of different doping types and concentrations is determined largely by several factors such as temperature uniformity and gas flow pattern. Factors that can negatively impact on the characteristics of the epilayer are trapped pockets (also known as dead space) of gas within the gas delivery lines; cells of recirculating gas above the sample; and outgassing from the susceptor, among others. The gas composition within the dead space and recirculating cells cannot change rapidly with any change in gas flow rate or gas composition.

The delay in responding to a change in flow rate or composition of the process gas causes a memory effect where preceding gas composition lingers for a while before gradually changing to a new value. The regions of recirculating gas, or recirculating cells, are due to the geometry of the reactor tube and the susceptor, reactor pressure, and gas flow velocity as well as the buoyancy force acting on the gas. Buoyancy is a phenomenon where a hot fluid rises and a cold fluid sinks. The unheated gas that enters the reactor will be heated when it impinges on the hot susceptor. Since density is inversely proportional to temperature, the hot gas will see a reduction in density and subsequently rises due to buoyancy force. On the other hand, the gas in contact with the cold wall of the reactor tube will have a higher density due to its lower temperature. This colder gas sinks along the cold wall due to gravitational force. The movement of the gas that transfers heat between regions of different temperatures is known as convection. There are three different types of convection, namely, free convection where buoyancy force is dominant, forced convection where buoyancy force is negligible or minimal, and mixed convection consisting of both forced and free convections. To determine the different modes of convection for a vertical reactor, a common practice is to calculate the ratio of Grashof (Gr) to Reynolds (Re) numbers [37]. The convection is considered as free or mixed convection if the ratio Gr/Re is within a certain range or exceeds a critical value. However, due to variations in reactor and susceptor design, this critical value is not precise and needs to be determined either by computational fluid dynamics (CFD) simulation or experimentally. Generally the smaller the value of the ratio Gr/Re the more likely it is that forced convection is dominant.

At low pressure, the flow pattern or velocity vector field is commonly that of forced convection where inertial force dominates. Performing the growth process at low pressure reduces the turbulence effects above the sample surface. This shows that buoyancy force is not a significant factor in a LPCVD process. As pressure increases, buoyancy force gradually grows in significance and becomes predominant at atmospheric pressure. At atmospheric pressure, it is common to see recirculating cells and turbulence due to buoyancy force. The consequence of this free convection can be seen in the thickness and doping nonuniformity across the epilayer.

There are a few remedies that can be implemented to resolve the issue of recirculating cells or mitigate its impact. Other than performing thin film growth at low pressure, the susceptor can be rotated to increase film uniformity. At atmospheric pressure, the rotation of the susceptor produces the suction force that counteracts the effect of buoyancy force and suppresses the occurrence of recirculating cells above the susceptor. Another technique is to increase the flow rate of the input gas. Increasing the velocity of the gas will increase Re and decrease the ratio Gr/Re, but this depends on whether the ratio Gr/Re allows for an increase in gas flow velocity without creating fluid instabilities. Modifying the reactor or susceptor geometry also helps, but this requires CFD simulations for guidance.

Since the gas flow pattern in LPCVD is usually that of a forced convection, it leads one to believe that epitaxial growth at low pressure is surface activated and not mass-flow limited. This means that the growth rate is not dependent on gas flow pattern. When a growth is surface-activated it means that any surface excitation can strongly affect epitaxial growth and that film uniformity is largely dependent on temperature uniformity across the sample. However, in the range of operating pressure (> 0.1 Torr) for most conventional CVD systems, the predominant growth mechanism is still mass-flow limited, which means mass transport and gas flow pattern are still important.

The disadvantage of LPCVD is the low growth rate. APCVD is preferred in cases where higher growth rate is required as in growth of power device structures. Temperature uniformity across the wafer is easier to maintain in an atmospheric pressure CVD process due to better heat conduction through the gas around the sample. Hence, APCVD gives better uniformity in thickness and doping due to a better temperature distribution across the wafer.

The presence of recirculating cells mentioned previously and the overall flow pattern of the gas are affected by reactor configuration, pressure, gas flow velocity, as well as the shape of the susceptor. The horizontal reactor configuration is more widely used although the vertical configuration is gaining popularity. Unlike the vertical reactor, the ratio to determine the type of convection in a horizontal reactor is Gr/Re^2 [37]. This ratio does not depend on the physical property of the gas. A horizontal reactor is cheaper than a vertical reactor. In the horizontal configuration, the sample has to be tilted with respect to the direction of the gas flow in order to increase the flow velocity at the far, or downstream, end of the sample. The increase in flow velocity compensates for the depletion of the gas species as gas flows over the sample. Without the tilt, the thickness and doping level of the epilayer tend to vary across the sample. However, the need for the tilt prevents the sample from being rotated. Figure 8 shows the computational fluid dynamics (CFD) modeling of the gas flow patterns for a horizontal reactor with tilted susceptor for both 100 Torr (LPCVD) and 760 Torr (APCVD) processes.

Figure 8 shows that gas flows across the surface of the substrate are laminar for both LPCVD and APCVD. In LPCVD, there is no recirculating cell or turbulence above the surface of the sample, which means that buoyancy does not play a significant role. However, for the case of APCVD under the conditions imposed in the simulation, there appears to be a minor gas flow instability next to the wall of the reactor tube above the susceptor. If buoyancy force were to be dominant, it would affect flow in the direction perpendicular to the horizontal axis of the reactor tube because that would be the direction of the gravitational force. Figure 9 shows the temperature distribution for the horizontal reactor at different reactor pressures. The temperature distribution mirrors the gas flow pattern. Since forced convection is predominant, a longer tail can be seen downstream from the susceptor in APCVD showing heat conduction by a larger quantity of gas moving downstream.

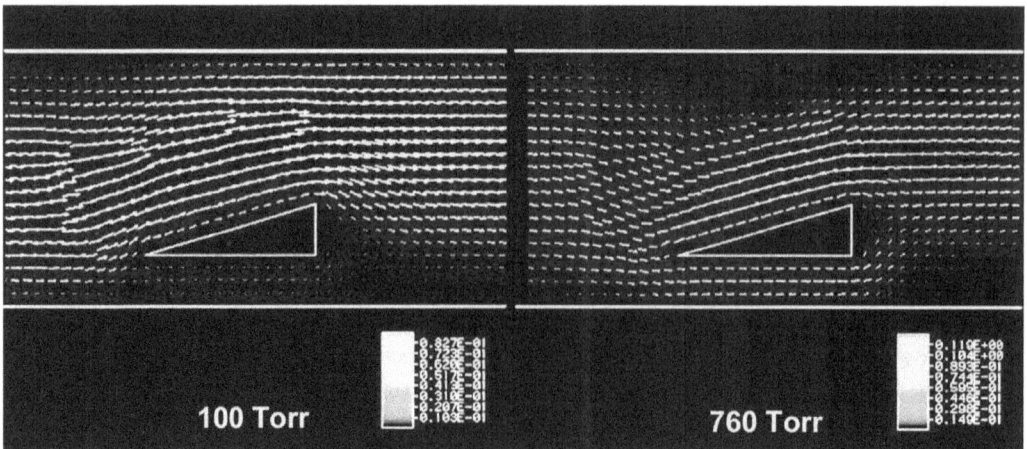

FIGURE 8

CFD simulations of gas velocity vector field for a horizontal reactor with tilted susceptor at 100 Torr and 760 Torr.

In the vertical configuration, the gas is incident perpendicularly on the surface of the sample. The gas flow patterns for the cylindrical susceptor in a vertical reactor for both LPCVD and APCVD processes are shown in Figure 10.

It can be seen from Figure 10 that for the same flow velocity, recirculating cells appear in the APCVD process but are absent in the LPCVD process. The effect of the buoyancy force in APCVD in Figure 10 is clearly seen. Input gas flowing over the cold wall sinks to the bottom due to higher density, but the gas that hits the hot susceptor heats up and rises to the top due to lower density. The rising hot gas subsequently meets the incoming cold gas and creates a stagnation point with two recirculating cells at both sides above the

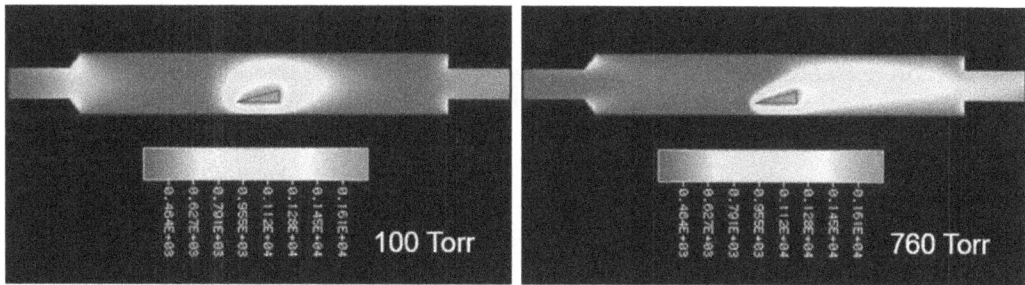

FIGURE 9
CFD simulations of temperature distribution around a tilted susceptor in a horizontal reactor at different pressures.

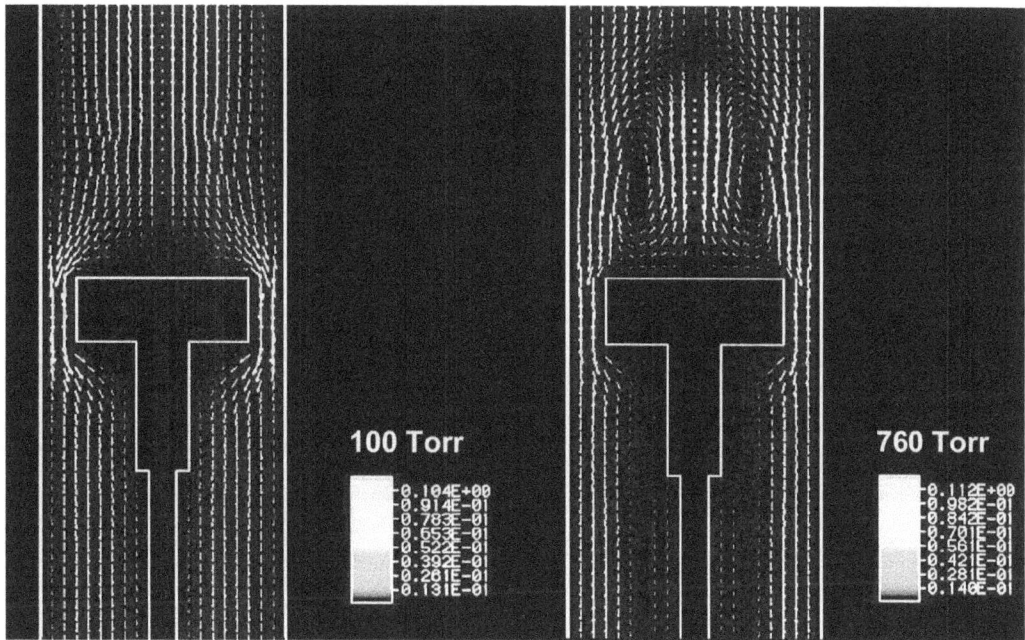

FIGURE 10
CFD simulations of gas velocity vector field at different reactor pressures for a vertical reactor with cylindrical susceptor.

surface of the susceptor. The recirculating gas cells above the flat surface of the susceptor in an APCVD process can be removed by using a higher gas flow to a certain limit, as previously mentioned above. The vertical configuration is also more versatile in that the cylindrical susceptor can be rotated to give another optional remedy to remove recirculating cells.

Figure 11a shows the corresponding temperature distribution around a cylindrical susceptor in a vertical reactor at different pressures.

The free convection by the rising hot gas due to the buoyancy force is efficient in transferring heat further upstream from the susceptor. This is clearly evident in Figure 11a. The buoyant hot gas transfers heat to the incoming cold gas causing the latter to heat up. This causes gas phase pyrolysis and reactions to occur at a distance further upstream from the surface of the sample at higher pressure (760 Torr) compared to that at lower pressure (100 Torr). A higher concentration resulting from higher gas pressure allows more gaseous species to transfer heat upstream from the susceptor to preheat the incoming gas. This preheating creates higher concentration of reactive gas species even before impinging on the hot susceptor, and accounts for the higher growth rate observed in an APCVD process. Figure 11b shows an actual APCVD process where this phenomenon of buoyancy force-induced turbulence can be seen as a plume extending

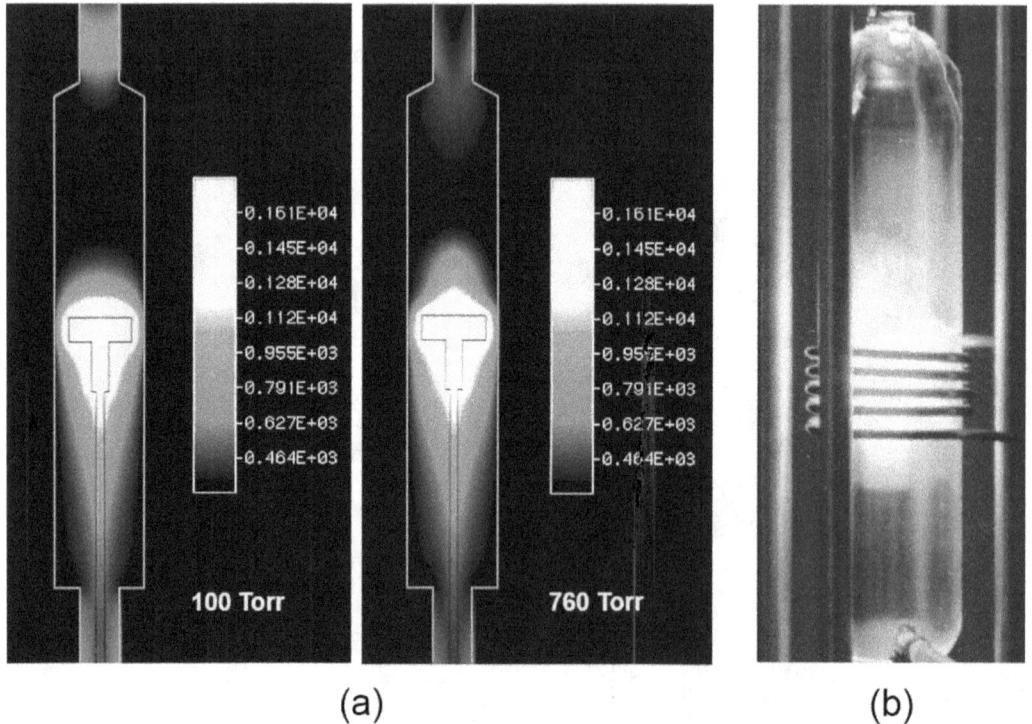

(a)   (b)

FIGURE 11

(a) CFD simulations of temperature distribution around a cylindrical susceptor in a vertical reactor at different pressures; (b) actual SiC CVD growth in APCVD mode showing turbulence.

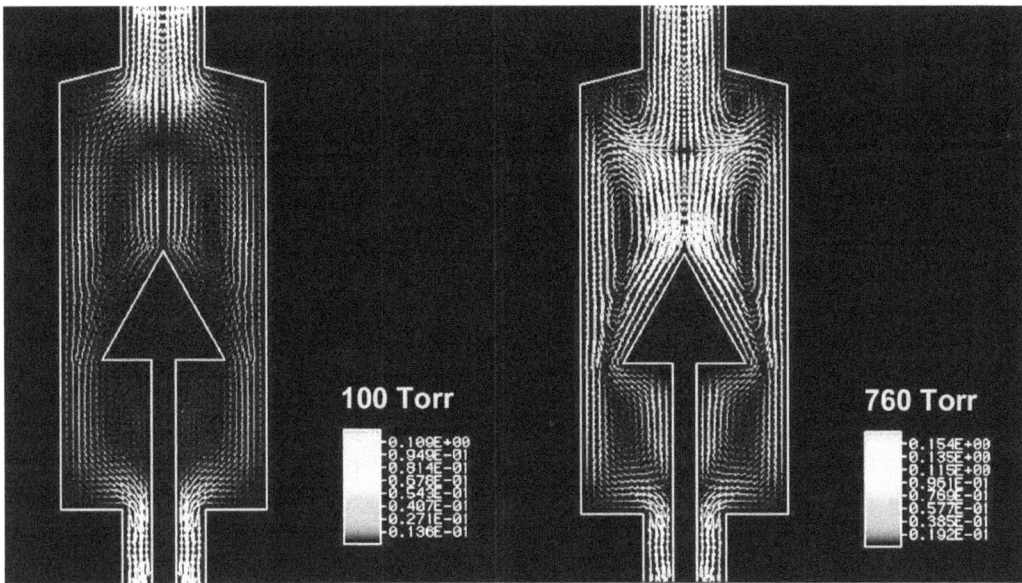

FIGURE 12
CFD simulations of gas velocity vector field for a vertical reactor with pyramidal shape susceptor at different pressures.

upstream from the surface of the sample, followed by deposition at the cold wall of the reactor tube.

For the sake of discussion to show impact of susceptor design, Figure 12 shows the gas flow patterns in the case of a pyramidal shape susceptor capable of holding four wafers on its four sides. In this case, there are more turbulence and recirculating gas cells appearing in both LPCVD and APCVD processes contrary to that seen in the cylindrical, pancake, or barrel susceptor design. Again, these recirculating gas cells diminish to a certain extent when the gas flow velocity is high enough. However, excessive increase in gas flow velocity exacerbates the problem of gas turbulence due to the increasing dominance of the buoyancy force.

In all cases, it can be seen that the gas pressure affects the temperature distribution around the susceptor with the heated gas acting as a good medium for thermal conduction and carrying the heat further from the susceptor.

The simulation results shown here are consistent with the expectations adduced from other simulation works in the literature [37–41].

Reactor geometry and susceptor design are important considerations for an optimal CVD system. Reactor configuration is the most important aspect because that is the heart of the system. Bearing in mind the focus on an affordable, simple, versatile, but yet productive CVD system in this discussion, a straight wall water-cooled vertical reactor, as shown in Figure 1, can be constructed using concentric large bore quartz tubings at the initial stage. Quartz glass-blowing expertise is still needed, and consideration must be given to the direction of the inlet and outlet for the cooling water supply to enable induction heating coil to slide in or out. In the past, large bore vacuum-grade stainless steel flanges were difficult to procure off-the-shelf because vendors prefer to sell a complete system instead of individual components. However, it is now possible to get suitable water-cooled flanges with many choices readily available off-the-shelf [42] and at a reasonable price. After more experience, the reactor configuration can later be modified to adopt some of the techniques in hot-wall CVD if one desires to venture into CVD growth of thick diffusion epilayers for high-power devices. If the focus is on niche areas in materials and low-power SiC device technology development, then the system described here is adequate. It also pays to study the many CFD simulations that have been carried out and published on various CVD systems of different dimensions and designs. But those dimensions and designs may not be identical to the design being considered. A concurrent research project on a CFD simulation of a design under consideration would be worthwhile because it can provide quantitative data on the optimal dimensions as well as additional insights even though the general principles of gas flow dynamics are generally known.

We have shown here, through computational fluid dynamics simulations, that the gas flow pattern and temperature distribution are affected by pressure, reactor configuration, susceptor design, and gas flow velocity. Knowing the effects of these parameters allows one to arrive at an optimal reactor configuration and susceptor design and proper growth conditions to avoid recirculating cells and turbulence.

2.5 Precursor Chemistry and Delivery

The most common precursor for silicon is silane, and that for carbon is propane. The basic reaction chemistry to produce SiC is described by the following reactions:

$$SiH_4 \rightleftharpoons SiH_2 + H_2$$

$$SiH_2 \rightleftharpoons Si\,(g) + H_2$$

$$2C_3H_8 \rightleftharpoons 3C_2H_2 + 5H_2$$

$$Si\,(g) + \tfrac{1}{2}C_2H_2 = SiC + \tfrac{1}{2}H_2$$

Although silicon can react with other carbon species such as CH_4 or C_2H_4, it has been determined that the reaction with C_2H_2 is the predominant reaction leading to the formation of SiC [43]. A recent simulation also shows that CH_3 and C_2H_2 are the main active hydrocarbon species in growth of SiC [44].

Silane and propane are usually supplied in diluted form, i.e. 2% silane in hydrogen and 2% propane in hydrogen. Other precursors for silicon are dichlorosilane (SiH_2Cl_2) [45], hexachlorodisilane Si_2Cl_6 [46], etc. Single precursors of both C and Si such as 1,3-disilacyclobutane ($C_2H_8Si_2$) [47], trimethylsilane ($Si(CH_3)_3H_3$) [48], methyltrichlorosilane (CH_3SiCl_3) [6,49–51], dichlorodimethylsilane [52], hexamethyldisilane ($(CH_3)_6Si_2$) [53,54], bis-trimethylsilylmethane ($(CH_3)_3Si)_2CH_2$) [55], etc. have also been used. The alternative sources of carbon are ethylene, C_2H_4 [23] and acetylene, C_2H_2 [28,56].

To ensure ultra-high purity throughout the gas and source delivery system, only grade 304 or 316 stainless steel components are used. Plumbing lines should be made of electropolished seamless 0.25 inch grade 304 stainless steel tubing. Cut ends should be free of burrs to prevent particulate from entering the gas lines. Fittings should be stainless steel grade 316. Since this is not an ultra-high vacuum system, Swagelok compression fittings would be adequate. Prior to final installation, tubing and fittings should be rinsed sequentially with acetone, methanol, and deionized water to remove organic contaminants.

3 Material Characteristics and Growth Procedures

3.1 Polytypes

SiC has over 250 polytypes. The most commonly used polytypes are the cubic 3C, and the hexagonal 6H- and 4H-polytypes.

The cubic 3C-SiC polytype is usually grown on a silicon substrate. Due to a large lattice and coefficient of thermal expansion mismatch, heteroepitaxial growth of 3C on silicon results in a large density of defects, such as stacking faults and dislocations, at the interface. The existence of these defects hinders the usefulness of the 3C-SiC/Si structure in device applications, although the quality of the 3C-SiC epilayer improves with thickness, i.e. the crystal quality improves as one gets further away from the interface.

At the interface between a thin film and a substrate of different lattice parameters, the lattice mismatch can be accommodated by a uniform elastic strain in the epilayer without the formation of misfit dislocations provided the thickness of the thin film is less than a critical value. Using the model of Mathews and Blakeslee [57] and the data for the 3C-SiC/Si interface provided by Cheng *et al.* [58], the critical thickness corresponding to the maximum strain for the $Si_{(1-x)}C_x$ system can be estimated [59] as shown in Figure 13. It can be seen from Figure 13 that the critical thickness reduces with increasing carbon content, which means that beyond the critical film thickness, misfit dislocation density increases to accommodate the misfit. For the 3C-SiC/Si system, the critical thickness is of the order of the lattice parameter. A thin carbon-rich interfacial layer is therefore required to accommodate the effect of the lattice misfit. Without this interfacial layer, polycrystalline 3C-SiC will be formed instead.

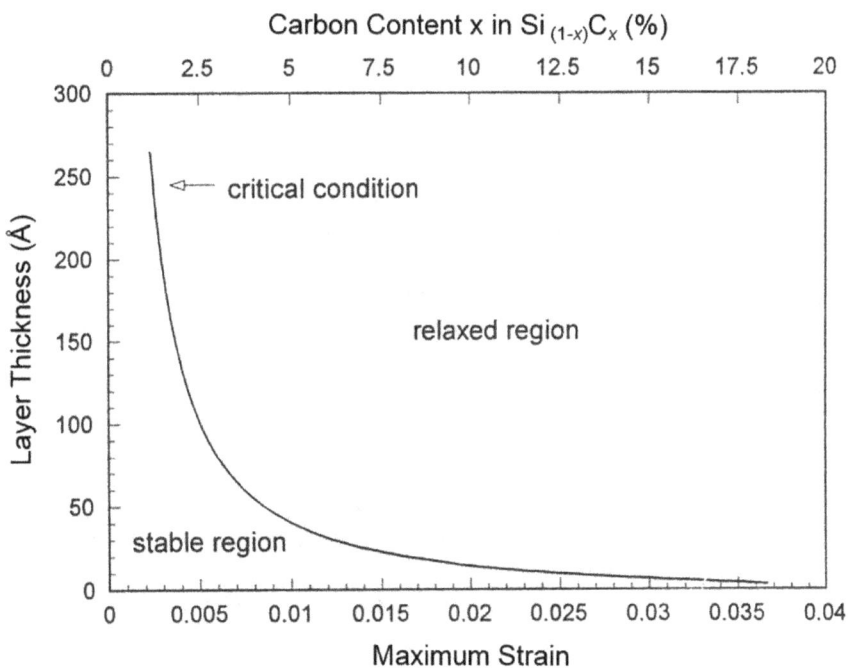

FIGURE 13
Maximum strain and critical thickness for different C content in $Si_{(1-x)}C_x$.

Heteroepitaxial growth of 3C-SiC on Si therefore requires the growth of an interfacial layer before normal growth of 3C-SiC epilayer. The growth of this interfacial layer, which is also known as the carbonization process, is a crucial step. Without a proper interfacial layer, the 3C-SiC epilayer will fracture and peel off from the Si substrate on cooling to ambient temperature. To grow the interfacial layer, the reactor is backfilled with propane/hydrogen gas mixture and the temperature of the sample is ramped rapidly from ambient to a growth temperature of about 1350°C within 2 min. Care must be taken to ensure that the temperature does not overshoot or else the silicon substrate would melt onto the susceptor if the temperature happens to reach the melting point of silicon. In APCVD, propane is necessary in the carbonization step, but silane is optional. But in LPCVD, silane is necessary as will be discussed below. Since the presence of silane gas during carbonization does not hurt the growth of the interfacial layer, it is normally included to make it more convenient to transition seamlessly from interfacial growth to normal growth.

Etch pits are frequently seen on 3C-SiC epilayers. The occurrence of etch pits in the 3C-SiC epilayers is more prevalent in low-pressure CVD techniques, such as LPCVD or MBE. The formation of etch pits originates at the silicon substrate as shown by a cross-sectional scanning electron micrograph in Figure 14.

Etch pits are caused partly by the loss of silicon atoms from the surface due to the high vapor pressure of silicon and hydrogen gas etching. To suppress the formation of these etch pits, silane is necessary in the carbonization step [31]. Another technique [60] that can be used to suppress etch pits' formation in low-pressure processes is to use a Si substrate that has been carbonized at atmospheric pressure and has a thin 3C-SiC epilayer previously grown at atmospheric pressure. The use of a Si wafer with an APCVD-grown 3C-SiC epilayer as the substrate prevents the evaporation of Si from the Si wafer during the initial stage of epitaxial growth at low pressure. In other words, for MBE growth, the substrate should be a Si wafer with a pre-grown 3C-SiC epilayer that has been grown by APCVD. However, in a variable-pressure LPCVD system, the carbonization and subsequent deposition of a thin 3C-SiC epilayer can be done *in situ* on Si wafer at atmospheric pressure. The reactor tube pressure is then reduced to the desired pressure and LPCVD of 3C-SiC is then carried out as usual. The resultant low-pressure-grown 3C-SiC epilayer will have reduced etch pits' density.

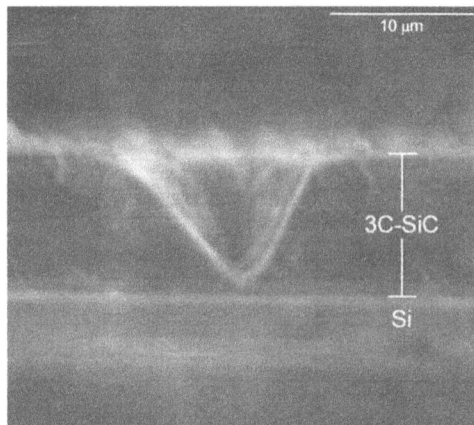

FIGURE 14
Scanning electron micrograph of cross-section of an etch pit in 3C-SiC/Si epilayer. From Figure 6 of [31], with permission of Elsevier.

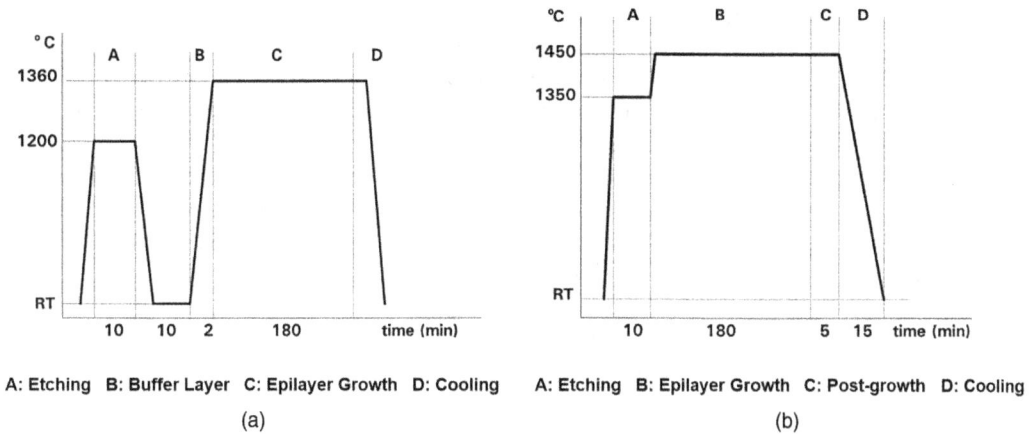

A: Etching B: Buffer Layer C: Epilayer Growth D: Cooling   A: Etching B: Epilayer Growth C: Post-growth D: Cooling

(a)   (b)

FIGURE 15
Different stages in epitaxial growth of (a) 3C-SiC and (b) 4H-SiC or 6H-SiC.

A slightly different technique [28,55] to reduce etch pits in LPCVD of 3C-SiC on silicon is to use a longer carbonization step that is slightly over 60 min duration using only C_2H_2 as the process gas. The end result is the same and that is an overgrowth of SiC layer over the etch pits formed in the silicon substrate.

The commercial availability of both 6H- and 4H-polytype substrates enables homoepitaxial growth of 6H- and 4H-epilayers on the respective substrates. With homoepitaxial growth, higher quality device-grade epilayers can be obtained. This leads to successful fabrication of myriad devices using both 6H- and 4H-SiC epilayers.

Epitaxial growths of 6H- and 4H-SiC epilayers are easier than that for 3C-SiC because there is no need for an interfacial layer and thus no carbonization step. The other difference is the higher growth temperature of 1450°C. The growth procedure for 6H-SiC epilayer is similar to that for 4H-SiC epilayer.

The outlines of the growth procedures for 3C-SiC and 4H-SiC or 6H-SiC are shown in Figure 15 for comparison.

3.2 Substrate Crystal Orientation

The crystal orientation of the most commonly used 6H-SiC substrate is 3° off the *c*-axis towards the 11$\bar{2}$0 plane, whereas the misorientation for the 4H-SiC substrate is 8° off the *c*-axis towards the 11$\bar{2}$0 plane. Misorientation is important as it reduces the growth temperature, i.e. a higher growth temperature is needed if on-axis substrate is used. Off-axis substrate provides more steps and kinks that enhance epitaxial growth. The mis-cut angle affects the morphology of the epilayer. A 4H-SiC substrate with a 3° mis-cut gives a lot of stacking faults whereas an 8° mis-cut gives a smooth mirror-like finish. The use of off-axis SiC substrate for growth of epitaxial layer is also commonly referred to as step-enhanced epitaxial growth.

The presence of micropipe defects, which are screw dislocations, in both on- and off-axis (0001) substrates has been a problem. The possibility that bulk material grown along different crystal direction may be micropipe-free has induced several wafer producers to grow SiC boules of both 6H- and 4H-polytypes in crystal directions other than the <0001> direction. The availability of wafers with the *a*-plane (11$\bar{2}$0) or the *p*-plane (1$\bar{1}$00) has enabled CVD epitaxial growth to be carried out and characterized. For vertical power devices, such as the UMOS or DMOS, which are fabricated on the *c*-plane of a hexagonal SiC wafer, charge carrier conduction is actually along the planes normal to the *c*-plane such as the *a*-plane (11$\bar{2}$0) or the *p*-plane (1$\bar{1}$00). For this reason, there is an interest in characterizing charge carrier transport in the planes perpendicular to the *c*-plane. The availability of *a*- or *p*-plane substrates allows such study to be undertaken. Another compelling reason to look at the *a*- or *p*-planes is that a vertical power device built on an *a*- or *p*-face substrate will have carrier conduction along the *c*-axis, which has higher carrier mobility. CVD epitaxial growth on the *a*-face gives specular surface morphology, similar to that of the *c*-face. Lateral 4H-SiC MOSFET devices grown on the *a*-face have been shown to give higher channel mobility compared to those fabricated on the *c*-face. In the case of the *p*-face substrate, the surface of the *p*-face epilayer is not a mirror-finish but consists of undulating features.

Another crystal orientation that is of interest is (03$\bar{3}$8). The morphology of the epilayer grown on the (03$\bar{3}$8) face is similar to that of the *a*-face. The nitrogen and boron doping characteristics between the two faces are also similar and both are different from those on off-axis (0001) substrates. These differences are attributed to different bonding configurations at the surface. It is also reported that although *n*-type doping of the epilayers grown on the (03$\bar{3}$8) and *a*-face is usually higher than an off-axis (0001) substrate, the background doping of epilayers grown on (03$\bar{3}$8) substrates is lower than that on off-axis (0001) substrate [61].

3.3 Pre-Growth Etching

Pre-growth *in situ* treatment of the sample is an important procedure to assure a good prestine surface before actual growth. There are basically two *in situ* etching techniques: hydrogen gas or hydrochloric acid etching. Etching by hydrogen gas requires longer etching time whereas use of hydrochloric acid is faster and thus requires shorter etching time. The concentration of hydrochloric acid in the gas and the temperature of the sample are two parameters that can be adjusted to control the etching rate. These two parameters should be optimized to produce a good surface because etching by HCl is a very aggressive process, which can create etch pits in the substrate. These etch pits would then cause the final epilayer to have more pits similar to an orange peel-like morphology. The different etching temperature of the substrate produces a different surface condition, which affects epitaxial growth. The use of different etching temperature in the polytype control of SiC epitaxial growth on 6H-SiC has been reported by Powell *et al.* [62].

The *in situ* etching rate of HCl in a horizontal hot-wall CVD reactor has been reported by Zhang *et al.* [63]. The etching data cover a temperature range of 1300–1550°C and can be extrapolated down to 1250°C, which is the optimum etching temperature for our cold-wall vertical CVD system.

3.4 Post-Growth Termination

The common procedure to terminate the growth process is to turn off the flow of silane and propane into the growth chamber. The sample temperature is still maintained at the growth temperature for several minutes with hydrogen gas continuing to flow into the chamber. Maintaining the growth temperature would allow silane gas to be purged out of the chamber before the sample is allowed to cool to ambient. The removal of the silane is necessary because if the sample is allowed to cool in the presence of silane, silicon droplets would form on the surface of the sample. Silicon droplets would also form if the purging of silane is incomplete, or if residual silane still remains in the chamber during cooling. Another effective post-growth termination technique is to shut off the flow of silane and propane, and replace hydrogen gas with argon. In this latter technique, the sample temperature must still be maintained at the growth temperature for several minutes while process gas is purged out of the chamber. The sample is subsequently cooled to ambient in the presence of argon flow. The absence of hydrogen suppress the decomposition of residual silane. In the event that silicon droplets are present, they can be easily removed by dipping the sample in aqua regia solution for a few seconds.

3.5 Susceptor Effect

The two common types of susceptor materials are high purity graphite and SiC-coated high purity graphite. The SiC coating helps to prevent outgassing of impurities and carbon from the graphite susceptor. It also prevents the graphite susceptor from being etched by the hydrogen gas during CVD growth. The main problem with SiC-coated susceptor is the short lifetime of the SiC coating, which tends to peel off to expose the graphite after a few CVD growth runs. Susceptors made from pure graphite without the SiC coating have also been used for SiC epitaxial growth. It should be noted however that the presence of carbon from the graphite affects the Si/C ratio of the gaseous precursors, which may be a source of inconsistency in doping characteristics. The fact that the pure graphite susceptor has been used satisfactorily in some CVD reactors shows that the presence of carbon does not really pose a big problem, especially since growth and doping conditions are characteristic of an individual reactor system.

The condition of the susceptor is one of the significant factors that controls the surface morphology of the sample. The detrimental effect of a degraded susceptor is shown in the Nomarski micrographs in Figure 16. Figure 16a shows a typical surface morphology of a 3C-SiC sample grown with a new susceptor whereas Figure 16b shows the surface morphology of a 3C-SiC sample grown with a bad susceptor. Numerous etch pits can be seen in the sample grown with a susceptor that has been used numerous times without reconditioning.

The cause of the etch pits in this case is the outgassing that occurs when the temperature of the susceptor is ramped to over 1300°C. As the susceptor begins to outgas, undesirable deposits would deposit on the substrate. The location of these deposits will give rise to

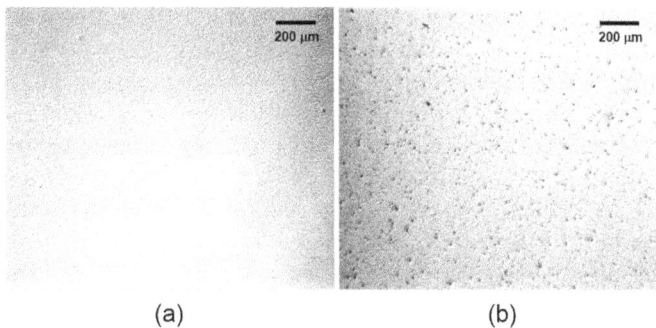

(a)   (b)

FIGURE 16
Surface morphology of 3C-SiC grown with (a) new susceptor and (b) degraded susceptor.

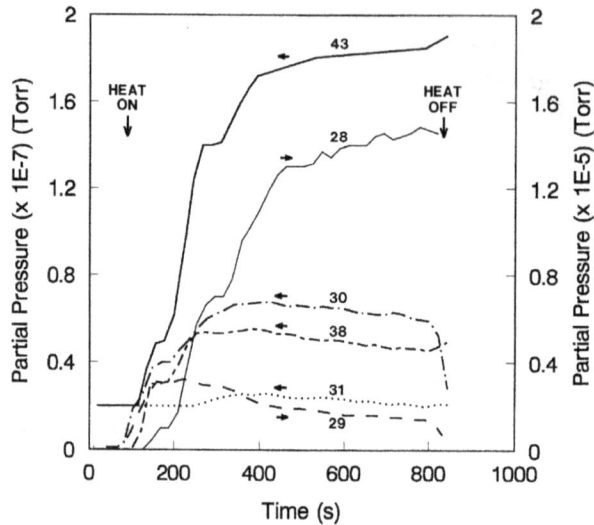

FIGURE 17

Time evolution of gas species detected by a quadrupole mass spectrometer during outgassing of a used susceptor at elevated temperature, with numbers indicating mass-to-charge ratio (m/z) of the gas species.

various crystallographic defects such as stacking faults and pits. Outgassing is especially serious in a susceptor that has been previously used many times. Figure 17 shows the gas species in the chamber during outgassing of the susceptor detected by a quadrupole mass spectrometer. The gas released from the susceptor contains mostly silicon and carbon-related species. To reduce outgassing from the susceptor during crystal growth, the susceptor must be conditioned. The conditioning process involves etching in HCl for 15–20 min to remove free silicon on the surface of the susceptor. The etching process is followed by heating in hydrogen gas at low pressure at a temperature above the growth temperature for about 30 min.

3.6 Doping

The applicability of any electronic materials in device fabrication depends crucially on the ability to introduce both n- and p-type dopants with controlled concentrations ranging from $< 1 \times 10^{15}$ cm^{-3} for Schottky contacts and high-power devices to $>5 \times 10^{18}$ cm^{-3} for low resistance ohmic contacts. In compound semiconductors of the form AB (such as GaAs, InP, etc.) the degree of dopant introduction can be controlled by adjusting the relative gas-phase concentrations of the A and B constituents during epitaxy. For instance, altering the flow rates of A and B precursors into the growth chamber changes the A:B ratio in the gas phase, which then favors the introduction of one type of dopant over another. The same concept can be applied to SiC doping. It has been shown first by Larkin *et al.* [64] that changing the Si:C ratio in the gas phase during epitaxial growth affects the efficiency of dopant introduction in the epilayers. This process is also known as site-competition epitaxy. Simply put, a high Si:C ratio favors the incorporation of nitrogen atoms in the lattice because N competes with C for the C sites to produce n-type doping. The converse is true of p-type doping where a low Si:C ratio favors aluminum doping because Al competes with Si for the Si sites.

The actual doping mechanism, however, is not so simple because a Si or C vacancy can be an acceptor or donor depending on whether they are on hexagonal or cubic sites. Another complication to the doping problem is the formation of defect complexes involving both dopants and native defects.

The dopant incorporation mechanism has been studied by several groups. Kimoto *et al.* [65] reported that the dopant incorporation generally followed the site-competition epitaxy model for the (0001) Si face but does not follow this model on the (0001) C face.

The most common dopants in SiC are nitrogen as *n*-type dopant and aluminum as *p*-type dopant. Other dopants used are phosphorous and boron. The nitrogen atom can be introduced as nitrogen gas or ammonia, although the former is more commonly used. Since nitrogen is a ubiquitous element that is present as a residual impurity in hydrogen gas supply, the epilayers will be nominally *n*-type with variable background doping level controlled by the Si:C ratio during epitaxial growth, as shown in Figure 18. This is an important concept that one needs to keep in mind when designing a doping protocol for epitaxial growth of SiC. Phosphorous and boron are introduced using phosphine [66,67] and diborane gas, respectively, whereas aluminum is introduced using an organometallic compound such as trimethylaluminum or triethylaluminum. The dopant incorporation as a function of precursor flow rate has been studied [66,67].

The common precursors for aluminum are trimethylaluminum and triethylaluminum. Hydrogen gas is used as the carrier gas and is allowed to bubble through the metalorganic liquid held at a controlled temperature. The volume flow rate of the precursor transported into the growth reactor is calculated according to the equation:

$$F = F_{H2} \times \frac{P_{TMA}}{\left(P_B - P_{TMA}\right)}$$

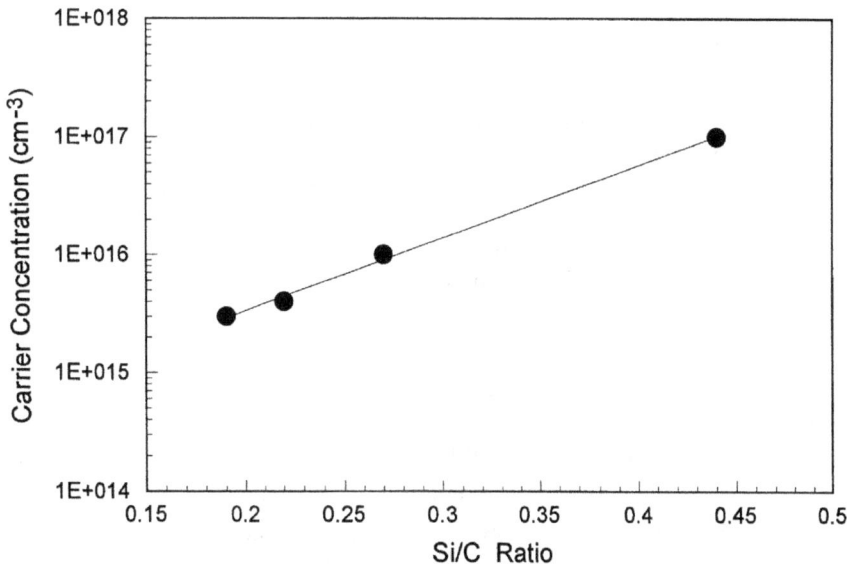

FIGURE 18
Background doping level affected by Si:C ratio in the process gas during SiC CVD.

where F_{H2} is the hydrogen gas flow rate through the bubbler, P_{TMA} is the vapor pressure of the metalorganic source, and P_B is the total pressure in the bubbler. The vapor pressure of the metalorganic source follows the equation:

$$\log\left(P_{TMA}\right) = A - \frac{B}{T}$$

where $A = 8.224$ and $B = 2134.83$ K for trimethylaluminum; whereas for triethylaluminum, $A = 10.784$ and $B = 3625$ K.

3.7 Dopant Activation

Dopant activation during epitaxial growth depends on the growth temperature. A higher growth temperature allows more dopants to be introduced into the epilayers. This explains the gradual increase in doping concentration with time as the growth process proceeds. This trend is especially obvious in the long-duration growth process in a quartz reactor, where the light intensity as seen by a pyrometer aimed at the sample surface is affected by light reflected from the opaque materials deposited on the wall of the tube. One way to avoid such spurious temperature readings is to use a second pyrometer aimed at the bottom of the sample holder as previously mentioned. This second pyrometer is used to control the temperature controller connected to the induction power supply. The accuracy of the temperature reading from the second pyrometer would not be affected by deposits on the wall of the growth chamber. A temperature difference of 100°C would affect dopant incorporation by a magnitude [66].

The degree of dopant activation in a normal CVD growth of SiC is about 2% for aluminum dopant and 6% for nitrogen dopant.

There is a limit to how much dopant that can be introduced without creating defects such as stacking faults and dislocations [68,69].

3.8 Porous Substrates

Porous SiC is prepared by photoelectrochemical etching in hydrofluoric acid [70]. The use of porous substrates for CVD growth of SiC has been carried out by several groups [71-73]. The purported reason for use of porous substrate is to reduce defects in the epilayer. Saddow *et al.* [69] reported that 4H-SiC epilayer grown by CVD on porous 4H-SiC substrate showed lower defect density and better structural characteristics than that grown on untreated substrate. They attributed the defect reduction to stress relief achieved by lateral epitaxial overgrowth of the pores.

4 Hot-Wall and High-Temperature CVD

An important benchmark characteristics of a high-power device is its ability to withstand high voltage and high current levels. A device with high breakdown voltage requires low doping concentration, which in turn produces a wider depletion layer for a given reverse-bias voltage. The electric field breakdown of a device depends on many factors such as crystalline defects, avalanche breakdown, and depletion layer breakthrough, i.e. the width of the depletion layer induced by a reverse-bias voltage exceeds the epilayer thickness. A 4H-SiC junction device designed to withstand a few kilovolts would require an epilayer thickness of over 50 μm. It should be mentioned that, in practice, breakdown voltage for any given epilayer thickness is even lower than calculated because of other factors such as edge termination, material quality, and device structure. Hence, the required epilayer thickness is even greater than theoretical value to compensate for deficiency in device design and fabrication.

The cold-wall CVD technique is able to produce high quality SiC epilayers suitable for device fabrication. The growth of epitaxial SiC layer is usually done with a substrate temperature at about 1400–1500°C. This commonly used cold-wall CVD technique has a relatively low growth rate of 1–3 μm/hr. The fastest growth rate using cold-wall CVD is about 6 μm/hr [74]. In our experience, the growth rate is typically about 1 μm/hr for CVD growth of high quality SiC epilayer with a low doping level of $\approx 10^{15}$ cm^{-3}. The low growth rate is due to the fact that in order to achieve such low doping concentration, it is necessary to use a low Si:C ratio in the gas phase based on the concept of site-competition epitaxy. At this low growth rate, it would take over 50 hours to grow a 50 μm layer using the conventional cold-wall CVD technique, which is commercially unviable. Gradual drift in system parameters and deterioration in growth conditions over such an extended period of time render this technique unreliable and inefficient for growth of thick device structures.

An alternative technique with a higher growth rate is required for growth of high-power SiC device structures. A variation of the CVD growth process called hot-wall CVD [27] allows the process gas to pass through an elongated heated susceptor before reaching the substrate. The aim is that preheating of the gas increases the pyrolytic cracking efficiency of the precursors. Thermal loss is reduced by using graphite foam thermal insulation around the susceptor. The growth temperature is about 1500–1650°C. This technique is also able to provide device-quality epilayers, but the growth rate is not much higher compared to cold-wall growth rate. An advantage of this technique is the improved long-term stability of the growth process due to better heating efficiency and temperature control. This allows for long growth period to produce thick epilayers with thicknesses up to 30–100 μm. However, growth rate is still relatively slow and inefficient. The growth rates for both cold- and hot-wall CVD reactors are substantially lower compared to those of another technique known as high-temperature CVD (HTCVD) [27,29,75].

The HTCVD technique uses basically the same graphite reaction chamber or susceptor concept that is used in hot-wall CVD. There are basically two versions, commonly known as the chimney and inverted chimney reactors. Process gas is preheated before reaching the samples or substrates. The difference between this technique and other CVD techniques lies in the fact that the growth temperature is about 1800–2300°C, which is much higher than that of either cold-wall or hot-wall CVD. The greatest advantage offered by the HTCVD growth technique is the high growth rate. The HTCVD process can give a growth rate as high as 500 μm/hour [27,29,75].

Due to the high growth temperature, the initial implementation of the HTCVD technique suffers from the same problems [70] as found in the sublimation growth or PVT method. The main problems are poor epilayer quality, autodoping, and polytype control. Since the inception of the hot-wall and HTCVD techniques in the 1990s the growth procedures have been refined and material quality has improved significantly [76,77,78].

The first HTCVD system used a graphite susceptor and graphite insulation [27]. According to Ellison *et al.* [75], the unintentional doping of HTCVD-grown 4H-SiC was found to be *p*-type due to boron impurities. This is not surprising because Al, Ti, V, and B are among the common impurities in graphite, which is the material widely used for susceptors, thermal insulation, and fixtures in both bulk and epitaxial SiC crystal growth techniques. Graphite makes the control of carbon in the gas phase more difficult and unpredictable. Etching of carbon by hydrogen and the outgassing of carbon at high temperature increase the C/Si ratio of the vapor in the growth chamber. This promotes the incorporation of *p*-type dopants following the 'site-competition' mechanism. A high carbon content is an advantage for growth of *n*-type epilayer with low *n*-type doping concentration but this is not true for *p*-type material. This explains the observation that low *p*-type doping concentration is not possible in some CVD systems.

Kushibe *et al.* [79] has reported that at a high substrate temperature of over 1620°C, there is a competitive process between etching and epitaxial growth. The etching process follows the reaction $2SiC + H_2 \rightleftharpoons 2Si\ (g) + C_2H_2$. The increase in etching rate by hydrogen causes a decrease in deposition rate. It is then necessary to increase the flow rate of the process gas to compensate for the etching effect [76].

5 Safety

It cannot be stressed enough that safety is of paramount importance in and around a SiC CVD system. A SiC CVD system involves high tension power, hazardous chemicals and gases, and high temperature. Laboratory personnel should be aware of the dangers and the precautions to take, and to be knowledgeable about the emergency standard operating procedures (SOPs) in force in the laboratory.

About 98% of the gas involved in SiC CVD growth process is hydrogen, which is flammable. Silane is pyrophoric. Hydrogen, propane, or other hydrocarbon gases such as ethylene and acetylene are also flammable. For safety precaution, it is better to use gas mixture for silane, such as 2% silane in hydrogen, instead of pure silane, which is both pyrophoric and toxic. The metalorganic source that is used as the dopant is also pyrophoric. Metalorganic source is highly flammable and reacts violently with water and oxygen in air. Therefore, water should not be used to extinguish a fire caused by a metalorganic source. A powder fire retardant should be used instead. Proper protocols should be followed in installing or replacing metalorganic source in the system both to protect the cleanliness and purity of the system as well as the safety of laboratory personnel .

The flammable gas cylinders should be in the gas cabinets and the SiC CVD system should be within an enclosure. Both these locations should be vented to the outside atmosphere so that any leaks would not escape into the laboratory. Since 98% of the gas used in the SiC CVD system is hydrogen, it is important that at the very least, several hydrogen gas detectors should be installed in the gas cabinets and system enclosure to set off the alarm at the slightest indication of a gas leak.

Every gas line carrying hazardous gas from the gas cabinet to the SiC CVD system should preferably be enclosed by a larger tubing to form an outer jacket around the gas delivery tube. Inert gas such as nitrogen can be used to continuously flush through the outer jacket thus forming a gas shield. The outer jackets are vented through the exhaust at the gas cabinets. Any leaks would then be confined within the outer jackets and exhausted through the exhaust outlet of the gas cabinet without endangering the laboratory personnel.

Another hazardous situation is the presence of both high-temperature and flammable hydrogen gas. It is important to make sure the exhaust is not plugged and the reactor pressure is not building up inside the reactor tube. For APCVD, the check valve to the vent should open when reactor pressure is at atmospheric pressure. The process should be stopped immediately if pressure does not stop rising beyond atmospheric pressure. A burst joint or cracked reactor tube during growth process would be a dangerous situation especially with the susceptor at high temperature and flammable or pyrophoric precursors being present.

The SiC CVD system should have a panic button or emergency switch that shuts down the system immediately when activated.

There are laboratory safety manuals setting out the laboratory SOPs in the handling of metalorganic chemicals and flammable gases at other institutions available on the Internet. It would be wise to look over such manuals to study useful practices and policies for adoption in the SiC CVD laboratory.

6 Summary

Great strides have been achieved in epitaxial growth of SiC epilayers by CVD in the last few decades, due principally to the commercial availability of bulk SiC materials. In this chapter, we have discussed the basic principles in SiC CVD growth and the procedures for obtaining high quality epilayers. The techniques of SiC CVD are similar to those employed in the epitaxial growth of other electronic materials such as silicon and the compound semiconductors, but with several subtle differences. We have discussed the basic principles of SiC CVD and highlighted the problems of SiC epitaxy. We have provided important pointers necessary for a successful in-house construction of a SiC CVD system that can give many years of productive research work in the growth of SiC epilayers and fabrication of high-performance and high-power SiC devices.

References

1. I.N. Pring and W. Fielding, J. Chem. Soc. 95, 1497 (1909).
2. E.B. Gasilova, M.A. Gurevich, T.N. Nistor, and M.B. Reifman, "Single-crystal structure types in the hexagonal modification of silicon carbide, α-SiC, prepared by vapor-phase recrystallization," in Silicon Carbide, ed. I.N. Frantsevich, Consultants Bureau, New York, 1970, p.31.
3. N.K. Prokof'eva and M.B. Reifman, "Preparation of doped beta silicon carbide," ibid, p. 156.
4. R.W. Brander, "Epitaxial growth of SiC layers," in Proceedings of the third International Conference on Silicon Carbide, ed. R.C. Marshall, J.W. Faust, Jr., C.E. Ryan, University of South Carolina Press, Columbia, 1974, p.8, and references therein.
5. B. Wessels, H.C. Gatos, and A.F. Witt, "Epitaxial growth of silicon carbide by chemical vapor deposition," ibid, p. 25.
6. S.N. Gorin and L.M. Ivanova, "Cubic Silicon Carbide (3C-SiC): structure and properties of single crystals grown by thermal decomposition of methyl trichlorosilane in yydrogen," Phys. Sta. Sol. (b) 202, 221 (1997), and references therein.
7. J.R. O'Connor and J. Smiltens (Editors), Silicon Carbide – A High-Temperature Semiconductor, Pergamon Press, London (1960).

8. S. Nishino, J.A. Powell, and H.A. Will, "Production of large-area single-crystal wafers of cubic SiC for semiconductor devices," Appl. Phys. Lett. 42, 460 (1983).
9. Yu. M. Tairov and V.F. Tsvetkov, J. Cryst. Growth 52, 146 (1981).
10. G. Augustine, H. McD. Hobgood, V. Balakrishna, G. Dunne, and R.H. Hopkins, "Physical vapor transport growth and properties of SiC monocrystals of 4H polytype," Phys. Sta. Sol. (b) 202, 137 (1997).
11. R.C. Glass, D. Henshall, V.F. Tsvetkov, and C.H. Carter, Jr., "SiC seeded crystal growth," Phys. Sta. Sol. (b) 202, 149 (1997).
12. J. Takahashi and N. Ohtani, "Modified-Lely SiC crystals grown in [1$\underline{1}$00] and [11$\underline{2}$0] directions," Phys. Sta. Sol. (b) 202, 163 (1997).
13. M. Syväjärvi, R.Yakimova, H.H. Radamson, N.T. Son, Q. Wahab, I.G. Ivanov, and E. Janzén, "Liquid phase epitaxial growth of SiC," Journal of Crystal Growth 197, 147–154 (1999).
14. V.A. Dmitriev, "LPE of SiC and SiC-AlN," in Properties of Silicon Carbide, Ed. G.L. Harris, (INSPEC, IEE, London, 1995) pp. 214–227; V.A. Dmitriev and M.G. Spencer, "SiC fabrication technology: growth and doping," Semiconductors and Semimetals, 52, 21 (1998). (and references therein).
15. T. Fuyuki, t. Hatayama, and H. Matsunami, "Heterointerface control and epitaxial growth of 3C-SiC on Si by gas source molecular beam epitaxy," Phys. Sta. Sol. (b) 202, 359 (1997).
16. R.S. Kern, K. Jarrendahl, S. Tanaka, and R.F. Davis, "Homoepitaxial SiC growth by molecular beam epitaxy," Phys. Sta. Sol. (b) 202, 379 (1997).
17. A.O. Konstantinov, "Sublimation growth of SiC," in Properties of Silicon Carbide, Ed. G.L. Harris, (INSPEC, IEE, London, 1995) pp. 170–203.
18. T. Kimoto, A. Itoh, and H. Matsunami, "Step-controlled epitaxial growth of high-quality SiC layers," Phys. Sta. Sol. (b) 202, 247 (1997).
19. R. Rupp, Yu. N. Makarov, H. Behner, and A. Wiedenhofer, "Silicon carbide epitaxy in a vertical CVD reactor: Experimental results and numerical process simulation," Phys. Sta. Sol. (b) 202, 281 (1997).
20. S. Nishino, "Chemical vapor deposition of SiC," in Properties of Silicon Carbide, Ed. G.L. Harris, (INSPEC, IEE, London, 1995) pp. 204–213.
21. D.J. Larkin, "An overview of SiC epitaxial growth," Mater. Res. Soc. Bull. 22, 36 (1997).
22. J.A. Powell, L.G. Matus, and M.A. Kuczmarski, "Growth and characterization of cubic SiC single-crystal films on Si," J. Electrochem. Soc. 134, 1558 (1987).
23. H.S. Kong, J.T. Glass, and R.F. Davis, "Chemical vapor deposition and characterization of 6H-SiC thin films on off-axis 6H-SiC substrates," J. Appl. Phys. 64, 5 (1988).
24. H. Matsunami, "Progress in epitaxial growth of SiC," Physica B 185, 65 (1993).
25. R. Rupp, P. Lanig, J. Volkl, and D. Stephani, "First results on silicon carbide vapor phase epitaxy growth in new type of vertical low pressure chemical vapor deposition reactor," J. Cryst. Growth 146, 37 (1995).
26. Z.C. Feng, A. Rohatgi, C.C. Tin, R. Hu, A.T.S. Wee, and K.P. Se, "Structural, optical, and surface science studies of 4H-SiC epilayers grown by low pressure chemical vapor deposition," J. Electron. Mater. 25, 917 91996.
27. O. Kordina, C. Hallin, A. Henry, J.P. Bergman, I. Ivanov, A. Ellison, N.T. Son, and E. Janzén, "Growth of SiC by Hot-Wall CVD and HTCVD," Phys. Sta. Sol. (b) 202, 321 (1997).
28. H. Nagasawa and K. Yagi, "3C-SiC single-crystal films grown on 6-inch Si substrates," Phys. Sta. Sol. (b) 202, 335 (1997).
29. O. Kordina, C. Hallin, A. Ellison, A.S. Bakin, I.G. Ivanov, A. Henry, R. Yakimova, M. Touminen, A. Vehanen, and E. Janzén, "High temperature chemical vapor deposition of SiC," Appl. Phys. Lett 69, 1456 (1996).
30. A. Ellison, J. Zhang, J. Peterson, A. Henry, Q. Wahab, J.P. Bergman, Y.N. Makarov, A. Vorob'ev, A. Vehanen, E. Janzén, "High temperature CVD growth of SiC," MaterialsScience and Engineering B61–62, 113–120 (1999).
31. C.C. Tin, R. Hu, R.L. Coston, and J. Park, "Reduction of etch pits in heteroepitaxialgrowth of 3C-SiC on silicon," J. Cryst. Growth 148, 116 (1995).

32. R. A. Matula, "Electrical resistivity of copper, gold, palladium, and silver," Journal of Physical and Chemical Reference Data 8, 1147 (1979).
33. J.A. Dean (ed), *Lange's Handbook of Chemistry* (15th Edition), McGraw-Hill, 1999.
34. Valery Rudnev, Don Loveless, Raymond L. Cook, *Handbook of Induction Heating*, CRC Press, 2017.
35. Spartaco Caniggia and Francescaromana Maradei, *Signal Integrity and Radiated Emission of High-Speed Digital Systems*, John Wiley & Sons, Ltd., 2008.
36. T. Kimoto, S. Nakazawa, K. Hashimoto, and H. Matsunami, "Reduction of doping and trap concentrations in 4H–SiC epitaxial layers grown by chemical vapor deposition," Appl. Phys. Lett. 79, 2761 (2001).
37. W.L. Holstein and J.L. Fitzjohn, "Effect of buoyancy forces and reactor orientation on fluid flow and growth rate uniformity in cold-wall channel CVD reactors," Journal of Crystal Growth 94, 145–158 (1989).
38. J. Zhang, A. Ellison, Ö. Danielsson, M.K. Linnarsson, A. Henry, E. Janzén, "Epitaxial growth of 4H SiC in a vertical hot-wall CVD reactor: Comparison between up- and down-flow orientations," Journal of Crystal Growth 241, 421–430 (2002).
39. C. Houtman, D.B. Graves, and K.F. Jensen, "CVD in stagnation point flow: An evaluation of the classical 1D treatment," J. Electrochem. Soc. 133, 961 (1986).
40. G. Evans and R. Greif, "A Numerical Model of the Flow and Heat Transfer in a Rotating Disk Chemical Vapor Deposition Reactor," Journal of Heat Transfer 109/929, 928–935 (1987).
41. R. Wang, R. Ma, and M. Zupan, "Modeling of chemical vapor deposition of large-area silicon carbide thin film," Crystal Growth & Design 6, 2592–2597 (2006).
42. MTI Corporation, Richmond, CA.
43. T. Kimoto and H. Matsunami, "Surface kinetics of adatoms in vapor phase epitaxial growth of SiC on 6H-SiC(0001) vicinal surfaces," J. Appl. Phys. 75, 850 (1994).
44. O. Danielsson, M. Karlsson, P. Sukkaew, H. Pedersen, and L. Ojamäe, "A systematic method for predictive *in silico* chemical vapor deposition," J. Phys. Chem. C, 124, 7725–7736 (2020).
45. Y. Ohshita, "Low temperature and selective growth of β-SiC using the SiH_2Cl_2/I-C_4H_{10}/HCl/H_2 gas system," Appl. Phys. Lett. 57, 605 (1990)
46. C. Jacob, M.-H. Hong, J. Chung, P. Pirouz, and S. Nishino, "Selective epitaxial growth of silicon carbide on patterned silicon substrates using hexachlorodisilane and propane," Materials Science Forum 338–342, 249 (2000)
47. A.K. Chaddha, J.D. Parsons, J. Wu, H-S. Chen, D.A. Roberts, and H. Hockenhull, "Chemical vapor deposition of silicon carbide thin films on titanium carbide using 1,3-disilacyclobutane," Appl. Phys. Lett. 62, 3097 (1993)
48. Golecki, F. Reidinger, and J. Marti, "Single-crystalline, epitaxial cubic SiC films grown on (100) Si at 750°C by chemical vapor deposition," Appl. Phys. Lett. 60, 1703 (1992).
49. M. Ganz, N. Dorval, M. Lefebvre, M. Pealat, F. Loumagne, F. Langlais, "In situ optical analysis of the gas phase during the deposition of silicon carbide from methyltrichlorosilane," J. Electrochem. Soc. 143 (5), 1654 (1996).
50. H. Sone, T. Kaneko, and N. Miyakawa, "In situ measurements and growth kinetics of silicon carbide chemical vapor deposition from methyltrichlorosilane," J. Cryst. Growth 219 (3), 245 (2000).
51. G.D. Papasouliotis and S.V. Sotirchos, "Experimental study of atmospheric pressure chemical vapor deposition of silicon carbide from methyltrichlorosilane," J. Mater. Res. 14(8), 3397 (1999).
52. T. Takeuchi, Y. Egashira, T. Osawa, H. Komiyama, "A kinetic study of the chemical vapor deposition of silicon carbide from dichlorodimethylsilane precursors," J. Electrochem. Soc., 145 (4), 1277 (1998).
53. H.-T. Chiu and J.-S. Hsu, "Low pressure chemical vapor deposition of silicon carbide thin films from hexamethyldisilane," Thin Solid Films 252 (1), 13 (1994).

54. C. Sartel, V. Souliere, J. Dazord, Y. Montiel, I. El-Harrouni, J.M. Bluet, and G. Guillot, "Epitaxial growth of 4H-SiC with hexamehyldisilane HMDS," Materials Science Forum 389–393, 263 (2002).

55. J.K. Jeong, M.Y. Um, H.J. Na, B.S. Kim, I.B. Song, and H.J. Kim, "Homoepitaxial growth of 4H-SiC on porous substrate using bis-trimethylsilylmethane precursor," Materials Science Forum 389–393, 267 (2002).

56. H. Nagasawa and Y. Yamaguchi, "Suppression of etch pit and hillock formation on carbonization of Si substrate and low temperature growth of SiC," J. Cryst. Growth 115, 612 (1991).

57. J.W. Mathews and A.E. Blakeslee, J. Cryst. Growth 27, 118 (1974).

58. T. Cheng and M. Aindow, Philosophical Magazine Letters 62, 239 (1990).

59. R. Hu, "Silicon carbide chemical vapor deposition, in-situ doping, and device fabrications," Ph.D. Thesis, Auburn University, 1996.

60. C.C. Tin, R. Hu, J. Park, T.F. Isaacs-Smith, and E. Luckowski, "Control of etch pit density in low-pressure-grown 3C-SiC/Si by variation of reactor pressure," J. Mater. Sci. Lett. 15, 823 (1996).

61. T. Kimoto, S. Nakazawa, K. Fujihara, T. Hirao, S. Nakamura, Y. Chen, K. Hashimoto, and H. Matsunami, "Recent Achievements and Future Challenges in SiC Homoepitaxial Growth," Materials Science Forum, vol. 389–393, pp. 165–170 (2002).

62. J.A. Powell, J.B. Petit, J.H. Edgar, I.G. Jenkins, L.G. Matus, J.W. Yang, P. Pirouz, W.J. Choyke, L. Clemen, and M. Yoganathan, "Controlled growth of 3C-SiC and 6H-SiC films on low-tilt-angle vicinal (0001) 6H-SiC wafers," Appl. Phys. Lett., Vol. 59, 334 (1991).

63. J. Zhang, O. Kordina, A. Ellison, and E. Janzen, "In-situ etching of 4H-SiC in H_2 with addition of HCl for epitaxial CVD growth," Materials Science Forum Vol. 389–393, pp 239–242 (2002).

64. D.J. Larkin, P.G. Neudeck, J.A. Powell, and L.G. Matus, "Site-competition epitaxy for superior silicon carbide electronics," Appl. Phys. Lett. 65 (13), 1659 (1994).

65. T. Kimoto, A. Itoh, and H. Matsunami, "Incorporation mechanism of N, Al, and B impurities in chemical vapor deposition of SiC," Appl. Phys. Lett. 67, 2385 (1995).

66. R. Wang, I.B. Bhat, and T. P. Chow, "Epitaxial growth of n-type SiC using phosphine and nitrogen as the precursors," J. Appl. Phys. 92, 7587 (2002).

67. R. Wang, I. Bhat, and T.P. Chow, "Vapor-phase epitaxial growth of n-type SiC using phosphine as the precursor," Materials Science Forum Vol. 389–393, pp 211–214 (2002).

68. J. Q. Liu, H. J. Chung, T. Kuhr, Q. Li, and M. Skowronski, "Structural instability of 4H–SiC polytype induced by n-type doping," Appl. Phys. Lett. 80, 2111 (2002).

69. M.K. Linnarsson, M.S. Janson, U. Zimmermann, B.G. Svensson, P.O.A. Persson, L. Hultman, J.Wong-Leung, S. Karlsson, A. Schoner, H. Bleichner, and E. Olsson, "Solubility limit and precipitate formation in Al-doped 4H-SiC epitaxial material," Appl. Phys. Lett. 79, 2016 (2001).

70. J.S. Shor, I. Grimberg, B.Z. Weiss, and B.D. Kurtz, "Direct observation of porous Sic formed by anodization in HF," Appl. Phys. Lett. 62, 2836 (1993).

71. S.E. Saddow, M. Mynbaeva, W.J. Choyke, R.P. Devaty, S. Bai, G. Melmychuck, Y. Koshka, V. Dmitriev, and C.E.C. Wood, "SiC defect density reduction by epitaxy on porous surfaces," Materials Science Forum 353–356, 115 (2001).

72. J.E. Spanier, G.T. Dunne, L.B. Rowland, I.P. Herman, "Vapo-phase epitaxial growth on porous 6H-SiC analyzed by Raman scattering," Appl. Phys. Lett. 76, 3879 (2000).

73. M. Mynbaeva, S.E. Saddow, G. Melnychuk, I. Nikitina, M. Scheglov, A. Sitnikova, N. Kuznetsov, K. Mynbaev, and V. Dimitriev, "Chemical vapor deposition of 4H-SiC epitxial layers on porous SiC substrates," Appl. Phys. Lett. 78, 117 (2001).

74. S. Nakamura, T. Kimoto, and H. Matsunami, "Fast growth and doping characteristics of α-SiC in horizontal cold-wall chemical vapor deposition," Materials Science Forum 389–393, pp. 183 (2002).

75. A. Ellison, J. Zhang, W. Magnusson, A. Henry, Q. Wahab, J.P. Bergman, C. Hemmingsson, N.T. Son, and E. Jansen, "Fast SiC epitaxial growth in a chimney CVD reactor and HTCVD crystal growth developments," Materials Science Forum 338–342, 131 (2000).

76. Y. Tokuda, E. Makino, Naohiro Sugiyama, I. Kamata, N. Hoshino, J. Kojima, K. Hara, H. Tsuchida, "Stable and high-speed SiC bulk growth without dendrites by the HTCVD method," Journal of Crystal Growth 478, 9–16 (2017).
77. D. Chaussende and N. Ohtani, "Silicon carbide," in *Single Crystals of ElectronicMaterials: Growth and Properties*, R. Fornari, Ed., Woodhead Publishing, 2018.
78. R.T. Leonard, Y. Khlebnikov, A.R. Powell, C. Basceri, "100 mm 4HN-SiC wafers with zero micropipe density," Mater. Sci. Forum 600–603, 7–10 (2009).
79. M. Kushibe, Y. Ishida, H. Okumura, T. Takahashi, K. Masahara, T. Ohno, T. Suzuki, T. Tanaka, S. Yoshida, and K. Arai, "Competitive Growth between deposition and etching in 4H-SiC CVD epitaxy using quasi-hot wall reactor," Materials Science Forum, Vol. 338–342, pp. 169–172 (2002).

6

Homo-Epitaxy of Thick Crystalline 4H-SiC Structural Materials and Applications in an Electric Power System

Yingxi Niu

Institute of Semiconductors, Chinese Academy of Sciences, Beijing, China
Wuhu Advanced Semiconductor Manufacturing Co., LTD, Anhui, China

1 Introduction

With the development and utilization of fossil energy, human society is facing global resource shortage, environmental pollution, climate change, and many other problems. The traditional method for energy development is difficult to sustain; clean energy to replace fossil energy will be the general trend. It is imperative to build a global energy Internet (GEI) to meet the global electricity demand in a clean and green way. The essence of GEI is to connect large-scale clean energy bases and various distributed power sources with ultrahigh-voltage (UHV) grid (above ±800 kV DC and above 1000 kV AC) as the backbone grid and clean energy transmission as the guide to deliver clean energy to users [1]. Clean energy grid access, long-distance transmission as well as electric vehicles, industrial motor control, and other terminal consumption are inseparable from the power conversion device with high-power power devices as the core. High-performance power devices play a decisive role in the performance of a power conversion system.

"One generation of materials, one generation of devices, one generation of equipment". Silicon power devices are the main power devices used in a power system. With the development of microelectronics technology, the structure design and manufacturing technology of silicon-based power devices are becoming mature. However, due to the limitation of material properties, the performance of silicon-based power devices is close to its theoretical limit, and the voltage and power capacity are limited. Silicon-based power devices have to adopt series connection, parallel connection technology, and complex circuit topology to meet the actual requirements, which leads to the increase of equipment failure rate and cost, and restricts the development of a power grid. Therefore, we need to rely on higher performance power devices to improve the performance of new power devices and systems to meet the needs of the future power grid. In recent years, the development of wide band gap semiconductor materials represented by silicon carbide (SiC) has opened a new era of semiconductor industry, and power device technology and industry ushered in new development opportunities [2, 3].

DOI: 10.1201/9780429198540-8

At present, silicon power devices, such as IGBT, thyristor, GTO, are widely used in high-power power electronic devices of a grid system to realize the control and conversion of electric energy. Due to the performance limitation of silicon power electronic devices, the realization of high-voltage and high-power devices, such as 100kVA solid-state transformer (SST), static synchronous compensator (STATCOM), voltage source inverter (VSC), need to use series voltage sharing, multi-level voltage sharing and other technologies, several or even dozens of power devices in series to achieve the required breakdown voltage and current conduction level. This will have a negative impact on the reliability, overall loss, and stability of the system. Therefore, in order to fundamentally improve the reliability and stability of high-power electronic equipment, reduce the overall loss of the system, and improve the control and conversion efficiency of energy, it is necessary to research and develop new power electronic devices with higher breakdown voltage, lower power consumption, and high temperature resistance, such as SiC-based power devices [4–6].

2 Brief History Review of Research and Development on the Epitaxy and Devices of High Voltage

2.1 4H-SiC epitaxy

In the standard process the growth rate was limited to 6–8 μm/h by the homogeneous silicon nucleation in the gas phase. In order to make 4H-SiC power electronic devices work at a voltage above 10 kV, the thickness of SiC epitaxial material should reach 100 μm above, one of the main goals of the SiC thick epitaxy development was to increase the epitaxial growth rate, One of the methods is the chlorine-based method, that is, adding HCl additive into the reaction gas or directly using the precursor containing Cl [7–11]. The chemical bond energy of Si-Si bond is 226 kJ/mol and that of Si-Cl bond is 400 kJ/mol. The bond energy of the Si-Cl bond is almost twice that of Si-Si. Therefore, Cl and Si are easier to combine than Si and Si. Moreover, the existence of the Cl element will inhibit the combination of Si and Si, hinder the formation of Si clusters or Si droplets, and improve the epitaxial quality. The research of Crippa *et al.* [12] shows that when other gas parameters remain unchanged, the addition of HCl gas helps to inhibit the formation of silicon droplets and increase the epitaxial growth rate. Figure 1 shows that the new processes with chlorine addition resulted in a very high growth rate (>100 μm/h) with silicon precipitation [13]. TCS is the typical precursor used in silicon epitaxy for its safety and stability in industrial processes and also has been used in 4H-SiC epitaxy for commercialization, such as Figure 2 (b) LPE1o6 and (c) Aixtron G5 WW.

At present, another important topic of SiC thick epitaxy is defect control. Figure 3 schematically depicts some of the SiC epilayer defects. These defects are generally divided into two categories. One is morphological defects, such as carrots, downfall, step bunching, and triangular defects, are the main concerns at present. Morphological defects generally belong to "killer" defects. These defects are formatted by polishing damage or non-optimized growth process. The downfall is generated by the falling of a SiC particle from the reactor upper wall. Okada [14] and Wahab [15] found that carrot defects will increase the reverse leakage current of the device. For step bunching, Kato's [16] research shows that it will mainly lead to the increase of reverse leakage current of SiC diode. In thick epitaxy, triangular defects are most likely to appear, and the size is especially large. Matsunami

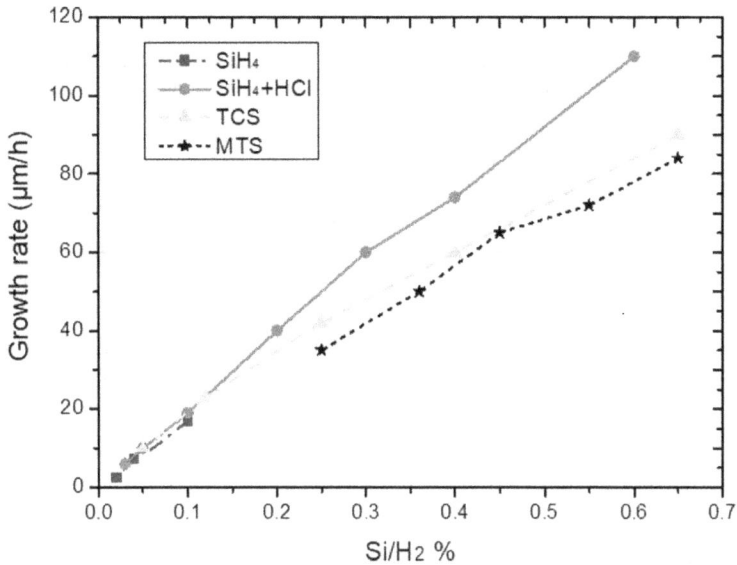

FIGURE 1
Growth rate vs. Si/H_2 ratio.

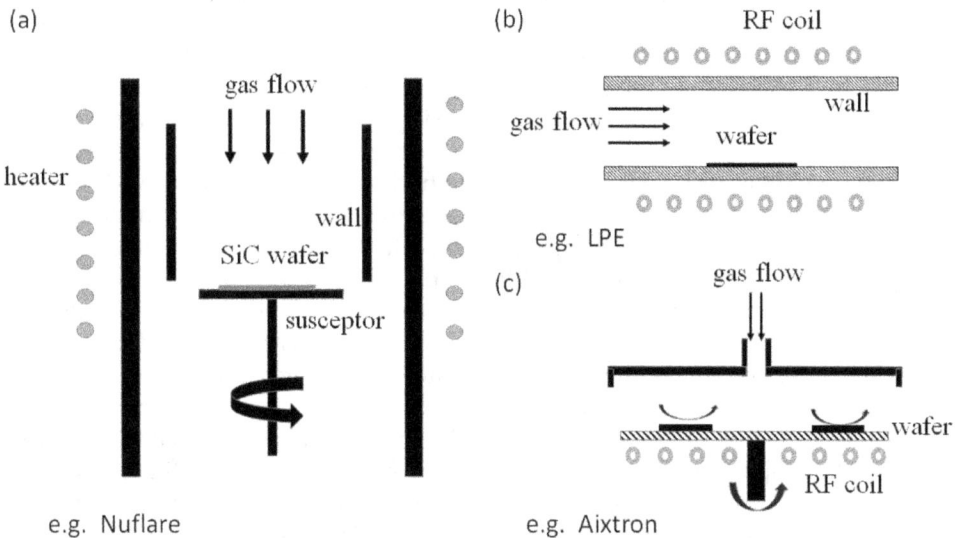

FIGURE 2
Schematic illustrations of several typical reactors of SiC epitaxy. (a) Vertical reactor, (b) horizontal reactor, (c) warm-wall reactor.

[17] found that the formation of triangular defects may be related to 2D nucleation caused by excessive step width on the substrate surface and defects in the substrate. According to Dong [18], low surface temperature, high C/Si, and low Cl/Si will promote the formation of triangular defects.

The other is crystal structural defects, such as basal dislocation, screw dislocation, edge dislocation, and stacking fault. A large number of studies have been carried out on the

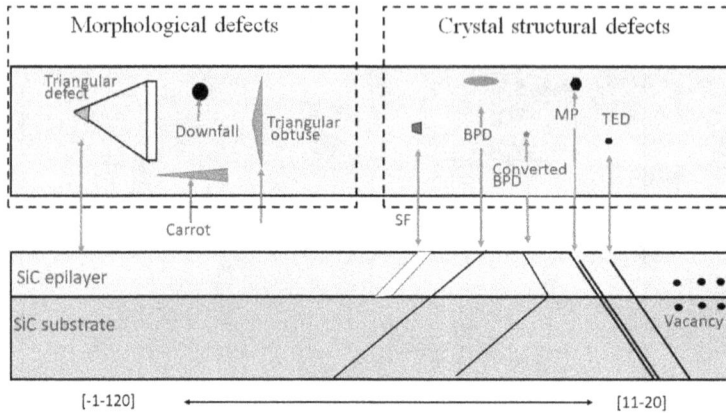

FIGURE 3
Some common SiC epilayer defects.

transformation of basal plane dislocation (BPD). The BPD in the structural defects is special, which has a great impact on the reliability of bipolar devices, mainly in the forward voltage drift and the failure of gate oxide. Most of the BPD that leads to the performance degradation of bipolar power devices comes from the substrate and penetrates into the epitaxial layer. At the same time, because the main structure of high-voltage devices for the power system in the future is bipolar devices, it is very necessary to control the basal dislocation density in epitaxy. Sumakeris [19] and Zhang [20, 21] reduced the BPD density to < 10 cm^{-2} by KOH etching the substrate before epitaxial growth, and characterized the evolution process of etch pits on the surface of silicon carbide epitaxial layer by AFM. It was found that when the lateral growth rate exceeded the growth rate of the step flow control method, the BPD turned. However, this method is not suitable for industrial production. In 2009, Stahlbush [22] developed the interrupted epitaxial growth method (cutting off the supply of silicon source during growth), which increased the conversion rate of BPD to 98% and reduced the density of BPD to about 7 cm^{-2}.

For thick film epitaxial materials, because the conductivity modulation affects bipolar devices, it needs a long minority carrier life to achieve low forward voltage drop, so it is also very necessary to improve the minority carrier life. The results show that the main defect affecting the minority carrier lifetime of 4H-SiC epitaxial material is $Z_{1/2}$ (Ec-0.6eV) [23–26]. Moreover, the formation of the $Z_{1/2}$ energy level is mainly related to the complex of carbon vacancy (V_C) [27, 28]. Therefore, in order to improve the minority carrier lifetime, it is necessary to reduce the $Z_{1/2}$ deep level defect concentration. At present, the mainstream methods are thermal oxidation treatment [29, 30] and C ion implantation and annealing treatment [31, 32]. Some people also reduce the concentration of $Z_{1/2}$ energy level in epitaxial materials by optimizing epitaxial growth process conditions (such as temperature, C/Si, etc.). Litton [33, 34] found that the $Z_{1/2}$ defect concentration will decrease with the increase of C/Si by changing C/Si. Danno [35] found that changing the growth rate will not affect the concentration of $Z_{1/2}$, and increasing the temperature will increase its concentration.

2.2 4H-SiC High-Voltage Devices

In the aspect of SiC high-voltage power devices, the breakdown voltage of 4H-SiC Metal Oxide Semiconductor Field Effort Transistor (MOSFET) and Insulated Gate Bipolar

Transintor (IGBT) has exceeded 20kV. Cree developed 10 kV 4H-SiC MOSFET with a 6×10^{14} cm^{-3} doped, 100 μm thick epilayer and 10 kV n-IGBT with a 3×10^{14} cm^{-3} doped, 120 μm thick epilayer in 2008, individually. At 20 V on the gate, The MOSFET has q linear I_D-V_D characteristic corresponding to a specific on-resistance of 127 mΩcm². The n-IGBT has a 3 V knee before turning on with a differential on-resistance of 14.3 mΩcm², both devices can maintain 10 kV breaking capability up to 200°C [36]. Cree developed 15 kV SiC p-IGBT with 2×10^{14} cm^{-3} doped, 140 μm thick p-type epilayer as the drift layer in 2012, The room temperature differential specific on-resistance of is 24 mΩ-cm² with a gate bias of -20 V ; 12.5 kV SiC n-IGBT also made and showed a room-temperature differential specific on-resistance of 5.3 mΩ-cm² with a gate bias of 20V [37]. In 2013, Cree, DOE and North Carolina State University jointly developed 4H-SiC n-IGBT with a breakdown voltage of 20.7 kV, with $2\times10^{14}cm^{-3}$ doped, 160 μm thick n-type epilayer. The leakage current of the device was 140 μA. The device showed a V_F of 6.4 V at an I_C of 20 A, and a differential Ron, sp of 28 mΩ-cm² [38]. In 2015, wolfspeed [39] reported 27 kV planar gate n-channel IGBT and 15 kV p-channel IGBT. In 2015, Jeffrey *et al.* [40] reported a new generation of 10 kV SiC MOSFET, whose on-resistance at room temperature was improved from 160 mΩ to 100 mΩ.

In terms of diodes, the breakdown voltage of SiC Schottky diodes has exceeded 10 kV also, and that of SiC PiN diodes has exceeded 20 kV. SiC bipolar devices are superior due to the reduced on-resistance owing to the conductivity modulation effect.

In 2003, Zhao *et al.* developed the first 4H-SiC Schottky barrier diode (SBD) blocking over 10 kV based on 115 μm n-type epilayer doped $5.6\times10^{14}cm^{-3}$, with multi-junction termination extension (MJTE) structure. The current density of 48 A/cm², was achieved with a forward voltage drop of 6V, and the specific on-resistance was 97.5 mΩ-cm² [41]. A 10 kV junction barrier Schottky diode (JBS) was developed by Cree in 2008, with the $6\times10^{14}cm^{-3}$

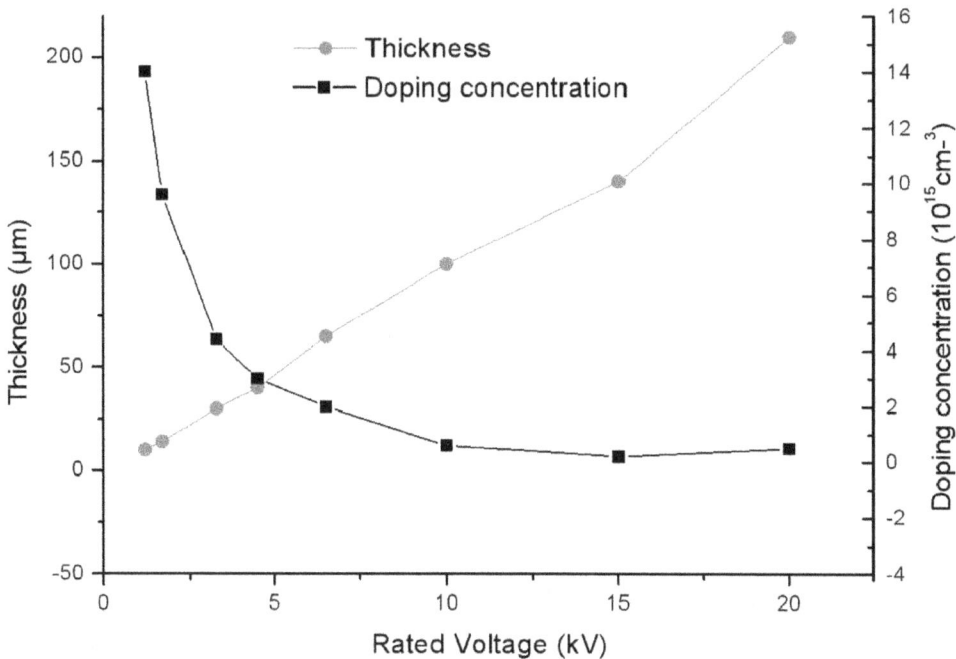

FIGURE 4
The thickness of epilayer and doping as a function of rated voltage.

MOSFET   n-IGBT

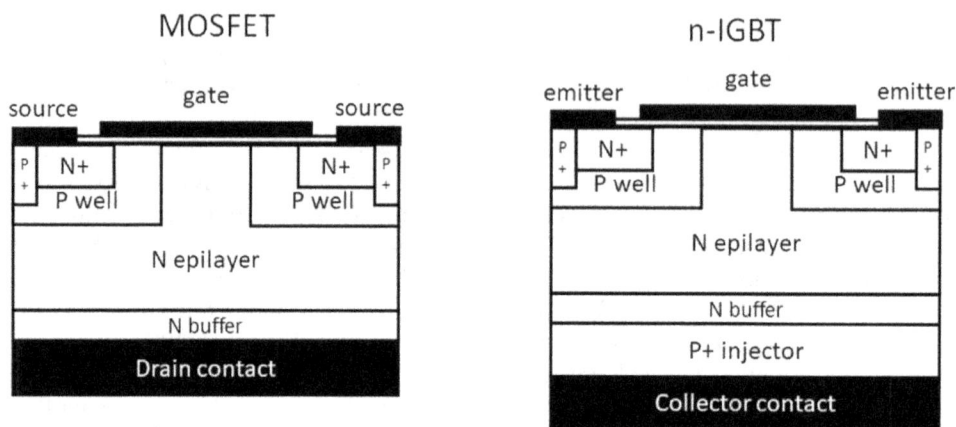

FIGURE 5
Structure of high-voltage SiC MOSFET and *n*-IGBT.

doped, 120 μm thick epilayer. The diodes show a positive temperature coefficient of resistance and a stable Schottky barrier height of up to 200°C [42].

In 2001, Kansai Electric Power and Cree jointly developed 12 kV 4H-SiC PiN diodes and 19kv 4H-SiC PiN diodes. The doping concentrations of epitaxial layers of the two devices are 2×10^{14} cm^{-3}, 8×10^{13} cm^{-3}, respectively, the thickness of epilayer is 120 μm and 200 μm, respectively. The reverse recovery time of the two devices is 34 ns and 43 ns, respectively, which is less than 1/30 of that of the commercial 6 kV Si diode [43]. In 2012, Kyoto University developed a 15 kV 4H-SiC PiN diode with 7×10^{14} cm^{-3} doped, 147 μm thick epilayer [44]. In the same year, Kyoto University released a 4H-SiC PiN diode with a breakdown voltage up to 21.7 kV, and the doping concentration of its epilayer was 2.3×10^{14} cm^{-3}. The thickness is 186 μm. The differential specific on resistance is 63.4 mΩ•cm^2 at 50 A/cm^2 [45]. In 2013, Kyoto University released 4H-SiC PiN diode with the highest voltage level at present. Its breakdown voltage reaches 26.9 kV, by using a 270 μm-thick, 1×10^{14} cm^{-3} doped epitaxial layer. The differential on-resistance of the device is 19.2 mΩ•cm^2 at 100 A/cm^2 [46].

In 2016, Hiroki Niwa *et al.* [47] reported an ultrahigh-voltage hybrid structure 10 kV-class SiC MPS diode with snapback-free operation. This study shows that MPS also has excellent forward conduction and reverse blocking characteristics, which has great potential in the field of ultrahigh voltage. In 2016, Huang Runhua *et al.* [48] of the 55th Institute of Electrical Science and Technology in China developed a 17 kV PiN diode. In 2018, Nakayama *et al.* [49] developed a 27.5 kV SiC PiN diode by using spacing modulation JTE and carrier injection technology, which is the highest breakdown voltage SiC power device reported so far.

3 Challenges of Ultrathick SiC Epitaxial Materials

3.1 Defect Engineer

In the grid system, the high-voltage devices' current is considerably higher, the device area is bigger than low voltage. High-density defects in ultrathick SiC epitaxial materials

TABLE 1

Effects of defects on SiC device performance and reliability

Defect	SBD	MOSFET, JFET	PiN, BJT, Thyristor, IGBT
TSD	No	Degradation of oxide reliability	Local reduction of carrier lifetime
TED	No	No	Local reduction of carrier lifetime
BPD	No	Degradation of body diode, oxide reliability	Bipolar degradation (Increase of on-resistance and leakage current)
In-grown SF	VB reduction (20–50%)	VB reduction (20–50%)	VB reduction (20–50%)
Carrot, triangular Defects	VB reduction (30–70%)	VB reduction (30–70%)	VB reduction (30–70%)
Downfall	VB reduction (50–90%)	VB reduction (50–90%)	VB reduction (50–90%)
Point defects (Z1/2)	No	No	Reduction of carrier lifetime

are the main problem restricting the development of high-voltage and high-power SiC devices. Once they appear, the device performance will be affected, which greatly reduces the performance of the device, such as Table 1. At present, the total density of typical droplets and triangular defects in 6 inch 10 μm thick epitaxial layers can reach the level of 0.1~1.0 cm^{-2}. Thomas [50] realized the defect level of 0.75 cm^{-2} on 6 inch 20–40 μm thick epitaxial wafer. At 2018, Niu [51] studied 70 μm thick homoepitaxial layers on 6 inch substrate, the density of triangular defects reduced from 1.01 cm^{-2} to 0.14 cm^{-2} by optimized process. However, the defect levels in 6 inch and thick (>70 μm) epitaxial materials have not been reported.

For the growth of thick-film SiC epitaxial, because the thickness required is thicker, the deposition on the inner wall of the reactor during the growth process is also thicker, in this way, the probability of the particles falling from the inner wall of the reactor into the surface during the growth process is also increased, and the particles falling from the reactor are also called nucleation points, which induce the triangular defects, etc., the defect size in SiC epitaxy is proportional to the thickness of the epitaxial layer, $L=d/\tan\theta$ (L is the characteristic size of the defect, d is the thickness of the epitaxial layer, and θ is the angle of the crystal plane). With the increase of thickness, the size of the defect is larger, and the characteristic size of the surface triangular defect can reach 2860 μm for the epitaxial layer with 200 μm thickness. So a large defect is very disadvantageous to the development of high-voltage devices, so how to realize thick film SiC epitaxial material with mirror surface and low defect is the key to manufacture high-voltage SiC power devices.

Improvement of the epitaxial process of SiC is strongly required to achieve a high yield and a high current in fabrication of large-area devices.

3.2 Carrier Lifetime

SiC bipolar power devices (such as IGBT) are the main device forms used in the field of an ultrahigh-voltage power grid due to the conductivity modulation effect. Different from unipolar devices, bipolar devices have high requirements for minority carrier lifetime, which directly affects the on-state and switching characteristics of bipolar devices. In order to obtain the ideal forward characteristic, the minority carrier life must be at least greater than 5 μs [52]. However, a long minority carrier lifetime may not necessarily obtain good conduction characteristics, for 200 μm, The minority carrier lifetime of epitaxial layer is greater than 5 μs, it basically did not contribute to the improvement of forward pressure drop [53]. At present, the primary minority carrier lifetime of silicon carbide epitaxial materials is about 1~2 μs [54], so it is very necessary to study the technology of enhancing minority carrier lifetime.

There are many intrinsic defects in SiC materials. These intrinsic defects are the main sources of deep level defects, which have a serious impact on the minority carrier lifetime of SiC materials. Among them, $Z_{1/2}$ (0.63 eV below the conduction band) [55] and $EH_{6/7}$ (1.55 eV below the conduction band) [56] are one of the main defects in 4H-SiC materials. Figure 6 illustrates the energy levels of a carbon vacancy defect in SiC. As a composite center, they are the main restrictive factors affecting the minority carrier lifetime in n-type 4H-SiC [58, 59].

A large number of studies show that the formation of $Z_{1/2}$ energy level is related to carbon vacancy (V_C) [61-63], which is clearly the "killer" of minority carrier lifetime in n-type materials. However, $EH_{6/7}$ cannot be considered as the "killer" defect of minority carrier life of n-type materials because $EH_{6/7}$ does not participate in the composite process [57]. Therefore, in order to improve the minority carrier life, it is necessary to repair the V_C and reduce the concentration in $Z_{1/2}$ below 3e12 cm^{-3}. At present, there are two methods. One method is to optimize the epitaxial process, because $Z_{1/2}$ is very dependent on C/Si ratio [64, 65] and growth temperature [64, 66]. Research shows that reducing the growth temperature and increasing C/Si during epitaxial growth can reduce the concentration of $Z_{1/2}$ center in epitaxial materials, but the reduction range is limited. Another method is post-treatment technology, mainly including C injection [67, 68], high-temperature oxidation [69, 70], high-temperature annealing [71, 72], and electron irradiation [73]. The main principle of C ion implantation method is that the implanted excess C ions will diffuse into the material during high-temperature annealing under the protection of Ar gas, so as to fill the V_C in the material. The main mechanism of high-temperature thermal oxidation is that during the high-temperature oxidation of SiC materials, C atoms will not be completely discharged in the form of CO and CO_2 gas, and C interstitials will be generated at the interface between SiO_2 oxide layer and SiC materials. At high temperature, these C interstitials will move towards the inside of SiC material, so as to fill the C vacancy

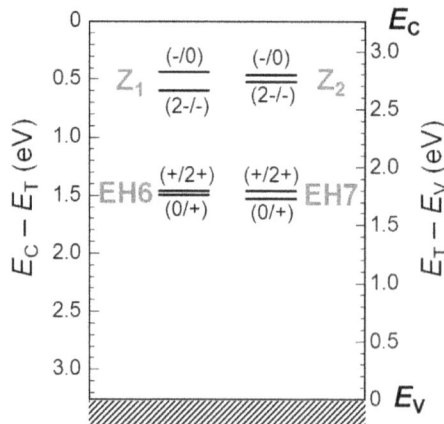

FIGURE 6
Energy levels of a carbon vacancy (V_c) defect in SiC. From [60] Figure 2, reproduced with permission of IOP.]

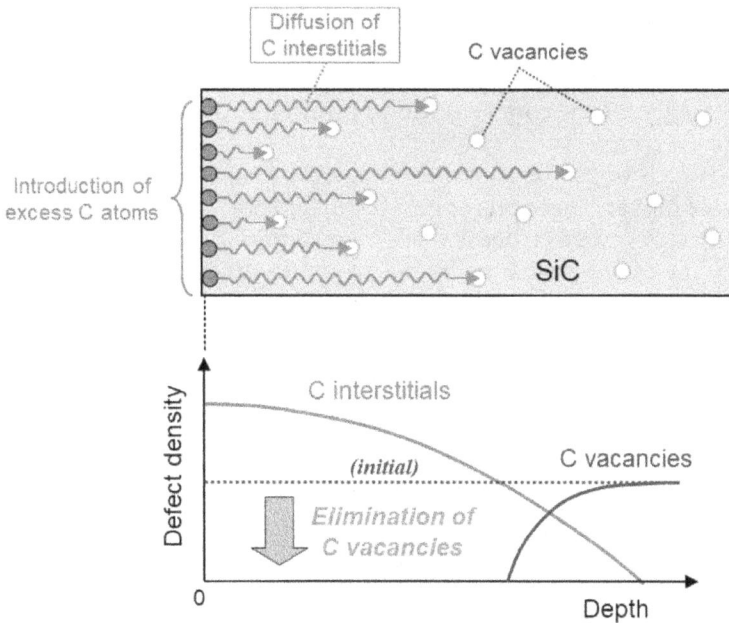

FIGURE 7
Schematic illustration of the basic idea for the elimination of V_c defects in SiC. From [60] Figure 4, reproduced with permission of IOP.

between the material and $Z_{1/2}$ deep level defect, as explained in Figure 7. The carrier lifetime was drastically enhanced from 2.1 µs for the as-grown sample to 48 µs after the V_c elimination.

In addition to the intrinsic defects analyzed above, some structural defects and dislocations are also called recombination centers, which affect the minority carrier lifetime [74, 75]. In thick film epitaxy, deposits will be generated on the inner wall of the cavity while epitaxial growth, and will fall on the surface, which will induce many defects. Therefore, it is very meaningful to analyze the influence of such defects on minority carrier lifetime in thick film.

4 Conclusion

With the development of the global energy Internet, the utilization of energy is developing in the direction of more intelligent and efficient, the power transmission and transformation technology is developing in the direction of longer distance, larger scale, more environmental protection and efficient, the intelligent power distribution is developing in the direction of more flexible interaction and higher self-healing ability, and the power network security defense is developing in the direction of more active and intelligent. In order to adapt to the change of power grid form in the future, the capacity of power electronic devices with power devices as the core will reach GW level, and the energy density per unit volume is expected to be hundreds of times higher than that at present. At the same time, it has lower cost, smaller heating capacity, faster response speed, higher artificial intelligence level, and better environmental compatibility. Therefore, the future power grid puts forward the overall demand of higher voltage, larger capacity, higher efficiency, and higher reliability for the core switching power electronic devices of power electronic equipment.

In the special needs of power grid, the voltage, current, and reliability of power devices are required to be higher. Therefore, epitaxial wafers with ultrathick thickness, low doping concentration, and low defect density are also required in the aspect of SiC epitaxial materials. Therefore, the research on high-quality ultrathick SiC epitaxial materials is the basis for the preparation of high-voltage SiC devices for the power grid, The realization of a single SiC device withstanding tens of thousands of volts will greatly reduce the number of series devices and auxiliary equipment, thus greatly reducing the loss, volume, and cost of power equipment, improving the reliability, flexibility, and applicability, which is of great significance to support the development of the grid.

With the continuous improvement of SiC epitaxial material quality, the fabrication process of SiC power electronic devices is constantly improved, and the high-voltage bearing capacity of a single device is constantly improved. Combined with its advantages in current conduction ability and reliability, it will play a great role in improving the efficiency of energy control and conversion, in flexible DC transmission, solid-state transformer, static reactive power compensator, and other fields of power electronic transformation devices.

References

1. Zhenya Liu, *Global energy Internet*. China Electric Power Press, Beijing, 2pp, 2015.
2. Nashida N, Hinata Y, Horio M, et al. "All SiC power module for photovoltaic power conditioner system", 2014 IEEE 26th International symposum on power semiconductor devices & IC's (ISPSD), 15–19 June, 2014, Waikoloa, USA: IEEE, 342–345 (2014).
3. Filsecker F, Alvarez R. Bernet S. "Evaluation of 6.5 kV SiC PiN diodes in a medium-voltage high power 3L-NPC converter". IEEE Transactions on Power Electroics, 29, 5148–5156 (2014).
4. Jingrong Yu, Yijia Cao, Yijun Wang, *et al.* "Influence of SiC Power Electronic Devices on the Power System". Micronanoelectronic Technology, 49, 503–509 (2012).
5. Kuang Sheng, Qing Guo. "Prospects of SiC Power Electronic Device Application in Power Grid". China Southern Power Grid, 10, 87–90 (2016).
6. Kuang Sheng, Qing Guo, Junming Zhang. "Development and Prospect of SiC Power Devices in Power Grid". Proceedings of the CSEE, 32, 1–7 (2012).
7. F. La Via, G. Izzo, M. Mauceri, G. Pistone, G. Condorelli, L. Perdicaro,G. Abbondanza, L. Calcagno, G. Foti, D. Crippa, "4H-SiC epitaxial layer grown by trichlorosilane (TCS)", J. Cryst. Growth, 311, 107–113 (2008).
8. F. La Via, M. Camarda, A. La Magna, "Mechanism of growth and defect properties of epitaxial SiC", Appl. Phys. Rev. 1, 031301 (2014).
9. G. Dhanaraj, M. Dudley, Yi Chen, B. Ragothamachar, B. Wu, H. Zhang, "Epitaxial growth and characterization of silicon carbide films", J. Cryst. Growth, 287, 344–348 (2006).
10. I. Chowdhury, M.V.S. Chandrasekhar, P.B. Klein, J.D. Caldwell, T. Sudarshan, "High growth rate 4H-SiC epitaxial growth using dichlorosilane in a hot-wall CVD reactor", J. Cryst. Growth, 316, 60–66 (2011).

11. H. Pedersen, S. Leone, A. Henry, F.C. Beyer, V. Darakchieva, E. Janzén, "Very high growth rate of 4H-SiC epilayers using the chlorinated precursor methyltrichlorosilane(MTS)", J. Cryst. Growth, 307, 334–340 (2007).

12. D. Crippa, G. L. Valente, A. Ruggiero, et al. "New achievements on CVD based methods for SiC epitaxial growth", Materials Science Forum, 483–485, 67–72 (2005).

13. F. La Via, G. Izzo, et al., "Thick epitaxial layers growth by chlorine addition", Materials Science Forum, 615–617, 55–60 (2009).

14. Okada T, Kimoto T, et al., "Correspondence between surface morphysical faults and crystallographic defects in 4H-SiC homoepitaxial film", Japanese Journal of Applied Physics, 41, 6320–6326 (2002).

15. Wahab Q, Ellison A,et al.,"Influence of epitaxial growth and substrate induced defects on the breakdown of high-voltage 4H-SiC Schootky diodes", Materials Science Forum, 338–342, 1175–1178 (2000).

16. Kato T, Kinoshita A, et al., "Morphology improvement of step-bunching on 4H-SiC wafers by polishing technique", Materials Science Forum, 645–648, 763–765 (2010).

17. Matsunami H, Kimoto T, "Step controlled epitaxial growth of SiC: High quality homoepitaxy", Materials Science and Engineering: R: Reports, 20, 125–166 (1997).

18. Dong L, Sun G S, et al., "Growth of 4H-SiC epilayers with low surface roughness and morphological defects density on 4° off-axis substrates", Applied Surface Science, 270, 301–306 (2013).

19. Sumakeris J. J., Bergman J. P., Das M. K., et al. "Techniques for minimizing the basal plane dislocation density in SiC epilayers to reduce Vf drift in SiC bipolar power devices". Materials Science Forum, 527–529, 141–146 (2006).

20. Zhang Z, Moulton E, Sudarshan T S, "Mechanism of eliminating basal plane dislocations in SiC thin films by epitaxy on an etched substrate" Applied Physics Letters, 89, 081910 (2006).

21. Zhang Z, Sudarshan T S, "Basal plane dislocation-free epitaxy of silicon carbide". Applied Physics Letters, 87, 151913 (2005).

22. Stahlbush R E, Vanmil B L, Myers-Ward R L, et al. "Basal plane dislocation reduction in 4H-SiC epitaxy by growth interruptions". Applied Physics Letters, 94, 041916 (2009).

23. Carlos W E, Glaser E R, et al., "The role of the carbon vacancy-carbon antisite defect in semi-insulating 4h silicon carbide". Bulletin of the American Physical Society, 48, 1322 (2003).

24. Klein P B, Shanabrook B V,et al., "Lifetime-limiting defects in n–n– 4H-SiC epilayers", Applied Physics Letter, 88, 052110 (2006).

25. Zhang J, Storasta L, et al., "Electrically active defects in n-type 4H-silicon carbide grown in a vertical hot-wall reactor". Journal of Applied Physics, 93, 4708 (2003).

26. Danno K, Nakamura D, et al., "Investigation of carrier lifetime in 4H-SiC4H-SiC epilayers and lifetime control by electron irradiation", Applied Physics Letter, 90, 202109 (2007).

27. Kawahara K, Suda J, et al., "Analytical model for reduction of deep levels in SiC by thermal oxidation", Journal of Applied Physics, 111, 053710 (2012).

28. Kawahara K, Thang T X, et al., "Investigation on origin of $Z_{1/2}$ center in SiC by deep level transient spectroscopy and electron paramagnetic resonance", Applied Physics Letter, 102, 112106 (2013).

29. Hiyoshi T, Kimoto T, "Reduction of deep levels and improvement of carrier lifetime in n-type 4H-SiC by thermal oxidation", Applied Physics Express, 2, 041101 (2009).

30. Hiyoshi T, Kimoto T, "Elimination of the major deep levels in n-type and p-type 4H-SiC by two-step thermal treatment", Applied Physics Express, 2, 091101 (2009).

31. Storasta L, Tsuchida H, "Reduction of traps and improvement of carrier lifetime in 4H-SiC epilayers by ion implantation", Applied Physics Letter, 90, 062116 (2007).

32. Storasta L, Tsuchida H, et al., "Enhanced annealing of the $Z_{1/2}$ defect in 4H–SiC epilayers", Journal of Applied Physics, 103, 013705 (2008).

33. Litton C. W., Johnstone D., Akarca-Biyikli S., et al, "Effect of C/Si ratio on deep levels in epitaxial 4H-SiC", Applied Physics Letter, 88, 121914 (2006).

34. Fujiwara H., Danno K., Kimoto T., et al, "Effects of C/Si ratio in fast epitaxial growth of 4H-SiC(0001) by vertical hot-wall chemical vapor deposition", Journal of Crystal Growth, 281, 370–376 (2005).

35. Danno K., H ori T., Kimoto T., "Impacts of growth parameters on deep levels in n-type 4H-SiC", Journal of Applied Physics, 101, 053709 (2007).

36. Mrinal K.Das, Robert Callanan, D. Craig Capell, et al. "State of the Art 10 kV NMOS Transistors", Proceedings of the 20th International Symposium on Power Semiconductor Devices & IC's, Orlando, FL, USA, 253–255 (2008).

37. Sei-Hyung Ryu, Allen Hefner, Subhashish Bhattacharya, et al., "Ultra High Voltage, High Performance 4H-SiC IGBTs". 24th International Symposium Power Semiconductor Devices and ICs (ISPSD), 257–260 (2012).

38. Ryu S., Capell C., Jonas C., et al. "Ultra High Voltage IGBTs in 4H-SiC".2013 IEEE Workshop on Wide Bandgap Power Devices and Applications (WiPDA), 36–39, (2013).

39. Van Brunt E, Cheng L, O'Loughlin M J, et al. "27 kV, 20 A 4H-SiC n-IGBTs". Materials Science Forum, 821–823, 847–850 (2015).

40. Casady J B, Pala V, Lichtenwalner D J, et al., "New generation 10 kV SiC power MOSFET and diodes for industrial applications", Proceedings of PCIM Europe 2015, Nuremberg, Germany: VDE, 1–8 (2015).

41. Jian H. Zhao, Petre Alexandrov, and X. Li. "Demonstration of the first 10-kV 4H-SiC Schottky barrier diodes". IEEE electron device letters, 24, 402–404 (2003).

42. Breet A. Hull, Joseph J. Sumakeris, Michael J. O'Loughlin, et al. "Performance and Stability of Large-Area 4H-SiC 10 kV junction barrier schottky rectifiers". IEEE Transactions on electron devices, 55, 1864–1870 (2008).

43. Sugawara Y., Takayama D., Asano K., et al. "12–19kV 4H-SiC pin Diodes with Low Power Loss". Proceedings of 2001 International Symposium on Power Semiconductor Devices & ICs, 27–30 (2001).

44. Hiroki Niwa, Gan Feng, Jun Suda, et al. "Breakdown Characteristics of 15-kV-Class 4H-SiC PiN Diodes with Various Junction Termination Structures". IEEE Transactions on electron devices, 59, 2748–2753 (2012).

45. Hiroki Niwa, Gan Feng, Jun Suda, et al. "Breakdown Characteristics of 12–20 kV-class 4H-SiC PiN Diodes with Improved Junction Termination Structures". Proceedings of the 2012 24th International Symposium on Power Semiconductor Devices and ICs, 381–384 (2012).

46. Kaji N., Niwa H., Suda J., et al. "Ultrahigh-Voltage(>20kV) SiC PiN Diodes with a Space-Modulated JTE and Lifetime Enhancement Process via Thermal Oxidation". International Conference on Silicon Carbide and Related Materials, 86 (2013).

47. Niwa H., Suda J., Kimoto T., "Ultrahigh-Voltage SiC MPS diodes with hybrid unipolar/bipolar operation", IEEE Transactions on Electron Devices, 99, 1-8 (2016).

48. Runhua Huang, et.al, "Development of 17kV 4H-SiC PiN diode", Journal of Semiconductors, 36, 084001 (2016).

49. K. Nakayama, et al., "27.5 kV 4H-SiC PiN diode with space-modulated JTE and carrier injection control", Proceedings of the 30th ISPSD, 395–398 (2018).

50. Thomas B., Zhang J. et al. "Homoepitaxial chemical vapor deposition of up to 150 um thick 4H-SiC epilayers in a 10x100mm batch reactor", Materials Science Forum, 858, 129–132 (2016).

51. Y.X. Niu, X.Y. Tang. et al. "Lowdefect thick homoepitaial layers growth on 4H-SiC wafers for 6500V JBS devices", Materials Science Forum, 954: 114–120 (2019).

52. Kimoto T., Yamada K., Niwa H., Suda J., "Promise and challenges of high-voltage SiC bipolar power devices", Energies, 9, 908 (2016).

53. Kimoto T., Yonezawa Y., "Current status and perspectives of ultrahigh-voltage SiC power devices", Materials Science in Semiconductor Processing, 78: 43–56 (2018).

54. Kimoto T., Tsunenobu. "Material science and device physics in SiC technology for high-voltage power devices". Japanese Journal of Applied Physics, 54, 040103 (2015).

55. Dalibor T., Pensl G., Matsunami H., et al. "Deep Defect Centers in Silicon Carbide Monitored with Deep Level Transient Spectroscopy". Physica Status Solidi, 162, 199–225 (1997).

56. Hemmingsson C., Son N.T., Kordina O., et al. "Deep level defects in electron-irradiated 4H SiC epitaxial layers", Journal of Applied Physics, 81, 6155 (1997).
57. Kimoto T. and Cooper J.A., "Fundamentals of Silicon Carbide Technology: Growth, Characterization, Devices and Applications". New York. Wiley-IEEE Press, 103 pp, 2014.
58. Klein P.B., Shanabrook B.V., et al., "Lifetime-limiting defects in n–n– 4H-SiC epilayers", Applied Physics Letter, 88, 052110 (2006).
59. Danno K., Nakamura D., et al., "Investigation of carrier lifetime in 4H-SiC epilayers and lifetime control by electron irradiation", Applied Physics Letter, 90, 202109 (2007).
60. Kimoto T. et al., "carrier lifetime and breakdown phenomena in SiC power device material", Journal of Physics D: Applied Physics, 51, 363001 (2018).
61. Hornos T., Gali A., et al., "Large-Scale Electronic Structure Calculations of Vacancies in 4H-SiC Using the Heyd-Scuseria-Ernzerhof Screened Hybrid Density Functional", Materials Science Forum, 2011, 679–680, 261–264 (2011).
62. Kawahara K., Suda J., et al., "Analytical model for reduction of deep levels in SiC by thermal oxidation", Journal of Applied Physics, 111, 053710 (2012).
63. Kawahara K., Thang T.X., et al., "Investigation on origin of $Z_{1/2}$ center in SiC by deep level transient spectroscopy and electron paramagnetic resonance", Applied Physics Letter, 102, 112106 (2013).
64. Lilja L., Booker I., et al., "The influence of growth conditions on carrier lifetime in 4H–SiC epilayers", 381, 43–50 (2013).
65. Litton C.W., Johnstone D., et al., "Effect of C/Si ratio on deep levels in epitaxial 4H-SiC", Applied Physics Letter, 88, 121914 (2006).
66. Danno K., Hori T., et al., "Impacts of growth parameters on deep levels in n-type 4H-SiC", Journal of Applied Physics, 101, 053709 (2007).
67. Storasta L., Tsuchida H., "Reduction of traps and improvement of carrier lifetime in 4H-SiC epilayers by ion implantation", Applied Physics Letter, 90, 062116 (2007).
68. Storasta L., Tsuchida H., et al., "Enhanced annealing of the $Z_{1/2}$ defect in 4H–SiC epilayers", Journal of Applied Physics, 103, 013705 (2008).
69. Hiyoshi T., Kimoto T., "Reduction of deep leves and improvement of Carrier lifetime in n-type 4H-SiC by thermal oxidation", Applied Physics Express, 2, 041101 (2009).
70. Hiyoshi T., Kimoto T., "Elimination of the major deep levels in n-type and p-type 4H-SiC by two-step thermal treatment", Applied Physics Express, 2, 091101 (2009).
71. Ayedh H.M, Nipoti R., Hallén, A., et al. "Elimination of carbon vacancies in 4H-SiC employing thermodynamic equilibrium conditions at moderate temperatures", Applied Physics Letters, 107, 252102 (2015).
72. Negoro Y., Katsumoto K., Kimoto T., et al. "Electronic behaviors of high-dose phosphorus-ion implanted 4H-SiC (0001)", Journal of Applied Physics, 96, 224–228 (2004).
73. Kimoto T., Danno K., Suda J., "Lifetime-killing defects in 4H-SiC epilayers and lifetime control by low-energy electron irradiation", Phys. Status Solidi (b), 245, 1327–1336 (2008).
74. Myers Ward R.L., Lew K.K., Vanmil B.L., et al. "Impact of 4H-SiC Substrate Defectivity on Epilayer Injected Carrier Lifetimes", Materials Science Forum, 600–603, 481–484 (2009).
75. Taishi T., Hoshikawa T., Yamatani M., et al. "Influence of crystalline defects in Czochralski-grown Si multicrystal on minority carrier lifetime", Journal of Crystal Growth, 306, 452–457 (2007).

7

Epitaxial Growth and Structural Studies of Cubic SiC Thin Films Grown on Si-face and C-face 4H-SiC Substrates

Hao-Hsiung Lin
Graduate Institute of Electronics Engineering, Graduate Institute of Photonics & Optoelectronics, and Department of Electrical Engineering, National Taiwan University, Taipei, Taiwan, ROC

Bin Xin
Department of Materials Science and Engineering, King Abdullah University of Science and Technology, Jeddah, Saudi Arabia

Zhe Chuan Feng and Ian T. Ferguson
Department of Electrical and Computer Engineering, Southern Polytechnic College of Engineering and Engineering Technology, Kennesaw State University, Kennesaw, Marietta, GA, USA

1 Introduction

Silicon carbide has many attractive mechanical and electrical properties superior over the traditional semiconductors, such as high thermal conductivity, high critical electric break-down field, excellent temperature stability, and good chemical inertness. These properties make it a promising material for advanced power semiconductor devices in harsh environment where the traditional semiconductors are not durable [1, 2]. Silicon carbide is an *IV-IV* semiconductor in which the basic Si-to-C tetrahedral structure forms close-packed planes stacking in different sequences, called polytypes. So far, more than 200 SiC polytypes have been identified [3]. Their different electronic energy band structures and invariant chemical composition allow a novel type band-gap engineering for device applications. Compared with traditional heterojunctions, consisting of different semiconductors, such as Si/SiC or AlGaN/GaN, the SiC polytype heterojunction avoids the problems of lattice mismatch and atomic interdiffusion in the heterojunction preparation. Electrical and optical properties of 3C-SiC/6H-SiC and 3C-SiC/4H-SiC polytype heterojunctions have been theoretically and experimentally studied. For the 3C-SiC/4H SiC heterojunction, the conduction band offset is as large as 1 eV, two-dimensional electron gas (2DEG) induced by the spontaneous field in C-face 4H SiC has been observed [4–8], indicating the great potential applications to novel high electron mobility devices [9–11].

Because SiC heterojunction consists of different polytypes, its preparation is no longer governed by the precision control of chemical compositions, but rather by optimizing growth conditions to achieve the required polytypes of SiC. In this circumstance, defects

DOI: 10.1201/9780429198540-9

become a bottleneck for the growth of SiC heterojunctions. In this chapter, we focus on the vapor phase epitaxy and structural properties of 3C-SiC grown on 4H-SiC. The structure of 3C-SiC is zincblende lattice, which is with the highest symmetry among the SiC polytypes. Thanks to this feature, 3C-SiC has the smallest energy gap, 2.3 eV, among the polytypes and is free from spontaneous polarization effect, which makes it the best channel material for electronic devices with polytype heterojunction [12–14]. In addition, the high symmetric also results in high mobility, high saturation velocity, and high intrinsic carrier density for 3C-SiC, which is beneficial to the electrical properties for applications to power devices [15, 16].

Unfortunately, 3C-SiC technology encounters a big difficulty, that is the absence of an appropriate bulk growth technique and thus it has no 3C-SiC seed for epitaxy [17]. For this reason, 3C-SiC technology is still far behind the more mature 4H-SiC technology and the 3C-SiC epitaxy usually uses 4H-SiC substrates. In this respect, most studies on 3C-SiC epitaxial layers were grown on the Si-face hexagonal substrates. Because of the difficult migration on the C-terminated surface, 3C-SiC grown on the Si-face shows higher crystalline quality and better surface morphology than those grown on the C-face [18, 19]. However, the spontaneous polarization field in C-face hexagonal SiC is in outward-pointing normal direction [12, 20], and the direction is reverse in Si-face hexagonal SiC. When 3C-SiC is on top of C-face hexagonal SiC, the discontinuity of the spontaneous polarization field at the interface may induce a sheet of electron gas [13, 21, 22]. For the 3C-SiC on Si-face hexagonal SiC structure, the spontaneous polarization direction is inversed, a two-dimensional hole gas at the interface has been predicted [8]. Because the polarization field affects the band structure of the heterojunction, the epitaxial technologies on both C-face and Si-face SiC are of great importance in polytype band-gap engineering.

Because the Si- and C-faces of hexagonal SiC have different surface free energies [23, 24], 3C-SiC layers formed on these faces display different surface morphology and epitaxial defects. This chapter presents the growth and structural properties of 3C-SiC grown on Si-face and C-face 4H-SiC substrates [25–27]. The details of the chemical vapor deposition are described in section 2 [25]. Defect structures and the formation mechanisms for 3C-SiC grown on Si-face 4H SiC are presented and discussed in section 3. For the epitaxial growth, the double position boundary (DPB) defect strongly affects the surface morphology of the hetero-epilayer. 3C-SiC has two types of stacking sequence with a 60° rotation difference, namely 3C-I type with an A–B–C stacking order and 3C-II type with an A–C–B stacking order, when grown on the basal plane of a hexagonal polytype [25, 26, 28]. The mismatched boundary between domains with the two different stacking sequences results in DPB defects. This implies that the formation of the DPB can be traced back to the nucleation stage of hetero-epitaxy. Although DPB-free 3C-SiC epitaxial layers grown on (0 0 0 1) face 4H-SiC has been reported very recently [29], the required growth conditions are still very harsh. For the 3C-SiC grown on C-face 4H SiC, there are some researches addressing the deposition on the C-face hexagonal [30]. However, the growth mechanism remains poorly understood because of a lack of systematic studies. In section 4, we present and discuss in details the super-V-shaped structure (SVSS) often observed on the surface of 3C-SiC films deposited on the C-face of 4H-SiC substrates by chemical vapor deposition (CVD) [27].

Our structural analyses are closely related to the epitaxial growth. In sections 4.3 and 4.4, two growth models for 3C-SiC grown on 4H-SiC are proposed and discussed, with details given in these two sections. Finally, section 5 is a conclusion.

2 Epitaxial Growth

All the 3C-SiC epi-films were grown on 4H-SiC substrates in an Epigress VP508 SiC hot-wall CVD reactor. The reactant sources were silane (SiH_4) and propane (C_3H_8) gases diluted with H_2. For the growth of Si-face SiC, the substrates were 2-inch Si-face 4H-SiC *n*-type wafers, produced by TankeBlue Company. The wafers were on-axis with less than 0.3° off the basal plane typically. Before growth, standard cleaning processes, including degreasing in acetone, cleaning in alcohol and rinsing in de-ionized water, were performed sequentially. The SiC substrate was first etched from room temperature to 1600°C in SiH_4 and H_2 flows, because a Si-rich etching environment is beneficial for growth on the Si face of the substrate [31]. The flow rates of H_2 and SiH_4 were 4 sccm and 80 slm, respectively. Subsequently, the SiH_4 flow was stopped for 3 min to remove silicon atoms in the chamber, and the precursor C_3H_8 was injected into the reactor for 30 s at a flow rate of 4.2 sccm. The usage of C_3H_8 precursor is because it reduces the density of DPB defects [32]. Then, homoepitaxial growth was conducted for 3 min at 1600°C to elongate the step terraces for subsequent 3C-SiC nucleation. Finally, the 3C-SiC epitaxial growth was carried out for 3 h with the temperature ramped down from 1600 to 1550°C. During the growth, including both the homoepitaxial and heteroepitaxial growth, the flow rates of SiH_4 and C_3H_8 were 21 and 7 sccm, respectively, and the flow rate of the carrier gas, H_2, was fixed at 80 slm. The chamber pressure was always kept at 100 mbar. The *in situ* etching step, homoepitaxy step, and a nucleation step before the hetero-epitaxy have previously studied and optimized [25].

For the growth of 3C-SiC on the carbon-terminated 4H-SiC. The substrate was a 3-inch *n*-type on-axis commercial wafer purchased from TankeBlue. Heteroepitaxial growth was conducted by keeping the growth temperature at 1550°C for 1 h. The source gases were SiH_4 and C_3H_8 with flow rates of 21 and 7 sccm, respectively. The carrier gas was H_2 kept at 80 slm and the pressure in the chamber was maintained at 100 mbar.

3 Twinning and Double Position Boundary Defects in 3C-SiC Grown on Si-face 4H-SiC

In the epitaxy of 3C-SiC on 4H-SiC substrates, the substrate plays the role of seed for the nucleation of the 3C polytype. The hexagonal tetrahedron concentration of 4H-SiC is 50%, i.e. stacking sequence of 4H-SiC contains 50% cubic sequences in which the two cubic stacking sequences, ABC and ACB, coexist in equal probability [6]. Basically, 3C-SiC could nucleate on substrate surface with equal probability at the substrate interface [25, 33], except the extreme conditions that the surface is atomic flat or tilted with the step bunching edge in coherent with the period of 4H-SiC. In these conditions, one of the sequences may be preferential. The two cubic sequences are a mirror reflection of each other and form a twinning pair when they are stacked along the normal of close packed plane. The transition of the two 3C twining pairs is just by rotating the lattice by 60° with respect to the growth direction. The interface is coherent, without bond distortion or breaking. In fact, the interface, i.e. the twin plane, is a single 2H-SiC layer. Note that 2H-SiC accounts for 50% in the bottom 4H-SiC substrate. However, when the nucleation develops to a domain on the growth plane, the boundary between two domains with two different 3C-SiC sequences is incoherent, which usually gives rise to DPB defects. The atomic bonds of DPBs are highly distorted, which is thought to result in the poor electronic properties of 3C-SiC films. Normally, the DPB defect is shallow grooves [34]. However, a more complex V-shaped structure is often found in deposited films and is attracting considerable attention [34–36]. This defect has a twinning complex structure resulting from the stress release along the {111} plane. In this section, we first discuss the surface morphology of 3C-SiC grown on 4H-SiC substrate. Then, we analyze a V-shaped twinning structure in 3C/4H-SiC. The defect region was characterized using cross-sectional transmission electron microscopy (TEM) and was found to have a polycrystalline complex structure. On the basis of the TEM analysis, we propose a kinetic mechanism of adatom migration to describe the linear prolongation of the V-shaped boundary between the regular and twinned domains, and the nonlinear DPB between two twinned domains.

3.1 Surface Properties

The electron backscatter diffraction (EBSD) mapping conducted on the surface of a 3C-SiC epilayer grown on Si-face 4H-SiC substrate is shown in Figure 1(a). The SEM image with an orange dashed ellipse, marking the scan region of EBSD measurement, is shown in left up inset. Corresponding to the SEM image, the region surrounded by the hill-like structure is in light-gray color with the pattern of pseudo-Kikuchi bands noted at right up corner inset. The region outside the surrounding hill-like structure is in deep gray color with its Kikuchi pattern noted at left down corner inset. From the resolved Kikuchi band, the two regions show a 60° angle difference, indicating they represent two different twinning regions. The black color of the boundary results from the unresolvable Kikuchi band, suggests that the hill-like boundary is with very complicated structures. From the SEM image shown in the left top inset of Figure 1(a), the twinning domain size of 3C-SiC is close to minimeter scale.

To examine the surface features on DPB defect morphologies, an atomic force microscopy (AFM) observation on the Si face 3C-SiC epitaxy layer was conducted. In our sample, the DPB defects show a protrusive hill-like structure and is visible to the naked eye with lower density. The largest area free from DPB defect is $500 \times 1000 \ \mu m^2$. A typical AFM image of a DPB defect with a scanning area of $100 \times 100 \ \mu m^2$ is shown in Figure 1(b). From the cross-section profile along the white line shown in the left-bottom inset, the hill-like structure has two peaks standing on a much smoother "hill", and there is a "canyon" embedded in between the two peaks. The height from the bottom to the top is ~0.7 μm and the width of the bottom is 22 μm. We believe that the "canyon" results from the original shallow groove of the DPB defects reported in Ref. [29], and the twin peaks can be ascribed to the gathering of redundant atoms due to the low local surface energy. An evidence is that the longitudinal direction of the "hill" is roughly parallel to the {1 1 −2 0} planes. Basically, the Si and C adatoms undergo anisotropic migration, and {1 1 −2 0} is the fast diffusion direction for adatoms on the Si-terminated surface. This conclusion is similar to the findings of Ref. [37]. The authors of the reference observed that the V-shaped mesa with arms oriented parallel to {1 −1 0 0} has a larger webbed SiC cantilever than that with the arms oriented parallel to {1 1 −2 0}. The result indicates that the migration of adatoms on the arm prefers {1 1 −2 0} directions. A schematic diagram drawn in Figure 2(a) shows the surface diffusion channels reserved by the topmost Si atoms on the (0 0 0 1) face. The green arrow direct to the normal of {1 1 −2 0} planes provides a straight channel for adatom migration. Other directions, like the normal of {1 −1 0 0}

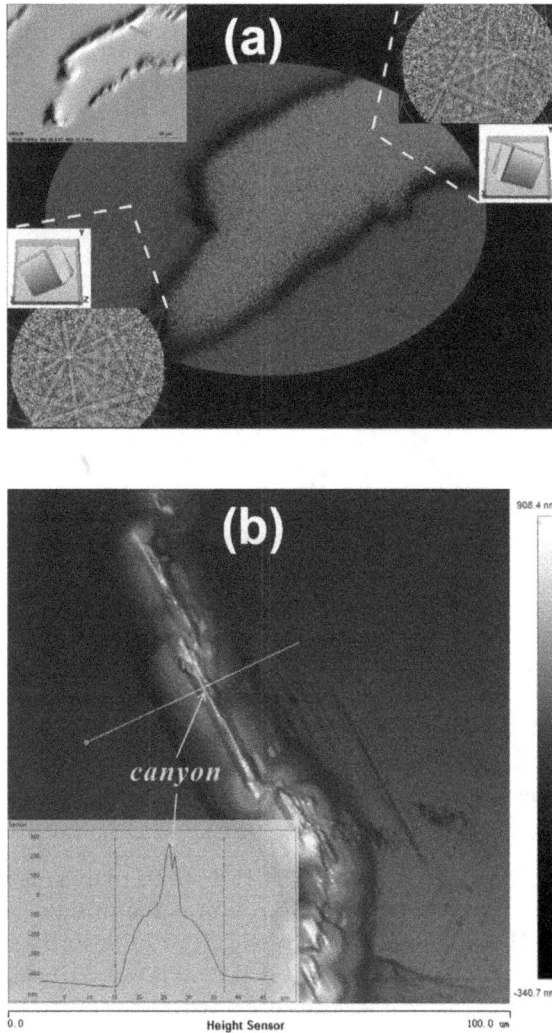

FIGURE 1

(a) Electron backscattered diffraction (EBSD) mapping of a 3C-SiC epilayer grown on Si-face 4H-SiC. Left-top inset shows the SEM image, corresponding to the central EBSD mapping. The right-top inset and left-bottom inset show the Kikuchi patterns of the light gray and deep gray regions, respectively. (b) A typical AFM image of a DPB defect with a cross-section profile along the white line shown in the bottom inset. (From [25] Figure 6, reproduced with permission of Elsevier.)

planes, represented by the red dotted arrow, have more probability to be blocked by the topmost Si atoms. Borysiuk *et al.* calculated the surface energy for adatoms on the Si-terminated surface of 4H-SiC at a high temperature close to 1600°C using density functional theory [24]. They showed that when the Si and C adatom on top of the topmost Si atoms, the surface energy reaches its maximum. However, the surface energy has six local minimum sites surrounding each top Si atom. Three of them are on top of the Si atoms in the bottom layer, while the other three sites are on top of the hollow site in the bottom layer. The adatoms hop from one minimum site to the nearest neighboring minimum

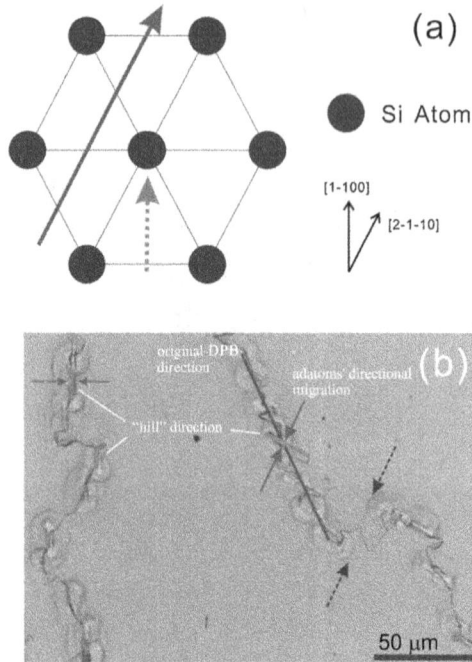

FIGURE 2
(a) Top view of Si atom on Si-terminated surface (0 0 0 1) plane of 4H-SiC, a straight channel represented by a solid arrow and an impassable path represented by a dotted arrow. (b) A typical surface morphology of DPB defects. (From [25] Figure 7, reproduced with permission of Elsevier.)

site, and the direction is along the normal of {1 1 −2 0} planes. Another unneglectable factor in considering the anisotropic lateral growth is the number of dangling bonds in the growth directions. The rate to absorb atoms [31] or to desorb adatoms by H_2 etching [38] in <1 1 −2 0> is higher than that in <1 −1 0 0>. Figure 2(b) shows a typical surface morphology of DPB defects. There is a series of DPB defects extending from the top center to the right bottom corner with the direction represented by the long blue line. The direction is parallel to {1−1 0 0} planes. However, the longitudinal direction of each DPB defect, represented by the short red lines, is parallel to the {1 1 −2 0} planes. A reasonable explanation is that the distribution of the DPB defects is along the coalescence boundary of the twinning domains, which is relevant to the original nucleation positions and lateral growth of the domains. However, the formation of the DPB defects, i.e. the gathering of adatoms is through the hopping migration and redirects the hill ridge to parallel to the {1 1 −2 0} planes. In the figure, a white dashed elliptical indicates an area free from the hill-like DPB defects. In this area, because the direction of adatom migration paths, indicated by the dashed black arrows, are blocked by the former formed hill-like DPB defects, hill-like DPB defects are not formed.

3.2 V-shaped Twinning Structure

In this section, we studied a V-shaped twinning structure in 3C-SiC epilayer grown on Si-face 4H-SiC. The defect was characterized using cross-sectional TEM and the core region was found in a polycrystalline complex structure.

A low-magnification, cross-sectional TEM image taken from the [-2110] zone axis is shown in Figure 3(a). The bright field image clearly shows the V-shaped structure. In preparing the TEM sample, we intentionally selected the zone axis to observe the twinning structure in cubic SiC. As mentioned in above section, the longitudinal direction of the hill-like DPB defects parallel to {11–20} planes. In the preparation, under the SEM of the focused ion beam system, we let the longitudinal direction of the DPB "hill" tilts the normal of the sample cross-section by ~22°. In other words, the hill-like structure is roughly along the [-1010] direction, i.e. parallel to the (-12–10) plane of the 4H-SiC. The selected area electron diffraction (SAED) patterns of the regions indicated in Figure 3(a) as 3C-I, 3C-II, and 4-H are shown in Figures 3(b), (c), and (d), respectively. From the SAEDs, it is clear that the deposited layer in either the left (3C-I) or the right (3C-II) of the V-shaped structure is a high-quality, single-crystalline 3C-SiC epitaxial film and well registered to the 4H-SiC substrate. While from the SAED of the 3C-I and 3C-II regions, they are mirror twins with respect to their common axis [111]. The relationships between 3C-I, 3C-I, and the bottom 4H substrate are described as follows: $[111]_{3C-I}$ || $[111]_{3C-II}$ || $[0001]_{4H}$; $[01–1]_{3C-I}$ || $[0-11]_{3C-II}$ || $[-2110]_{4H}$; and $[-211]_{3C-I}$ || $[2-1-1]_{3C-II}$ || $[01-10]_{4H}$, where 3C's directions use cubic Miller indices and 4H's directions use hexagonal Miller indices. In here, 3C-II is the 180° rotation variant of 3C-I. To elucidate the structure of the twinning complex, the 3C-I in the left region is hereafter defined as the standard coordinate system.

The V-shaped structure shown in Figure 3(a) is composed of Twin-I, Twin-II, and the DPB defect. The SAED pattern taken from the tip region of the V-shaped structure, indicated

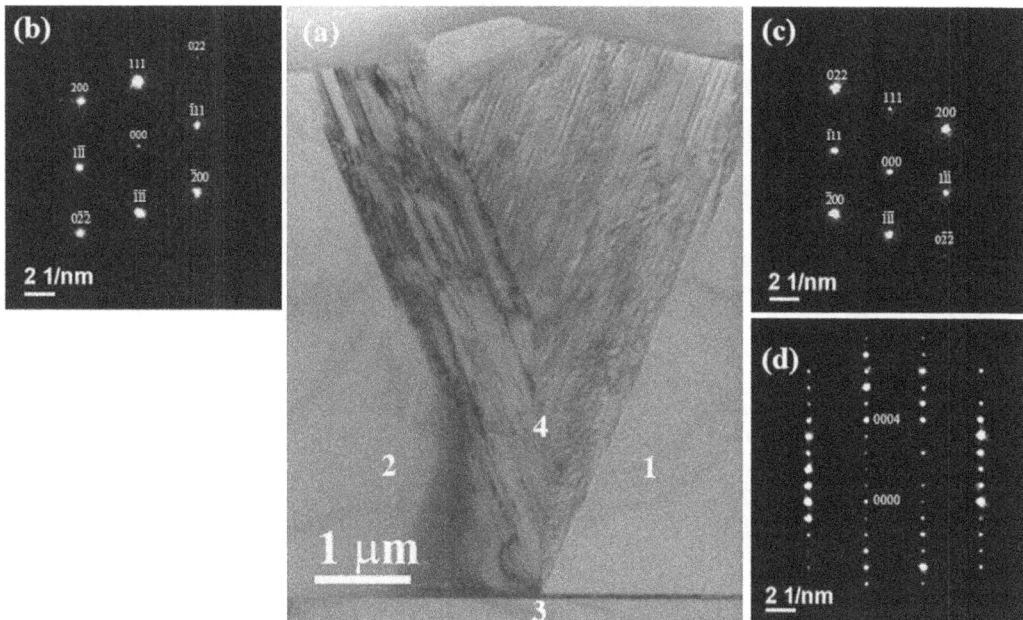

FIGURE 3

(a) Low-magnification cross-sectional TEM image of 3C-SiC grown on a Si-face 4H-SiC substrate measured along the [-2110] axis. SAED patterns of the epitaxial layer taken from regions 1, 2, and 3 are shown in (c), (b), and (d), respectively. The SAED patterns of (b) and (d) indicate the regions are cubic 3C-SiC and the mirror symmetry with respect to [111] suggests they are a twin pair. They are named 3C-I and 3C-II. The SAED of (d) shows a hexagonal lattice, which is the pattern of 4H-SiC substrates. The SAED pattern of region 4 is very complicated and is shown in Figure 4.

by the number 4 is shown in Figure 4(a). The diffraction pattern is very complicated and contains four different reciprocal lattices, belonging to 3C-I, Twin-I, Twin-II, and 3C-II regions, respectively. To clearly identify the lattice points of each of these four reciprocal lattices, the SAED pattern is redrawn schematically in Figure 4(b)–(e), in which individual reciprocal lattice points are high-lighted by larger black spots. In Figure 4(b) and 4(e), the reciprocal lattices of 3C-I and 3C-II are highlighted, respectively. Their twinning axis [111] is indicated in the figures. In Figure 4(c), the highlighted are the lattice points of Twin-I. The twinning axis $[11\text{-}1]_{3C\text{-}I}$ is represented by the solid line in both Figure 4(b) and 4(c), and the reflections between the lattice points of Twin-I and their twining counterparts are connected by the dash lines shown in Figure 4(c). Through the dash lines, one can find that the counterparts belong to 3C-I reciprocal lattice. In the similar way, one also can find that the lattice points of Twin-II highlighted in Figure 4(d) are the twining points from those of 3C-II, highlighted in Figure 4(e), and the twining axis is $[11\text{-}1]_{3C\text{-}II}$.

To recognize the ranges of Twin-I, Twin-II in the V-shaped structure, we took the dark-field images using the diffraction beams from the lattice points belonging to Twin-I and Twin-II reciprocal lattices. For comparison, we also took the diffraction beam belonging to 3C-II. The bright field images of the whole V-shaped structure including the tip area is shown in Figure 5(a) and the three dark images are shown in Figure 5(b), 5(c), and 5(d). The lattice points for the diffraction beams are {022} and are marked with circles in Figure 4(c),

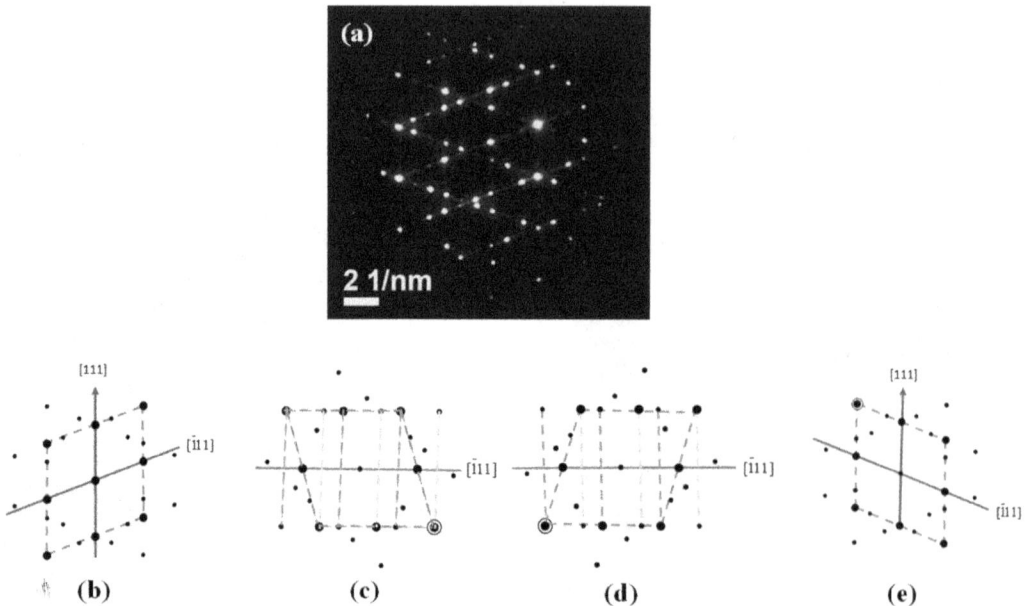

FIGURE 4
(a) SAED pattern of region 4 indicated in Figure 3(a). The pattern contains four different reciprocal lattices, belonging to 3C-I, Twin-I, Twin-II, and 3C-II regions, respectively. Schematic diagrams of the SAED pattern are redrawn using larger spots to represent the highlight reciprocal lattices. (b) is for 3C-I lattice, (c) is for Twin-I lattice, (d) is for Twin-II lattice, and (e) is for 3C-II lattice. In (b) and (c), the twinning axis between 3C-I and Twin-I, $[\text{-}111]_{3C\text{-}I}$, is indicated. In (d) and (e), the twinning axis between 3C-II and twin-II, $[1\text{-}1\text{-}1]_{3C\text{-}II}$, is indicted. In (c), (d), and (e), each figure has a circled spot. The circled spots are {022} planes, which are selected to generate dark-field images, which are shown in Figure 5.

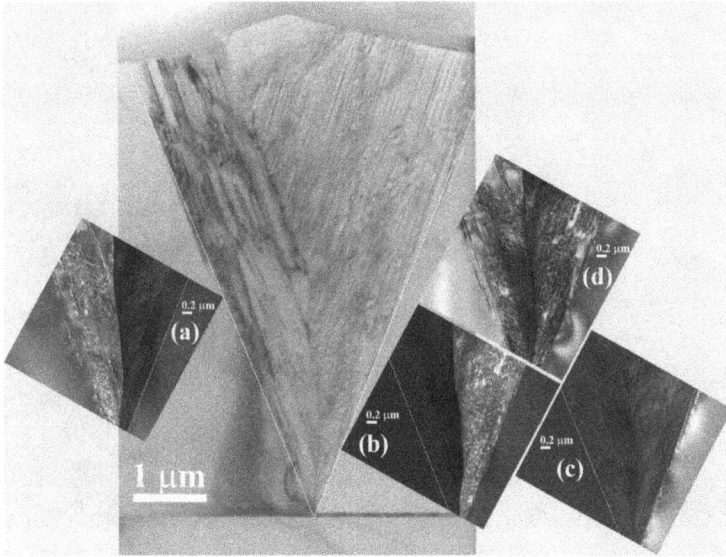

FIGURE 5

(a) Bright-field image of the whole V-shaped structure including the tip area for three dark-images from the {022} diffraction beams. (b) Dark field image of twin-I region, showing that the region occupies the left part of the tip area. (c) Dark field image of twin-II region, showing that the region occupies the right part of the tip area. (d) Dark field image of 3C-II region. (e) Bright field image of the tip area. The diffraction spots for the dark-field images shown in (b), (c), and (d) are indicated in Figures 4(c), (d), and (e), respectively.

4(d), and 4(e), respectively. In the dark-field mode, only the area belonging to the selected diffraction lattice is displayed. From Figure 5, we can easily identity the area of Twin-I and Twin-II in the V-shaped structure. Twin-I and Twin-II occupy the left part and the right part of the V-shaped structure, respectively, and the DPB defects locate at their boundary.

A high-resolution cross-sectional TEM image of the V-shaped twinning complex is shown in Figure 6(a). The image reveals a 64-nm-thick 3C-SiC layer existing in between the bottom of the DPB defects and the 4H-SiC layer, implying that the DPB defect does not originate from the 3C/4H interface. In this layer, one can see several planar defects or coherent twin bands, i.e. the stacking sequence of the depositing layer switches between 3C-I type to 3C-II type and forms a thin 3C-II type layer. Note that the interface between the two types can be considered as a single 2H layer. The type changing near the substrate is not usual, because the 4H-SiC substrate contains 50% 2H hexagonal component. Once the type switching cannot synchronize in the whole growth plane, the boundary between the two 3C types becomes incoherent in the lattice and thus results in the DPB defect. Moreover, the developing of the DPB defects further prompts the twinning with respect to the (-111) plane in 3C-I to form Twin-I, and the twinning with respect to the (5-1-1) plane in 3C-II to form Twin-II. The (5-1-1) plane is in the coordinate system of 3C-I. In fact, it is equivalent to the (-111) plane in the coordinate system of 3C-II. From the higher magnification image shown in the right panel of Figure 6(a), it is apparent that the planar defects in the left panel are series of coherent, alternating twin bands parallel either to the (111) plane in the 3C-I region or to the (-111) plane in the Twin-I region. Zhang *et al.* proposed a model of strain relaxation during hetero-epitaxial growth to explain the periodic array

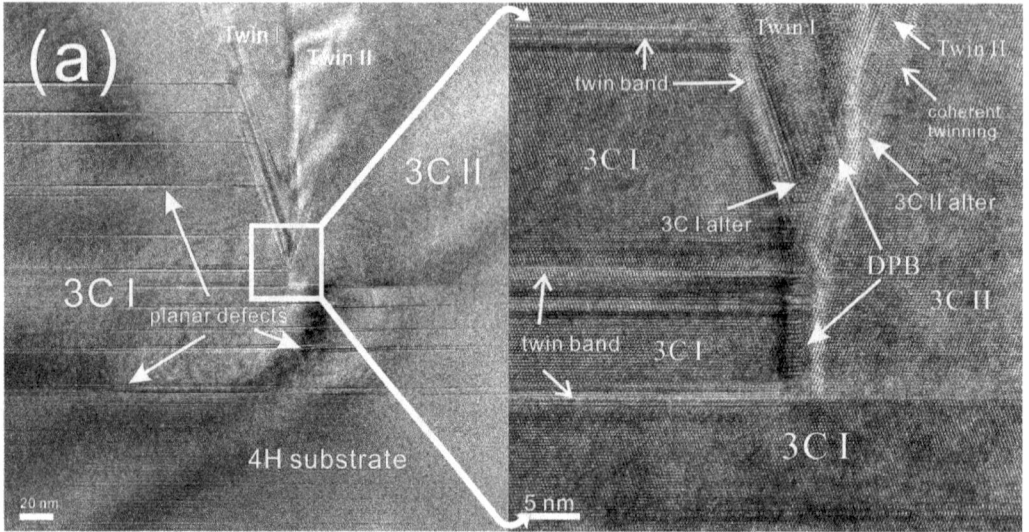

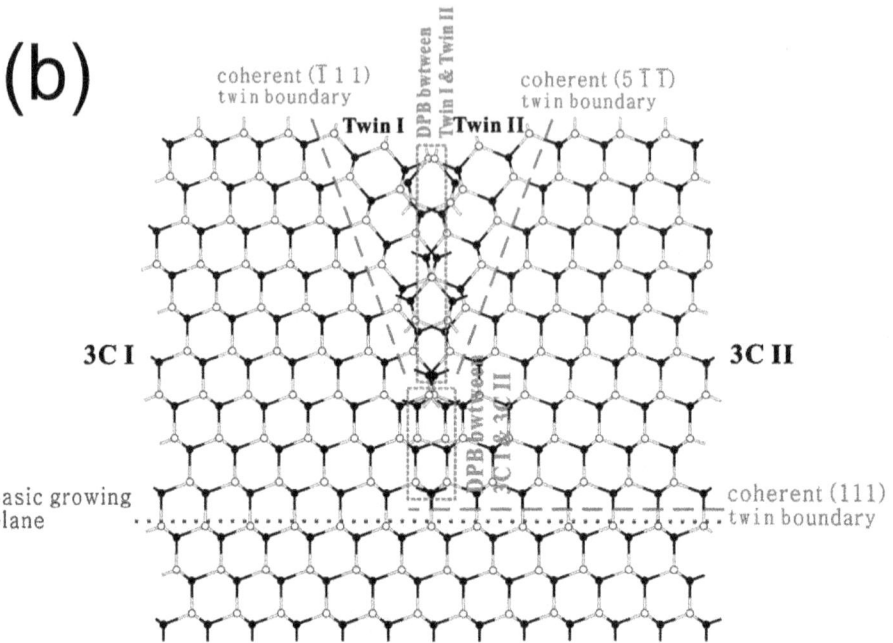

FIGURE 6
(a) High-resolution cross-sectional TEM image showing the tip area of the V-shaped twinning defect. Defects paralleled (111) plane or (-111) plane are multi-bilayer coherent twins. (b) Ball-and-stick model of the V-shaped twinning complex defect. (From [26] Figure 4, reproduced with permission of the American Vacuum Society.)

of misfit twins [39]. According to their model, the DPB defect is deemed to have a larger lattice size than a perfect crystal has. The existence of twin bands in Twin-I or Twin-II domains helps release the strain energy of the DPB. Figure 6(b) depicts a simplified ball-and-stick model for the V-shaped structure based on the TEM analysis. The dashed line represents the twin boundary, and the pink dotted rectangle indicates the distorted bonds of the DPB. The blue dotted line describes the basic growing plane where the DPB defect begins to emerge.

3.3 Dynamics of Adsorb Atoms near the DPB Defects

From the TEM analysis in section 3.2, we observed that the two coherent twin boundaries are linear and forms a V-shape, but the incoherent DPB region is composed of irregular piecewise linear sections. To explain the linear prolongation of the V-shaped boundary between the regular and twinned domains, and the nonlinear DPB between two twinned domains, we proposed a kinetic migration model for adatoms crossing different domains. The potential energy diagram of adatom interaction with the substrate is illustrated in the top of Figure 7(a). As the lattice of the Twin-I domain is that of the 3C-I domain after a rotation of 180° with respect to the [-111] axis, the vertical direction of Twin-I region is [-5-1-1] in Ttwin-I's coordinated system. Note that [5-1-1] is very close to [-100]. The angle between them is only 15.8°. In Figure 7(a), the normal of the titling surface of Twin-I is just [-100], one can see that the surface atom has two dangling bonds, while the normal of surface in 3C-I is [111] and the surface atom has only one dangling bond. As a result, it is more difficult for adatoms to desorb from site 2 in the Twin-I surface than to desorb from site 1 in the 3C-I region, even when the Twin-I surface turns to (-5-1-1). Therefore, adatoms in the Twin-I domain have a lower adsorption energy (E_{d2}) than that in the 3C-I domain (E_{d1}). Once a carbon atom is absorbed on the SiC surface, it will preferentially migrate toward the Twin-I domain if it has sufficient thermal energy to overcome the activation barrier between the 3C-I and Twin-I domains. If the epitaxy is dominated by step flow, the lateral growth rate will be much faster than the longitudinal growth rate. The dynamic

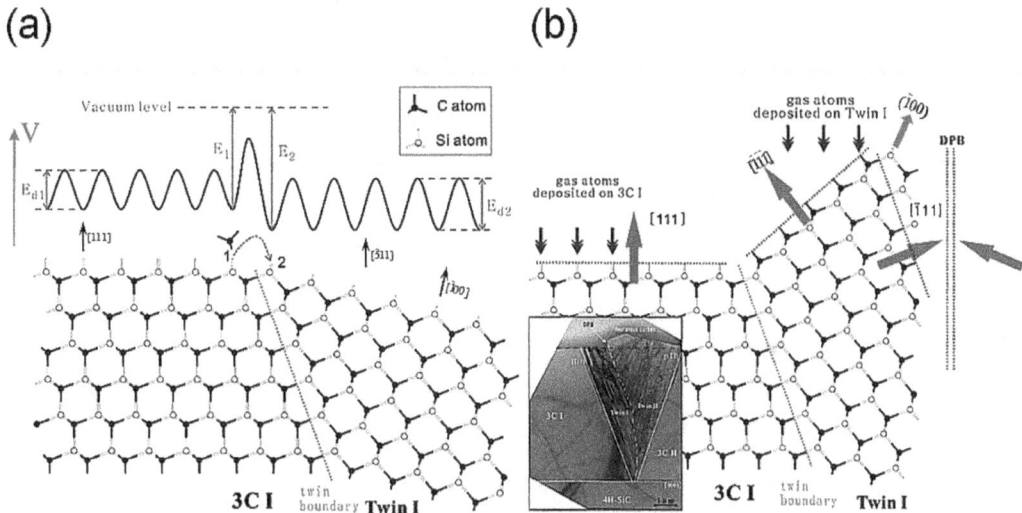

FIGURE 7

(a) Schematic representation of adatom migration from 3C-I domain to Twin-I domain. The potential surface energy of the adatom is shown in the top part of the panel. (b) The growth direction of the 3C-I and Twin-I domains restricts their boundary to the (-111) plane. The low magnification cross-sectional TEM image in the bottom left corner is used to compare the surface of the V-shaped twinning region and the atom model. The down-direction arrows double-fold-arrows indicate the gas atoms deposited on different domains. Deposition rate on the (-5-1-1) plane is slower than on the (111) plane. (From [26] Figure 5, reproduced with permission of the American Vacuum Society.)

morphology of the epitaxy surface tends to tilt to the low dangling bond close-packed plane, as illustrated in Figure 7(b), because of the lower surface energy. However, the preferred growth orientations for the 3C-I and Twin-I domains are still the (111) and (-5-1-1) planes, respectively. Because adatoms migrate easily between 3C-I and Twin-I domains, the equal speed of growth in the two directions restricts growth of the twin boundary to the (-111) plane.

Similarly, the crystal growth of the Twin-I domain along the $[-111]_{3C-I}$ axis competes with the growth of the Twin-II domain along the $[1-1-1]_{3C-II}$ axis. A different situation is that the distorted bonds of the DPB have much higher potential surface energy than the V-shaped coherent boundary. Thus, the adatoms have a low probability to overcome the high activation barrier, and adatom migration cannot balance the growth rates of the Twin-I and Twin-II domains. If different numbers of atoms are absorbed on the different domains, the side with more adatoms will grow with a faster rate, which causes the extension of the DPB region to deviate from the [111] direction, as can be seen in the low-magnification TEM image shown in Figure 7(b). According to the ball-and-stick model shown in Figure 7(b), when the DPB region deviates from the [111] direction, a height difference of the growing layer at the DPB is built, as shown in the inset of TEM image in Figure 7(b). However, this height difference cannot increase without limits. The increment in height difference may change the surface orientation and the surface potential to allow the adatoms to overcome the activation barrier to balance the growth rate difference between the Twin-I and Twin-II region.

The measured angle between the surface of the Twin domain and the (0001) plane is smaller than that in the model shown in Figure 7(b). It is believed that the macroscopic growth factors, such as the heat transfer and source gas flow, may weaken the preference for a particular growth direction in a step-flow epitaxy. For example, consider an on-axis wafer is placed horizontally in the chamber of the CVD reactors. When gas atoms are evenly and perpendicularly deposited on the surface of the substrate, the (111) plane has the highest deposition rate in the epitaxy. Other planes, such as the (511) plane in the Twin-I domain, have higher densities of dangling bonds projected on the (111) plane, which results in a lower deposition rate per unit area.

The twinning defect leads to different local growth rates around the DPB. In here, the AFM profile image in the inset of Figure 1(b) and the TEM cross-section in Figure 3(a) are combined together for comparison in Figure 8. Note that the cross-sectional AFM profile is along the [-12-10] direction, i.e. perpendicular to the in-plane direction of the DPB defect, as discussed in section 3.1. The angle between the direction and the zone axis of the TEM image is not 90° but ~30°. In other words, the AFM profile and the TEM image are slightly tilted by ~30°. Though they are not along the same path, a qualitative comparison is still reasonable. From the figure, the hill-like region is divided into three parts. The first region, called the "DPB region" is very thin and contains many distorted bonds. Because the distorted bonds extend to the surface of the deposited film, it is difficult for adatoms to migrate across that region, which results in a lower local growth rate. This explanation is consistent with previous reports of DPB causing a groove-like morphology and of a lower local growth rate in the DPB region [29, 40]. The second region is designated as the "twin region," which comprises the Twin-I and Twin-II domains. This region has much steeper surfaces because of its proximity to the close-packed plane, as discussed earlier in this section. The last region, named the "hill region," corresponds to the normal 3C crystal and comprises 3C I and 3C II domains. The surfaces of this region are not as steep as those of the twin region. It is believed that adatoms can easily be incorporated into DPBs via any migration path because DPBs are a closed cycle from the top view of the surface. We therefore attribute the higher local growth rate to excess gathered adatoms.

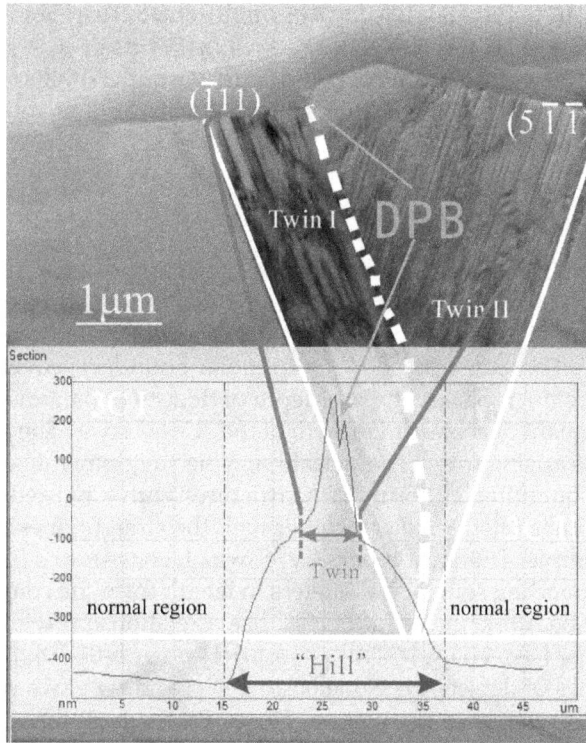

FIGURE 8

A comparison between the profile AFM image shown in the inset of Figure 1(b) and the low-magnification TEM bright field image shown in Figure 3(a) to elucidate the growth in local regions around the DPB defect. (From [26] Figure 6, reproduced with permission of the American Vacuum Society.)

3.4 Summary

We studied a V-shaped twinning defect in 3C-SiC grown on a Si-face (0001) 4H-SiC substrate. On the basis of TEM analysis, an atomic-structure model is proposed to explain the observed twinning complex. The growth rates along the (111) plane in the 3C-I domain and the (511) plane in the Twin-I domain were found to be balanced by adatom intermigration. The synchronous growth rates cause the coherent boundary to grow linearly. Conversely, the growth rates in the Twin-I domain and in the Twin-II domain are not balanced and the asynchronous growth rates, leading to the formation of a nonlinear boundary.

4 Super-V-Shaped Structure on 3C-SiC Grown on the C-Face 4H-SiC

Si- and C-faces of hexagonal SiC have different surface free energies [23, 24]. According to the DFT study [24], either Si adatom or C adatom encounter significant higher energy barrier in the surface of C-face 4H-SiC than in the surface of Si-face 4H-SiC. Because of the different surface properties, 3C-SiC layers grown on C-face 4H-SiC display different surface morphology and epitaxial defects. Although some researches indicated the promising application of C-face 3C-SiC to electronic devices [30], its growth mechanism remains poorly understood. This section presents and discusses a super-V-shaped structure (SVSS) often observed on the surface of 3C-SiC films deposited on the C-face 4H-SiC substrates by chemical vapor deposition (CVD).

4.1 Defects in C-face 3C-SiC

DPB defects discussed in the previous section are no longer the major defects on the surface of 3C-SiC films grown on the C-face 4H-SiC due to the different surface properties [25, 29]. Generally, the major defects on C-face 3C-SiC can be categorized into three types. The first one is called "basin". These defects show an irregular boundary, cover an area of several to hundreds of square microns, and have a depth of dozens of nanometers. The basins are thought to be the remnant of coalesced growth of the 3C-SiC layer. The second type is polycrystalline complexes arising from SiC nucleation, which agglomerate on the surface of the 3C-SiC film in large quantities to form black structures with a twisted linear topography. These black twisted lines often appear together with the straight lines of SVSSs or emerge on the edge of the sample. The third type is SVSS, which consists of a high rising point and two straight lines, extending several millimeters in length from the common point.

Normally, the epitaxial orientation of 3C-SiC has a relationship with the C-face 4H-SiC substrate of $[000\text{-}1]_{4H}$ || $[\text{-}1\text{-}1\text{-}1]_{3C}$, $[\text{-}2110]_{4H}$ || $[01\text{-}1]_{3C}$, and $[01\text{-}10]_{4H}$ || $[\text{-}211]_{3C}$ [26, 41]. Among the three types of defects, type one basin defect and type two polycrystalline twist line have no distinct preferential orientation. However, the type three SVSS defect shows an orientation-dependent morphology. Figure 9 shows several SVSS defects. For convenience in the following text, the high point and two straight lines of the SVSS are hereafter called the "vertex" and "arms", respectively. As shown in the figure, the angle between the two arms is $60°$, and the angle bisector is along the same direction $[1\text{-}100]$, which is the step terrace direction on this sample. Although the substrate is on-axis nominally, step terrace due to the slight uncertain misorientation has been confirmed by AFM measurement and the direction was also determined [26]. The consistency in the direction of step terrace and the angle bisector suggests that the formation of SVSS is relevant to the step flow in the growth. Figure 9(a) shows SVSS defects and their interaction. Step-flow direction is also indicated in the figure. One arm of SVSS 1 is interrupted and ended by an existing arm belonging to SVSS 2. Figure 9(b) shows another case, in which two arms belonging to different SVSS defects contacts and stops their extensions. In this case, the arms grew at equal speeds, reached the meeting point at the same time, and blocked each other. On the other hand, the upper arm of SVSS 1 and the lower arm of SVSS 2 extend to a much longer distance and out of the image.

A series of 3C-SiC samples were grown on the C-face of 4H-SiC under different growth conditions. The angular bisectors of SVSSs have the same orientation and the angles between two arms are equal for the same sample. These regularly arranged straight arms are connected at their meeting points to segment the sample into small prismatic-like pieces, particularly when viewed from the inverse direction to the refractive light. Different samples have different angular bisector orientations and arm angles. According to the statistical results, the angle ranges from $50°$ to $90°$. The main reasons for the different arm angles in different samples could be the orthogonal misorientation of the wafer and different ambient temperatures. These two factors influence the migration length as well as the step-flow direction of the adatoms. For the same sample, the two factors are constant, which guarantees that the SVSSs in a single sample have the same angular bisector orientation and arm angle.

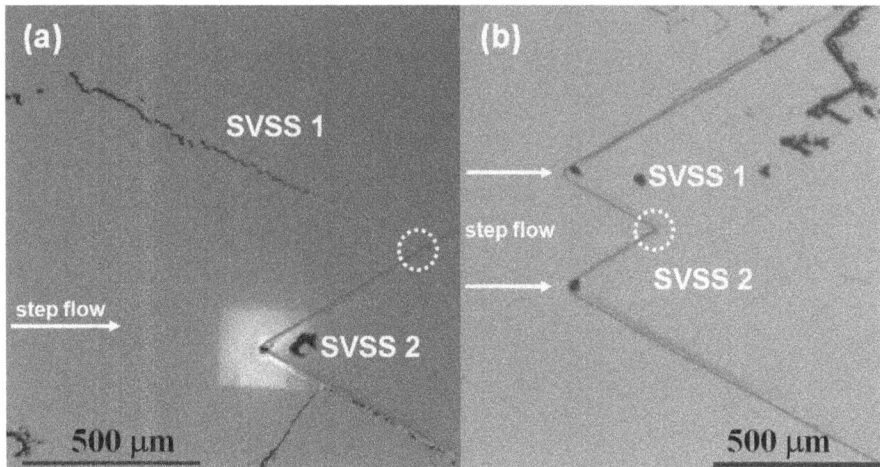

FIGURE 9
(a) Two SVSS defects and their interaction. One arm of SVSS 1 is interrupted by an arm of SVSS 2. The rectangular bright spot surrounding the vertex of SVSS 2 is caused by the FIB technology, and the result of the prepared TEM sample is shown in Figure 10. (b) Another pair of SVSS defects. The arms of SVSS 1 and SVSS 2 are interrupted by each other because they reach their meeting point, denoted by the circle, at the same time. (From [27] Figure 8, reproduced with permission of IOP Publishing.)

4.2 TEM Characterization

To characterize the inner structure of the SVSS, two TEM samples were cut from the vertex region and midpoint of an arm by FIB technique, respectively. The position of the first sample is highlighted by the white solid line marked in the scanning electron microscopy (SEM) image shown in Figure 10(a). As can be seen in the figure, the protruding vertex is a triangular-like mesa containing two groups of stripes, denoted by two white dashed lines along [-1-120] and [-2110] directions, respectively. The white line is along [1-100] and cuts the stripe group along [-1-120], and the zone-axis is [11-20]. The full cross-sectional TEM image with a width of 10 μm and a height of 4 μm is shown in the inset surrounded by a white dashed rectangle in Figure 10(a) and the left bottom inset in Figure 10(b). A low-resolution TEM image is displayed in Figure 10(b) to reveal the evolution of the defects. In the bottom of the image, one can see a polycrystalline inclusion, from which several planar defects, belonging to the [-1-120] stripe group, extend to the surface. A selected-area electron diffraction image with an objective aperture of 200 nm, indicated by a white solid circle shown in Figure 10(b), was obtained across the boundary of the planar defect, and is depicted in Figure 3(c). Based on the results of our analysis in section 3.2, the SAED pattern is composed of a 3C-SiC fundamental pattern and its twin pattern along the coherent (−1 1 1) plane. The twining lattice points indicated by white dotted circles and their original counterparts are connected by the dotted arrows. A higher magnification image of the black agglomerations is illustrated in Figure 3(d). The four inserts are SAED images taken from four different positions indicated in the figure. The diffraction patterns were obtained by rotating the sample to different angles, revealing that the black agglomerations are a cluster of tiny 3C-SiC crystals with random orientations. We believe that they originated from a polycrystalline SiC nucleation inclusion. Because each of them has different preferred growth directions, some residual voids were left during the growth and are thus observed in the figure. From the SEM image and the superimposed full TEM image in Figure 10(a), the inclusion is located at the center of a deeper position under the triangular-like mesa.

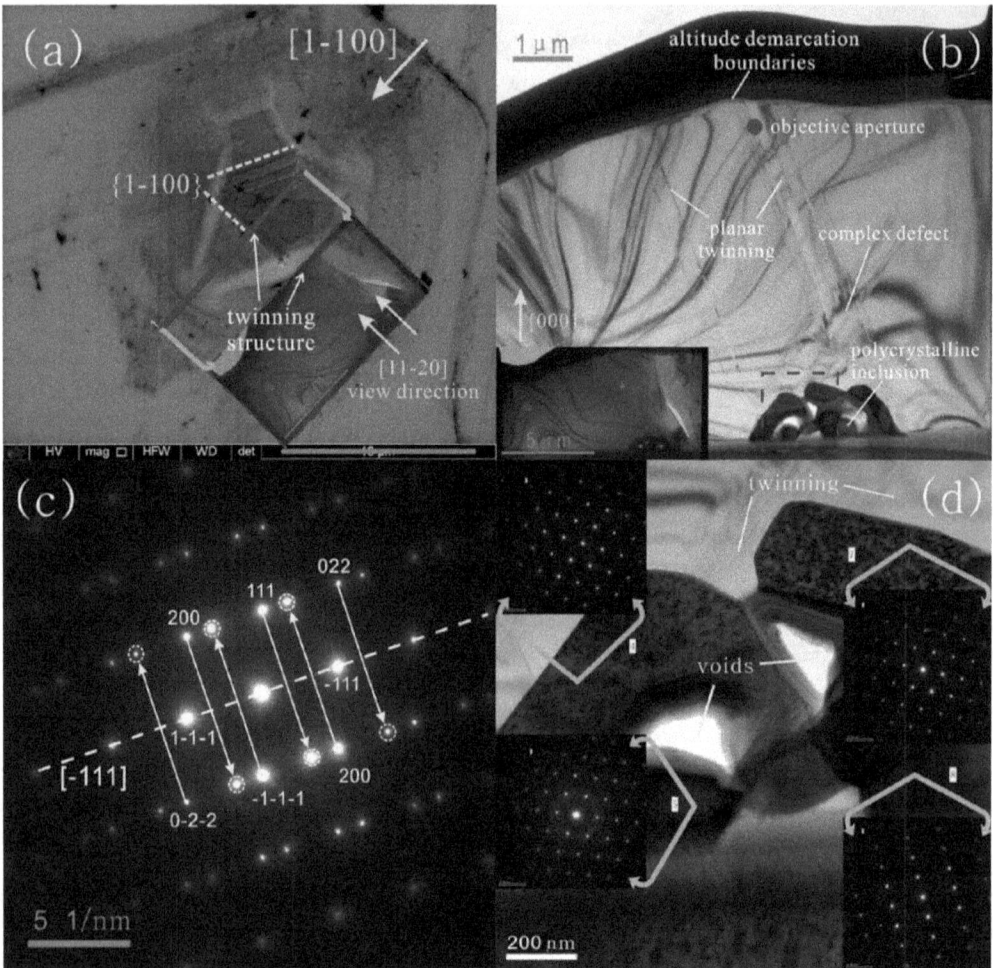

FIGURE 10
(a) SEM image of a typical vertex. The thick line shows where the cut was performed to obtain the TEM sample. The inset surrounded by the dashed rectangle is the full cross-sectional TEM image. (b) TEM image revealing the inner structure of the triangular mesa. (c) SAED pattern taken from the objective aperture marked with a solid circle in (b). (d) TEM image of a 3C-SiC polycrystalline nucleation inclusion in the deposited film. Insets are the SAED patterns of the marked areas. (From [27] Figure 3, reproduced with permission of IOP Publishing.)

The above analysis indicates that the polycrystalline inclusion induces a complex defect, and subsequently the complex defect converts into the twinned structures. Twin bands can be easily formed in the parallel {1 1 1} plane because twinning in this coherent plane only involves a twist of atomic bonds with negligible additional energy. From Figure 10(b), it is apparent that several twin bands extend to the surface and form a series of straight stripes in Figure 10(a). Another group of stripes is believed also composed of twin bands. As indicated in Figure 10(b), the stripes are also the altitude demarcation boundaries of the mesa along the [0 0 0 -1] direction. Thus, we concluded that the top of the polycrystalline inclusion has a preferred local growth rate on the (0 0 0 -1) plane.

The second TEM sample was 9 μm wide and 3 μm high, crossing the arm of an SVSS perpendicularly, with the TEM image shown in Figure 11. The inset HRTEM image and SAED pattern indicate that the cut region is composed of a high-quality 3C-SiC crystal without any defects. Since the width of the sample covers more than half of the 15-μm-wide arm, we believe that the SVSS arm is a single crystal with linear protruding morphology.

4.3 Growth Model

To explain the formation mechanism of SVSS, a model considering adatom migration affected by the protruding point defects on the substrate surface is proposed. As illustrated in Figure 12, adatoms tend to migrate toward lower terraces under step-flow dominant growth. The presence of a point defect changes the surface morphology, blocking the migration path, so that some adatoms blend into the protruding point while the rest will

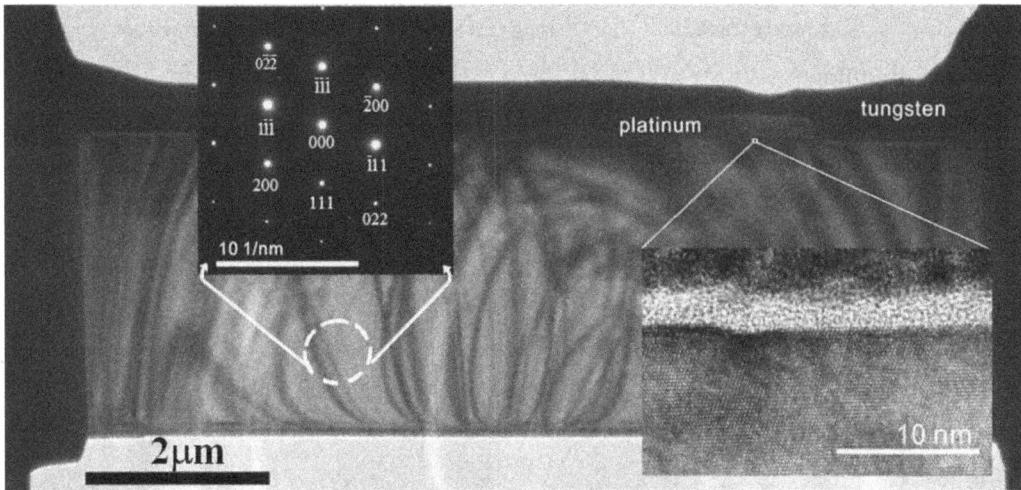

FIGURE 11
TEM image of a sample containing more than half of one arm of an SVSS. The HRTEM image, SAED pattern, and the inset HRTEM lattice image show that the arm is 3C-SiC without any special defects. The bright part in the inset HRTEM lattice image between the platinum and 3C-SiC is a thin SiO_2 interfacial layer resulting from natural oxidation. (From [27] Figure 4, reproduced with permission of IOP Publishing.)

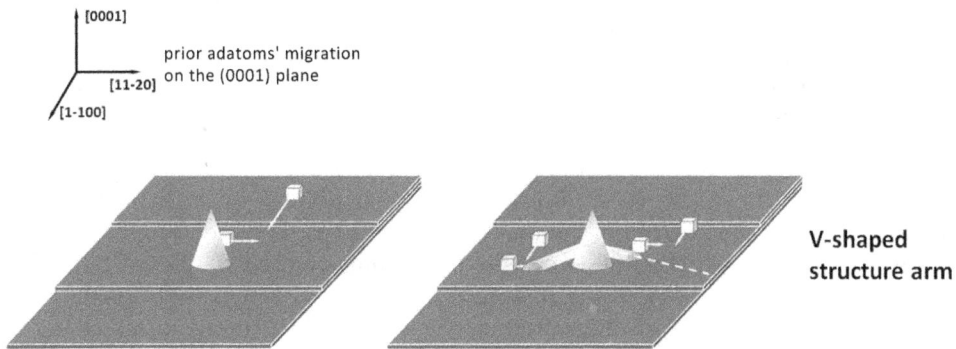

FIGURE 12
Schematic diagrams outlining the proposed formation mechanism of an SVSS, in which the cubes represent adatoms, the arrows show their migration direction, i.e. step-flow direction, and the cones indicate the protruding points (mesa morphology in Figure 10(a)). The protruding point blocks the migration of adatoms and forms the two arms.

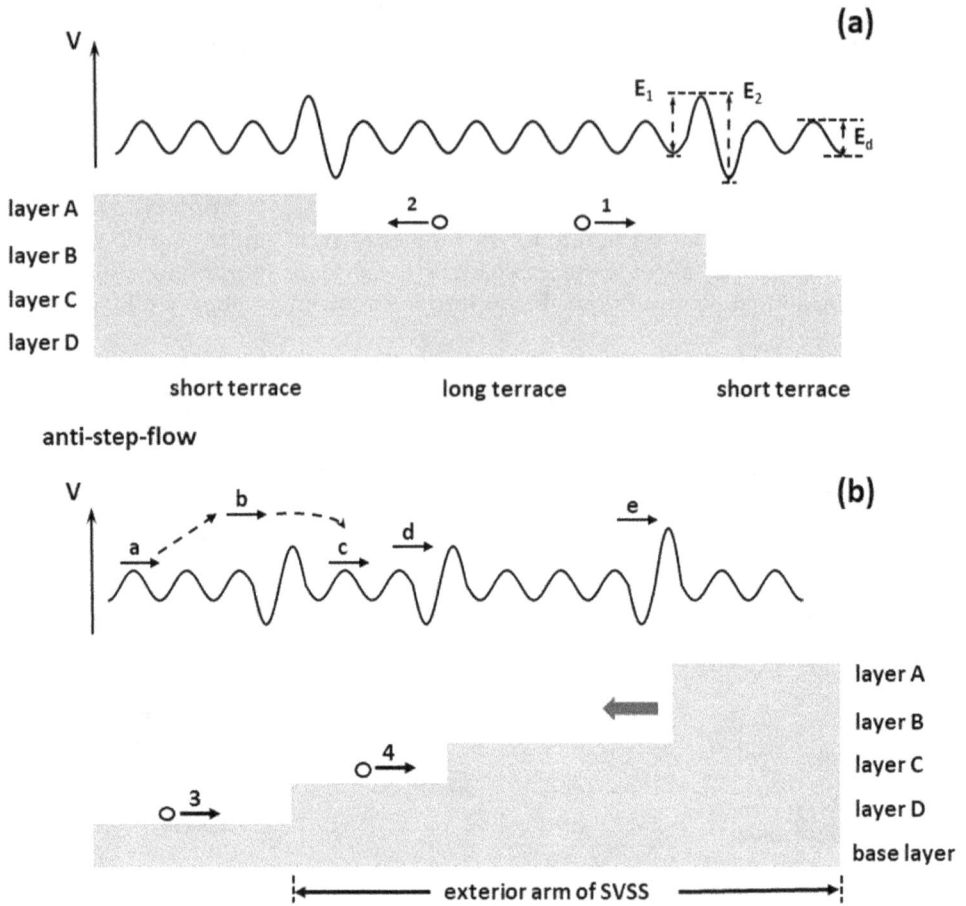

FIGURE 13
(a) Schematic diagram illustrating the migration of adatoms under step-flow dominant conditions. This situation also applies to the interior of the arm. The potential energy diagram illustrates the interaction between adatoms and substrate; E_d is the activation energy for diffusion on a terrace, E_1 the activation energy for an adatom to cross the step edge from an upper terrace to a lower terrace; E_2 is the activation energy from a lower terrace to an upper one. (b) Schematic diagram illustrating the migration of adatoms under anti-step-flow. This situation applies to local protruding structures, such as the exterior of an arm. In the upper potential energy diagram, the height of the arrows stands for the kinetic energy of the adatom. An adatom can cross the step edge to upper tracer only if its kinetic energy is higher than the activation barrier.

detour around it. Those absorbed into the protruding point form a distinct boundary, which may be observed in front of the mesa in Figure 10(a). The others, diverted around the protruding point, induce a higher density of adatoms on the two "downstream" sides compared with that at other locations. It is well known that 3C-SiC growth has the features of easy nucleation and difficult migration on the C-face [18]. The two arms of the SVSS stretch out along the two downstream sides of the protruding point because the higher adatom density increases the probability of nucleation and crystallization. It should be noted that this process would accumulate progressively, with the newly added lengths blocking other adatoms from migrating "upstream" to the protruding point. As a result, the outstretched arms will elongate continuously. We observe that the lengths of these arms may elongate to hundreds of microns.

In the previous section, we mentioned the twisted black lines of the second type poly-crystalline complex that accompanies the arm of an SVSS. The formation of polycrystalline particles is also driven by the higher density of adatoms nearby. The biggest difference between the protruding arms and polycrystalline particles along an SVSS is that the latter have a random nucleation direction. Moreover, SVSSs and polycrystalline nucleation are never observed on our Si-face under the same growth conditions. It is well known that adatoms on Si-face have much longer migration length than on C-face. To locally confine adatoms on Si-face is more difficult than on C-face. The protruding point is the origin of the SVSS. However, examples of such points are not limited to the mesa illustrated in Figure 10. Many other kinds of defects, such as micropipes, screw dislocations, and poly-crystalline nucleation complexes formed on the wafer surface, could also bring protruding points. The observations in our experiments confirm that polycrystalline nucleation is the most frequent type of protruding point defect. For example, the vertices in Figure 9 are composed of polycrystalline nucleation complexes. One possible reason for this is that micropipes or screw dislocations would be released during epitaxial growth, as deduced from our previous study of 3C-SiC grown on the Si-face [29].

4.4 Step-flow and Anti-step-flow in the Growth Model of SVSSs

Figure 13(a) is a schematic cross-section diagram showing multi-step structure with different terrace lengths. The top portion of the figure shows the surface potential energy function of a traditional interaction model for adatoms [42]. On the terraces, the potential energy reaches minimum at each adsorption site on the surface lattice. To escape a site and hop to the neighboring site, the adatom needs to overcome an energy barrier E_d. However, the potential energy is modulated at each step edge. At the adsorption site at the edge bottom, the potential energy is deeper than those at the ordinary sites on the terrace, which is because that adatom at this edge site can bind the atoms in the upper layer and thus has larger binding energy. At the site at the edge top, the potential energy is higher than E_d by an additional Enrlich-Schwoebel barrier [42]. This modulation makes different energy barriers seen from the site at edge bottom and the site at edge top. As shown in the figure, the former become E_2, while the latter becomes E_1. Both are higher than E_d.

Under the step-flow dominant conditions, there are two types of adatom migration, which are denoted mode 1 and mode 2 as illustrated on layer B in Figure 13(a). The migra-tion direction of mode 1 is the same as the step-flow direction. On the step edge, mode 1 adatom needs to overcome an additional Schwoebel barrier to migrate from the upper terrace to the lower terrace. If the adatom incorporates into the crystal at the edge bottom site, this process elongates the terrace of layer B. Mode 2 migration occurs in the opposite direction to that of the step-flow, thus an adatom can be directly incorporated into the step edge, which shortens layer B and elongates the upper layer A. Suppose that the long-terrace adsorbs more adatoms than the short-terrace, then different migration modes would result in different step distributions. When mode 1 dominates the kinetics, the terrace length is subject to positive feedback, leading to step-bouncing or macro-step behavior in crystal growth [43]. In contrast, if mode 2 dominates, the terrace lengths becomes uniform, resulting in a smooth crystal surface.

Considering the adatoms at the step edges, it is more difficult for adatoms to hop from a lower terrace to an upper terrace than vice versa because the activation barrier E_2 is larger than E_1. Thus, take this effect into account, it is plausible that the adatoms in mode 1 have higher average kinetic energy than those in mode 2. Given a group of adatoms with an equal distribution of mode 1 and mode 2, adatoms in the latter will change their direction more easily than the former when they collide. As a result, mode 1 dominates the step-flow kinetics after continuous collisions. The kinetic mode in the whole substrate can thus be summarized as follows. Most adatoms migrate toward lower terraces with higher kinetic energy. A few adatoms migrate toward upper terraces with lower kinetic energy. Adatoms with lower kinetic energy are easily trapped by the deep potential at the step edge.

Nucleation is strongly related to the density and the kinetic energy of adatoms. A protruding point or arm could slow down the speed of adatoms from upstream (mode 1) and gather them together, creating a suitable environment for nucleation. In contrast, adatoms from downstream (mode 2) are difficult to gather and nucleate because of their small number. This explains the formation mechanism proposed in the growth model of SVSSs.

However, step-flow theory does apply well to the exterior side of the protruding arms, where the adatoms from upstream accumulate in front of the ascending steps of the arms. The increment in the population enhances the flux of adatoms hoping to upper terrace, which may be superior to the adatoms' flux descending from the arm. The latter is mainly from the gas phase atoms adsorbing on the arms. An anti-step-flow migration model is thus proposed and shown in Figure 13(b). In this model, two cases are considered: adatoms on a long terrace (shown as adatom "mode 3") and adatoms on a short terrace (adatom "mode 4"). Adatoms usually have a low kinetic energy when they are adsorbed on the surface or hop to an upper terrace. In the case of "mode 3" on a long terrace, the adatom has enough time in a high-temperature environment to accelerate to a high speed (a to b), which could overcome the activation barrier E_2 (b to c). The kinetic energy of the adatom will be consumed in this process, and after that increased again in the high-temperature environment until the final kinetic energy is lower than the activation barrier E_2 and the adatom is trapped in the deep step edge potential. In this process, all the layers above the long terrace have the possibility to elongate. In the other case of "mode 4" on a short terrace, the initial low-speed adatom on the short terrace cannot accelerate to a high enough speed to overcome the step edge barrier, which results in the elongation of the upper nearest-neighbor layer (layer C). This process corresponds to the unique result of the short terrace becoming shorter.

Since the adatoms on the long terrace induces an uncertain result on its terrace length (mode 3) and the short terrace definitely is shortened (mode 4), the adatoms that come from upstream and hop to short terraces determine the evolution of the exterior arm. The short terrace will eventually disappear so that finally two layers combine to form one double-layer or even several layers combine to produce a multi-layer. The double-layer has a higher barrier than E_2. Most adatoms cannot cross this barrier and are trapped at the double-layer step edge, which results in the double-layers extending together. Overall, the exterior protruding arm of the SVSS has a driving force to grow in the opposite direction to that of the adatoms' migration direction. This also causes the exterior of the arm to be steeper than the interior as we observed from AFM results [27].

4.5 Summary

The growth and properties of SVSSs on 3C-SiC films deposited on (000-1) C-face 4H-SiC wafers have been studied. Optical images show that the SVSS consists of two protruding straight lines with a common higher protruding point; TEM images indicate that the protruding point is induced by polycrystalline nucleation and the straight lines are 3C-SiC crystals without any special defects. A growth model combining nucleation and the kinetic mechanism of adatom migration is proposed to explain the SVSS formation mechanism. The driving force, steepness of the sloping sides of the arms, interruption of arm growth, and the V-angle of the SVSSs have been discussed, and results of their characterization coincide well with the growth model. The effects of step-flow and anti-step-flow in the formation model have also been discussed.

5 Conclusion

3C-SiC layers have been successfully deposited onto Si-face 4H-SiC (0 0 0 1) and C-face (0 0 0-1) substrates using high temperature CVD. For the 3C-SiC layer deposited on Si-face 4H-SiC, twinning and DPB defects are observed. EBSD mapping was used to characterize the orientations and boundaries of the twinning domains. Through the EBSD measurement, the twinning domain near mini-meter scale was observed. Optical microscopy and AFM were used to characterize the morphology of the defects in the samples. The longitudinal direction of DPB defects is parallel to {2 -1 -1 0} planes. The anisotropic migration rate of adatoms has been proposed to explain the formation of the DPB defects. TEM study on the V-shaped DPB twinning defects shows that the defect originates from the boundary between two domains of the twining with respect to (111) planes. Further twinning with respect to (-111) and (1-1-1) planes constitutes the V-shaped structure of the defect. Based on TEM analysis, an atomic-structure model was proposed to describe the observed twinning structure. The growth rates along the (111) plane in the 3C-I domain and the (511) plane in the Twin-I domain were found to be balanced by adatom intermigration. The synchronous growth rates cause the coherent boundary to grow linearly. Conversely, the growth rates in the Twin-I domain and in the Twin-II domain are not balanced and the asynchronous growth rates, leading to the formation of a nonlinear boundary. SVSS, an orientation-dependent defect in 3C-SiC films grown on (000-1)C-face 4H-SiC wafers, has been studied. Optical and AFM images reveal the structure of SVSS defect, which consists of a protruding point with extended protruding arms. The angle bisector of the two arms is along the descending step direction. TEM study indicates that the protruding point is induced by polycrystalline nucleation and the arms are 3C-SiC with good crystallinity. A growth model is proposed to explain the formation and orientation dependency of the SVSS defect. Step-flow and anti-step-flow considering Schwoebel barrier are proposed to support the formation model.

Acknowledgements

The authors would like to acknowledge the financial support from the National Natural Science Foundation, China (Grant No. 51272202 and No. 61234006), the Science Project of State Grid, China (Grant No. SGRI-WD-71-14-004) and the Ministry of Science and Technology, Taiwan (Contract No. NSC 102-2221-E-002-191-MY3). H.H. Lin would like to thank Dr. H.M. Wu for the TEM measurements and Dr. S.J. Tsai for helping in preparing the manuscript.

References

1. P. G. Neudeck, R.S. Okojie, L.-Y. Chen, "High-temperature electronics: a role for wide bandgap semiconductors?", Proc. IEEE 90, 1065 (2002).
2. J. R. Waldrop, R.W. Grant, Y. C. Wang, R. F. Davis, "Metal Schottky barrier contacts to alpha 6H-SiC", J. Appl. Phys. 72, 4757 (1992).
3. N. Churcher, K. Kunc, V. Heine, "Calculated ground-state properties of silicon carbide", J. Phys. C: Solid State Phys. 19. 4413 (1986).
4. A. Fissel, "Artificially layered heteropolytypic structures based on SiC polytypes: molecular beam epitaxy, characterization and properties", Phys. Rep. 379, 149 (2003).
5. A. Lebedev, G. N. Mosina, I. P. Nikitina, N. S. Savkina, L. M. Sorokin, A. S. Tregubova, "Investigation of the structure of (p)3C-SiC-(n)6H-SiC heterojunctions", Tech. Phys. Lett. 27, 1052 (2001).
6. A. Lebedev, "Heterojunctions and superlattices based on silicon carbide", Semicond. Sci. Technol. 21, R17 (2006).
7. M. V. S. Chandrashekhar, C. I. Thomas, J. Lu, M. G. Spencer, "Observation of a two-dimensional electron gas formed in a polarization doped C-face 3C/4H SiC heteropolytype junction", Appl. Phys. Lett. 91, 033503 (2007).
8. M. V. S. Chandrashekhar, C. I. Thomas, J. Lu, M. G. Spencer, "Electronic properties of a 3C/4H SiC polytype heterojunction formed on the Si face", Appl. Phys. Lett. 90, 173509 (2007).
9. J. Lu, M. V. S. Chandrashekhar, J. J. Parks, D. C. Ralph, M. G. Spencer, "Quantum confinement and coherence in a two-dimensional electron gas in a carbon-face 3C-SiC/6H-SiC polytype heterostructure", Appl. Phys. Lett. 94, 162115 (2009).
10. A. Pérez-Tomás, M. Placidi, N. Baron, S. Chenot, Y. Cordier, J.C. Moreno, J. Millan, P. Godignon, "2DEG HEMT mobility vs inversion channel MOSFET mobility", Mater. Sci. Forum 645–648, 1207 (2010).
11. Henry, X. Li, H. Jacobson, S. Andersson, A. Boulle, D. Chaussende, E. Janzén, "3C-SiC heteroepitaxy on hexagonal SiC substrates", Mater. Sci. Forum 740–742, 257 (2013).
12. Qteish, V. Heine, and R. J. Needs, "Polarization, band lineups, and stability of SiC polytypes", Phys. Rev. B 45, 456534 (1992).
13. V. M. Polyakov and F. Schwierz, "Formation of two-dimensional electron gases in polytypic SiC heterostructures", J. Appl. Phys. 98, 025709 (2005).
14. S. Bai, R. P. Devaty, and W. J. Choyke, "Determination of the electric field in 4H/3C/4H-SiC quantum wells due to spontaneous polarization in the 4H SiC matrix", App. Phys. Lett. 83, 3171 (2003).
15. M. Bhatnagar, and B. J. Baliga, "Comparison of 6H-SiC, 3C-SiC, and Si for power devices", IEEE Trans. Electron Dev. 40, 645 (1993).
16. M. Ruff, H. Mitlehnor, and R. Helbig, "SiC devices: physics and numerical simulation", IEEE Trans. Electron Dev. 41, 1040 (1994).
17. G. Ferro, "3C-SiC epitaxial growth on α-SiC polytypes", a book chapter in Silicon carbide epitaxy, edited by F. La Via, Research Signpost, Kerala/India, Ch. 9, pp. 213–215, 2012.
18. H. S. Kong, J. T. Glass, and R. F. Davis, "Growth rate, surface morphology, and defect microstructures of β-SiC films chemically vapor deposited on 6H-SiC substrates", J. Mater. Res. 4, 204 (1989).
19. L. Latu-Romain, D. Chaussende, P. Chaudouet, F. Robaut, G. Berthome, M. Pons, and R. Madar, "Study of 3C-SiC nucleation on (0 0 0 1) 6H-SiC nominal surfaces by the CF-PVT method", J. Cryst. Growth 275, e609 (2005).
20. A. Qteich, V. Heine, and R. J. Needs, "Electronic-charge displacement around a stacking boundary in SiC polytypes", Phys. Rev. B 45, 6376 (1992).

21. J. Lu, C. I. Thomas, M. V. S. Chandrashekhar, and M. G. Spencer, "Measurement of spontaneous polarization charge in C-face 3C-SiC_6H-SiC heterostructure with two-dimensional electron gas by capacitance–voltage method" J. Appl. Phys. 105, 106108 (2009).

22. O. Kim-Hak, G. Ferro, J. Dazord, M. Marinova, J. Lorenzzi, E. Polychroniadis, P. Chaudouet, D. Chaussende, and P. Miele, "Study of the 3C-SiC nucleation from a liquid phase on a C face 6H-SiC substrate", J. Cryst. Growth 311, 2385–2390 (2009).

23. M. Sabisch, P. Kruger, and J. Pollmann, "Ab initio calculations of structural and electronic properties of 6H-SiC (0 0 0 1) surfaces", Phys. Rev. B 55, 10561–10570 (1997).

24. J. Borysiuk, J. Sołtys, R. Bozek, J. Piechota, S. Krukowski, W. Strupinski, J. M. Baranowski, and R. Stepniewski, "Role of structure of C-terminated 4H-SiC surface in growth of graphene layers—transmission electron microscopy and density functional theory studies", Phys. Rev. B 85, 045426 (2012).

25. Xin, R. X. Jia, J. C. Hu, C. Y. Tsai, H. H. Lin, and Y. M. Zhang, "A step-by-step experiment of 3C-SiC hetero-epitaxial growth on 4H-SiC by CVD", Appl. Surf. Sci. 357, 985–993 (2015).

26. Xin, Y. M. Zhang, H. M. Wu, Z. C. Feng, H. H. Lin, R. X. Jia, "Kinetic mechanism of V-shaped twinning in 3C/4H-SiC heteroepitaxy", J. Vac. Sci. Technol. A 34, 031104 (2016).

27. Xin, R. X. Jia, J. C. Hu, and Y. M. Zhang, "Super-V-shaped structure on 3C-SiC grown on the C-face of 4H-SiC", J. Phys. D: Appl. Phys. 49, 335305 (2016).

28. H. S. Kong, B. L. Jiang, J. T. Glass, G.A. Rozgonyi, K. L. More, "An examination of double positioning boundaries and interface misfit in beta-SiC films on alpha-SiC substrates", J. Appl. Phys. 63, 2645–2650 (1988).

29. X. Li, H. Jacobson, A. Boulle, D. Chaussende, and A. Henry, "Double-position-boundaries free 3C-SiC epitaxial layers grown on on-axis 4H-SiC", ECS J. Solid State Sci. Technol. 3, P75–P81 (2014).

30. R. Vasiliauskas, S. Juillaguet, M. Syvajarvi, and R. Yakimova, "Cubic SiC formation on the C-face of 6H-SiC (0 0 0 1) substrates", J. Cryst. Growth 348, 91 (2012).

31. T. Kimoto, H. Matsunami, "Surface kinetics of adatoms in vapor phase epitaxial growth of SiC on 6H-SiC{0 0 0 1} vicinal surfaces", J. Appl. Phys. 75, 850–859 (1994).

32. Turnbull, "Kinetics of heterogeneous nucleation", J. Chem. Phys. 18, 198–202 (1950).

33. K. Nishino, T. Kimoto, and H. Matsunami, "Reduction of double positioning twinning in 3C-SiC grown on α-SiC substrate", Jpn. J. Appl. Phys., Part 1 36, 5202–5207 (1997).

34. M. Marinova, A. Mantzari, A. Andreadou, J. Lorenzzi, G. Ferro, and E. K. Polychroniadis, "Influence of Ga doping on the microstructure of 3C-SiC layers grown on 4H-SiC substrates by VLS mechanism", Phys. Status Solidi C 10, 72–75 (2013).

35. M. Marinova, A. Andreadou, A. Mantzari, and E. K. Polychroniadis, "On the twin boundary propagation in (111) 3C-SiC layers", Mater. Sci. Forum 717–720, 419–422 (2012).

36. N. Jegenyes, M. Marinova, G. Zoulis, J. Lorenzzi, A. Andreadou, A. Mantzari, V. Soulière, S. Juillaguet, J. Camassel, E. K. Polychroniadis, and G. Ferro, "Influence of C/Si ratio on the dopant concentration and defects in CVD grown 3C-SiC homoepitaxial layers", AIP Conf. Proc. 1292, 31–34 (2010).

37. P. G. Neudeck, A. J. Trunek, D. J. Spry, J. A. Powell, H. Du, M. Skowronski, N. D. Bassim, M.A. Mastro, M. E. Twigg, R. T. Holm, R. L. Henry, C. R. Eddy Jr., "Recent results from epitaxial growth on step free 4H-SiC mesas", Mater. Res. Soc. Symp. Proc. 911, 0911-B08–03 (2006).

38. A. Nakajima, H. Yokoya, Y. Furukawa, and H. Yonezu, "Step control of vicinal 6H–SiC (0 0 0 1) surface by H_2 etching", J. Appl. Phys. 97, 104919 (2005).

39. Y. Zhang, L. Liu, and T. Y. Zhang, "Strain relaxation in heteroepitaxial films by misfit twinning: II. Equilibrium morphology", J. Appl. Phys. 101, 063502 (2007).

40. M. Soueidan, G. Ferro, B. Nsouli, F. Cauwet, J. Dazord, G. Younes, and Y. Monteil, "Effect of growth parameters on the heteroepitaxy of 3C-SiC on 6H-SiC substrate by chemical vapor depostion", Mater. Sci. Eng. B 130, 66–72 (2006).

41. P. Pirouz, and J. Yang, "Anti-site bonds and the structure of interfaces in SiC", Mater. Res. Soc. Symp. Proc. 183, 173 (1990).

42. R. L. Schwobel, and E. J. Shipesy, "Step motion on crystal surface", J. Appl. Phys. 37, 3682 (1966).

43. N. Ohtani, M. Katsuno, J. Takahashi, H. Yashiro, and M. Kanaya, "Evolution of macrosteps on 6H-SiC (0 0 0 1): impurity-induced morphological instability of step trains", Phys. Rev. B 59, 4592–4595 (1999).

8

SiC Thermal Oxidation Process and MOS Interface Characterizations: From Carrier Transportation to Single-Photon Source

Yasuto Hijikata

Graduate School of Science and Engineering, Saitama University, Japan

Yu-ichiro Matsushita

Tokyo Tech Academy for Convergence of Materials and Informatics, Tokyo Institute of Technology, Japan

Takeshi Ohshima

Quantum Beam Science Research Directorate, National Institutes for Quantum and Radiological Science and Technology, Japan

1 Introduction

There are a lot of mysteries in the SiC MOS interfaces. For example, the oxidation rate and the carrier transportation characteristics at the oxide–SiC interface are quite different from each other between 4H-SiC [0001] (Si-face) and [000–1] (C-face) [1]. To obtain the intrinsic low-on-resistance of SiC MOSFET, a lot of efforts have been performed for reducing the interface electron traps, which reduce the channel mobility and increase the on-resistance. Some oxide growth methods, such as nitrogen-included oxidation [2] and high-temperature oxidation [3], exhibit an enormous effect. However, side reactions such as threshold voltage instability [4,5] occurs instead. Moreover, for more than ten years, alkaline-metal-doped (sodium, potassium, vanadium, etc.) oxides have been regarded as the best way to improve SiC MOS characteristics [6,7], despite such oxides causing severe deterioration in Si MOS developments. After all, it is inferred that little is known on its oxidation mechanism compared to Si, making an improvement in MOS interface characteristics more difficult.

Another mysterious phenomenon on the SiC MOS interface is that bright and electrically controllable single-photon sources (SPSs) form on the substrate surface by oxidation [8,9]. We have recently been focusing specifically on this SPS because if such an MOS-type single-photon emitter is realized it has a great potential to integrate to MOS-based devices as well as Si-integrated circuits.

In this chapter, we introduce some recent notable results on SiC MOS interface characteristics, as follows:

I. SiC oxidation mechanism and its correlation to the carrier transportation at the MOS interface.

II. Creation of single-photon sources on the oxidized SiC surface.

DOI: 10.1201/9780429198540-10

In Section I, a summary of the results from macroscopic simulations of the SiC oxidation process based on "Si and C emission model [10]" will be introduced. Theoretical studies on the carrier transportation at the MOS interface performed by *ab initio* calculations will also be exhibited.

In the next section, basic properties and structure analyses of the SPSs formed on the surface of an oxidized SiC substrate will be introduced. Besides, some examples of demonstration on the electric control of single-photon emission from the device-embedded SPSs will be shown.

2 SiC Oxidation Mechanism and the Carrier Transportation at the MOS Interface

2.1 SiC Oxidation Process and Characteristics of MOS Interfaces

2.1.1 SiC Oxidation Model

We have attempted to establish an optimized oxidation procedure that reduces the carrier traps at the MOS interface by elucidating the SiC oxidation mechanism. For this purpose, we develop an *in situ* spectroscopic meter for observing the SiC oxidation process in real time. By utilizing the obtained real-time oxide growth rate data, we proposed a SiC oxidation model, termed "Si and C emission model [10]", by which we succeeded in the reproduction of the observed growth rate data. According to this oxidation model, oxide on SiC grows through the following four processes (Figure 1) [11]:

- Stage (i): Si atoms are emitted into the growing oxide and reach the oxide surface, resulting in the oxide growth at the oxide surface.
- In addition to the surface oxide growth mentioned above, another oxide growth is also generated at the oxide–SiC interface. Namely, these oxide growths simultaneously occur, and hence, the oxide growth rate is expressed by

$$\frac{dX}{dt} = \frac{dX}{dt}\Big|_{\text{surface}} + \frac{dX}{dt}\Big|_{\text{surface}} \tag{1}$$

  where $dX/dt\,|_{\text{surface}}$ and $dX/dt\,|_{\text{interface}}$ denote surface growth rate and interface growth rate, respectively.
- Stage (ii): As oxide layer is grown, the emitted Si atoms cannot reach the oxide surface. Therefore, in this oxide thickness region, the interface growth becomes dominant. On the other hand, since the Si atoms accumulate inside oxide near the interface and these atoms prevent the oxidation reaction at the interface, the oxide growth rate reduces as oxide growth. In the case of SiC oxidation, it is considered that C accumulation also occurred, which also reduces the interface oxide growth rate.
- Stage (iii): The interface oxide growth is still dominant in this stage. The difference from Stage (ii) is that the Si and C accumulation is already in a saturated state and these atoms start a diffusion into the SiC substrate side.
- Stage (iv): In this stage, the oxide has been very thick. Hence, since the oxidants diffusing inside oxide gradually reduce, their amount becomes insufficient for oxidation. Therefore, in Stage (iv), oxide growth rate can be expressed as

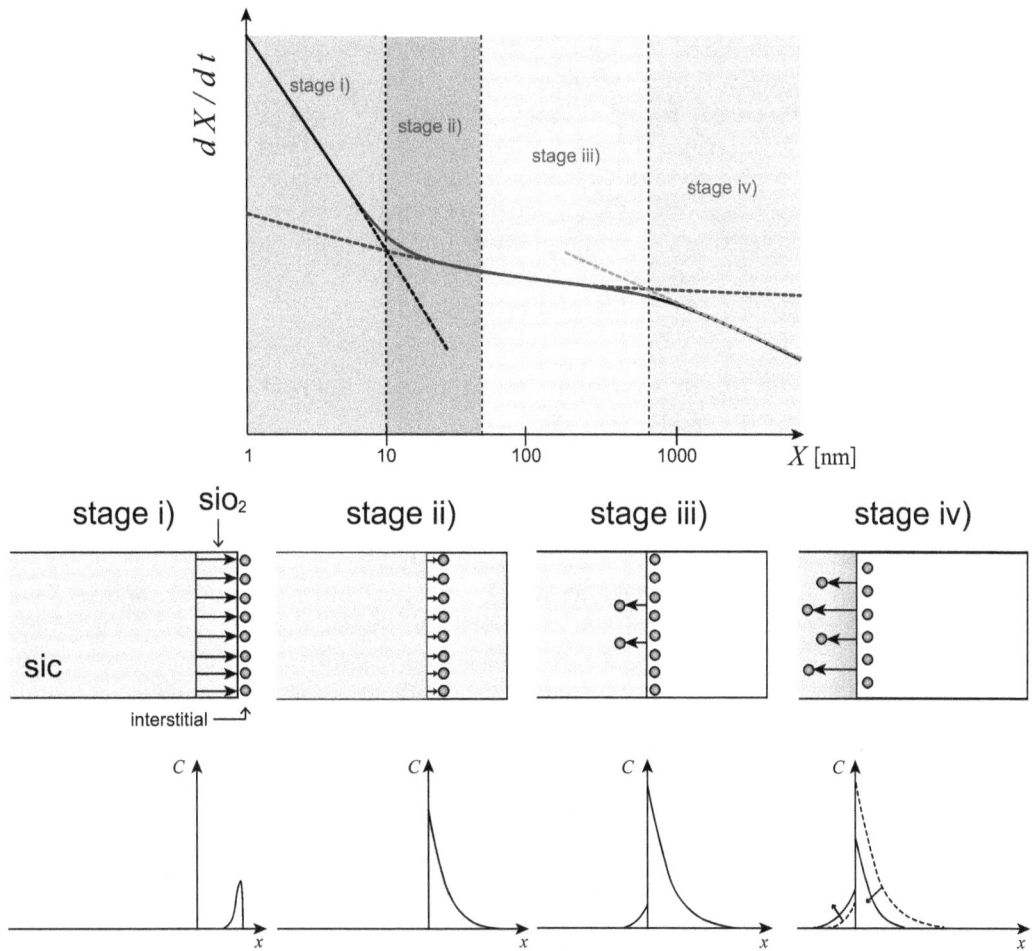

FIGURE 1
The four-stage oxidation process deduced by "Si and C emission model." Reproduced from [11]. CC BY 3.0.

$$\frac{dX}{dt}^{-1} \propto \left(\frac{d[O]_{ox}}{dt}\right)^{-1} + \left(\frac{d[O]_{int}}{dt}\right)^{-1} \tag{2}$$

where $[O]_{ox}$ and $[O]_{int}$ mean the amount of oxidant consumed by oxidation and that arriving at the interface via diffusion inside oxide, respectively. Since $[O]_{int}$ decreases with increasing oxide thickness, the term $[O]_{int}$ term becomes a rate-limiting process. As a result, the oxidation rate reduces, and then, Si and C emission into the substrate is further enhanced.

2.1.2 Carrier Traps and Point Defects' Generation at the MOS Interface

Our oxidation model performs simulations of the concentrations of oxidant, Si and C interstitial, as well as oxide growth rate. These concentrations provide hints especially on the interface structure that includes point defects. According to several articles, point defects such as Si vacancies (V_{Si}) [12] or C di-interstitials [13] are promising candidates of the origin of electron traps at the MOS interface. Therefore, we believe that such a point defect evaluation is very significant to improve the carrier transportation at the MOS interface. In addition, it has been clearly revealed that C vacancies (V_C) are filled with C interstitials generated at the MOS interface during oxidation. V_Cs are one of the most unpleasant defects in the bipolar high-power devices because they act as recombination centers in the depletion layer. Hence, our model is also valuable in the developments of bipolar devices. In Section II, we will describe the single-photon source formed on the surface of SiC by oxidation. Again, it is believed that the generation of point defects during oxidation is important for the formation of a single-photon source.

2.2 Theoretical Studies on the Carrier Transportation at the MOS Interface Performed by *Ab Initio* Calculations

Power losses in switching devices consist of two major components: electrical resistance losses in the on-state (on-resistance losses) and power losses during switching transients (switching losses). The on-resistance loss is dominant in the low operating frequency region, while the switching loss is dominant in the high operating frequency region. There is a trade-off between on-resistance and switching loss. Until now, power devices have been created primarily based on Si, and MOSFETs and IGBT devices have been widely used. However, with Si-MOSFETs, the on-resistance increases significantly as the withstand voltage increases, making it particularly difficult to create power device elements with high withstand voltages exceeding 1 kV. In addition, the application of Si-IGBTs is currently limited to the low operating frequency region in order to reduce switching losses.

In this context, SiC is attracting significant attention as an alternative power semi-conductor material to Si [14]. SiC has a tetrahedral structure composed of Si and C as a building unit, and is composed of stacked units in the direction of one bonding axis. It is known that there are more than 200 types of structural polytypes due to differences in the stacking sequences (see Figure 2). In order to distinguish these crystalline polytypes, they are designated by specifying the repetition period n in the stacking direction and the crystal symmetry (index, C, H, or R for cubic, hexagonal, and rhombohedral crystals, respectively). According to this nomenclature, the wurtzite structure is called the 2H structure

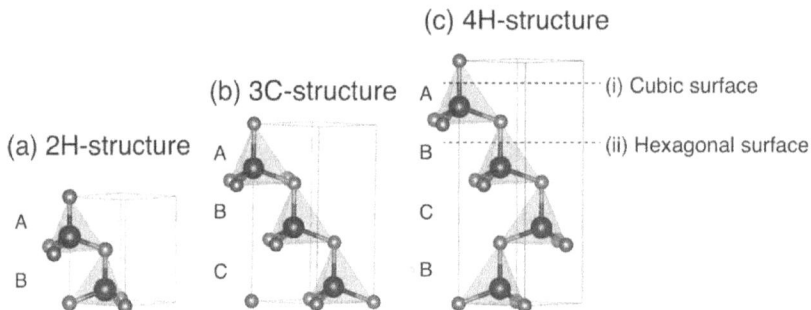

FIGURE 2

Atomic structures of (a) 2H-SiC, (b) 3C-SiC, and (c) 4H-SiC polytypes. Different stacking sequences of atomic bilayers are depicted by the letters, A, B, and C. In (c), two dotted lines show (i) the cubic and (ii) the hexagonal surface planes. When SiO_2 films contact with the SiC substrate at the position (i) or (ii), we call it the cubic or the hexagonal interface. Reproduced with permission of ACS [15].

TABLE 1

Physical properties of Si, 4H-SiC, Gan, β-Ga$_2$O$_3$, and diamond. [14, 16–20]

	Si	4H-SiC	GaN	β-Ga$_2$O$_3$	Diamond
Band gap (eV)	1.12	3.26	3.42	4.9	5.5
Electron Mobility (cm²/Vs)	1400	1000($\perp$c) 1200(//c)	1300(Bulk) 2000(2DEG)	300	4000
Hole Mobility (cm²/Vs)	450	120	30		3800
Breakdown Electric Field (MV/cm)	0.3	2.5–2.8	2.5–2.8		
Thermal Conductivity (W/cmK)	1.5	4.9	2.0	0.2	20

(stacking sequence: AB) and the zincblende structure is called the 3C structure (stacking sequence: ABC). As shown in Figure 2, 4H-SiC (stacking sequence: ABCB), known as one of the most stable structures among the polytypes of SiC, has 50% cubic and 50% hexagonal stacking structure [15], which has already been put into mass production process and is now available in 150-mm diameters (6-inch) are available on the market. At the same time, 4H-SiC is known to have excellent bulk properties: band gap three times larger, breakdown electric field ten times larger, and thermal conductivity three times larger than Si. Table 1 compares the bulk physical property values. Among them, the breakdown electric field is the most important physical property value for low on-resistance. The theoretical limit for the on-resistance R_{on} can be calculated by the following equation:

$$R_{on} = \frac{4V_B^2}{\varepsilon \mu E^3},$$ (3)

where V_B is withstand voltage, ε relative permittivity, μ carrier mobility, and E breakdown electric field strength. In other words, the higher the breakdown electric field, the lower the on-resistance can be realized. 4H-SiC has a breakdown electric field ten times higher than that of Si, which means that SiC can theoretically reduce the on-resistance by two orders of magnitude. 4H-SiC has extremely excellent physical properties as a power semiconductor. Henceforth in this section, unless otherwise specified, the term "SiC" will refer to 4H-SiC.

Despite the excellent physical properties of SiC, the actual SiC-MOSFET device characteristics are far from what is expected from the physical properties [14, 16–20]. This is due to the high density of interface levels at or near the SiC/SiO$_2$ interface in SiC-MOSFET devices. High-low method experiments have shown that the density of interfacial defects increases exponentially as one approaches the conduction-band minimum, typically reaching 1×10^{12} to 10^{13} cm^{-2} eV^{-1}, which is two orders of magnitude larger than that at the Si/SiO$_2$ interface [14]. Although attempts have been made to identify defects near the SiC/SiO$_2$ interface from both experimental and theoretical perspectives, they have not yet been fully elucidated. In this section, we review the recent progress of theoretical studies on interface defects and focus on a recent topic: a new mechanism of interface defects that has not been considered before.

The following are the bullet points that have recently been clarified by theoretical calculations. Two types of interface defect candidates have been proposed.

- One of the candidates is a fluctuation at the conduction-band minimum, which is peculiar to SiC.
- The other candidate for interface defects is residual carbon defects at the interface.

In the following, we will explain the microscopic nature of each interface defect candidate, how the experimental facts can be explained in terms of these interface defect candidates, and introduce recent experimental facts that suggest the validity of the interface defect candidates. In addition, I will introduce a proposal of a method to reduce the interface defects from the viewpoint of these interface defect candidates and a recent experimental research report on the effectiveness of the method.

Before introducing the first SiC/SiO$_2$ interface defect candidate, I will explain a little more about the bulk properties of SiC and its strange electronic properties. It is the large band gap dependence of SiC on the polytypes. According to experimental results, the band gap of 4H-SiC (stacked sequence: ABCB) is 3.26 eV, while 3C-SiC (stacked sequence: ABC) is 2.36 eV and 6H-SiC (stacked sequence: ABCACB) is 3.02 eV [14]. From the viewpoint of chemical bonding, all of these polytypes consist of sp^3 bonds, and there is no significant difference in the local atomic structure. Nevertheless, the fact that the difference in stacking structure alone produces a large change in band gap is known to be a phenomenon that is difficult to understand from the viewpoint of chemical bonding. Theoretical calculations based on the density-functional theory (DFT) have provided a microscopic clarification of this phenomenon [21, 22]. As a result, it was found that the microscopic mechanism originates from the peculiar character of the wave function at the conduction-band minimum of SiC. The peculiarity is the appearance of a "floating state" in the nano internal space within the structural channel (the [110] channel in the 3C structure), rather than an atomic orbital character as usually thought [21]. The floating state is an electronic state that is not in the vicinity of the atoms, but is spread out in a floating manner within the structural channel. Therefore, the wave function at the conduction-band minimum is sensitive to the deformation of the structural channel, and if the length of the structural channel changes among structural polytypes, the quantum confinement effect changes accordingly, and the electronic level is significantly modulated. In fact, theoretical calculations have shown that 24 SiC polytypes have been calculated and the correlation between their band gaps and structural channel lengths shows a clear relationship, which can be well fitted by the formula for the following equation as an electronic level of a 1D quantum well (see Figure 3) [22]:

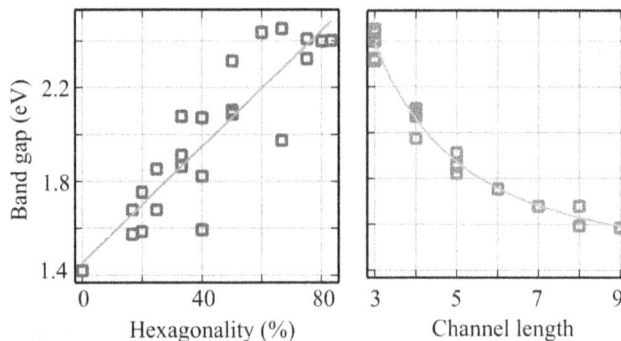

FIGURE 3

Band gaps for 24 representative SiC polytypes calculated in generalized gradient approximation as a function of the hexagonality (left panel) and a function of the channel length (right panel). In each panel, the fitting function (see text) is also shown. Reproduced with permission of APS [22].

$$\varepsilon = \varepsilon_{3C} + \frac{\pi^2 \hbar^2}{2m^* \left(l + \Delta\right)^2} , \qquad (4)$$

where ε is the band gap in each polytype, ε_{3C} the band gap in 3C-SiC, $\hbar$ Planck's constant, m^* the effective mass of the electron in the [110] direction, l the channel length normalized by the building unit length, and Δ the outgrowth of the wave function at the conduction band minimum from the channel space. For example, in 3C-SiC, the length of the structural channel can be regarded as infinite, and the band gap is 2.36 eV, the smallest value among SiC, because there is no quantum confinement effect. On the other hand, in 4H-SiC, the length of the structural channel is 3 in the unit of primitive structural building unit, and the wave function at the conduction-band minimum is confined in a narrow internal space with a finite length, resulting in a larger band gap of 3.23 eV. 6H-SiC has a structural channel length of 4, leading to a weaker quantum confinement effect and smaller band gap than 4H-SiC, i.e., 3.02 eV. The anisotropy of effective mass, etc. can also be explained in a straightforward manner by considering the peculiar character of the wave function at the conduction-band minimum.

Let us move on to the main topic, the first interface defect candidate. It has been explained that the difference of stacking structure in bulk SiC has a significant effect on the SiC band gap. Next, we will discuss how the stacking structure in 4H-SiC near the 4H-SiC(0001)/SiO$_2$ "interface" can affect the electronic state near the interface. For simplicity, we consider an ideal 4H-SiC(0001)/SiO$_2$ interface with no interface defects (neither dangling bonds nor defect structures). In the ideal 4H-SiC(0001)/SiO$_2$ interface, two types of possible interface stacking structures can be found by considering the stacking structure of 4H-SiC sides (i) (hereafter called cubic interface) and (ii) (hereafter called hexagonal interface) in Figure 2(c). Figure 4 schematically shows the band alignment of the 4H-SiC(0001)/SiO$_2$ interface; the stacking sequence of the 4H-SiC is BCBA, and the cubic interface in Figure 4(a) shows the case where the SiO$_2$ film is attached to layer A. On the other hand, the hexagonal interface in Figure 4(b) shows the case where the SiO$_2$ film is bonded to layer B. Furthermore, Figure 4(c) shows the stacking structure when the topmost surface stacking sequence changes from B to A in Figure 4(b) (hereinafter referred to as stacking-fault interface). In fact, it has been experimentally reported that such a stacking

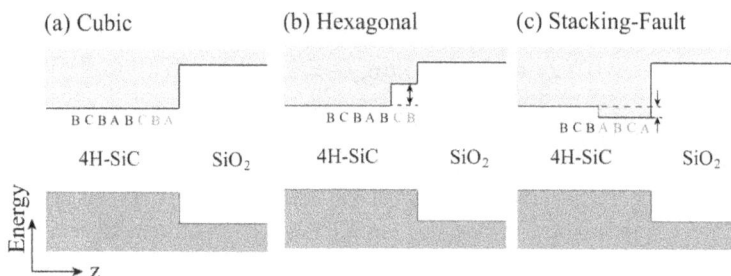

FIGURE 4

Schematic pictures of the band alignment along the direction perpendicular to the interface (z-direction) for the three possible interface structures of SiC/SiO$_2$. (a) The cubic interface (BCBA-stack/SiO$_2$), (b) the hexagonal interface (ABCB-stack/SiO$_2$), and (c) the stacking-fault interface (ABCA-stack/SiO$_2$). The stacking sequence near the interface is shown by the letters. The red letters denote the region in which the interstitial channel is connected. Reproduced with permission of ACS [15].

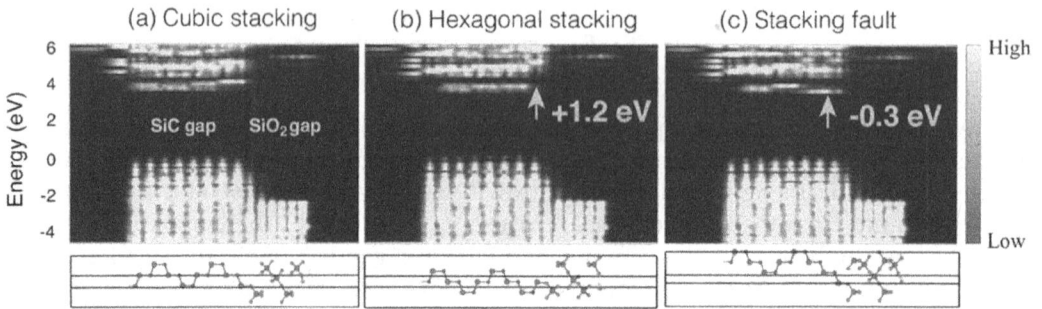

FIGURE 5

Calculated localized density of state (LDOS) by the Heyd-Scuseria-Ernzerhof (HSE) functional for the cubic-stacking (BCBA-stack/SiO$_2$) (a), the hexagonal-stacking (ABCB-stack/SiO$_2$) (b), and the stacking-fault (ABCA-stack/SiO$_2$) (c). SiC/ SiO$_2$ interfaces as functions of the energy and the z-coordinate are perpendicular to the interface. The magnitude of LDOS is presented by the color code shown in the legend. The left and right sides correspond to the SiC and SiO$_2$ regions. The origin of the energy is set to be the valence band top of SiC at the interface. Below each LDOS, the corresponding atomic configuration is illustrated where blue, brown, red, and white balls depict Si, C, O, and H atoms, respectively. Reproduced with permission of ACS [15].

fault structure appears on the 4H-SiC(0001) surface [23, 24]. It is important to note that even in the ideal 4H-SiC(0001)/SiO$_2$ interface with no defect structure, the interface structure is not unique, and several interface stacking sequence models are possible based on the "stacking structure difference" of 4H-SiC. We have explained that the SiC band gap in bulk SiC is strongly dependent on the stacking structure. In the same way, we will discuss how the stacking structure affects the interfacial electronic properties at the interface for the three interface structure models shown in Figure 4. In the hexagonal interface, the stacking sequence is …ABCB/SiO$_2$, and we can treat the stacking structure as 2H-SiC locally up to the third layer near the interface. On the other hand, at the stacking-fault interface, the stacking sequence is …BABCA/SiO$_2$, and is treated as 6H-SiC appeared locally up to the fourth layer near the interface. These differences in the stacking structure cause modulation of the electronic level at the conduction-band minimum in the vicinity of the interface. In other words, the electronic state near the interface can behave very sensitively to the modulation of the structural channel near the interface, just as it does in the bulk. In fact, this has been confirmed by DFT-based calculations as shown in Figure 5 [15]. Indeed, the conduction-band minimum is found to shift upwards by 1.2 eV at the hexagonal interface and shift downwards by 0.3 eV at the stacking-fault interface. Although the ideal interface is considered here, in reality, the conduction-band minimum of 4H-SiC itself can change continuously, from 0.3 eV below to 1.2 eV above, by considering the additional strain and other effects [25–28]. In other words, the conduction-band minimum of SiC itself fluctuates at the 4H-SiC/SiO$_2$ interface. This means that the stacking structure sensitivity of the band gap in bulk SiC also appears at the interface, and even if there are no defects at the interface, the SiC(0001)/SiO$_2$ interface is inherently prone to conduction band fluctuations, which can cause effective potential fluctuations from the electron carrier's point of view [16]. The behavior of conduction carriers in a fluctuating effective potential is a physical phenomenon known for a long time as Anderson localization and can be considered as a new scattering mechanism at the SiC(0001)/SiO$_2$ interface. Considering the behavior of electron carriers in terms of effective potential fluctuation or Anderson localization at the interface, the fluctuation of the stacking structure near the interface pushes down the edge of the conduction band, which can be regarded as the interface defect density observed in our experiments. An attempt has also been made to reconsider the experimental results in terms of Anderson localization [29]. By comparison with the experimental results, a length scale of about 4.3 nm was estimated for the in-plane direction fluctuations. This new scattering mechanism needs to be discussed in more detail from a theoretical point of view.

Recently, the strange behavior of SiC/SiO_2 interface defects was reported experimentally by Ito *et al.* [30]. In their experiments, when the conduction band levels were shifted relative to the vacuum level using the quantum confinement effect of a gated electric field, the interface defect levels also shifted with the conduction band levels. This is a behavior that cannot be understood for conventional interface defects, such as impurities. Conventional interface defects should create an electronic level independent of the quantum confinement effect, and should have an electronic level that is uniquely defined with respect to the vacuum level. If we understand that the nature of the interface defect level is the "conduction band itself," which is Anderson-localized due to fluctuations in the effective potential, it is not difficult to understand this experiment qualitatively. Further detailed investigation is needed in the future.

Now, let us move on to the second candidate for interface defects. Another candidate for interface defects is residual carbon defects at the interface, where SiO_2 and CO_2 are produced as end products when SiC is oxidized fully. On the other hand, the intermediate products are still poorly understood. Experiments have shown that carbon atoms are released and diffused into the SiC substrate, are emitted into the SiO_2 film, and deposit carbon residues at the interface, but it is not known which of these is the stable main product. Which ones are candidates for interface defects is still under active debate. Although there have been many reports on the electronic states of these defects based on theoretical approaches, there has been no discussion on which of them is the main product. Recently, a theoretical

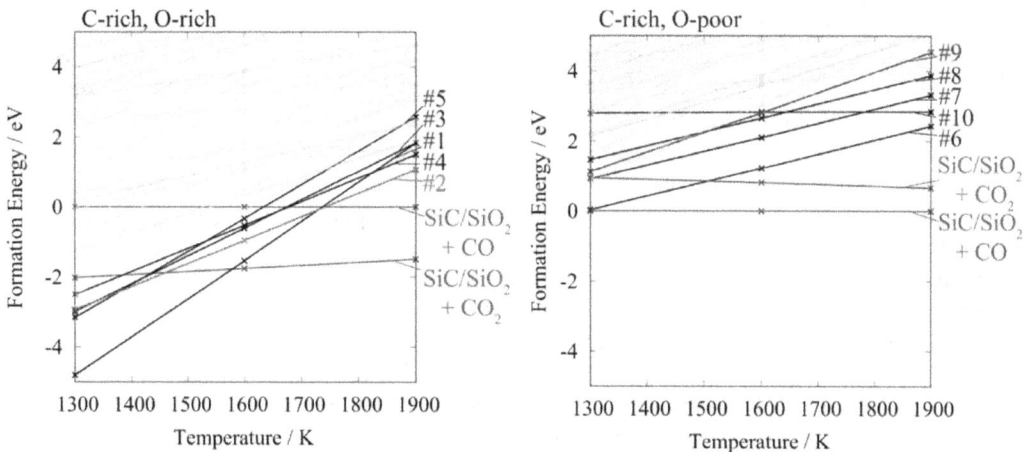

FIGURE 6

Temperature dependence of defect formation energies in comparison with the energies of gaseous CO and CO_2 molecules at the (a) C-rich O-rich and (b) C-rich O-poor limits. The zeros of the formation energies are set at the energy of a CO molecule. The temperature dependence of defect formation energies was calculated by correcting the chemical potentials of atomic species at finite temperatures by referring to thermochemical tables. Lines in dark contrast (red, blue, and black) indicate the defects stable at 1600 K, which is about the typical experimental temperature of SiC oxidation, The red, blue, and black solid lines correspond to the energies of defects in SiC, SiO_2, and SiC/SiO_2 interfaces, respectively. Green lines represent the energies of gaseous CO and CO_2 molecules. Reproduced with permission of AIP [31].

(a) 4H-SiC(0001)

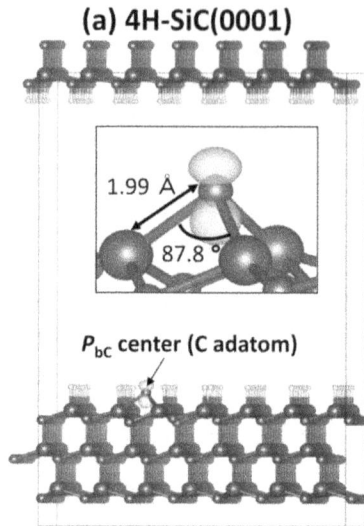

P_{bc} center (C adatom)

FIGURE 7

488-atom H-terminated 4H-SiC(0001) unit cell including a single P_{bc} center used for first-principles calculations. Reproduced with permission of AIP [33].

study has been performed to determine which of the three regions (SiC side, SiO_2 side, and just at the SiC/SiO_2 interface) is the most energetically favorable for carbon-related defects (see Figure 6). The study reported that most of the carbon-related defects distributed in the region just at the SiC/SiO_2 interface are energetically stable. Comprehensive DFT calculations revealed that carbon tends to precipitate once at the interface, i.e., as a mid-product [31]. Among the carbon-related defects revealed, there are carbon adatoms (P_{bc} centers) with dangling bonds at the interface (see Figure 7). This is a defect that has recently been identified by electron-spin resonance (ESR) measurements [32, 33], and certainly shows that theoretical calculations are consistent with experiments. Unfortunately, the P_{bc} center itself has a level near the valence band, not an interface defect that creates a level near the conduction band edge, but the agreement between experiment and theory is remarkable. Theoretical calculations report that carbon double bonds at the interface create a level near the conduction band, and these are candidates for interface defects [31, 34].

The presence of residual carbon defects deposited at the interface is also consistent with recent secondary-ion mass spectrometer (SIMS) experimental results [35]. In fact, although carbon defects in the SiO_2 film are too small for the experimental detection limit in the oxidized interface as it is, annealing under Ar atmosphere succeeded in pulling out the residual carbon defects accumulated at the interface from the interface, and a large amount of carbon defects were directly observed in the SiO_2 film after Ar annealing by SIMS.

We have reported two candidates for interface defects from the theoretical calculation approach. Reviewing the results in Figure 6 again reveals another hidden message. It is the fact that as long as SiC is thermally oxidized, it inevitably goes through interface residual carbon defects as intermediate products of carbon. This might mean that it is better to give up thermal oxidation of SiC if we want to avoid carbon-related defects at the interface. One recent experiment to note is that of epitaxial Si crystals on a SiC substrate, thermally oxidized at a relatively low temperature where only the Si crystals are oxidized, forming a SiC/SiO_2 interface (see Figure 8) [36]. Here, it should be noted that if NO gas is used as the interface treatment, the interface will be oxidized by the oxygen atoms contained in the

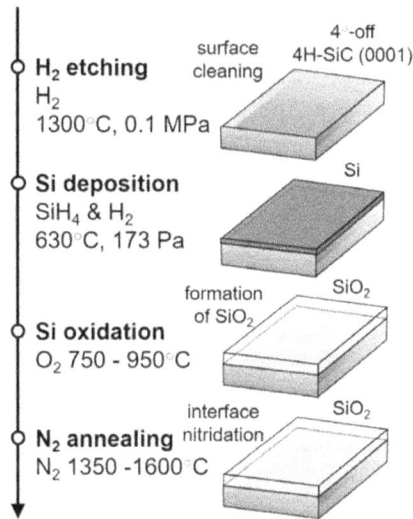

FIGURE 8

Novel schematic process flow of forming SiC/SiO$_2$ structures in the present study. The conditions (i.e., tempera-ture and pressure) and roles of each process are described in the figure. Reproduced from [36]. CC BY 4.0.

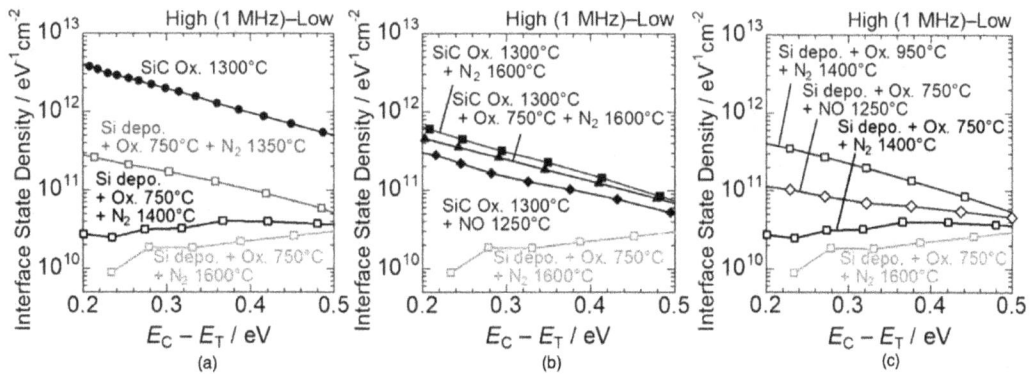

FIGURE 9

Energy distribution of density of interface density (D_{IT}) for SiC MOS structures obtained by a high (1 MHz)–low method; (a) impact of N$_2$ annealing temperature, (b) comparison with typical methods (NO and N$_2$ annealings), and (c) impacts of oxidation temperature and post-oxidation treatment. The possible error in the D_{IT} values is estimated as about ± (2–3) × 10^{10} eV^{-1} cm^{-2}. Reproduced from [36]. CC BY 4.0.

NO gas. Therefore, it is important to use N$_2$ gas as the interface treatment. As a result, it was shown that the interface defect density can be reduced by about one order of magni-tude compared to the conventional method as shown in Figure 9 [36,37]. This is thought to be an effect of the reduction of carbon-related defects as interface defects.

Theoretical calculations have revealed the SiC/SiO$_2$ interface. We have also explained how experiments supporting the theoretical results are being obtained. However, it is still true that there are many experimental results that cannot be fully explained from the the-oretical viewpoint, and it is important to clarify the interface structure from both theoret-ical and experimental viewpoints and its effect on devices.

3 Creation of Single-Photon Sources at the MOS Interface

3.1 Basic Properties and Structure Analyses of the SPSs

In recent years, our group found that oxidation of SiC substrate forms bright single-photon sources (SPSs) at the MOS interface [9] (hereafter termed "surface SPS"). There are notable features in the surface SPS and one of them is that it has an extremely high emission-rate. Another one is that single-photon emission is electrically controllable at room temperature by both current injection and electric field [9, 38–39] (will be mentioned in detail next section), which enables us to expect an innovative MOS-based SP emitting device.

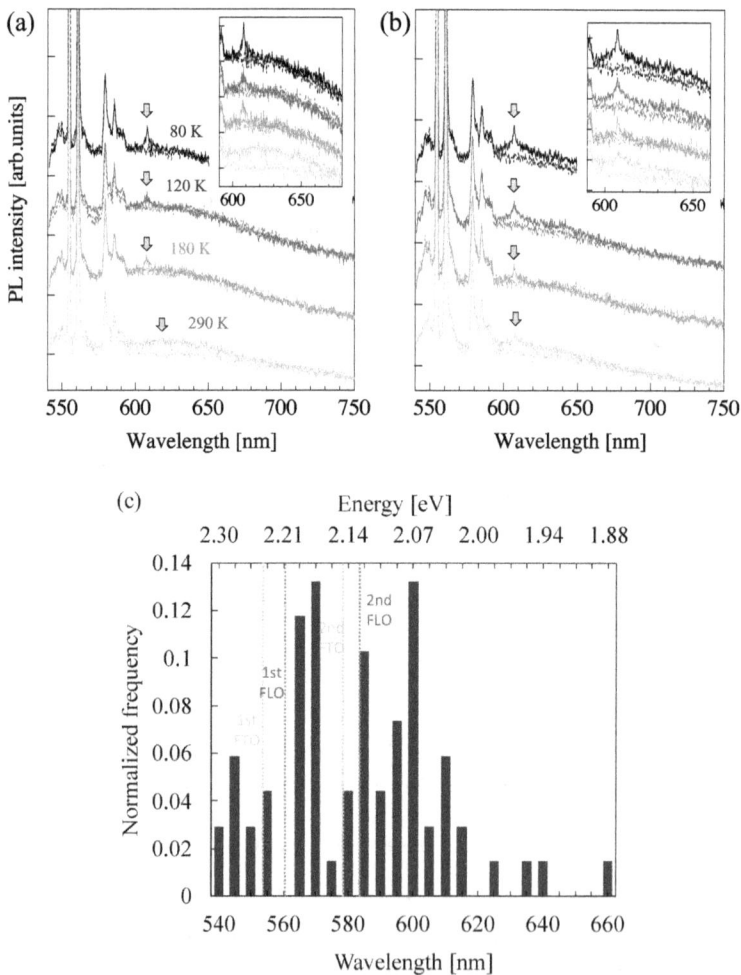

FIGURE 10

PL spectra from surface SPSs at various temperatures. (a) Broad peak at 290 K, and (b) sharp peak at 290 K. The arrows indicate the center wavelength and the solid and broken lines are spectra from SPS and background (BG), respectively. The measurements were carried out at temperatures from 290 to 80 K. Each spectrum is offset for clarity. The insets show spectra at a wavelength range around emission peak. (c) Histogram of ZPL obtained from PL measurements at 80 K. The broken lines in orange and in green represent the positions of the 1st and 2nd FTO phonon lines and those of the 1st and 2nd FLO phonon lines, respectively. Reproduced from [41]. CC BY 4.0.

Figure 10 shows photoluminescence (PL) spectra from typical two types of the surface SPS and histogram of zero-phonon line (ZPL) obtained from PL measurements at 80 K for SPSs more than 50. As shown in (a) and (b), it is found that there are two types of SPS that have broad spectrum and sharp spectrum at 290 K, respectively [40, 41]. The histogram indicates that the radiation wavelength is different from one another for each SPS. Since such a variation in radiation properties might hinder the application of surface SPS, some solution method should be found.

However, since the defect structure of surface SPS is unknown, it is difficult to find a control method of the specification of surface SPS. According to Lohnman *et al.* [9], even oxide with thickness below 1 nm produces the surface SPSs and most of the SPSs disappear after removal of oxide. Nevertheless, it is considered that the SPSs are located in the SiC-side because the optical polarization property of them is linear polarization and its orientation angle agrees with the crystal axis of SiC [9]. Based on this information, we inferred that oxygen was the origin or one of the compositions in complex of the surface SPSs. To verify this assumption, the surface SPSs were formed using isotope oxygen (^{18}O).

Figure 11 shows the schematic of electronic transitions along the configurational coordinate for an SPS [42]. In general, phonon side-band (PSB) appears in the longer wavelength region of PL spectrum because electronic transition usually induces coupling with phonons. Therefore, in the case that surface SPSs are formed using ^{18}O with heavier mass, the PSB should be shrunk and the ZPL appears in the shorter wavelength side compared with natural oxygen.

According to Hijikata *et al.* [42], the observed data from the surface SPSs formed with ^{18}O surely agreed with those expected. In addition, Matsushita *et al.* revealed that several oxygen-included defects have a highest-occupied molecular orbital to lowest-unoccupied molecular orbital (HOMO-LUMO) gap that is nearly equal to the photon energy of surface SPSs by performing *ab initio* calculations [43]. Therefore, we believe that the origin of the surface SPS includes oxygen.

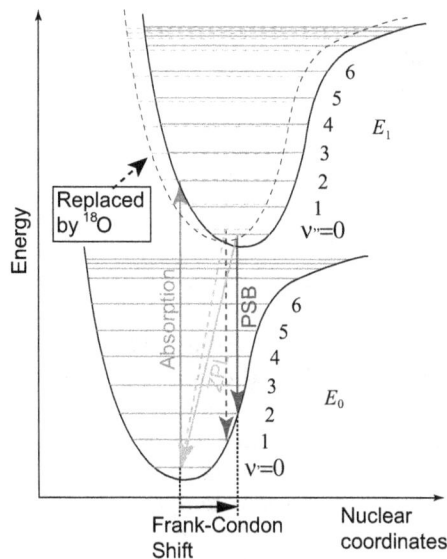

FIGURE 11

Schematic of electronic transitions with phonon coupling along the configurational coordinate for an SPS. Reproduced from Ref. [42]. CC BY 3.0.

Another issue, i.e., the variation in radiation wavelength, was also solved by the-oretical and experimental studies. For theoretical study, Matsushita *et al.* performed *ab initio* calculations for HOMO-LUMO gaps with different defect locations with respect to amorphized interlayer. As a result, HOMO-LUMO gaps varied in the range of 0.35 eV [43], which is in good agreement with the observed data [40]. In addition, the annealing tem-perature that changes radiation wavelength of the surface SPSs was 600 °C, which agrees with the temperature that changes the MOS interface structure and its carrier traps' density [41]. Therefore, it is inferred that the variation in radiation wavelength is attributed to the MOS structural change around the surface SPS.

3.2 Electrical Control of the Single-Photon Sources Formed at the MOS Interface

Controlling photon emission by device operation is important from the point of view of the development of single-photon emitter devices. In this section, controlling photon emission from surface SPS embedded in electronic devices is described [39]. Figure 12(a) shows the schematic drawing of an in-plane SiC p^+nn^+ diode used in this study. The diode was fabricated on an n-type 4H-SiC epitaxial layer grown on an n-type 4H-SiC substrate (Si face, 4° off). The donor concentration of the epitaxial layer is 4.7×10^{14} /cm^3. The n^+ (donor: 2.0×10^{20} /cm^3) and p^+-type (acceptor: 5.0×10^{19} /cm^3) regions were formed by phos-phorus and aluminum ion implantation at a temperature of 800 °C and subsequent thermal annealing at 1800 °C for 10 min in argon atmosphere. The surface of the diode was covered with an oxide layer, which was grown by pyrogenic oxidation (H$_2$:O$_2$ = 1:1) at a tempera-ture of 1100 °C. The thickness of the oxide layer was estimated to be 41 nm. Aluminum is used for metal electrodes on the n^+ and p^+-type regions. Figure 12(b) shows current-voltage (*IV*) characteristics of the diode with and without laser irradiation (wavelength: 532 nm, power: 1.0 mW).

Luminescence properties of surface SPSs in the SiC diode under forward bias voltage was investigated at room temperature. Here, no excitation laser was used since SPSs in the diode emit photons due to carrier injection (EL). EL maps of a region near the p^+n junction with three different forward bias voltages are shown in Figure 13(a). The EL intensity of the n-region near the p^+n junction obviously increases with increasing forward bias voltage. However, this luminescence is not only from the surface SPS but also from residual defects (recombination centers) in the SiC bulk region. Thus, minority carriers injected into the n-region were trapped and radiatively recombine in color centers [38]. Here, we focus on a surface SPS named "X", which is encircled by a black dashed line in Figure 13(b). This surface SPS shows forward bias dependence of EL intensity. Figure 13(b) shows the EL spectra for the surface SPS "X" under different applied voltage. The shape of EL spectra is not affected by applied voltage, and the intensity increases with increasing voltage. The antibunching behavior for the surface SPS at 7.5 V obtained from second-order autocorrel-ation function $g^2(\tau)$ measurement using HBT interferometry is shown in Figure 13(c). The value of $g^2(\tau = 0)$ is estimated to be 0.34 and as a result, it can be concluded that this color center has single-photon emission characteristics.

Next, photoluminescence (PL) properties for surface SPSs under revise vias are mentioned. In this study, a 532 nm laser was used for excitation of surface SPSs (PL), and all measurements were carried out at room temperature. Figure 14(a) shows a PL map obtained from a SiC diode using a home-built confocal microscope. Many bright spots are observed in all n^+-, n-, and p^+-regions although the overall luminescence intensities in the n^+-region is higher than that in the other regions. However, this result is not correlated to surface SPS but to D1 center. Thus, n^+-region was formed by ion implantation and its fluence was 4 times higher than that of the p^+-region. As a result, the larger amount of D1

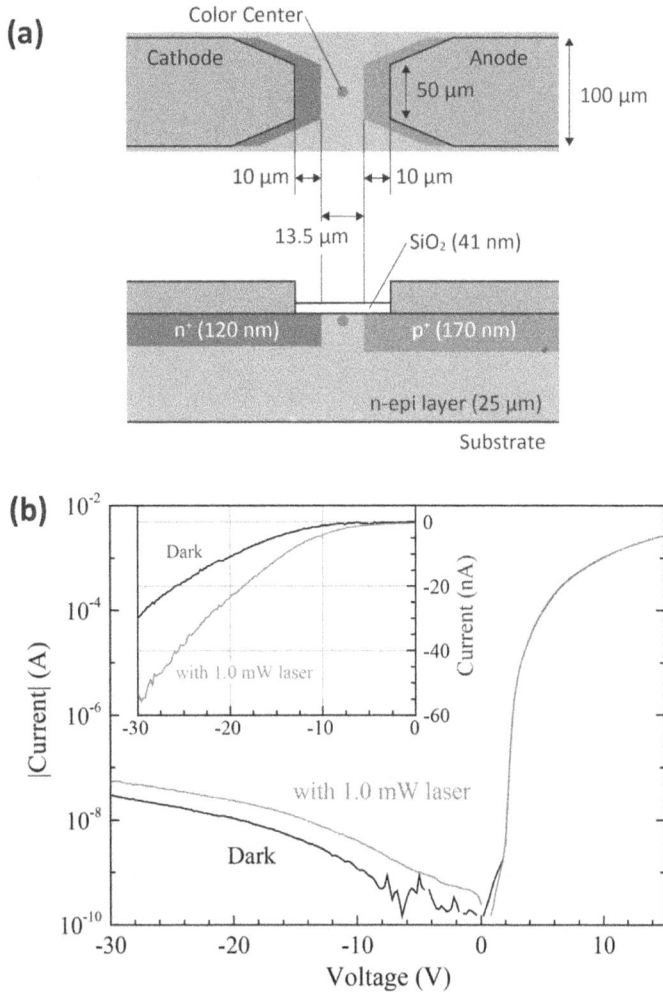

FIGURE 12

(a) A schematic drawing of planar-type 4H-SiC diode used in this study and (b) the *IV* characteristics under dark condition (black) and with 532 nm, 1.0 mW laser illumination (red) at room temperature. The ordinate is the absolute value of current. The inset depicts the same data with linear ordinate. Reproduced with permission of ACS [39].

centers remained even after the thermal annealing at 1800 °C since the D1 center is thermally quite stable. However, we can distinguish between the D1 center and surface SPS since the spectral range of the D1 center is 450–650 nm at room temperature and different from surface SPS. Here we focus on a surface SPS created in the *n*-region close to the boundary of *p*+-region (shown as the black open square in Figure 14(a)). Hereinafter, this surface SPS is called as "Y". Figure 3(c) shows PL maps in the black open square shown in Figure 14(a) with and without a reverse bias of −30 V. The PL intensity of "Y" increases by applying −30 V although no significant change in PL intensity is observed for surface SPS named "S" as shown in Figure 14(b). PL spectra for the surface SPSs "Y" and "S" are plotted in Figure 14(c). For the surface SPS "Y", the obvious zero phonon line is not observed and a broad spectrum with a peak is about 630 nm is shown. This broad PL spectrum can be explained in terms of the phonon sideband. The color center "S" also

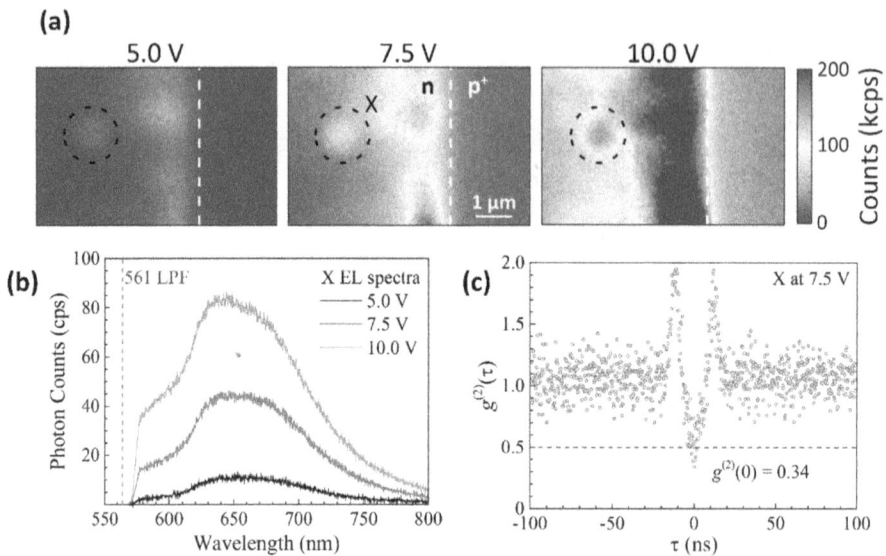

FIGURE 13
(a) EL-CFM maps with different forward bias voltages. White dashed lines show the boundary between the epi-region (*n*-region) and the *p*⁺-region. (b) EL spectra of a color center "X" in (a) with different bias voltages. (c) Measurement of the second-order autocorrelation function of the color center "X" at 7.5 V. The uncorrelated background was corrected and the value of $g^{(2)}(0)$ is shown in the figure. Reproduced with permission of ACS [39].

shows the broad spectrum with a peak around 670 nm. Since the PL peak wavelength for "Y" is slightly shorter than that for "S", the structure of these color centers might not be exactly the same. By applying the revise bias of –30 V, the intensity of "Y" increases by 11% without any change in the spectrum shape. This indicates that the emission of photon from "Y" simply increases without any structural changes. To confirm single-photon emission properties for "Y", the second-order autocorrelation function measurement was carried out (Figure 14(d)). Since the value of $g^{(2)}(\tau = 0)$ for "Y" is estimated to be 0.46 (below 0.5), the surface SPS "Y" is concluded to be a single-photon emitter. In addition to antibunching behavior, since bunching behavior is also observed between 10 and 20 ns, "Y" is not a single-photon source with two simple energy levels. Thus, the metastable level exists in addition to ground and excited states.

Here we apply the three-level system. In the three-level system, the antibunching characteristics can be described to be the double exponential function:

$$g^2(\tau) = 1 - (1+\alpha) \cdot \exp\left(-\frac{\tau}{\tau_1}\right) + \alpha \cdot \exp\left(-\frac{\tau}{\tau_2}\right), \qquad (5)$$

where τ is the delay time. The values of α, τ_1, and τ_2 are fitting parameters relating to the nonradiative decays and transitions via metastable state, the transition between ground and excited state, and the behavior of the metastable state, respectively [44]. The fitting result is plotted as the solid line in Figure 14(d). In this fitting, the values of 3.4, 3.2, and 6.4 ns are obtained for α, τ_1, and τ_2, respectively. It should also be mentioned that the antibunching characteristics of EL shown in Figure 13(c) cannot be fitted by Equation (5). This suggests that the mechanism of EL for surface SPS is not the same as that of PL, and another level should be considered to understand the mechanism of EL.

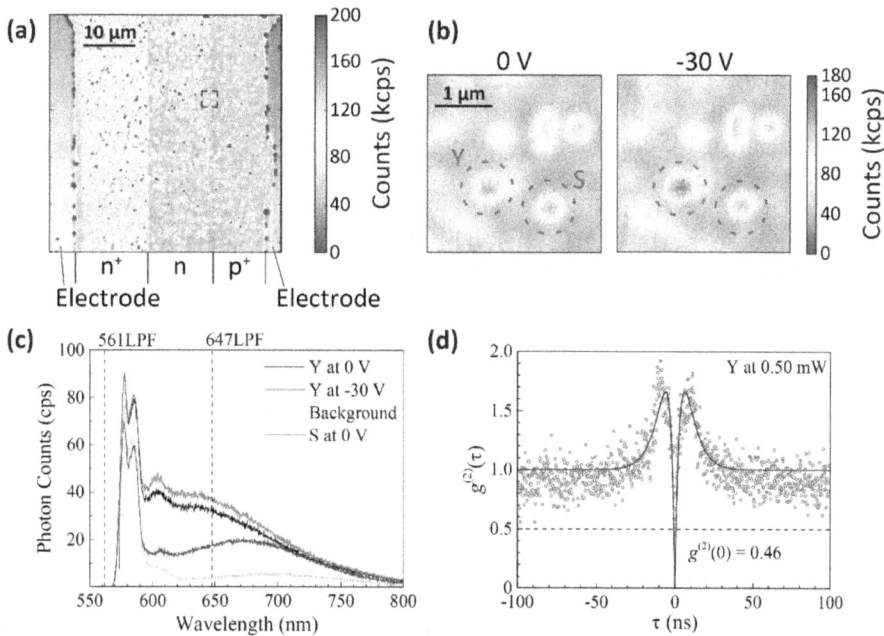

FIGURE 14

(a) A wide scan PL-CFM map of the diode with 532 nm laser excitation at 1.0 mW. (b) Focused PL-CFM maps without reverse bias voltage (left) and with –30 V (right). The color centers Y and S, which respectively responded and did not respond to the applied bias voltage, are encircled by dashed lines. (c) PL spectra of the color centers Y and S, and the background. The strong peaks at around 580 nm are the second-order Raman shift. A small peak at around 600 nm is thought to be an artifact, which often appeared probably due to the stray light in the measurement setup since the peak is observed for the background spectrum. (d) Measurement of the second-order autocorrelation function of the color center Y at 0.5 mW. The uncorrelated background was corrected and the experimental value of $g^{(2)}(0)$ is shown in the figure. The fitting curve by Equation (1) is also drawn. Reproduced with permission of ACS [39].

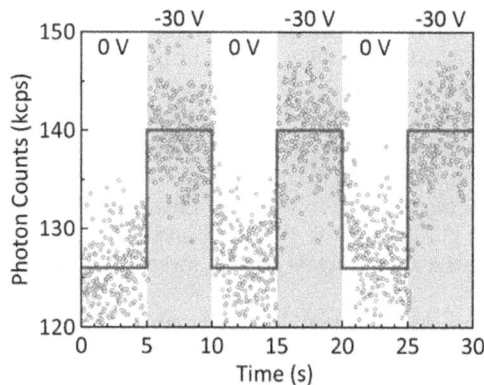

FIGURE 15

Photon count variations of the color center Y by switching the applied bias voltage. ON and OFF in the figure denote the applied bias voltages were 0 and –30 V, respectively. Lines in the figure are drawn to guide the eye. Reproduced with permission of ACS [39].

The electrical control of photon emission from a surface SPS in SiC *pn* diode was also demonstrated. In this study, the reverse bias of –30 V was cyclically applied to the diode and photon emission from the surface SPS "Y" was measured. The duration of bias was 5 seconds. Photons from "Y" increases from 126 to 140 kcps by applying reverse bias of –30 V as shown in Figure 15. On the other hand, no significant change in photon emission from the surface SPS "S" is observed between with and without applying bias. For the stability of photon emission from the surface SPS "Y", no blinking was observed at least for 2 hours. However, other surface SPS showed blinking of photon emission. At present, since we do not understand the exact mechanism of photon emission from the surface SPSs, the origin of this blinking has not yet been interpreted. Further investigations are necessary to clarify this.

4 Summary

In this chapter, we focused on some peculiar phenomena induced by the oxidation of SiC, and introduced their characteristics and findings obtained from recent research results. Regarding carrier transport at the MOS interface, the findings obtained by comparing theoretical calculations and experimental results were outlined. In addition, we described the current analysis results of the defect structure of the recently discovered single-photon sources formed at the MOS interface by oxidation of the SiC substrate, and introduced demonstrations of their electrical drive. Although there are many mysteries on the SiC MOS interface characteristics, if utilized well, we expect some enormous potential to further improve the performance of existing MOS devices (e.g., power MOSFET), or to lead to the realization of epoch-making devices such as a non-classical light-emitting device.

References

1. Suzuki, H. Ashida, N. Furui, K. Mameno and H. Matsunami, *Thermal Oxidation of SiC and Electrical Properties of Al–SiO₂–SiC MOS Structure*, Jpn. J. Appl. Phys. 21, 579–85 (1982).
2. S. Dimitrijev, H.-F. Li, H. B. Harrison, D. Sweatman, *Nitridation of silicon-dioxide films grown on 6H silicon carbide*, IEEE Electron. Dev. Lett. 18, 175 (1997).
3. T. Hosoi, D. Nagai, M. Sometani, Y. Katsu, H. Takeda, T. Shimura, M. Takei, H. Watanabe, *Ultrahigh-temperature rapid thermal oxidation of 4H-SiC(0001) surfaces and oxidation temperature dependence of SiO₂/SiC interface properties*, Appl. Phys. Lett. 109, 182114 (2016).
4. J. Rozen, S. Dhar, M. E. Zvanut, J. R. Williams, and L. C. Feldman, *Density of interface states, electron traps, and hole traps as a function of the nitrogen density is SiO₂ on SiC*, J. Appl. Phys. 105, 124506 (2009).
5. H. Sakata, D. Okamoto, M. Sometani, M. Okamoto, H. Hirai, S. Harada, T. Hatakeyama, H. Yano and N. Iwamuro, *Accurate determination of threshold voltage shift during negative gate bias stress in 4H-SiC MOSFETs by fast on-the-fly method*, Jpn. J. Appl. Phys. 60, 060901 (2021).
6. E. Ö. Sveinbjörnsson, F. Allerstam, H. Ö. Ólafsson, G. Gudjónsson, D. Dochev, T. Rödle, R. Jos, *Sodium Enhanced Oxidation of Si-Face 4H-SiC: A Method to Remove Near Interface Traps*, Mater, Sci. Forum 556–557, 487–492 (2007).

7. D. J. Lichtenwalner, L. Cheng, S. Dhar, A. Agarwal, and J. W. Palmour, *High mobility 4H-SiC (0001) transistors using alkali and alkaline earth interface layers*, Appl. Phys. Lett. 105, 182107 (2014).

8. A. Lohrmann, N. Iwamoto, Z. Bodrog, S. Castelletto, T. Ohshima, T. J. Karle, A. Gali, S. Prawer, J. C. McCallum and B. C. Johnson, *Single-photon emitting diode in silicon carbide*, Nat. Commun. 6, 8783 (2015).

9. Lohrmann, J. R. Klein, T. Ohshima, M. Bosi, M. Negri, D. W. M. Lau, B. C. Gibson, S. Prawer, J. C. McCallum, and B. C. Johnson, *Activation and control of visible single defects in 4H-, 6H-, and 3C-SiC by oxidation*, Appl. Phys. Lett. 108, 021107 (2016).

10. Y. Hijikata, H. Yaguchi and S. Yoshida, *A Kinetic Model of Silicon Carbide Oxidation Based on the Interfacial Silicon and Carbon Emission Phenomenon*, Appl. Phys. Express 2, 021203 (2009).

11. D. Goto and Y. Hijikata, *Unified theory of silicon carbide oxidation based on the Si and C emission model*, J. Phys. D.: Appl. Phys. 49, 225103 (2016).

12. J. Cochrane, P. M. Lenahan, and A. J. Lelis, *Identification of a silicon vacancy as an important defect in 4H SiC metal oxide semiconducting field effect transistor using spin dependent recombination*, Appl. Phys. Lett. 100, 023509 (2012).

13. X. Shen and S. T. Pantelide, *Identification of a major cause of endemically poor mobilities in SiC/SiO2 structures*, Appl. Phys. Lett. 98, 053507 (2011).

14. T. Kimoto, J.A. Cooper, *Fundamentals of Silicon Carbide Technology*; Wiley (2014).

15. Y. Matsushita and A. Oshiyama, *A Novel Intrinsic Interface State Controlled by Atomic Stacking Sequence at Interfaces of SiC/SiO$_2$*, Nano Lett. 17, 6458–6463 (2017).

16. A. Lidow, M.D. Rooij, J. Strydom, D. Reusch, and J. Glaser, GaN *Transistors for Efficient Power Conversion*, Wiley, New York (2015).

17. H. Amano, *et al.*, *The 2018 GaN power electronics roadmap*, J. Phys. D.: Appl. Phys. 51, 163001 (2018).

18. S. Fujita, *Wide-bandgap semiconductor materials: For their full bloom*, Jpn. J. Appl. Phys. 54, 030101 (2015).

19. M. Higashiwaki, A. Kuramata, H. Murakami and Y. Kumagai, *State-of-the-art technologies of gallium oxide power devices*, J. Phys. D: Appl. Phys. 50, 333002 (2017).

20. S. Koizumi, H. Umezawa, J. Pernot, and M. Suzuki, *Power Electronics Device Applications of Diamond Semiconductors*, Woodhead Publishing, Duxford (2018).

21. Y. Matsushita, S. Furuya, and A. Oshiyama, *Floating Electron States in Covalent Semiconductors*, Phys. Rev. Lett. 108, 246404 (2012).

22. Y. Matsushita and A. Oshiyama, *Interstitial Channels that Control Band Gaps and Effective Masses in Tetrahedrally Bonded Semiconductors*, Phys. Rev. Lett. 112, 136403 (2012).

23. U. Starke, J. Schardt, J. Bernhardt, M. Franke, K. Heinz, Phys. Rev. Lett., *Stacking Transformation from Hexagonal to Cubic SiC Induced by Surface Reconstruction: A Seed for Heterostructure Growth*, 82, 2107–2110 (1999).

24. J. B. Hannon, R. M. Tromp, N. V. Medhekar, V. B. Shenoy, *Spontaneous formation and growth of a new polytype on SiC(0001)*, Phys. Rev. Lett., 103, 256101 (2009).

25. K. Chokawa and K. Shiraishi, *Theoretical study of strain-induced modulation of the bandgap in SiC*, Jpn. J. Appl. Phys. 57, 071301 (2018).

26. S. Iwase, C. J. Kirkham, and T. Ono, *Intrinsic origin of electron scattering at the 4H-SiC(0001)/SiO$_2$ interface*, Phys. Rev. B 95, 041302 (2017).

27. T. Ono, C. J. Kirkham, S. Saito, Y. Oshima, *Theoretical and experimental investigation of the atomic and electronic structures at the 4H-SiC(0001)/SiO$_2$ interface*, Phys. Rev. B 96, 115311 (2017).

28. J. Kirkham and T. Ono, *First-Principles Study on Interlayer States at the 4H-SiC/SiO$_2$ Interface and the Effect of Oxygen-Related Defects*, J. Phys. SoC. Jpn. 85, 024701 (2016).

29. H. Yoshioka and K. Hirata, *Characterization of SiO$_2$/SiC interface states and channel mobility from MOSFET characteristics including variable-range hopping at cryogenic temperature*, AIP Advances 8, 045217 (2018).

30. K. Ito, T. Kobayashi, and T. Kimoto, *Effect of quantum confinement on the defect-induced localized levels in 4H-SiC(0001)/SiO$_2$ systems*, J. Appl. Phys. 128, 095702 (2020).

31. T. Kobayashi and Y. Matsushita, *Structure and energetics of carbon defects in SiC (0001)/SiO₂ systems at realistic temperatures: Defects in SiC, SiO₂, and at their interface*, J. Appl. Phys. 126, 145302 (2019).

32. T. Umeda, G.-W. Kim, T. Okuda, M. Sometani, T. Kimoto, and S. Harada, *Interface carbon defects at 4H-SiC(0001)/SiO₂ interfaces studied by electron-spin-resonance spectroscopy*, Appl. Phys. Lett. 113, 061605 (2018).

33. T. Umeda, T. Kobayashi, M. Sometani, H. Yano, Y. Matsushita, and S. Harada, *Carbon dangling-bond center (carbon P_b center) at 4H-SiC(0001)/SiO₂ interface*, Appl. Phys. Lett. 116, 071604 (2020).

34. K. Endo, *et al.*, 10th European Conference on Silicon Carbide & Related Materials (ECSCRM-2014).

35. T. Kobayashi and T. Kimoto, *Carbon ejection from a SiO₂/SiC(0001) interface by annealing in high-purity Ar*, Appl. Phys. Lett. 111, 062101 (2017).

36. T. Kobayashi, T. Okuda, K. Tachiki, K. Ito, Y.-i. Matsushita, and T. Kimoto, *Design and formation of SiC (0001)/SiO₂ interfaces via Si deposition followed by low-temperature oxidation and high-temperature nitridation*, Appl. Phys. Express 13, 091003 (2020).

37. K. Tachiki, *et al.*, 13th European Conference on Silicon Carbide & Related Materials (ECSCRM-2021).

38. F. Fuchs, V. A. Soltamov, S. Väth, P. G. Baranov, E. N. Mokhov, G. V. Astakhov, V. Dyakonov, *Silicon carbide light-emitting diode as a prospective room temperature source for single photons*, Sci. Rep. 3, 1637 (2013).

39. S.-i. Sato, T. Honda, T. Makino, Y. Hijikata, S.-Y. Lee, T. Ohshima, *Room Temperature Electrical Control of Single Photon Sources at 4H-SiC Surface*, ACS Photonics 5, 3159–3165 (2018).

40. H. Tsunemi, T. Honda, T. Makino, S. Onoda, S.-I. Sato, Y. Hijikata and T. Ohshima, *Various single photon sources observed in SiC pin diodes*, Mater. Sci. Forum 924 204–207 (2018).

41. Y. Hijikata, S. Komori, S. Otojima, Y.-i. Matsushita, T. Ohshima, *Impact of formation process on the radiation properties of single-photon sources generated on SiC crystal surfaces*, Appl. Phys. Lett. 118, 204005 (2021).

42. Y. Hijikata, T. Horii, Y. Furukawa, Y.-i. Matsushita and T. Ohshima, *Oxygen-incorporated single-photon sources observed at the surface of silicon carbide crystals*, J. Phys. Commun. 2, 111003 (2018).

43. Y.-i. Matsushita Y. Furukawa, Y. Hijikata, and T. Ohshima, *First-principles study of oxygen-related defects on 4H-SiC surface: The effects of surface amorphous structure*, Appl. Surf. Sci. 464, 451–454 (2019).

44. Lienhard, T. Schröder, S. Mouradian, F. Dolode, T. T. Tran, I. Aharonovich, D. Englund, *Bright and photostable single-photon emitter in silicon carbide*, Optica 3, 768–774 (2016).

Part III

SiC Materials Studies and Characterization

9

Multiple Raman Scattering Spectroscopic Studies of Crystalline Hexagonal SiC Crystals

Ian T. Ferguson
Southern Polytechnic College of Engineering and Engineering Technology,
Kennesaw State University, Marietta, GA, USA

Zhi Ren Qiu
State Key Laboratory of Optoelectronic Materials and Technologies and
School of Physics, Sun Yat-Sen University, Guangzhou, China

Lingyu Wan
Center on Nano-Energy Research, Laboratory of Optoelectronic Materials & Detection
Technology, Guangxi Key Laboratory for the Relativistic Astrophysics, School of
Physical Science & Technology, Guangxi University, Nanning, China

Jeffrey Yiin
Southern Polytechnic College of Engineering and Engineering Technology,
Kennesaw State University, Marietta, GA, USA

Benjamin Klein
Southern Polytechnic College of Engineering and Engineering Technology,
Kennesaw State University, Marietta, GA, USA

Zhe Chuan Feng
Southern Polytechnic College of Engineering and Engineering Technology,
Kennesaw State University, Marietta, GA, USA; Science Exploring
Lab, Arbour Glenn Drive, Lawrenceville, GA, USA

1 Introduction

Silicon carbide (SiC) has been recognized as an important material for a wide variety of high-power and high-temperature electronic applications [1–5]. SiC exhibits a large number (>250) of polytypes with different structural and physical properties, as well as possesses a wide range of applications [6–21]. Above, Refs. [1–5] are excellent SiC books since 2004 and Refs. [6–21] are good SiC review articles published within the last three decades. Of course, some good SiC books and review articles exist, beyond our above cited ones.

The SiC polytypes have the same chemical composition but exhibit different crystallographic structures and stacking sequences along the principal crystal axis. Several important polytypes of SiC such as 4H and 6H have C_{6v} crystallographic symmetry. In the "a" direction 4H- and 6H-SiC are almost identical (<1%) change; however, the 4H polytype

DOI: 10.1201/9780429198540-12

consists of four units in the c direction and the 6H consists of six units. Different polytypes have different band gaps, electron mobility, and other physical properties. In particular, 4H-SiC has attracted significant attention due to its high electron mobility and excellent thermal properties.

Since the late 1980s, high-quality and large-size wafers of both 6H- and 4H-SiC have come into the industry production. Gradually, the wafer size has become bigger and bigger, up to 8-inches recently, and the wafer crystalline quality performs better and better, and is very widely used for device applications. Wafers of SiC are also promising substrates for nitride semiconductor growth due to their compatible lattice structure and similar thermal expansion coefficients, as well as for other materials.

Raman scattering spectroscopy (RSS) has merits of non-destructiveness, fast analysis, and being free from special preparation for samples. It has been proven to be a useful and informative tool for the investigation and identification of SiC polytype materials. Furthermore, the Raman efficiency of SiC is high because of its strong covalence of the bonding, so the Raman signals can be easily detected. The Raman parameters such as intensity, width, and peak frequency provide plenty of information on the crystal quality. Early in 1968, Feldman et al. had reported Raman scattering studies in 6H-SiC [22], 4H-SiC [23] and phonon dispersion curves for several SiC polytypes [23]. To date, Raman spectroscopy has been widely applied in the SiC bulk and epitaxial materials, in particular, 6H-SiC [22–40] and 4H-SiC [23, 27, 32, 36, 41–53], although beyond these, there exist more in the literature.

In this chapter, we introduce and review some important topics in our Raman scattering studies on hexagonal wurtzite 4H-SiC and 6H-SiC crystal bulk materials, such as, semi-insulating and n-type (nitrogen) doped 4H-SiC and 6H-SiC wafers [54], Second-order Raman scattering from semi-insulating and doped 4H-SiC and 6H-SiC [55], resonance enhancement of electronic Raman scattering from nitrogen defect levels in 4H-SiC and 6H-SiC [56], temperature dependence of Raman scattering in bulk 4H-SiC with different carrier concentration [57], and anisotropic properties of 4H-SiC studied by angular-dependent (or rotation) Raman scattering [58].

2 Experimental Details

2.1 Materials

All 6H-SiC and 4H-SiC experimental samples are pieces at size of 5×5 mm or 10×10 mm, cut of commerce wafer materials (typically 1 3/8 to 2 inch) purchased from the CREE Company. The sample parameters, especially the semi-insulation and doping properties are mentioned in the corresponding place within this chapter. For example, the doped wafers of SiC were n type, nitrogen doped, with concentrations ranging from 2.1×10^{18} to 1.2×10^{19} cm^{-3} [54] and over a nominal range of 10^{18} to 10^{19} cm^{-3} [55, 56]. The semi-insulating sample was from a 1 3/8 inch wafer with a resistance of greater than 10^5 Ohm-cm [54, 55].

Three used bulk 4H-SiC samples in [57] and Section 6, with different doping levels, purchased from CREE Company, were prepared by physical vapor transport technique: S1 undoped, S2 medium-doped, and S3 heavy-doped, with carrier concentration of 9.09×10^{14}, 2.18×10^{18}, and 4.87×10^{18} cm^{-3}, respectively [57].

In [58] and Section 7, a semipolar "a" face (1120) undoped 4H-SiC at size of 5×5mm, cut from a 2 inch wafer – also purchased from CREE, was used.

2.2 Raman Spectrometer Systems

A variety of Raman spectrometer systems were used, and are also described in the subsections presenting the Raman spectra within this chapter. In Section 3 and 4 [54, 55], Raman spectra were recorded at room temperature using a Coherent Model INNOVA 90 Ar/Kr laser, and a SPEX Model 1877E triple monochromator with a charge-coupled detector (CCD) cooled by liquid nitrogen. The excitation source was the 514.5 nm laser light. Calibration was performed with atomic emission lamps, so that the spectrum is accurate to approximately 0.5 cm^{-1}.

In Section 5 [56], Raman spectra were recorded using confocal Raman microscopy. A Dilor LabRam system and a Renishaw Series 1000 Raman microscope were both used. All micro-Raman data were collected at room temperature. Micro-Raman spectra were obtained with laser excitation at 785 nm (1.58 eV) and 633 nm (1.96 eV), which was compared with data taken at 514 nm (2.41 eV) with both a Raman microscope and a bulk Raman spectrometer. The bulk Raman spectrometer was also used to collect data using 568 nm (2.18 eV) and 647 nm (1.92 eV) laser excitation, and to take low-temperature data using a crystal cooled with liquid nitrogen. This system has been previously described and used to examine SiC wafers [11]. The spectral resolution is approximately 1 cm^{-1} for the micro-Raman spectra. The experiments were all done in a backscattering geometry with the light collected along or close to the c axis of the SiC [56].

In Section 6 [57], a confocal micro-Raman spectroscope system, Jobin Yvon T64000 with a 2400 l/mm grating, was employed to study Raman mode shift of 4H-SiC samples with temperature varying from 80 K to 873 K. Spectral resolution of the system is up to 0.6 cm^{-1}. A Nd: YOV4 532 nm laser was utilized for excitation [57].

In Section 7 [58], Raman spectra were measured in backscattering geometry at room temperature (RT), by a Raman micro-spectrometer, named "Finder One", made by Zolix Company in China, under the excitation of 532 nm line of the solid state laser [58].

---

3 RT Raman Scattering and Line Shape of Doped 4H-SiC and 6H-SiC

3.1 RT Raman Spectra of 4H-SiC and 6H-SiC Crystalline Wafers

Group-theoretical analysis shows that the Raman-active modes of a wurtzite structure, which has C$_{6v}$ symmetry, are the A1, E1, and E2 modes [59]. The A1 and E1 phonon modes, which are also infrared (IR) active, are split into LO and transverse optical (TO) modes. In a backscattering geometry, where the incident and collected light are parallel to the c axis of the sample, the A$_1$(LO), E$_1$(TO), and E$_2$ phonons are expected to be seen in the Raman spectra at excitation energies, which are non-resonant [59, 60]. Since the band gap of hexagonal SiC is approximately 3.3 eV for 4H-SiC and near 3.0 eV for 6H-SiC [8] and our excitation energy within this subsection is no greater than 2.6 eV, the non-resonant selection rules are appropriate. We note that the 4H-SiC wafers studied were cut 8° off the c axis; however this is too small to significantly change the Raman selection rules.

As noted earlier, the existence of SiC polytypes has a major impact on the Raman spectra [29]. The one-phonon Raman spectra of 6H-SiC, 4H-SiC, and other polytypes of SiC can be explained by the folding of the Brillouin zone due to the polytype behavior of SiC [22, 23]. Because different SiC polytypes only differ by the length of the c axis, only the Brillouin zone in the direction of Γ-L is affected. This folding makes modes away from the Γ point visible in the one-phonon Raman spectrum. Group theory can be used to identify the

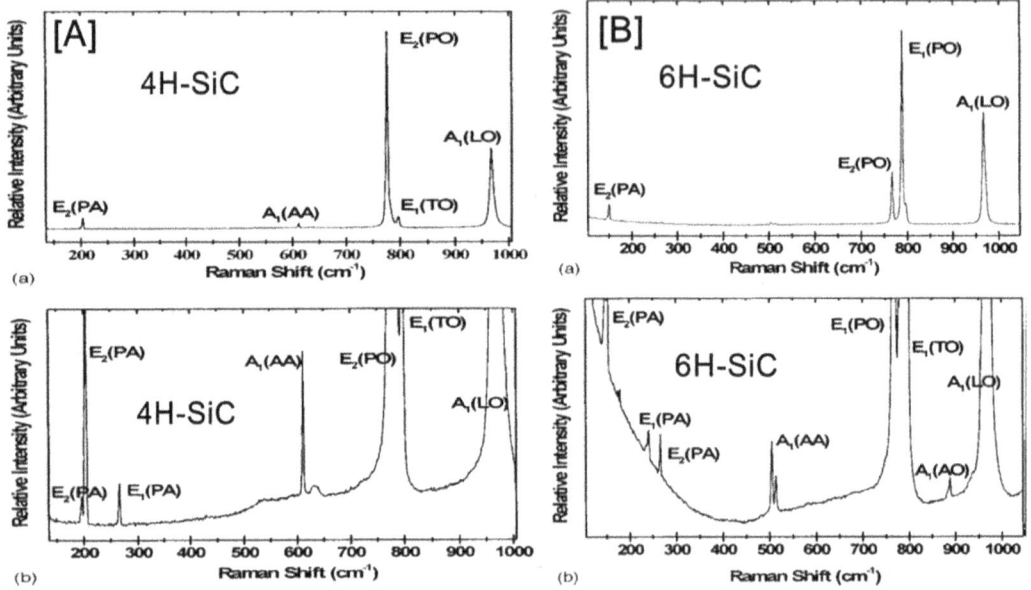

FIGURE 1

[A] (a) Raman spectrum of 4H–SiC wafer taken with 514.5 nm laser light at room temperature (nominal nitrogen concentration = 2.1x10^{18} cm^{-3}). The strongest peaks are E$_2$(TA), 203.5 cm^{-1}; A$_1$(LA), 610.5 cm^{-1}; E$_2$(TO), 777.0 cm^{-1}; E$_1$(TO), 797.5 cm^{-1}; A$_1$(LO) 967.0 cm^{-1}. (b) Same data as (a) but with an expanded y scale; [B] (a) Raman spectrum of 6H–SiC wafer taken with 514.5 nm laser light at room temperature (nominal nitrogen concentration = 2.1×10^{18} cm^{-3}. The strongest peaks are E$_2$(TA), 150.5 cm^{-1}; E$_2$(TO), 767.5 cm^{-1}; E$_1$(TO), 788.0 cm^{-1}; A$_1$(LO), 966.5 cm^{-1}. (b) Same data as (a) but with an expanded y scale. From [54] Figures 1 and 2, reproduced with permission of AIP.

symmetry of these additional modes. Phonons with atomic motion parallel to the c axis are designated axial and phonons perpendicular to the c axis are planar.

In Figures 1 [A] (a) and (b) we show typical Raman data for 4H-SiC, taken at room temperature. Figures 1 [B] (a) and (b) show data taken under the same conditions for 6H-SiC. The major peaks in the 4H-SiC Raman spectra are identifiable from previous studies [4,5]. The peak at 203.5 cm^{-1} is an E$_2$ planar or transverse acoustic (TA) mode, 610.5 cm^{-1} is A$_1$ axial or longitudinal acoustic (LA), 777.0 cm^{-1} is E$_2$ planar optical, 797.5 cm^{-1} is an E$_1$ mode, and 967.0 cm^{-1} is A$_1$(LO). The primary peaks in 6H-SiC are an E$_2$ planar acoustic mode at 150.5 cm^{-1}, two planar or TO modes of E$_2$ symmetry at 767.5 and 788.0 cm^{-1}, and an A$_1$(LO) phonon at 966.5 cm^{-1}. The mode at 796.0 cm^{-1} is a planar optical mode of E$_1$ symmetry. We note that with different experimental arrangements, different symmetry modes may be seen. For example, A$_1$ and E$_1$ modes may cause a peak at approximately 788 cm^{-1} in the 6H-SiC spectrum in addition to, or instead of, the E$_2$ mode to which we have assigned that peak, depending on experimental geometry [22,23]

The high quality of the data obtained in our experiments allows for the observation of several weaker peaks present in the Raman spectra. These additional peaks, with symmetry identification, are also listed in Table 1. In the 4H-SiC Raman spectra, two small additional features are worth noting. A pedestal is seen in the 4H-SiC Raman spectra, starting around 500 cm^{-1}. This pedestal appears at the same location in the Raman spectra as the laser wavelength is varied; therefore we conclude that this feature is Raman scattering and not luminescence. A similar feature is also seen in the 6H-SiC Raman spectrum. The Si–Si

TABLE 1

Peak assignments for peaks shown in Figures 1 and 2

4H-SiC peak	Mode	6H-SiC peak	Mode
Raman shift (cm^{-1})	Symmetry	Raman shift (cm^{-1})	Symmetry
195.5	E2 planar acoustic	146.0, 150.5	E2 planar acoustic
203.5	E2 planar acoustic	235.0, 240.0	E1 planar acoustic
266.0	E1 planar acoustic	266.0	E2 planar acoustic
610.5	A1 axial acoustic	505.0, 513.5	A1 axial acoustic
777.0	E2 planar optic	767.5, 788.0	E2 planar optic
797.5	E1(TO)	796.0	E1(TO)
		888.5	A1 axial optic
967.0	A1(LO)	966.5	A1(LO)

bond appears at approximately 519 cm^{-1} [62]; however, the feature in our experiments is sufficiently broad that we believe that this explanation is unlikely. We have attributed this feature to the acoustic branch of the second-order Raman spectra. Theoretical calculations support this assignment [63, 64]. The experimental second-order Raman spectra of 4H- and 6H-SiC will be discussed in another article [55].

Around 630 cm^{-1} another feature is clearly seen. In many samples, this feature appears as a doublet. Variation of the laser wavelength has also verified that this peak is due to Raman scattering. Because the amplitude of this peak does not significantly change with nitrogen doping concentration, and is still observed in the Raman spectrum of the semi-insulating sample, we discount the possibility that this peak is due to a local mode of the nitrogen dopant atom. This peak is the subject of further investigation.

Two peaks at around 500 cm^{-1} are visible in the 6H-SiC Raman spectra, as shown in Figure 1 [B] (b). In addition, there appears to be a broad peak underneath the two sharper features. The two peaks at 505 and 513 cm^{-1} are A$_1$ LA modes associated with the $q=0.67$ point ($q = k/k_{max}$) of the reduced Brillouin zone. The broad background signal increases in magnitude and sharpens when the sample is cooled to 77 K. This behavior is consistent with the assignment of this feature by Klein and co-workers as Raman scattering from electronic defects [25].

3.2 A$_1$(LO) Phonon Line Shape versus Nitrogen Doping Concentration

In Figure 2, we show the A$_1$(LO) phonon line shape as a function of nitrogen doping concentration. As the doping concentration is increased, the phonon increases in frequency and asymmetrically broadens. The change in nitrogen concentration from 2.1×10^{18} to 1.2×10^{19} cm^{-3} in 4H-SiC produces a dramatic change in the position and shape of the A$_1$(LO) phonon. This large change in the frequency of the A$_1$(LO) phonon makes its position a sensitive probe of doping in this concentration range. We note that none of the other strong Raman peaks shift or broaden in the doping concentration range studied.

3.3 Theoretical Calculation of A$_1$(LO) Phonon Line Shape

The behavior of the A$_1$(LO) phonon can clearly be attributed to plasmon–phonon coupling. The strong dependence of the A$_1$(LO) peak position on carrier concentration allows us to use the position of the A$_1$(LO) phonon as a measure of the carrier concentration. Previous studies have used this approach to examine other wide gap materials such as GaN at high doping concentrations [65].

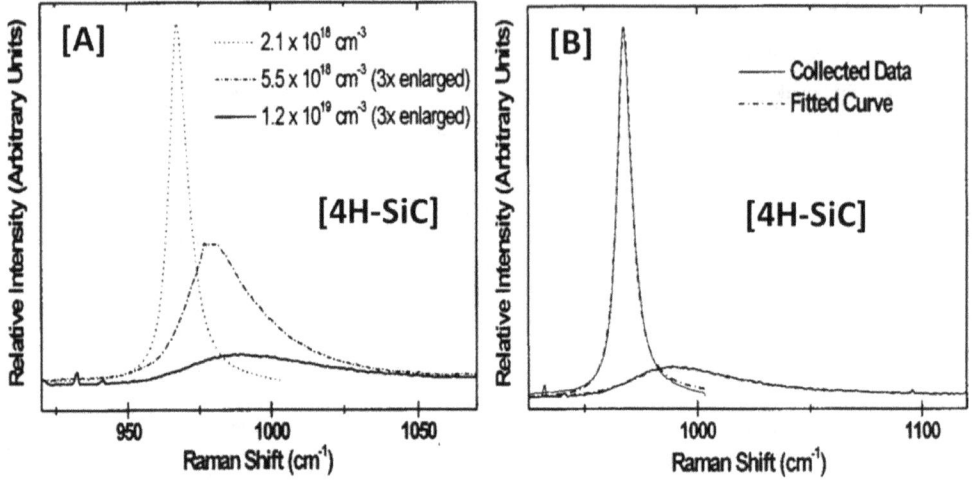

FIGURE 2

High-frequency optical phonon A_1(LO) in 4H-SiC at several nitrogen concentrations. The dotted line is 2.1×10^{18} cm^{-3}; dot-dash line 5.5×10^{18} cm^{-3}; solid line 1.2×10^{19} cm^{-3}. Experimental conditions as in Figure 1. The A_1(LO) phonon increases in frequency and broadens with increasing doping and carrier concentration. From [54] Figure 3, reproduced with permission of AIP.

Physically, at sufficiently large carrier concentration, longitudinal oscillations of the associated plasma modify the dielectric constant and consequently the Raman cross section. Klein and collaborators developed the first theory for the effects of plasmon–phonon coupling on Raman line shapes [25]. This theory has been refined by other research groups. For example, Irmer et al. extended the theory to include phonon dampening [61]. These works have established that in wide band gap semiconductors, the dominant mechanisms for Raman scattering are electro-optical and deformation potential, and not fluctuations in carrier density. The line shape of the LO phonon is given by

$$I(\omega) = SA(\omega)\mathrm{Im}[-1/\omega\varepsilon(\omega)], \tag{1}$$

where $A = 1 + 2C\dfrac{\omega_T^2}{\Delta}\left[\omega_p^2\gamma\left(\omega_T^2 - \omega^2\right) - \omega^2\eta\left(\omega^2 + \gamma^2 - \omega_p^2\right)\right] + C^2\left(\dfrac{\omega_T^4}{\Delta\left(\omega_L^2 - \omega_T^2\right)}\right)$

$$\times\left\{\omega_p^2\left[\gamma\left(\omega_L^2 - \omega_T^2\right) + \eta\left(\omega_p^2 - 2\omega^2\right)\right] + \omega^2\eta\left(\omega^2 + \gamma^2\right)\right\} \tag{2}$$

and

$$\Delta = \omega_p^2\gamma\left[\left(\omega_T^2 - \omega^2\right)^2 + \left(\omega\eta\right)^2\right] + \omega^2\eta\left(\omega_L^2 - \omega_T^2\right)\left(\omega^2 + \gamma^2\right) \tag{3}$$

ω_L and ω_T are the longitudinal and transverse optical phonon frequencies, γ is the plasmon dampening constant, η (i.e., Γ) is the phonon dampening constant, and C is the Faust–Henry co-efficient. S is a proportionality factor. The dielectric function $\varepsilon(\omega)$ contains a phonon and a plasmon contribution:

$$\varepsilon = \varepsilon_\infty\left(1 + \frac{\omega_L^2 - \omega_T^2}{\omega_T^2 - \omega^2 - i\omega\eta} - \frac{\omega_p^2}{\omega(\omega + i\gamma)}\right) \tag{4}$$

where ω_p, the plasma frequency, is given in cgs units by

$$\omega_p^2 = \frac{4\pi n e^2}{\varepsilon_\infty m^*}. \tag{5}$$

This model has been previously used to examine plasmon–phonon coupling in SiC [10,12] and other wide band semiconductors [66].

Using Eqs. (1–4), we fixed ω_T, ω_L, and C to accepted literature values (C=0.43; ω_T= 783 cm^{-1}; ω_L=965 cm^{-1}) [27]. The value for ω_L was confirmed by our measurements of semi-insulating 4H-SiC (964.5 cm^{-1}). We then allowed the program to find the values for Γ, γ, and ω_p that gave the best agreement between the fit and the experimental line shape. We were then able to relate these values to the nominal doping concentrations of each wafer, which enables us to make an estimate of nominal doping level for any given value of ω_p. The theoretical fitting results to all the data of Figure 2 are almost perfect, not shown here. A typical fitting example for the A$_1$(LO) phonon of 4H-SiC can be seen from Figure 4 in [54]. Similar quality fits are found for the 6H-SiC.

The high frequency coupled plasmon–phonon mode in the limit of no dampening can be expressed as

$$(\omega^+)^2 = (1/2)\left\{\omega_L^2 + \omega_p^2 \left[\left(\omega_L^2 + \omega_p^2\right)^2 - 4\omega_p^2\omega_T^2\right]^{1/2}\right\}, \tag{6}$$

where ω_L, ω_p, and ω_T are defined as above. Therefore, we can use the position of the A$_1$(LO) phonon to empirically estimate the carrier concentration [54].

In brief summary, we have used Raman spectroscopy to investigate wafers of 4H- and 6H-SiC. Theoretical models of plasmon–phonon coupling successfully predict the experimental A$_1$(LO) phonon line shape at dopant concentrations up to 10^{19} cm^{-3}. The phonon–plasmon coupling of the A$_1$(LO) phonon can be used as a measure of carrier concentration. The results suggest that Raman scattering of the phonon can be used as an *in situ*, non-contact diagnostic of doping uniformity in semiconductor production. This is in contrast to electrical conductivity measurements, which require contacts with the samples.

---

4 The Second-Order Raman Scattering of 4H-SiC and 6H-SiC

One-phonon Raman spectra are sensitive mainly to phonons at the Γ point (k=0) of the Brillouin zone. However, in SiC the polytype structure in the large zone allows us to see other phonons that have a pseudo-momentum of zero in the Brillouin zone [23]. The polytype dependence of one-phonon Raman spectra for doped SiC has been extensively studied [22,23,27,29,67,68]. Second-order Raman spectra reflect the entire Brillouin zone because any two phonons with opposite wave-vectors will have a total momentum of zero [62]. Two-phonon Raman spectra are a sensitive test of theoretical models of the lattice dynamics [62,63,69–72], because typically they are continuous spectra with peaks corresponding to regions or points of the dispersion curve with zero gradient and hence maxima in the phonon density of states. The second-order Raman spectrum of 3C-SiC has previously been studied [54,63].

We present here two-phonon Raman spectra for semi-insulating 4H-SiC, and *n*-type doped 4H- and 6H-SiC. The optical branch of the second-order Raman spectra for SiC was

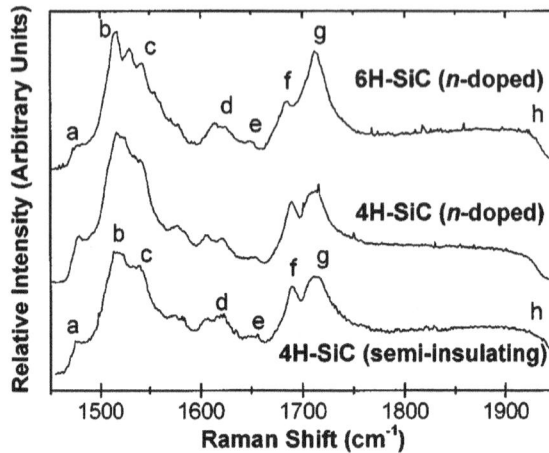

FIGURE 3

Second-order Raman spectra of semi-insulating 4H-SiC and *n*-type doped 4H- and 6H-SiC. The nominal nitrogen concentration was 2.1×10^{18} cm^{-3} for both doped samples. From [55] Figure 2, reproduced with permission of APS.

found to be polytype-dependent and much more complex than the (cubic) 3C-SiC polytype. These observations reflect clear changes in the phonon density of states with polytype, while the frequencies of the Γ point phonons vary by less than 5%. The identification of spectra was obtained by reference to experimental studies and theoretical calculations.

The spectra between 1450 cm^{-1} and 1950 cm^{-1} in SiC are shown in Figure 6, for semi-insulating and *n*-type doped 4H-SiC, and *n*-type doped 6H-SiC. The peak frequencies are listed in Table 2. We have assigned these peaks to the optical branch of the second-order Raman spectrum. This part of both the 4H- and 6H-SiC spectra is much more complex than for the cubic (3C) polytype, as observed by Windl et al. [63] and by Olego and Cardona [73]. Figure 3 clearly demonstrates the differences in the two-phonon spectra between polytypes. The smaller features visible are reproducible in samples of different origin.

There is no significant change in the spectra between the semi-insulating and doped 4H-SiC, nor is there any variance with doping level. Therefore we have concluded that none of the observed peaks in this region are due to local mode vibrations of the nitrogen.

The optical branch of the overtone spectrum begins at 1476.0 cm^{-1} for both 4H- and 6H-SiC. This peak is labeled "*a*" in the figures and in Table 2. The sharp edge of the overtone spectrum is due to a gap in the phonon density of states. Theoretical calculations of the phonon dispersion curve have shown that the lowest point in the optical phonon branch is the *K* point phonon [74]. The frequency of the *K* point phonon has been measured for 2H-SiC as 737 cm^{-1} [75]. We expect this value should be very similar for both 4H- and 6H-SiC. Our experimental value of 738 cm^{-1}, calculated from the second-order Raman spectra, confirms this hypothesis.

The high-frequency tail of the overtone spectra is at approximately 1930 cm^{-1} (peak h). A similar tail is observed in the experimental data for 3C-SiC [63] This feature is due to the overtone of the Γ point A_1(LO) phonon at 965 cm^{-1}. This is the highest point in the three-dimensional vibrational band structure. This feature is well predicted by theoretical calculations [74].

The region between 1514 cm^{-1} and 1580 cm^{-1} is the most prominent region in both the 4H- and 6H-SiC second-order Raman spectra. There is a clear dependence of the second-order Raman spectra on the polytype. The most recent data on 3C-SiC showed one near

TABLE 2

Peak frequencies for spectra in Figure 3

Peak semi-insulating 4H-SiC	Frequency (cm⁻¹) n-type doped 4H-SiC	n-type doped 6H-SiC	Label
1475	1478	1476	a
1515	1515	1516	b
	1526		
		1532	
1542	1544	1542	c
1606	1605		
		1614	
1622	1621	1626	d
1654	1655	1651	e
1689	1689	1686	f
1714	1713	1714	g
1925/1930	1925	1925/1930	h

featureless peak in this region [63]. The peak observed in 3C-SiC at 1519 cm⁻¹ or 1520 cm⁻¹ has been assigned [73] as the overtone of the TO(X) phonon at 761 cm⁻¹. This is consistent with theoretical phonon dispersion curves for 3C-SiC, which predict a high density of states in this region [74]. Bechstedt et al. [16] have calculated the high symmetry point phonon frequencies for both 4H- and 6H-SiC. We observe that the frequency of the peak labeled "b" in Figure 3 is consistent with possible values for an overtone of the L point phonon. Likewise, the 1542 cm⁻¹ peak (c), which is present in both the 4H- and 6H-SiC polytypes is possibly due to an overtone of the M symmetry phonon at 771 cm⁻¹. The additional peak at 1526 cm⁻¹ in the 6H-SiC spectrum is analogously due to a phonon from another point on the L-M branch, such as the U point. Calculations of the phonon dispersion curves [74] clearly show the optical L-M branch to be quite flat, indicating a high density of states.

The peak observed at 1714 cm⁻¹ (g) in 6H-SiC and a broader peak at the same location in 4H-SiC is the same as the high frequency peak reported in investigations of cubic SiC at approximately 1712 cm⁻¹. Olego and Cardona have pointed out that this frequency does not correspond to either 2LO(Γ) (1946 cm⁻¹), 2LO(X) (1658 cm⁻¹), or 2LO(L) (1676 cm⁻¹) [73]. The authors hypothesized that this peak is due to an overtone of an M point phonon, which had not been directly measured by Raman or IR. Absorption measurements, which are not as accurate as Raman, have placed this M point phonon in 6H-SiC to be at 863 cm⁻¹ [76]. This is close to the value we obtain (857 cm⁻¹) by using Olego and Cardona's hypothesis with our two-phonon Raman measurements. A similar analysis can be used for the peak at 1688 cm⁻¹ (f), which would attribute this peak to the overtone of an M point phonon previously measured to be 840 cm⁻¹ by analysis of the exciton absorption phonon sidebands [76].

Because of the gap in the phonon density of states, the numerous peaks that occur in the region between 1600 cm⁻¹ and 1670 cm⁻¹ for both 4H- and 6H-SiC must be combination bands, by analogy to the observations of Windl et al. [63] and Karch et al. [70] in 3C-SiC. We note that these peaks appear to be more intense than those observed in 3C-SiC and are seen here to be polytype dependent.

In brief summary, we have measured the first and second order Raman spectra of semi-insulating 4H-SiC. The second-order Raman spectra of n-type doped and semi-insulating 4H-SiC were found to be quite similar. The optical branches of the second-order Raman spectra for 4H- and 6H-SiC were found to be polytype dependent and much more complicated than the two-phonon Raman spectra for cubic 3C-SiC. Precise values for high symmetry point phonons in SiC have been obtained by the analysis of the second-order Raman spectra. Our experimental measurements are consistent with calculations of the phonon dispersion and calculated phonon frequencies for the *L-M* branch [70-72, 74]. Both overtones and combination bands are observed in the second-order Raman spectra of 4H- and 6H-SiC.

5 Electronic Raman Scattering from Nitrogen Defect Levels in 4H- and 6H-SiC

5.1 Theory of Plasmon–Phonon Coupling in N-doped 6H-SiC

Klein et al. established a theory of plasmon–phonon coupling in semiconductors, which successfully predicted Raman line shapes for *n*-type nitrogen-doped 6H-SiC [24]. Furthermore, Klein and Colwell were the first to observe electronic Raman scattering from nitrogen-doped 6H-SiC at low temperature [25]. Following Ref. [25] there have been experimental studies of electronic Raman scattering from nitrogen donor levels in other SiC polytypes [27, 29]. In contrast to 6H-SiC, the electronic Raman scattering in 4H-SiC is quite weak and quality experimental data is scarce [29]. It is important to note that these Raman studies used green or blue laser excitation. Under these conditions, electronic Raman scattering in nitrogen-doped silicon carbide is only clearly observed at low temperatures. It is generally accepted that nitrogen occupies carbon sites [77] in the SiC. In 4H-SiC, there are two inequivalent sites, one hexagonal and one quasicubic; for 6H-SiC, there are three sites, two quasicubic, and one hexagonal [68].

Here we demonstrate that electronic Raman scattering from nitrogen defect levels can be resonantly enhanced with red or near-IR laser excitation at room temperature. The resonantly enhanced electronic Raman scattering is found to be polytype dependent. We also present similar data taken at low temperature that shows this effect with sharper, more intense peaks. In 4H-SiC, the resonantly enhanced peaks are seen to shift with doping concentration. We note that IR Raman scattering was found to be a valuable tool for the characterization of diamond films [78].

5.2 Raman Scattering of N-doped 4H-SiC, Excited in Visible to NIR

Figure 4 shows room-temperature Raman spectra of a single nitrogen-doped 4H-SiC wafer ($n=5.5\times10^{18}$ cm^{-3}), which were taken at different laser excitation wavelengths: 514 nm (2.41 eV), 633 nm (1.98 eV), and 785 nm (1.58 eV). There is a very significant change in the Raman spectra upon changing the laser wavelength. When the laser excitation is in the red or near IR, clear resonant enhancements are observed for peaks at approximately 400, 530, and 570 cm^{-1}, enlarged and labeled Na, Nb, and Nc in the inset. At 620 cm^{-1}, a Fano resonance is observed. We note that we have not found any changes in the major peaks of the spectrum. The intensity and Raman shift of these peaks is unaffected by the change in excitation to red light. The $A_1(LO)$ mode behaves in the same manner as we have previously observed [54].

The strongest peak at 530 cm^{-1} (Nb) is broad and asymmetric; this asymmetry is possibly due to another peak near 500 cm^{-1}. The peaks at 530 and 570 cm^{-1} are consistent with

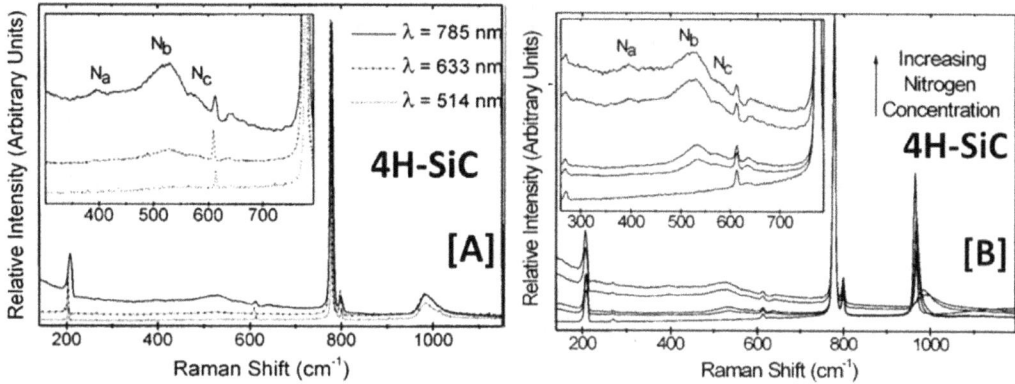

FIGURE 4

Raman spectra from a single *n*-type 4H-SiC sample taken at room temperature with different laser excitation wavelengths: 514 nm (2.41 eV), 633 nm (1.98 eV), and 785 nm (1.58 eV). Note the clear appearance of several additional peaks, shown enlarged in the inset, labeled Na, Nb, and Nc, as the laser wavelength is tuned to the near IR. We attribute these peaks to electronic Raman scattering from the nitrogen defect levels. The spectra are normalized to the peak at 777 cm^{-1}. From [56] Figure 1, reproduced with permission of AIP.

previous measurements of *n*-type 4H–SiC at low temperature [29]. At low temperatures, a sharp peak at 57 cm^{-1} has also been measured [29]. The three high frequency peaks and the low frequency mode at 57 cm^{-1} make for a total of four peaks that can be attributed to nitrogen donors in 4H-SiC. This is more than the number of inequivalent sites, two, for 4H-SiC.

Previously, Raman spectra for *n*-type 4H-SiC at different doping concentrations were taken with 785 nm (1.58 eV) excitation at room temperature, as shown in Figure 2 of [56]. As the *n*-type nitrogen doping concentration increases from semi-insulating to 7.1×10^{18} cm^{-3}, there was an increase in the intensity of the peaks at approximately 400, 530, and 570 cm^{-1}. The absence of peaks Na, Nb, and Nc in the semi-insulating sample demonstrates that these peaks are associated with nitrogen doping. In previous works [54, 55] we have drawn attention to a pedestal in the Raman spectrum beginning at around 500 cm^{-1}, which we attributed to a possible second-order scattering. This is also visible in the spectrum of the semi-insulating sample at 785 nm (1.58 eV), which shows further that the pedestal is unlikely to be due to nitrogen doping, and moreover is not a probable cause of the resonant effect that we report here. Careful inspection of the peak (Nb) near 530 cm^{-1} determines that the absolute peak position shifts to smaller values of Raman shift as the nitrogen concentration is increased. Similar resonantly enhanced electronic Raman scattering from nitrogen defect levels in 6H-SiC were also observed under visible to near-IR laser excitation [54].

5.3 Comparative Electronic Raman Scattering of N-doped 4H-/6H-SiC

Figure 5 [A] compares Raman spectra with 647 nm (1.92 eV) excitation for two 4H-SiC samples at low temperature and room temperature. This illustrates the differences between the Raman spectra at different doping levels and temperatures. The semi-insulating sample shows no trace of resonant peaks at either temperature. It is evident that the resonant peaks in the *n*-type doped sample are more intense at low temperature. The physics mechanism on this phenomenon is worthy to penetrate further.

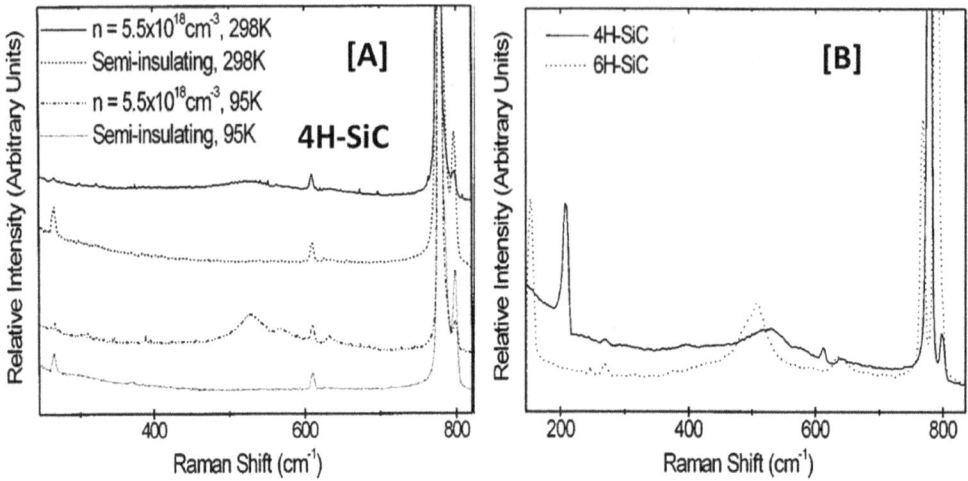

FIGURE 5

[A] Raman spectra of 4H–SiC wafers using laser excitation at 647 nm (1.92 eV). Low-temperature data are contrasted with room temperature data for a moderately high doping concentration n-type wafer and a semi-insulating wafer. The semi-insulating sample shows no peaks in the region of interest. There is clearly a resonant peak in both spectra of the *n*-type wafer, which is enhanced at low temperatures. [B] Comparison of Raman spectra of 4H-SiC and 6H-SiC taken at room temperature using 647 nm (1.92 eV) laser excitation. The spectra are normalized by their most intense peaks. The resonance effect is clearly larger in the 6H-SiC sample, even though the 4H-SiC sample has a higher doping concentration. From [56] Figures 6 and 7, reproduced with permission by AIP.

We have also observed that, for comparable nitrogen concentrations, it appears the electronic Raman scattering from nitrogen donor levels in 6H-SiC to be stronger than the 4H-SiC polytype. We demonstrate this in Figure 5 [B], which showed the resonant peaks are much more intense in the 6H-SiC sample than the 4H-SiC sample. We have already predicted in the last sub-section that the intensity of the resonant peaks increases with doping level. It is noted that the 4H-SiC sample ($n=5.5 \times 10^{18}$ cm^{-3}) actually has a higher concentration of *n*-type doping than the 6H-SiC ($n=1.7 \times \times 10^{18}$ cm^{-3}) [56], so the increased intensity in the 6H-SiC sample is not merely due to a doping effect. The peak at about 200 cm^{-1} in the 4H-SiC is a clear example of a Fano resonance, an asymmetry and distortion caused by overlap and interference between a broad electronic state and a discrete phonon state [29]. We note that the interference appears to increase as the laser excitation wavelength is tuned from the green to the near-IR, which suggests that the scattering from the electronic continuum is enhanced.

The electronic spectrum of donors in SiC is of fundamental importance. In addition to electronic Raman scattering, Fourier-transform infrared measurements have also been made of the electronic energy levels of nitrogen donors in SiC [79-81]. The IR and electronic Raman transitions obey different selection rules and are therefore complementary measurements. Previous works have identified the low-frequency electronic Raman scattering peak as due to valley-orbit splitting of the electronic ground state, because of the anisotropic SiC band structure [24,29]. As noted earlier and by other works, the electronic spectrum of donors in SiC is complicated by the existence of inequivalent sites for both 6H- and 4H-SiC. It is also possible that some of the Raman peaks observed are nitrogen local modes, this will be the subject of further investigation.

In brief summary of Section 5, we have observed electronic Raman scattering from nitrogen donor levels in both 4H- and 6H-SiC. We have found that the electronic Raman scattering is enhanced with red or near-IR laser excitation. The resonant enhancement is affected by temperature and polytype. This resonance is due to the near-IR absorption, typical of nitrogen-doped SiC, which has been attributed to deep defects inside the SiC wafer materials. More mechanisms related to these interesting Raman phenomena are left to explore and for further penetrative studies.

6 Temperature Dependence of Raman Scattering in Bulk 4H-SiC with Different Carrier Concentration

6.1 Experiment Temperature-Dependent RSS of Doped 4H-SiC

A confocal micro-Raman spectroscope system, Jobin Yvon T64000, was employed to study Raman mode shift of 4H-SiC samples with temperature varying from 80 K to 873 K. A Nd:YOV$_4$ 532 nm laser was utilized for excitation. Three bulk 4H-SiC samples of different doping levels were prepared: S1 undoped, S2 medium doped, and S3 heavy-doped, with carrier concentration of 9.09×10^{14}, 2.18×10^{18}, and 4.87×10^{18} cm^{-3}, respectively.

4H-SiC has wurtzite structure, which belongs to hexagonal symmetry. According to Raman selection rules of hexagonal SiC crystal (space group), 4H-SiC has 24 lattice modes ($6A_1 + 6B + 6E_1 + 6E_2$). Only ten ($3A_1 + 3E_1 + 4E_2$) of the 24 modes are Raman active, and the B modes are silent. The A_1 mode and E_1 mode are also infrared active and they split into longitude optical and transverse optical (TO) branches [29, 54, 82]. 4H-SiC presents phonon bands with A_1, E_1, and E_2 mode. In the backscattering geometric configuration ($z | | c, x,y$ in c-plane), A_1, E_1 and E_2, mode can be observed. Raman spectra of each sample are shown in Figure 6. Monotonous down shift of E_2(TO) and E_1(TO) mode with increasing temperature are clearly observed.

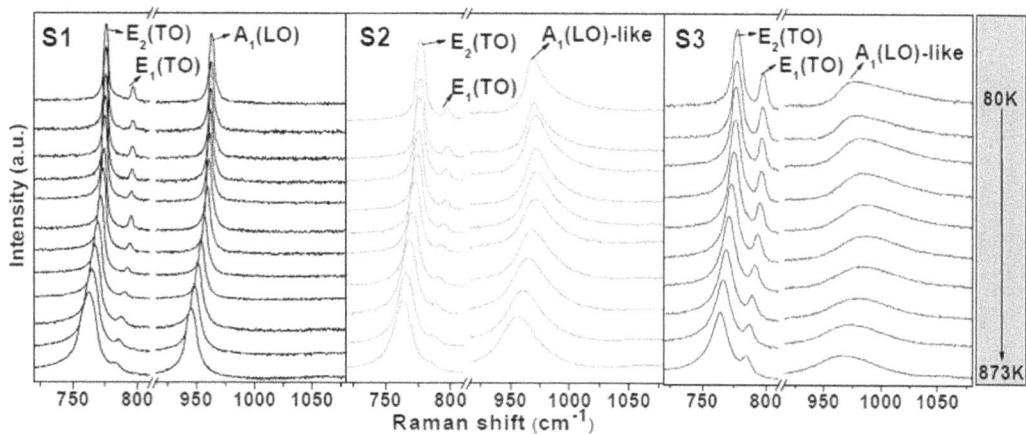

FIGURE 6

Raman spectra of three 4H-SiC samples at temperature varying from 80 K to 873 K. From [57] Figure 1, reproduced with permission from [copyright@osa.org 09.23.2021] © The Optical Society.

Raman mode located at about 970 cm^{-1} should be A$_1$(LO) mode theoretically, and the variation of peak position with temperature for undoped sample S1 is also monotonous. However, the two doped samples show a non-monotonous variation of this peak at around 970 cm^{-1} (named as A$_1$(LO)-like mode temporarily). Variation of A$_1$(LO)/A$_1$(LO)-like mode peak positions from three 4H-SiC samples can be seen later in Figure 8.

6.2 Theoretical Calculation of TO Phonon Frequency

Down shift of phonon scattering mode with temperature stems from linear thermal expansion and multi-phonon coupling processes. According to the Raman scattering model, temperature dependent of phonon frequency $\omega(T)$ can be given by [82]

$$\omega(T) = \Omega_0 + \omega^{(1)}(T) + \omega^{(2)}(T), \tag{7}$$

where $\Omega_0 = \omega_0 - M_1 - M_2$, ω_0 is the harmonic frequency of optical mode phonon at temperature near absolute zero. $\omega^{(1)}(T)$ is the term caused by linear thermal expansion, and $\omega^{(2)}(T)$ denotes anharmonic phonon coupling contribution to Raman shift. The second term $\omega^{(1)}(T)$ can be expressed as

$$\omega^{(1)}(T) = \Omega_0\{\exp[-\kappa\int(\alpha_c(t)+2\alpha_a(t))dt]-1\}, \tag{8}$$

where k is the Gruneisen parameter for optical Raman mode, $\alpha_c(t)$ and $\alpha_a(t)$ are coefficients of linear thermal expansion along direction of c and a axis. The third term $\omega^{(2)}(T)$ in Eq. (7) can be written as

$$\omega^{(2)}(T) = M_1\{1+\textstyle\sum_{i=1}^2 1/(e^{xi}-1)\} + M_2\{1+\textstyle\sum_{j=1}^3 [1/(e^{yj}-1)+1/(e^{yj}-1)^2]\}, \tag{9}$$

where M_1 and M_2 are fitting parameters. The value of the exponent x_i and y_j are given as: $\sum x_i = \sum y_j = \hbar\omega_0$. The first term of $\omega^{(2)}(T)$ describes the three-phonon process and the second denotes the four-phonon process. The split phonons of the same order are assumed to be of the same frequency. Therefore, it is supposed that $x_1=x_2=\hbar\omega_0/2$ and $y_1=y_2=y_3=\hbar\omega_0/3$.

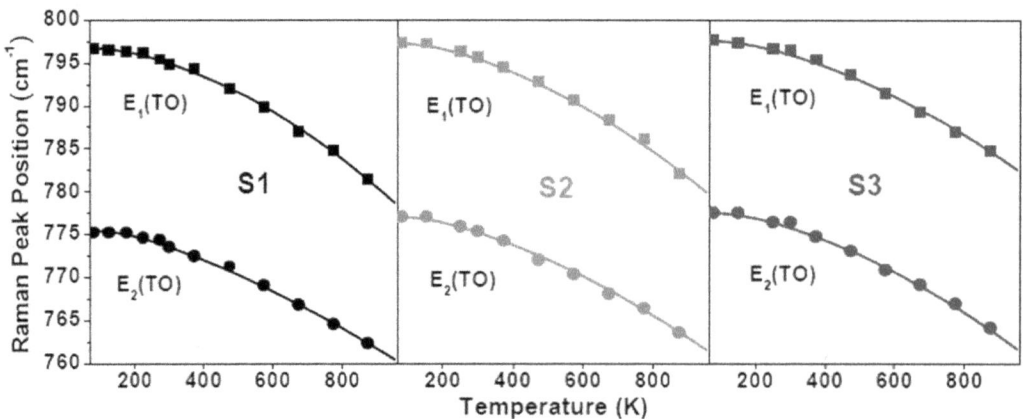

FIGURE 7

Fitting result of E$_1$(TO) and E$_2$(TO) mode peak positions of three 4H-SiC samples. From [57] Figure 3, reproduced with permission from [copyright@osa.org 09.23.2021] © The Optical Society.

Fitting results of the E_2(TO) and E_1(TO) mode shift dependent on temperature are shown in Figure 7. Curves of three 4H-SiC wafers are similar. The effect from free carrier concentration on phonon scattering modes is negligible. Therefore, 4H-SiC samples with different doping levels have the same variation of TO phonon scattering modes with temperature.

6.3 Theoretical Simulation on Temperature Dependence of LOPC Mode

In a doped polar semiconductor, such as SiC, free carrier (plasmon) interacts with the longitudinal optical (LO) phonon to form the LO phonon-plasma coupled (LOPC) mode, and its Raman scattering intensity can be expressed as [29,54,82],

$$I_{LOPC} = d^2S/d\omega d\Omega \,|_A = (16\pi h n_2 / V_0^2 n_1)(\omega_2^4/C^4)(d\alpha/dE)(n_\omega+1)A\text{Im}[-1/\varepsilon], \tag{10}$$

where $A = 1 + 2C\dfrac{\omega_T^2}{\Delta}\left[\omega_p^2\gamma\left(\omega_T^2 - \omega^2\right) - \omega^2\eta\left(\omega^2 + \gamma^2 - \omega_p^2\right)\right] + C^2\left(\dfrac{\omega_T^4}{\Delta\left(\omega_L^2 - \omega_T^2\right)}\right)$

$$\times\left\{\omega_p^2\left[\gamma\left(\omega_L^2 - \omega_T^2\right) + \eta\left(\omega_p^2 - 2\omega^2\right)\right] + \omega^2\eta\left(\omega^2 + \gamma^2\right)\right\} \tag{11}$$

$$\Delta = \omega_p^2\gamma\left[\left(\omega_T^2 - \omega^2\right)^2 + (\omega\eta)^2\right] + \omega^2\eta\left(\omega_L^2 - \omega_T^2\right)\left(\omega^2 + \gamma^2\right) \tag{12}$$

and ω_L is the longitudinal optical mode frequency; ω_T is transverse optical mode frequency; η is phonon damping constant; γ is plasma damping constant. Dielectric function is described as

$$\varepsilon = \varepsilon_\infty\left(1 + \frac{\omega_L^2 - \omega_T^2}{\omega_T^2 - \omega^2 - i\omega\eta} - \frac{\omega_p^2}{\omega(\omega+i\gamma)}\right) \tag{13}$$

$$\omega_p^2 = \frac{4\pi n e^2}{\varepsilon_\infty m^*} \tag{14}$$

where ω_p is plasma frequency; n is free carrier concentration; m^* is effective mass; e is unit charge; ε_∞ is high-frequency dielectric constant.

Anomalous variation is from temperature behavior of the LOPC mode. Peak position of LOPC mode can be calculated by [54]

$$\omega_{LOPC}^2 = (1/2)\{(\omega_p^2+\omega_{LO}^2)+[(\omega_p^2+\omega_{LO}^2)^2-4\omega_p^2\omega_{TO}^2]^{1/2}\}, \tag{15}$$

where ω_p is plasma frequency. ω_{LO} is the LO phonon frequency, using the A_1(LO) value of undoped 4H-SiC, and ω_{TO} is the TO phonon frequency, calculated from the Lyddane-Sachs-Teller (LST) relation of $\omega_{LO}^2/\omega_{TO}^2 = \varepsilon_0/\varepsilon_\infty$, where ε_0 and ε_∞ are the static permittivity and the high-frequency permittivity, respectively [83].

Eq. (15) reveals the interaction between LO phonons and plasma. Increasing temperature leads to a down shift of LO phonon mode and an increase of the degree of impurity ionization, which results in the increment of plasma frequency until complete ionization. Coupling between plasma frequency and LO phonon frequency contributes to the non-monotonous variation: up shift first and down shift later with temperature increasing.

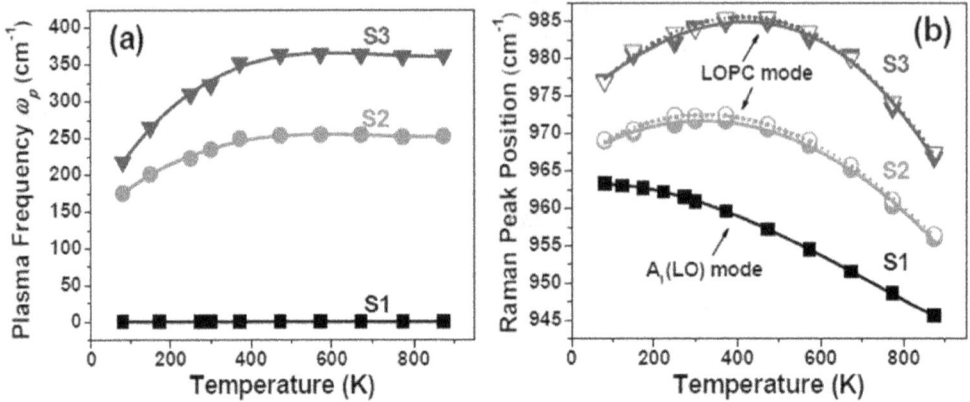

FIGURE 8
Temperature dependence of (a) plasma frequency ω_p (b) experimental results (solid line) and calculation (short dotted line) of LOPC mode peak positions. From [57] Figure 7, reproduced with permission from [copyright@osa.org 09.23.2021] © The Optical Society.

LOPC mode peak positions can be calculated according to the description of Eq. (15), as shown in Figure 8 (b) (short dotted lines). Calculation result is consistent with the variation of LOPC mode in experiment (solid lines in Figure 8 (b)).

In brief summary, temperature-dependent Raman scattering spectra on three pieces of bulk 4H-SiC wafers, undoped, medium doped, and high doped were measured from 80 K to 873 K. Three Raman modes, including E_2 (TO), E_1 (TO) and A_1(LO) (or A_1(LO)-like) mode, are clearly observed. Down shift of E_2 (TO) and E_1 (TO) phonon scattering modes is explained through phonon frequency function with temperature. Anomalous variation of A_1(LO) mode in doped 4H-SiC stems from the coupling between LO phonons and plasma, which forms the LOPC mode. The non-monotonous variation of blue-red shifts with temperature for LOPC mode from doped 4H-SiC could be explained by the influence from ionization process of impurities on the process of Raman scattering. A quantitative description on temperature-dependent Raman spectra for doped 4H-SiC is achieved, matching well to experimental data.

7 Rotation Raman Scattering Study on Anisotropic Property in Wurtzite 4H-SiC

7.1 Phonon Anisotropy Characteristics

The phonon anisotropy property of the wurtzite crystal is studied using angular-dependent Raman spectroscopy both theoretically and experimentally. The angle dependence of optical phonon modes and structural properties of *a*-plane 4H-SiC bulk crystal, respectively, is investigated by polarized Raman spectroscopy at room temperature. Corresponding Raman selection rules are derived according to measured scattering geometries to illustrate the angle dependence. The angle-dependent intensities of phonon modes are discussed and compared to theoretical scattering intensities, yielding the Raman tensor elements of A_1, E_1, and E_2 phonon modes.

In this Section 7, we report an investigation on the 4H-SiC phonon anisotropy employing angular-dependent Raman spectroscopy. The hexagonal wurtzite 4H-SiC is a tetrahedral coordinated semiconductor compound and belongs to the space group $C^4_{6v}(P6_3mc)$ with 6 f.u. in the primitive cell. All the atoms occupy sites of symmetry C_{3v}. Main Raman active modes include E_2(TA), 203.5 cm^{-1}; A_1(LA), 610.5 cm^{-1}; E_2(TO), 777.0 cm^{-1}; E_1(TO), 797.5 cm^{-1}; and A_1(LO) 967.0 cm^{-1}, as shown in previous Figure 1 of this chapter.

Owing to the different vibrating types, the phonon intensity induced by electric vector will vary from the polarization of incident laser beam and will cause the mani-fold scattered light. The polarizability tensor of the wurtzite crystal [84-88] has been used to verify the Raman selection rule. But the nonvanishing components of the susceptibility tensor should include the effect from different vibrating components. It is needed to further identify and quantify the anisotropic distribution of Raman signals.

Rahman scattering spectroscopy (RSS) is an effective non-destructive method to characterize materials, to analyze the crystal structure, electronic band structure, phonon energy dispersion, and electron-phonon coupling [84, 85]. By help of analyzing Raman spectra of wide band gap semiconductor (WGS) materials, such as SiC [22-58], GaN [84], and AlN [85, 86], ones are able to get knowledge on their crystal structure, carrier concentration, internal strain/stress and distribution, and other characteristic properties.

Angular-dependent, i.e., rotation Raman scattering has been applied to investigate WGS epitaxial GaN [84] and bulk AlN crystals [85–87]. In this chapter and [58], we employed the rotation Raman scattering (RRS) to study the angular-dependent polarized RSS on the "*a*" face 4H-SiC crystal by adjusting the polarized vector of incident and scattered laser light. The corresponding theoretical calculation on Raman selection role, Raman tensor elements of wurtzite 4H-SiC, and the variations of Raman spectral intensities are present in details and in comparison with experimental results.

7.2 Experimental Arrangements

In the current section of this chapter, we present the RT backscattering Raman spectra of the semi-polar *a*-face (1120) 4H-SiC crystal for the relationship between the lattice phonon vibration modes and the rotation angles of the wurtzite structure, which shows the crystal anisotropy characteristics. The experimental sample is a piece of un-doped *a*-face 4H-SiC with the size of 5×5 mm, cut from a wafer purchased from the CREE Company.

For experimental RSS measurements on the 4H-SiC sample, a Raman micro-spectrometer, connected with a charge coupled device (CCD) for signal detection, named "Finder One" from the Zolix Company, was used at room temperature (RT). The 532 nm excitation laser incident light, through the ×50 object, was focusing on the "*a*" surface of the 4H-SiC crystal sample, while the scattered light was collected via this same object. To make the scattered light parallel with or perpendicular to the incident light, a rotatable polarizer is inserted within the scattered light path. The sample stage could be rotated with 360°, i.e., the incident light could be rotated with 360° to get the angular-dependent Raman spectra.

Figure 9 shows the hexagonal wurtzite crystal structure and the experimental arrangement. The *XYZ* is in cordinance with the crystalline axis of the 4H-SiC crystal. The Z-axis is along the *c*-axis of the wurtzite 4H-SiC crystal, i.e., (0001) direction. The Y-axis is along the normal direction of the *a*-face of 4H-SiC. The rotation angle θ is the angle of inclusion between the ξ-axis and X-axis. The incident light is going along the Y-axis, the

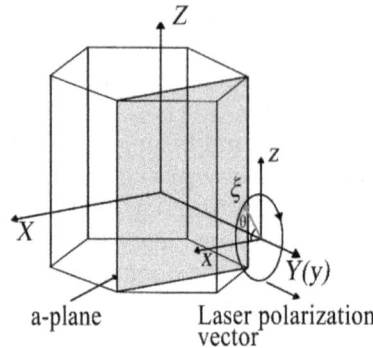

FIGURE 9
The hexagonal wurtzite crystal structure and the experimental arrangement. The Z-axis is along the *c*-axis of 4H-SiC crystal, i.e., (0001) direction. The *a*-plane is shown. The array ξ represents the incident light polarizing vector. The θ is the rotation angle with respect to the X-axis. From [58] Figure 1, reproduced with permission of JLS.

SiC sample is rotated around the Y-axis, and rotation angle θ lies between the incident light polarizing vector and X-axis.

7.3 Raman Selection Rules

4H-SiC crystal is a compound semiconductor with the hexagonal wurtzite structure, and belongs to the space group C^4_{6v} (*P63mc*) with eight atoms in the primitive cell. In its *Brillouin zone*, the lattice vibration produces nine optic branches, and three acoustic branches, including the A_1, B_1, E_1 and E_2 vibration mode. According to the group theory, A_1 and E_1 acoustic modes are Raman and infrared (IR) active. The A_1 phonon is polarized along the Y-axis direction, while the E_1 phonon is polarized within the XZ plane (see the Coordinate System in Figure 9). The E_2 mode is only Raman active, while B_1 belongs to non-active mode. Due to the creation of polar bonds, there appear the frequency shifts for A_1 and E_1 Raman modes with symmetries. The A_1 and E_1 Raman modes are split into the longitudinal phonon and transverse phonon modes, forming the A_1(TO, LO) and E_1(TO, LO) modes.

The Raman scattered light intensities can be expressed as [84–87]

$$I \sim \left| e_s \cdot R \cdot e_i \right|^2 \tag{16}$$

where e_i and e_s present the polarization vector for incident and scattered light, respectively. R is the second-order Raman tensor with the form of a 3×3 vector matrix [84–87]. This vector matrix represents the characteristics of Raman scattering phonon modes. The A_1, E_1, and E_2 Raman active modes in wurtzite structural materials are expressed as below [84–87]

$$R[A_1] = \begin{bmatrix} a & 0 & 0 \\ 0 & a & 0 \\ 0 & 0 & b \end{bmatrix}, \tag{17}$$

$$R[E_1] = \begin{bmatrix} 0 & 0 & -c \\ 0 & 0 & c \\ -c & c & 0 \end{bmatrix}, \tag{18}$$

$$\text{and} \quad R[E_2] = \begin{bmatrix} d & d & 0 \\ d & -d & 0 \\ 0 & 0 & 0 \end{bmatrix}, \tag{19}$$

where a, b, c, and d represent Raman tensor elements. In Raman backscattering measurements on the a-plane of 4H-SiC single crystalline material, referring to the hexagonal axis structure, the polarized vectors for incident and scattered light can be expressed as

$$e_i = \begin{pmatrix} \sin(\theta) \\ 0 \\ \cos(\theta) \end{pmatrix} \tag{20}$$

$$\text{and} \quad e_s = \begin{pmatrix} \sin(\theta_s) \\ 0 \\ \cos(\theta_s) \end{pmatrix}, \tag{21}$$

where θ (θs) represent the angle between the sample Y-axis and the incident (scattered) light vector, respectively.

Here, we discuss the case of the parallel and perpendicular configuration, from which, according to the calculation based upon Eq. (16), the Raman intensity at the surface of a-plane from the 4H-SiC crystal can be expressed as

$$I_{\parallel}(A_1) \sim |a|^2 \sin^4\theta + |b|^2 \cos^4\theta + \frac{|a||b|}{2}\sin^2 2\theta \cos(\varphi_{a-b}) \tag{22}$$

$$I_{\perp}(A_1) \sim \left(\frac{|a|^2 + |b|^2}{4} + \frac{|a||b|}{2}\cos(\varphi_{a-b}) \right) \sin^2 2\theta, \tag{23}$$

$$I(A_1) \sim I_{\parallel}(A_1) + I_{\perp}(A_1) \tag{24}$$

$$I_{\parallel}(E_1) \sim |c|^2 \cos(\varphi_{a-b}) \sin^2(2\theta), \tag{25}$$

$$I_{\perp}(E_1) \sim |c|^2 \cos(\varphi_{a-b}) \cos^2(2\theta), \tag{26}$$

$$I(E_1) \sim I_{\parallel}(E_1) + I_{\perp}(E_1) \tag{27}$$

$$I_{\parallel}(E_2) \sim |d|^2 \cos^4\theta, \tag{28}$$

$$I_{\perp}(E_2) \sim |d|^2 \sin^2\theta\cos^2\theta, \tag{29}$$

$$I(E_2) \sim I_{\parallel}(E_2) + I_{\perp}(E_2) \tag{30}$$

From these formalism, we can know that the phase difference φ_{a-b} between the Raman tensor element a and b, possesses obvious influence on the Raman scattering intensity of

A_1 mode. From Eq. (22), the values of $|a|$, $|b|$ and phase difference φ_{a-b} in the scattered light intensity $I_{//}(A_1)$ are certain lonely. They can be determined through the values of $I_{//}(A_1)$ at $\theta=0°$ and $90°$, respectively. But, the values of $|a|$, $|b|$ and phase difference φ_{a-b} in the scattered light intensity $I_\perp(A_1)$ are different from that of $I_{//}(A_1)$. As known from Eq. (23), the values of $|a|$, $|b|$ and phase difference φ_{a-b} in the scattered light intensity $I_{//}(A_1)$ have still influences on the scattered light intensity $I_\perp(A_1)$.

From these formalisms, we can perform the theoretical calculation and experimental measurements on the 4H-SiC anisotropy Raman scattering. Based upon the atomic dynamics theory, different elements of the vibration modes could display variations related to their space structures. In the following, we can discuss the anisotropy characteristics and the effects on the Raman tensor of various vibration modes.

7.4 Rotation Raman Spectra and Analyses

Figure 10 shows the rotation Raman spectra at the *a*-plane of 4H-SiC, in the wavenumber range of 580–820 cm^{-1} and with the angle rotated from 0° to 180° (step size of 5°), under (a) parallel and (b) perpendicular polarization, respectively. Within Figure 10, the A_1(LA), E_2(TO) and E_1(TO) modes are observed at 610.03 cm^{-1}, 776.49 cm^{-1}, and 796.88 cm^{-1}, respectively. The intensity and phase variations of these Raman modes, and their dependences on the rotation angle are clearly shown.

Figure 11 exhibits the experimental Raman peak values versus the rotation angle in symbols, and the theoretical curves in solid lines, calculated by Eqs. (22)–(30). These reveal clearly the interaction relationship of Raman intensity and rotation angle. As shown in Figure 11 (a), under the case of parallel polarization 〖y(xx)y-y(zz)y〗, with the rotation angle increased from 0° to 45°, the Raman intensity of A_1 mode decreases continually; while with the rotation angle increased from 45° to 90°, the Raman intensity of A_1 mode increases continually. At the rotation angles of 0°, 90°, 180°, 270°, and 360°, the A_1 mode has the maxima intensity; while at 45°, 135°, 225°, and 315°, the A_1 mode disappears. Under the case of perpendicular polarization 〖y(xz)y-y(zx)y〗 in Figure 11 (b), with the rotation angle increases from 0° to 45°, the Raman intensity of A_1 mode increases continually; while with the rotation angle increases from 45° to 90°, the Raman intensity of A_1 mode decreases continually. At the rotation angles of 0°, 90°, 180°, 270°, and 360°, the A_1 mode is disappeared; while at 45°, 135°, 225°, and 315°, the A_1 mode has the maxima intensity. These indicate the sinus-like variation of the A_1 Raman mode intensity depending on the rotation angle, and the strong anisotropic characteristics of 4H-SiC crystal. In addition, in order to determine the Raman tensor for the A_1 mode, it can employ the Raman selection rules to perform the theoretical modeling. Both the A_1 and E_1 phonon modes are both Raman and IR active, as shown in Figures 11 (c)–(d), with similar intensity variation trends for both A_1 and E_1 modes. However, the E_2 mode, with only Raman active, is different from A_1 and E_1 modes.

As shown in Figures 11 (e)–(f), under the case of parallel polarization 〖y(xx)y-y(zz)y〗, with the rotation angle increased from 0° to 90°, the Raman intensity of E_2 mode increases continually; while with the rotation angle increased from 90° to 180°, the Raman intensity of A_1 mode decreases continually. At the rotation angles of 0°, 180°, and 360°, the A_1 mode has the minima intensity; while at 90° and 270°, the A_1 mode is strongest. Under the case of perpendicular polarization 〖y(xz)y-y(zx)y〗, the Raman intensity variation of E_1 mode has the same variation trend as the A_1 mode.

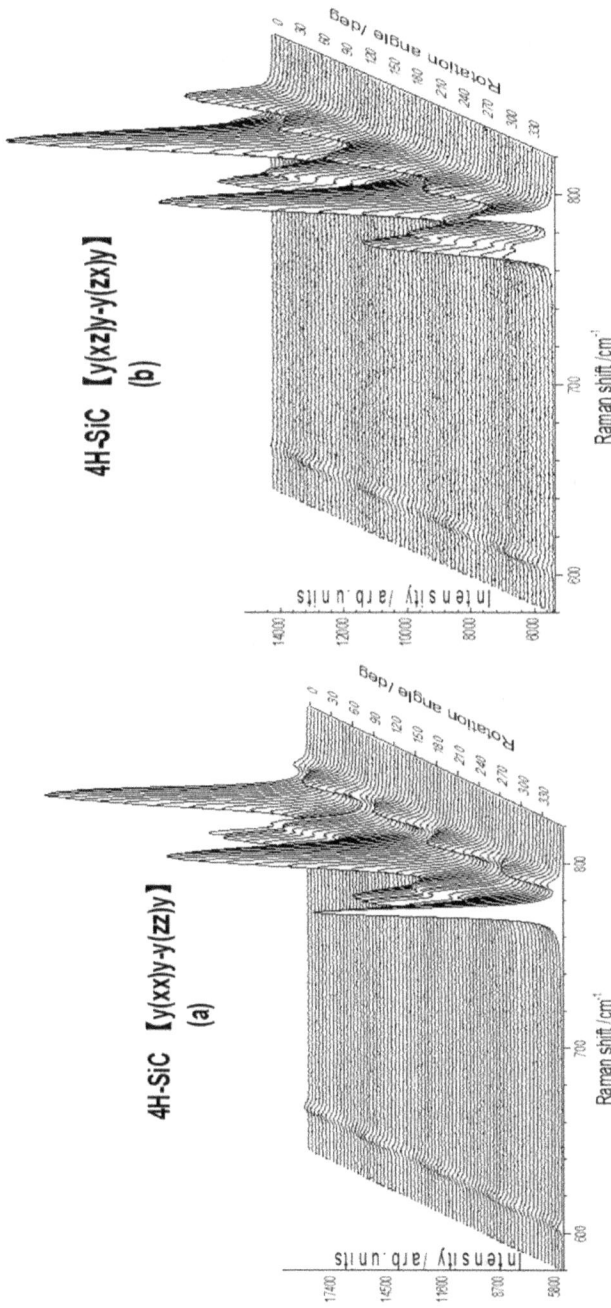

FIGURE 10

The rotation Raman spectra at the *a*-plane of 4H-SiC, with the angle varied from 0° to 360° (step size of 5°), under (a) parallel and (b) perpendicular polarization, respectively. From [58] Figure 2, reproduced with permission of JLS.

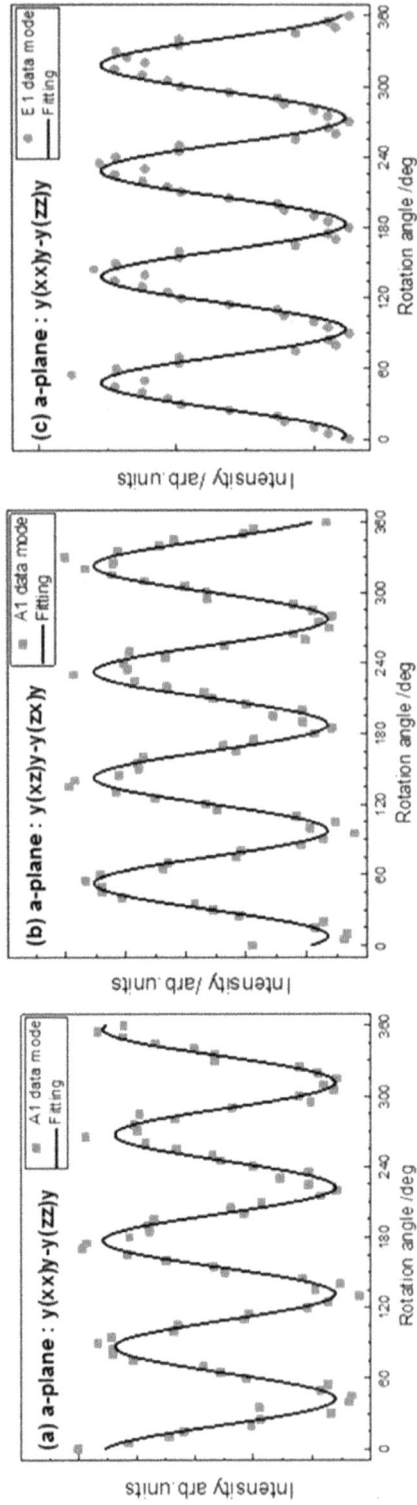

FIGURE 11

The relationship of Raman intensity and ration angle for the A_1 mode at the a-plane of 4H-SiC sample with the incident and scattered light vectors under (a) parallel and (b) perpendicular polarization. (c) and (d) are for the E_1 mode, while (e) and (f) are for the E_2 mode, respectively. The solid lines are fitting results based upon Eqs. (7)–(15), from which the values of $|a|$, $|b|$ and phase difference φ_{a-b} are determined. From [58] Figure 3, reproduced with permission of JLS.

7.5 Raman Tensor Element Analyses

Raman intensities of A_1, E_1, and E_2 phonon modes under the polarization states of parallel 【$y(xx)y-y(zz)y$】 and perpendicular 【$y(xz)y-y(zx)y$】 were shown as in Figure 11. The fitting results for the relationship of Raman intensity and ration angle under two polarization cases are expressed in Figures 11 (a)–(f), with small errors. To determine the Raman tensor elements, the fitting procedure by the least-squares method and formulas (22)–(30) with angle change of function fitting is applied. The best fitting curves are basically coincident with the experimental data, as shown in Figure 11. The relative values of Raman tensor elements for 4H-SiC crystal are calculated by way of the ratio calculation of fitting parameters are overviewed and shown in Table 3. It is found that different Raman modes are completely different. The phase difference and anisotropic ratio of each Raman mode characterizes similar information, but there exists phase shift among different elements. Also, the values of phase shift and anisotropic ratio are determined by the corresponding Raman mode. Each element of Raman tensor represents its single directional vibration and correlated with another element. Therefore, from experimental and theoretical results, the anisotropics of the 4H-SiC sample can be determined.

The 4H-SiC material belongs to a wide band gap semiconductor compound, is widely applied in the field of optoelectronics, and possesses the wurtzite structure with the hexagonal space group o C^4_{6v} ($P6_3mc$). Based upon the atomic dynamics, by way of Raman scattering measurements, different vibration elements can display the sensitivities correlated to their own space group [88]. The A_1, E_1, and E_2 modes in wurtzite 4H-SiC are Raman active. Among these modes, the A_1 polarization direction is coincident with the Y-axis. In doped semiconductor, the A_1 mode, by way of the resonant coupling between polarized phonon and plasma (free carriers), exerts an influence on the electronic transport properties [89]. As expressed from Table 3, the relative values of Raman tensor elements for 4H-SiC crystal are calculated by help of the ratio calculation on fitting parameters. From Raman scattering spectra under the parallel and perpendicular polarization, the phase difference φ_{a-b} of Raman tensor elements $|a|$ and $|b|$ (i.e., the parallel and perpendicular vectors) for 4H-SiC crystal, are determined as 117.37° and 92.86°, respectively.

It can be seen from the article of Strach et al. [90] that the selection of parameters a, b and phase difference φ_{a-b} is not unique for the perpendicular vectors of incident and scattered light. So, the ratio of perpendicular polarized vectors in Table 3 is not well defined. But this ratio is comparatively coincident with the result estimated from incident and scattered polarized vectors in parallel alignment. For wurtzite material, by re-doing measurements

TABLE 3

Relative values of Raman tensor elements for A_1, E_1, and E_2 modes of wurtzite 4H-SiC crystal, which are theoretical fitting results for the incident laser (e_i) and scattered light (e_s) under parallel and perpendicular polarization cases, respectively

a-plane tensor element		$\vec{e}_i \parallel \vec{e}_s$ $y(xx)\bar{y} \leftrightarrow y(zz)\bar{y}$	$\vec{e}_i \perp \vec{e}_s$ $y(xz)\bar{y} \leftrightarrow y(zx)\bar{y}$
A_1 mode	$\|a/d\|$	1.003±0.024	0.198±0.019
A_1 mode	$\|b/d\|$	1.004±0.023	0.198±0.020
	φ_{a-b}	111.37°	92.86°
E_1 mode	$\|c/d\|$	0.388±0.029	0.292±0.013
E_2 mode	$\|d/d\|$	1.000±0.007	1.000±0.006

on Raman scattering spectra, it can confirm the relationship between the anisotropy ratio and the phase difference of Raman tensor elements in wurtzite crystal.

In brief summary of Section 7, within this part of the work, for wurtzite 4H-SiC single crystalline material, the rotation Raman scattering spectra was studied in the backscattering at RT. The anisotropy characteristic properties are revealed. By adding in the polarization apparatus, the variation functional relationship of the angle between the incident laser polarization direction and the parallel (perpendicular) polarization direction was studied. By way of the parametrization on incident light and scattered light polarization vectors, the selection rules of wurtzite SiC are calculated. Based upon the selection rules, the intensity variations of the A_1, E_2, and E_1 modes dependent on the rotation angle are also calculated, which lead to the determination of Raman tensor elements of various modes.

8 Conclusion and Summary

This chapter offers a summarized review of some important topics in our Raman scattering studies on hexagonal wurtzite 4H-SiC and 6H-SiC crystal bulk materials, such as Raman scattering spectra in semi-insulating and n-type (nitrogen) doped 4H-SiC and 6H-SiC wafers, second-order Raman scattering from semi-insulating and doped 4H-SiC and 6H-SiC, resonance enhancement of electronic Raman scattering from nitrogen defect levels in 4H-SiC and 6H-SiC, temperature dependence of Raman scattering in bulk 4H-SiC with different carrier concentration, and anisotropic properties of 4H-SiC studied by angular-dependent (or rotation) Raman scattering.

The following main features are worthy to notice and useful for the R&D of SiC materials. (i) Theoretical models of plasmon–phonon coupling successfully predict the experimental A_1(LO) phonon line shape at dopant concentrations up to 10^{19} cm^{-3} in 4H- and 6H-SiC, which can be used as a measure of carrier concentration and as an *in situ*, non-contact diagnostic of doping uniformity in semiconductor production. This is in contrast to electrical conductivity measurements, which require contacts with the samples. (ii) Both overtones and combination bands are observed in the second-order Raman spectra of 4H- and 6H-SiC, which are quite similar for n-type doped and semi-insulating hexagonal SiC. The optical branches of the second-order Raman spectra for 4H- and 6H-SiC are polytype dependent and much more complicated than those for cubic 3C-SiC. Precise values for high symmetry point phonons in SiC have been obtained by the analysis of the second-order Raman spectra. Our experimental measurements are consistent with calculations of the phonon dispersion and calculated phonon frequencies for the L-M branch. (iii) Electronic Raman scattering from nitrogen donor levels in both 4H- and 6H-SiC are observed under the below band gap excitations of visible to near-IR. The resonance is due to the near-IR absorption, typical of nitrogen-doped SiC, attributed to deep defects inside the SiC wafer materials. (iv) Temperature-dependent (80–873 K) Raman scattering in bulk 4H-SiC with different doping levels is studied in combination with detailed theoretical simulation. T-variations of E_2 (TO) and E_1 (TO) phonon modes are well described through the three- and four-phonon process function with temperature. Anomalous variation of A_1(LO) mode in doped 4H-SiC stems from the coupling between LO phonons and plasma, forming the LOPC mode. The non-monotonous variation of blue-red shifts with temperature for LOPC mode from doped 4H-SiC could be explained by the influence from ionization process of impurities on the process of Raman scattering. A quantitative description on temperature-dependent Raman spectra for doped 4H-SiC is achieved, matching well to experimental data. (v) Rotation Raman scattering was studied in the combination of experiments and theories, revealing the anisotropy characteristic properties of 4H-SiC. The variation functional relationship of the angle between the incident laser polarization direction and the parallel (perpendicular) polarization direction, and the selection rules of wurtzite SiC are well established. The intensity variations of the A_1, E_2 and E_1 phonon modes dependent on the rotation angle are also calculated, which lead to the determination of Raman tensor elements of various modes.

The above penetrating investigations are helpful to better understand the mechanisms and find ways to further improve the material design and growth of hexagonal SiC as well as their wide range applications.

Acknowledgements

We acknowledge Profs. W.J. Choyke, F.H. Long and Weijie Lu, Drs. J.C. Burton and M. Pophristic, H.Y. Sun and D.-S. Zhao, for various help and supports on the works in this chapter. Different funding supports on the work are grateful.

References

1. W. J. Choyke, H. Matsunami, G. Pensl (Eds.), Silicon Carbide: Recent Major Advances, Springer, Berlin (2004).
2. Z. C. Feng (Ed.), SiC Power Materials – Devices and Applications, Springer, Berlin (2004).
3. Michael Shur, Sergey Rumyantsev and Michael Levinshtein (Eds.), SiC Materials and Devices, Vol. 1 and 2, World Scientific Publishing, Singapore (2006).
4. Peter Friedrichs, Tsunenobu Kimoto, Lothar Ley and Gerhard Pensl (Editors), Silicon Carbide: Volume 1: Growth, Defects, and Novel Applications (2009); Silicon Carbide: Volume 2: Power Devices and Sensors (2011) WILEY-VCH.
5. Tsunenobu Kimoto and James A. Co, Fundamentals of Silicon Carbide Technology: Growth, Characterization, Devices and Applications (Wiley – IEEE) (2014).
6. G. Pensl and W. J. Choyke "Electrical and optical characterization of SiC", Physica B 185, 264–283 (1993).
7. J. B. Casady and R. W. Johnson, "Status of silicon carbide (SiC) as a wlde-bandgap semiconductor for high-temperature applications: a review", Solid-State Electronics 39, 1409–1422 (1996).
8. W. J. Choyke and G. Pensl, "Physical properties of SiC", MRS Bulletin, 22, 25–29 (1997).
9. V. A. Dmitriev and M. G. Spencer, "Sic Fabrication Technology: Growth and Doping", in Semiconductors and Semimetals, Vol. 52, Ch. 2, p.21–75 (1998).
10. C. H. Carter, Jr., V. F. Tsvetkov, R. C. Glass, D. Henshall, M. Brady, St. G. Mu̇ller, O. Kordina, K. Irvine, J. A. Edmond, H.-S. Kong, R. Singh, S. T. Allen, J. W. Palmour, "Progress in SiC: from material growth to commercial device development", Materials Science and Engineering B 61–62, 1–8 (1999).

11. N. G. Wright and A. B. Horsfall, "SiC sensors: a review", J. Phys. D: Appl. Phys. 40, 6345–6354 (2007).

12. S. E. Saddow, C. L. Frewin, M. Nezafatil, A. Oliveros I, S. Afroz, J. Register, M. Reyes, and S. Thomas, "3C-SiC on Si: A Bio- and Hemo-compatible Material for Advanced Nano-Bio Devices", IEEE Nanotechnology Materials and Devices Conference (NMDC), p.46–53 (2014).

13. Masayuki Furuhashi, Shingo Tomohisa, Takeharu Kuroiwa and Satoshi Yamakawa, Practical applications of SiC-MOSFETs and further developments, Semicond. Sci. Technol., 31, 034003 (2016).

14. Raksha Adappa, Suryanarayana K, Swathi Hatwar H and Ravikiran Rao M, "Review of SiC based Power Semiconductor Devices and their Applications", 2019 2nd International Conference on Intelligent Computing, Instrumentation and Control Technologies (ICICICT), **IEEE**, 1197–1202 (2019).

15. W. M. Klahold, W. J. Choyke, and R. P. Devaty, "Band structure properties, phonons, and exciton fine structure in 4H-SiC measured by wavelength-modulated absorption and low-temperature photoluminescence", Physical Review B 102, 205203 (2020).

16. Dang-Hyok Yoon and Ivar E. Reimanis, "A review on the joining of SiC for high-temperature applications", Journal of the Korean Ceramic Society, 57, 246–270 (2020).

17. Jingxin Jian and Jianwu Sun, "A Review of Recent Progress on Silicon Carbide for Photoelectrochemical Water Splitting", Sol. RRL 4, 2000111 (2020).

18. Julian A. Michaels, Lukas Janavicius, Xihang Wu, Clarence Chan, Hsieh-Chih Huang, Shunya Namiki, Munho Kim, Dane Sievers, and Xiuling Li, "Producing Silicon Carbide Micro and Nanostructures by Plasma-Free Metal-Assisted Chemical Etching", Advanced Functional Materials, 2021, 2103298 (2021).

19. Xingliang Xu, Lin Zhang, Peng Dong, Zhiqiang Li, Lianghui Li, Juntao Li and Jian Zhang, Nanoscale Res. Lett. 16, 141 (2021).

20. Run Tian, Chao Ma, Jingmin Wu, Zhiyu Guo, Xiang Yang, and Zhongchao Fan, "A review of manufacturing technologies for silicon carbide superjunction devices", Journal of Semiconductors 42, 061801 (2021).

21. Ying Chang, Aixia Xiao, Rubing Li, Miaojing Wang, Saisai He, Mingyuan Sun, Lizhong Wang, Chuanyong Qu and Wei Qiu, "Angle-Resolved Intensity of Polarized Micro-Raman Spectroscopy for 4H-SiC", Crystals 11, 626 (2021).

22. D. W, Feidman, James H. Parker, Jr., W. J. Choyke, and Lyle Patric, "Raman Scattering in 6H SiC", Physical Review, 170, 698–704 (1968).

23. D. W, Feidman, James H. Parker, Jr., W. J. Choyke, and Lyle Patric, "Phonon Dispersion Curves by Raman Scattering in SiC, Polytypes 3C, 4H, 6H, 15R, and 21R", Physical Review, 173, 787–793 (1968).

24. Priscilla J. Colwell and Miles V. Klein, "Raman Scattering from Electronic Excitations in n-Type Silicon Carbide", Phys. Rev. B 6, 498–515 (1972).

25. Miles V. Klein, B. Ganguly, and Priscilla J. Colwell, "Theoretical and Experimental Study of Raman Scattering from Coupled LO-Phonon-Plasmon Modes in Silicon Carbide", Phys. Rev. B 6, 2380–2388 (1972).

26. J. Liu and Y. K. Vohra, "Raman Modes of 6H Polytype of Silicon Carbide to Ultrahigh Pressures: A Comparison with Silicon and Diamond", Phys. Rev. Lett. 72, 4105 (1994).

27. H. Harima, S.-I. Nakashima, and T. Uemura, "Raman scattering from anisotropic LOphonon–plasmon–coupled mode in ntype 4H– and 6H–SiC", J. Appl. Phys. 78, 1996 (1995).

28. Z.C. Feng, C.C. Tin, R. Hu, and K.T. Yue, "Combined Raman and luminescence assessment of epitaxial 6H-SiC films grown on 6H-SiC by low pressure vertical chemical vapor deposition", Semicond. Sci. & Tech. 10, 1418–1422 (1995).

29. S. Nakashima and H. Harima, "Raman investigation of SiC polytypes", Phys. Status Solidi A 162, 37 (1997).

30. Z. C. Feng, S. J. Chua, K. Tone and J. H. Zhao, "Recrystallization of C-Al Ion Co-implanted Epitaxial 6H-SiC", Appl. Phys. Lett. 75, 472–474 (1999).

31. Z.C. Feng, S.C. Lien, J.H. Zhao, X.W. Sun and W. Lu, "Structural and Optical Studies on Ion-implanted 6H-SiC Thin Films", Thin Solid Films, 516, no.16, 5217–5226 (2008).

32. Michael Bauer, Alexander M. Gigler, Andreas J. Huber, Rainer Hillenbrand, and Robert W. Stark, "Temperature-depending Raman line-shift of silicon carbide", J. of Raman Spectroscopy, 40, 1867–1874 (2009).

33. Xiang-Biao Lin, Zhi-Zhan Chen, and Er-Wei Shi, "Effect of doping on the Raman scattering of 6H-SiC crystals", Physica B 405, 2423–2426 (2010).

34. Shenghuang Lin, Zhiming Chen, Lianbi Li, and Chen Yang, "Effect of Impurities on the Raman Scattering of 6H-SiC Crystals", Materials Research 15, 833–836 (2012).

35. Yao Huang, Run Yang, Shijie Xiong, Jian Chen and Xinglong Wu, "Strong Coupling of Folded Phonons with Plasmons in 6H-SiC Micro/Nanocrystals", Molecules 23, 2296 (2018); doi:10.3390/molecules23092296.

36. Zongwei Xu, Zhongdu He, Ying Song, Xiu Fu, Mathias Rommel, Xichun Luo, Alexander Hartmaier, Junjie Zhang and Fengzhou Fang, "Topic Review: Application of Raman Spectroscopy Characterization in Micro/Nano-Machining", Micromachines 9, 361 (2018).

37. Aikaterini Flessa, Eleni Ntemou, Michael Kokkoris, Efthymios Liarokapis, Marko Gloginjić, Srdjan Petrović, Marko Erich, Stjepko Fazinić, Marko Karlušić and Kristina Tomić, "Raman mapping of 4-MeV C and Si channeling implantation of 6H-SiC", J Raman Spectrosc. 50, 1186–1196 (2019).

38. K. Kamalakkannana, R. Rajaramanb, B. Sundaravelb, G. Amarendrab, and K. Sivaji, "Effect of nitrogen ion implantation in semi insulating 6H-SiC and recrystallization probed by Raman scattering", Nuclear Inst. and Methods in Physics Research B 457, 24–29 (2019).

39. M. Mastellone, A. Bellucci, M. Girolami, R. M. Montereali, S. Orlando, R. Polini, V. Serpente, E. Sani, V. Valentini, M. A. Vincenti, and D. M. Trucchi, "Enhanced selective solar absorption of surface nanotextured semi-insulating 6H–SiC", Optical Materials 107, 109967 (2020).

40. N. Daghbouj, B.S. Li, M. Callisti, H.S. Sen, J. Lin, X. Ou, M. Karlik, and T. Polcar, "The structural evolution of light-ion implanted 6H-SiC single crystal: Comparison of the effect of helium and hydrogen", Acta Materialia 188, 609–622 (2020).

41. C. C. Tin, R. Hu, J. Liu, Y. Vohra and Z. C. Feng, "Raman microprobe spectroscopy of low-pressure-grown 4H-SiC epilayers", J. Crystal Growth 158, 509–513 (1996).

42. M. Chafai, A. Jaouhari, A. Torres, R. Antón, E. Martín and J. Jiménez, and W. C. Mitchel, "Raman scattering from LO phonon-plasmon coupled modes and Hall-effect in n-type silicon carbide 4H–SiC", Journal of Applied Physics, 90, 5211–5215 (2001). https://doi.org/10.1063/1.1410884.

43. S. Nakashima, T. Kitamura, T. Mitani, H. Okumura, M. Katsuno and N. Ohtani, "Raman scattering study of carrier-transport and phonon properties of 4H-SiC crystals with graded doping", Phys. Rev. B 76, 245208 (2007).

44. Yang Yin-Tang, Han Ru, and Wang-Ping, "Raman analysis of defects in n-type 4H-SiC", Chinese Physics B 17, 3459–3463 (2008).

45. R. Han, B. Han, D. H. Wang, and C. Li, "Temperature dependence of Raman scattering from 4H-SiC with hexagonal defects", Appl. Phys. Lett. 99, 011912 (2011).

46. N. Piluso, M. Camarda, and F. La Via, "A novel micro-Raman technique to detect and characterize 4H-SiC stacking faults", Journal of Applied Physics 116, 163506 (2014).

47. Yi-Chuan Tseng, Yu-Chia Cheng, Yang-Chun Lee, Dai-Liang Ma, Bang-Ying Yu, Bo-Cheng Lin, and Hsuen-Li Chen, "Using Visible Laser-Based Raman Spectroscopy to Identify the Surface Polarity of Silicon Carbide", J. Phys. Chem. C, 120, 18228–18234 (2016).

48. Lingyu Wan, Dishu Zhao, Fangzhe Wang, Gu Xu, Chin-Che Tin, Tao Lin, Zhaochi Feng and Zhe Chuan Feng, "Efficient quality analysis of homo-epitaxial 4H-SiC thin films by the forbidden Raman Scattering Mode", Optical Materials Express, 8,119–127 (2018).

49. Tao Liu, Zongwei Xu, Mathias Rommel, Hong Wang, Ying Song, Yufang Wang and Fengzhou Fang, "Raman Characterization of Carrier Concentrations of Al-implanted 4H-SiC with Low Carrier Concentration by Photo-Generated Carrier Effect", Crystals 9, 428, (2019).

50. S. M. Tunhuma, M. Diale, J. M. Nel, M. J. Madito, T. T. Hlatshwayo, and F. D. Auret, "Defects in swift heavy ion irradiated n-4H-SiC", Nuclear Inst. and Methods in Physics Research B 460, 119–124 (2019).

51. Ying Song, Zongwei Xu, Tao Liu, Mathias Rommel, Hong Wang, Yufang Wang and Fengzhou Fang, "Depth Profiling of Ion-Implanted 4H–SiC Using Confocal Raman Spectroscopy", Crystals 10, 131 (2020).

52. Alessandro Meli, Annamaria Muoio, Antonio Trotta, Laura Meda, Miriam Parisi and Francesco La Via, "Epitaxial Growth and Characterization of 4H-SiC for Neutron Detection Applications", Materials 14, 976 (2021).

53. Ying Chang, Aixia Xiao, Rubing Li, Miaojing Wang, Saisai He, Mingyuan Sun, Lizhong Wang, Chuanyong Qu and Wei Qiu, "Angle-Resolved Intensity of Polarized Micro-Raman Spectroscopy for 4H-SiC", Crystals 11, 626 (2021).

54. J. C. Burton, L. Sun, M. Pophristic, F. H. Long, Z. C. Feng and I. Ferguson, "Spatial characterization of doped SiC wafers by Raman spectroscopy", J. Appl. Phys. 84, 6268–6273 (1998). https://doi.org/10.1063/1.368947

55. J. C. Burton, L. Sun, F. H. Long, Z. C. Feng and I. Ferguson, "First- and second-order Raman scattering from semi-insulating 4H-SiC", Phys. Rev. B 59, 7282–7284 (1999). https://doi.org/10.1103/PhysRevB.59.7282.

56. J. C. Burton, F. H. Long and I. Ferguson, "Resonance enhancement of electronic Raman scattering from nitrogen defect levels in silicon carbide", J. Appl. Phys. 86, 2073–2077 (1999). https://doi.org/10.1063/1.371011.

57. Hua Yang Sun, Siou-Cheng Lien, Zhi Ren Qiu, Hong Chao Wang, Ting Mei, Chee Wee Liu, and Zhe Chuan Feng, "Temperature dependence of Raman scattering in bulk 4H-SiC with different carrier concentration", Optics Express 21, 26475–26482 (2013). https://doi.org/10.1364/OE.21.026475.

58. Di-shu Zhao, Fang-ze Wang, Ling-yu Wan, Qing-yi Yang, and Zhe Chuan Feng, "Raman scattering study on anisotropic property in wurtzite 4H-SiC", J. Light Scattering, 30 (6), 37–42 (2018). doi:10.13883/j.issn1004–5929.201802007

59. W. Hayes and R. Loudon, Scattering of Light by Crystals, 1st ed. Wiley, New York, (1978).

60. M. Cardona and G. Guntherodt (ed.), Light Scattering In Solids II, Springer, Vol. 50, New York, (1982).

61. G. Irmer, V. V. Toporov, B. H. Bairamov, and J. Monecke, "Determination of the charge carrier concentration and mobility in n-gap by Raman spectroscopy", Phys. Status Solidi A 119, 595 (1983). https://doi.org/10.1002/pssb.2221190219.

62. P. Y. Yu and M. Cardona, Fundamentals of Semiconductors, Springer, New York, (1996).

63. W. Windl, K. Karch, P. Pavone, O. Schutt, D. Strauch, W. H. Weber, K. C. Hass, and L. Rimai, "Second-order Raman spectra of SiC: Experimental and theoretical results from *ab initio* phonon calculations", Phys. Rev. B 49, 8764 (1994). https://doi.org/10.1103/PhysRevB.49.8764.

64. A. Zywietz, K. Karch, and F. Bechstedt, "Influence of polytypism on thermal properties of silicon carbide", Phys. Rev. B 54, 1791 (1996). https://doi.org/10.1103/PhysRevB.54.1791.

65. P. Perlin, C. Jauberthie-Carillon, J.-P. Itie, A. S. Miguel, I. Grzegory, and A. Polian, "Influence of heteroepitaxy on the width and frequency of the E_2 (high)-phonon line in GaN studied by Raman spectroscopy", Phys. Rev. B 45, 83 (1992). https://doi.org/10.1063/1.1347406.

66. P. Perlin, J. Camassel, W. Knap, T. Taliercio, J. C. Chervin, T. Suski, I. Grzegory, and S. Porowski, "Investigation of longitudinal-optical phonon-plasmon coupled modes in highly conducting bulk GaN", Appl. Phys. Lett. 67, 2524 (1995). https://doi.org/10.1063/1.114446.

67. A. Pe´rez-Rodrı´guez, Y. Pacaud, L. Calvo-Barrio, C. Serre, W. Skorup, and J. R. Morante, "Analysis of ion beam induced damage and amorphization of 6H-SiC by raman scattering", J. Electron. Mater. 25, 541 (1996). https://doi.org/10.1007/BF02666633.

68. Z. C. Feng, A. Rohatgi, C. C. Tin, R. Hu, A. T. S. Wee, and K. P. Se, "Structural, optical, and surface science studies of 4H-SiC epilayers grown by low pressure chemical vapor deposition", J. Electron. Mater. 25, 917 (1996). https://doi.org/10.1007/BF02666658.

69. A. Zyweitz, K. Karch, and F. Bechstedt, "Influence of polytypism on thermal properties of silicon carbide", Phys. Rev. B 54, 1791 (1996). https://doi.org/10.1103/PhysRevB.54.1791.

70. K. Karch, P. Pavone, W. Windl, D. Strauch, and F. Bechstedt, "Ab initio calculation of structural, lattice dynamical, and thermal properties of cubic silicon carbide", Int. J. Quantum Chem. 56, 801 (1995). https://doi.org/10.1002/qua.560560617.

71. K. Karch, F. Bechstedt, P. Pavone, and D. Strauch, "Pressure-dependent properties of SiC polytypes", Phys. Rev. B 53, 13 400 (1996). https://doi.org/10.1103/PhysRevB.53.13400.

72. F. Bechstedt, P. Kackell, A. Zyweitz, K. Karch, B. Adolph, K. Tenelsen, and J. Furthmüller, "Polytypism and properties of Silicon Carbide", Phys. Status Solidi B 202, 35 (1997).

73. Olego and M. Cardona, "Pressure dependence of Ratnan phonons of Ge and 3C-SiC",Phys. Rev. B 25, 1151 (1982)

74. M. Hofmann, A. Zywietz, K. Karch, and F. Bechstedt, "Lattice dynamics of SiC polytypes within the bond-charge model", Phys. Rev. B 50, 13 401 (1994). https://doi.org/10.1103/PhysRevB.50.13401.

75. L. Patrick, D. R. Hamilton, and W. J. Choyke, "Growth, Luminescence, Selection Rules, and Lattice Sums of SiC with Wurtzite Structure", Phys. Rev. 143, 526 (1966). https://doi.org/10.1103/PhysRev.143.526.

76. R. G. Humphreys, D. Bimberg, and W. J. Choyke, "Wavelength modulated absorption in SiC", Solid State Commun. 39, 163 (1981). https://doi.org/10.1016/0038-1098(81)91070-X.

77. H. H. Woodbury and G. W. Ludwig, "Electron Spin Resonance Studies in SiC", Phys. Rev. 124, 1083 (1961). https://doi.org/10.1103/PhysRev.124.1083.

78. E. Wörner, J. Wagner, W. Müller-Sebert, C. Wild, and P. Koidl, "Infrared Raman scattering as a sensitive probe for the thermal conductivity of chemical vapor deposited diamond films", Appl. Phys. Lett. 68, 1482 (1996). https://doi.org/10.1063/1.116261.

79. W. Suttrop, G. Pensl, W. J. Choyke, R. Stein, and S. Leibenzender, "Hall effect and infrared absorption measurements on nitrogen donors in 6H-silicon carbide", J. Appl. Phys. 72, 3708 (1992). https://doi.org/10.1063/1.352318.

80. W. Götz, A. Schöner, G. Pensl, W. Suttrop, W. J. Choyke, R. Stein, and S. Leibenzender, "Nitrogen donors in 4*H*-silicon carbide", J. Appl. Phys. 73, 3332 (1993). https://doi.org/10.1063/1.352983.

81. W. J. Moore, P. J. Lin-Chung, J. A. Freitas, Jr., Y. M. Altaiskii, V. L. Zuev, and L. Ivanova, "Nitrogen donor excitation spectra in 3C-SiC", Phys. Rev. B 48, 12289 (1993). https://doi.org/10.1103/PhysRevB.48.12289.

82. W. S. Li, Z. X. Shen, Z. C. Feng, and S. J. Chua, "Temperature dependence of Raman scattering in hexagonal gallium nitride films," J. Appl. Phys. 87, 3332–3337 (2000).

83. X. B. Li, Z. Z. Chen, and E. W. Shi, "Effect of doping on the Raman scattering of 6H-SiC," Physica B 405, 2423–2426 (2010).

84. H. C. Lin, Z. C. Feng, M. S. Chen, Z. X. Shen, I. T. Ferguson and W. Lu, "Raman Scattering Study on Anisotropic Property of Wurtzite GaN", J. Appl. Phys. 105, 036102 (2009).

85. Wei Zheng, Rui Sheng Zheng, Hong Lei Wu, and Fa Di Li, "Strongly anisotropic behavior of A_1(TO) phonon mode in bulk AlN", Journal of Alloys and Compounds, 584, 374–376 (2014).

86. L. Jin a, H. L. Wu, Y. Zhang, Z. Y. Qin, Y. Z. Shi, H. J. Cheng, R. S. Zheng, and W. H. Chen, "The growth mode and Raman scattering characterization of m-AlN crystals grown by PVT method", Journal of Alloys and Compounds, 584, 374–376 (2014).

87. Wei Zheng, Rui Sheng Zheng, Feng Huang, Hong Lei Wu, and Fa Di Li, "Raman tensor of AlN bulk single crystal", Photonics Research, 3, 38–43 (2015); and "Raman tensor of AlN bulk single crystal: erratum", Photonics Research, 8, 412–413 (2020).

88. T. Inui, Y. Tanabe, and Y. Onodera, Group Theory and Its Applications in Physics. Springer, Berlin (1996).

89. D. Olego and M. Cardona, "Raman scattering by coupled LO-phonon-plasmon modes and forbidden TO-phonon Raman scattering in heavily doped p-type GaAs", Phys. Rev. B 24, 7217–7232 (1981).
90. T. Strach, J. Brunen, B. Lederle, J. Zegenhagen, and M. Cardona, "Determination of the phase difference between the Raman tensor elements of the A_{1g}-like phonons in $SmBa_2Cu_3O_{7-\delta}$", Phys. Rev. B 57, 1292–1297 (1998). https://doi.org/10.1103/PhysRevB.57.1292.

10

Near-Infrared Luminescent Centers in Silicon Carbide

Ivan G. Ivanov and Nguyen T. Son

Department of Physics, Chemistry and Biology, Linköping University, Linköping, Sweden

1 Introduction

In recent decades, considerable effort has been spent on developing of single photon sources (SPSs) in various material platforms for applications in quantum communications [1]. Among these quantum dots (QDs) and photoluminescent (PL) centers (often referred as color centers) in semiconductors are particularly attractive due to their excellent optical properties and scalability.

Semiconductor QDs offer high photon generation rates (high brightness) and ability of tuning the spectral range. A QD involving a trapped electron or hole can act as an optical spin quantum bit (qubit) that combines single photon emission from excitons and a correlated spin qubit. Quantum entanglement between distant qubits – an important feature for quantum networks – has been demonstrated using these systems [2]. However, short spin coherence (below μs) caused by magnetic noise from the nuclear spin bath of III-V semiconductors remains a major challenge for QDs [3].

Many point defects in wide-bandgap semiconductors have both the ground and excited states within the bandgap and, hence, are color centers. In crystal hosts with low natural abundance of nonzero nuclear spins, such as diamond, silicon carbide (SiC) and II-VI compounds, spins associated with color centers can have long coherence times [4]. Optically addressable spins with long coherence time even at ambient conditions, the availability of nuclear spins for quantum memories, as well as high-fidelity spin-to-photon interfaces are distinct advantages of color centers for applications in quantum information processing and quantum sensing [1,5].

Among these systems, the negatively charged nitrogen-vacancy (NV) center in diamond, i.e., a complex between a substitutional N and a nearest lattice vacancy, is currently the leading contender. Its recent major advances in quantum information processing include high-fidelity (92%) entanglement of photons from two single NV emitters separated by ~1.3 km [6]. However, the high loss in fibers (~8 dB/km) hinders the application of the NV center in long-distance quantum communications using the fiber networks. Entanglement-preserving quantum frequency conversion of single NV photons at 637 nm into telecom C-band (1550 nm) with ~17% efficiency has been demonstrated [7]. However, down frequency conversion which more than doubles the wavelength comes at the cost of extra noise photons due to spontaneous scattering processes of the strong laser pump with a

DOI: 10.1201/9780429198540-13

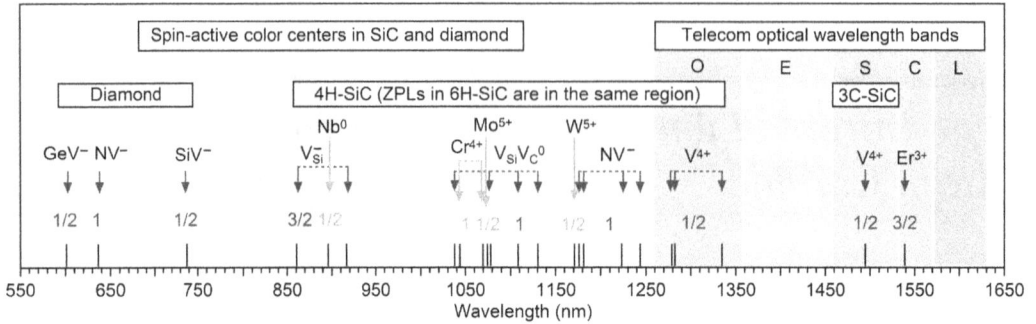

FIGURE 1
Emission wavelengths of spin-active color centers in diamond and SiC. The ZPLs of color centers in 4H- and 6H-SiC are in the same spectral region. The emission of Er^{3+} center is at ~1540 nm for 3C-, 4H-, and 6H-SiC.

wavelength shorter than that of the converted signal [8]. Other color centers in diamond also have emissions in the visible spectral region (Figure 1).

Silicon carbide is a wide-bandgap semiconductor which can exist in many polytypes with the hexagonal 4H-SiC (bandgap: 3.23 eV) and 6H-SiC (3.0 eV) and cubic 3C-SiC (2.36 eV) polytypes being the most common. The material has long been developed for energy-efficient power electronics and is the only wide-bandgap semiconductor with established large-scale industrial production methods, controlled doping, well-developed integrated circuit processing as well as high-quality nanofabrication. Recently, SiC has been emerging as promising material platform for optical spin qubits [4,9]. While sharing most of advantages of wide-bandgap material with diamond, such as spin qubits with long spin coherence times and room-temperature operation [10,11], SiC hosts various spin-active color centers emitting light near or at telecom wavelengths as shown in Figure 1. The development of optical spin qubits in SiC has been rapidly progressing, but so far concentrated mainly on the neutral divacancy ($V_C V_{Si}^0$), i.e., an uncharged complex consisting of a C vacancy (V_C) and a nearest Si vacancy (V_{Si}), and the negative Si vacancy (V_{Si}^-) [12]. In this Chapter, we give an overview on PL centers, which emit light near and at telecom wavelengths, and provide updated information on their identification and optical properties.

Both 4H- and 6H-SiC possess inequivalent lattice sites, two per host atom in 4H-SiC, and three in 6H-SiC. The notations used throughout this chapter for the inequivalent sites in 4H- and 6H-SiC are defined in Figure 2. We notice that while for 4H-SiC the notations for the inequivalent sites termed hexagonal and cubic are unambiguous, in 6H-SiC there exist some ambiguity in the literature concerning the (quasi)cubic sites k_1 and k_2. Most authors choose to label the Si and C sites in the first Si-C bilayer above the hexagonal sites in 6H-SiC as k_1 and the one on top of it as k_2, but then the notations k_1 and k_2 have no particular meaning. We choose the definition used in Ref. [13] and outlined in Figure 2. In this definition, k_1 denotes the quasicubic sites in 6H for which the first and the second neighbors mimic the environment in the cubic polytype (3C-SiC), but the third-nearest neighbors deviate from it. For the k_2 site the first-, second-, and third-nearest neighbors are arranged as in the cubic polytypes and only the fourth neighborhood deviates from it. Two pair defects, the divacancy and the NV pair, have four inequivalent configurations in 4H-SiC and six in 6H-SiC. The pairs corresponding to these inequivalent configurations are also outlined in Figure 2. Notice that each pair has two equivalent positions in the unit cell but only one is outlined: the lower half shows the axial configurations and the upper half – the basal ones.

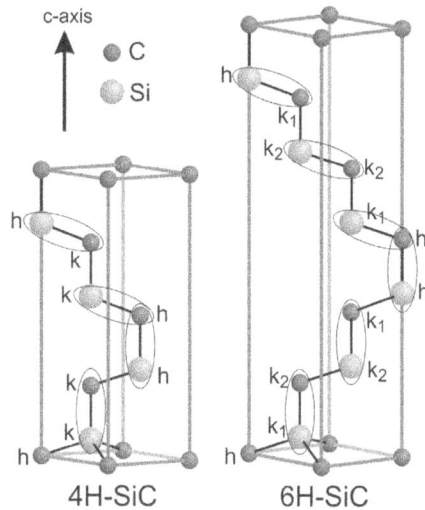

FIGURE 2

Unit cells of 4H- and 6H-SiC with the notations for the inequivalent lattice sites used throughout this chapter. Notice that the notations for 4H-SiC are common in the literature, but for 6H-SiC different definitions for the two cubic sites k_1 and k_2 are encountered in the literature. The present definition is the same as in Ref. [13]. Inequivalent configurations for nearest-neighbors pair defects are outlined with ovals but notice that each outlined configuration in the lower (upper) part of the unit cell has an equivalent counterpart in the upper (lower) part, which is not outlined.

2 Intrinsic Defects

The silicon and carbon vacancies as well as their nearest-neighbor combination, the divacancy are among the simplest defects in SiC. Another simple intrinsic pair defect with the same stoichiometry as V_{Si} is the carbon antisite–carbon vacancy pair ($C_{Si}V_C$), which can be obtained from V_{Si} if one of the nearest-neighbor C atoms jumps into the silicon vacancy site. Of these four defects, only the silicon vacancy in its negative charge state and the divacancy in the neutral charge state have well established infrared PL spectra and will be considered here in some detail.

2.1 The Silicon Vacancy

The Si vacancy has firstly been identified by electron paramagnetic resonance (EPR) in n-type 3C-SiC [14]. An effective spin of S=1/2 was assigned for this cubic (T_d) center. The same defect has been observed later in 4H- and 6H-SiC and its negative charge state with high spin configuration S=3/2 have been confirmed by electron nuclear double resonance (ENDOR) [15]. However, EPR experiments were performed within a short magnetic field range (40 G) and the low- and high-field lines of the S=3/2 center were not detected. Therefore, the negatively charged Si vacancy (V_{Si}^-) in the hexagonal polytypes 4H- and 6H-SiC was suggested to have the same symmetry as in the cubic polytype 3C-SiC [15]. The same center was observed later in 4H- and 6H-SiC but including the low- and high-field lines [16]. Being much weaker compared to the central line, which has already been assigned to V_{Si}^- [15], these lines were therefore attributed to the neutral Si vacancy with spin S=1 [16]. These EPR centers were studied by optically detected magnetic resonance (ODMR) with monitoring the near-infrared photoluminescence (PL) centers V1 and V2 in 4H-SiC and V1-V3 in 6H-SiC [17]. With selective excitations, ODMR spectra of V1-V3 show only the low- and high-field lines without the central line. This led to the initial assignment of V1-V2 in 4H-SiC and V1-V3 in 6H-SiC to the neutral Si vacancy at different inequivalent lattice sites [17].

It has been shown later that the center responsible for V2 has spin S=3/2 and therefore should be related to the negative Si vacancy but with axial symmetry [18]. The observation of different ^{13}C hyperfine (hf) interactions with nearest C neighbors of the central line leads to the suggestion of two different negatively charged Si vacancies, one with T_d-like symmetry often called undistorted V_{Si}^- without zero-field splitting (ZFS) and the other, i.e., the V1-V3 centers, with C_{3v} symmetry being isolated Si vacancies disturbed by a defect along the c-direction [19]. This disturbing defect was later suggested to be a distant C vacancy [20]. Combining high-resolution EPR and high-precision first-principles calculations, Ivády and co-workers [21] confirmed that these centers in 4H-SiC are the negative Si vacancy and reassigned V1 and V2 to V_{Si}^- at the hexagonal (h) and quasicubic (k) site, respectively. Similarly, identification has been reported for 6H-SiC with the V3 and V2 related to V_{Si}^- at the two quasicubic sties k_1 and k_2, respectively, and V1 to V_{Si}^- at the h-site [22].

Recently, with using isotopically purified 4H-^{28}SiC and proper annealing, EPR signals of V1 and V2 could be well separated from other overlapping spectra of other defects (Figure 3) and their ^{13}C hf interactions could be determined, confirming that the so-called

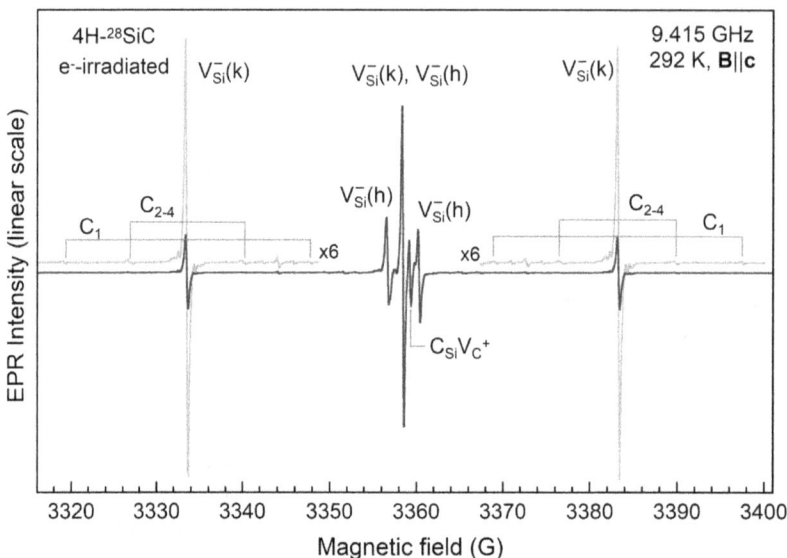

FIGURE 3

EPR spectra of the V_{Si}^- center in irradiated and annealed (at 700 °C) 4H-^{28}SiC measured at room temperature for the magnetic field along the c-axis (**B** | | **c**). In this isotopically enriched material (~99.85% of ^{28}Si), the ^{29}Si hf structures from the interaction with 12 Si in the next nearest neighbors are absent. The hf structures due to the interaction between the electron spin and the nuclear spins of nearest ^{13}C neighbors (along the c-axis, C_1, and in basal plan, C_{2-4}) for the low- and high-field components of V_{Si}^-(k) can be seen in the extended intensity (×6) parts of the spectrum.

undistorted negative Si vacancy center is indeed the central line of the V1 and V2 centers [23]. Those overlapping signals are predicted to be V_{Si}^- related complexes [24].

Recent hybrid functional theory calculations suggest that the charge transition level $(0|-)$, i.e., the ground state of V_{Si}^- in 4H-SiC, is at ~1.25 eV above the valence band maximum (VBM) (E_V + 1.25 eV), and its $(-|2-)$ and $(2-|3-)$ acceptor levels at ~0.7 and ~0.4 eV below the conduction-band minimum (CBM) (or ~E_C − 0.7 eV and ~E_C − 0.4 eV), respectively [25]. This defect is therefore an effective carrier compensation center in *n*-type materials and one of the main intrinsic defects used for creation of high-purity semi-insulating (HPSI) SiC [26,27]. However, since V_{Si} becomes mobile at ~700 °C [28,23] the SI-properties are not stable at high temperatures [26,27].

Due to short C dangling bonds, V_{Si}^- keeps the C_{3v} symmetry and has a high-spin ground state $^4A_2$ with an electron spin S=3/2 and the lowest excited state $^4A_2$. The ZPLs of V1 (861.6 nm) and V2 (916.5 nm) in 4H-SiC and V1 (864.7 nm), V2 (996.3 nm) and V3 (907.1 nm) in 6H-SiC [17] are related to the optical transitions $^4A_2 \rightarrow ^4A_2$, which have the optical dipole moment oriented along the c-axis of the crystal (E||c) [30]. Another ZPL, V1' (858.8 nm), which is often detected in 4H-SiC, was assigned to the transition from the higher-lying electronic excited state ^4E of the V1 center ($^4E \rightarrow ^4A_2$), which places the excited state ^4E at ~4.5 meV above the first excited state $^4A_2$ [30,31]. The corresponding transition for the V2 center has not been observed. The transition $^4A_2 \rightarrow ^4A_2$ responsible for the V1, V2 ZPLs in 4H-SiC and V1-V3 lines in 6H-SiC is allowed with parallel to the c-axis polarization (E||c), whereas the $^4E \rightarrow ^4A_2$ transition is allowed with perpendicular to c-axis (E⊥c) polarization (the V1' lines in both polytypes). Low-temperature PL spectra of V_{Si}^- in 4H- and 6H-SiC are shown in Figure 4.

However, in a recent calculation, the second excited state ^4E was found to be at ~342 and ~568 meV above the first excited state $^4A_2$ for V1 and V2, respectively, and cannot be related to the V1' ZPL [32]. An alternative explanation for the appearance of V1' is suggested in this work. Based on the obtained calculated polaronic spectrum of the V1 center, the V1' ZPL was attributed to the transition that connects the first polaronic excited state of predominant E electronic character (~5 meV above the excited $^4A_2$ state) to the ground $^4A_2$ state. Similarly, its counterpart V2' is predicted to be ~22 meV higher in energy than V2 [32], which may explain why it is not observed at low temperatures.

Due to small deviation from the quasitetrahedral symmetry of the spin density, the ZFS values (2D) are relatively small for the ground state of the V_{Si}^- centers (5 and 70 MHz for V1 and V2, respectively, in 4H-SiC [23], and 26.6, 128, and 27.8 MHz for V1, V2 and V3, respectively, in 6H-SiC [22]). For the excited electronic state, the ZFS is much larger and strongly depends on the temperature as shown from ODMR and level-anticrossing (LAC) measurements for V2 [33] and from calculations and PL measurements for V1 and V2 in 4H-SiC [32]. The ZFS is ~1.06 GHz for V2 and ~1 GHz for V1 at low temperatures. For the V2 center in 4H-SiC, the temperature dependence of the ZFS of the excited state can be described by a linear relation,

$$2D_{ES}(T) = 2D_{ES}(0) + \beta T, \tag{1}$$

with $2D_{ES}(0) = 1.06 \pm 0.02$ GHz denoting the ZFS in the limit T→0 and $\beta = -2.1 \pm 0.1$ MHz/K being the thermal shift coefficient [33].

Since the beginning of the last decade, EPR and ODMR studies of V_{Si} ensembles have suggested that defect is promising for near-infrared quantum bit (qubit) [34-36] and sensing [37-41]. While the C vacancy is always present in SiC, including pure films grown by chemical vapor deposition (CVD), the Si vacancy is usually absent in CVD materials

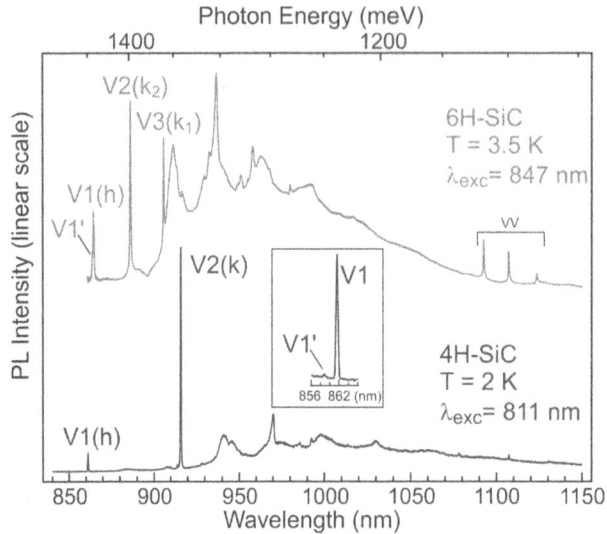

FIGURE 4

Low-temperature PL spectra of the V_{Si}^- centers in 4H- and 6H-SiC measured in conditions allowing detection of both E||c and E⊥c polarizations. Weak contribution of the divacancy in the 6H spectrum is denoted "VV". The ZPLs (V1, V2 in 4H, and V1-V3 in 6H) have E||c polarization. The rest of the spectra comprises the associated phonon sidebands (PCBs), which may have both polarizations. Thus, the observed ratio between the ZPLs and the PSBs depends on the experimental conditions. The assignment of hexagonal or cubic inequivalent lattice sites to the ZPLs follows [21] for 4H and [22] for 6H. The V1 ZPL in both polytypes exhibits a second component V1' associated with a second excited state (seen as a shoulder in 6H). V1' has E⊥c polarization and is observable at slightly higher temperature. Inset: V1' observed in 4H-SiC at 2 K but with higher excitation power causing local heating in the sample.

due to its considerably higher formation energy (~7.5 eV [25]). This makes it possible to use electron irradiation for well-controlled realization of V_{Si} with low concentrations suitable for observation of single-emitter studies [11]. It has been shown that the spin of the single V2 center has long spin coherence time in the millisecond range and can be optically controlled by ODMR at room temperature [11]. Coherent control of V2 spin ensembles by photocurrent detected magnetic resonance (PDMR) at ambient conditions has recently been demonstrated [42].

In addition to radiative transitions that give rise to the PL emission, nonradiative recombination via metastable states to the $m_S =\pm3/2$ ground state doublet is predicted to be efficient [43]. This allows spin-selective initialization of V_{Si}^- to be achieved in the ground state by optical excitation up to ~97% and a high-fidelity spin-photon interface [44]. The robust spin initialization of V_{Si}^- on the $m_S =\pm3/2$ spin ground states under optical excitation results in negligible population on the $m_S =\pm1/2$ states. This explains why the central line corresponding to the transition between $m_S =\pm1/2$ states of V_{Si}^- is not detected in ODMR [17] or EPR under optical excitation [35]. While enhancing the spin initialization that helps to increase the optical contrast of spin readout in Rabi oscillation experiments, the nonradiative recombination channel also reduces the PL intensity considerably, lowering the count rate of single V_{Si}^- emitters without using solid immersion lens (SIL) to typically a few kcounts/s [11]. The ODMR contrast of V_{Si}^- under off-resonant excitation is low (<1%) but can be increased to ~97% under resonant excitation as shown in Rabi oscillation of single V_{Si}^- emitters [44].

Like other color centers in semiconductors, the V_{Si}^- emission has a strong phonon side-band and the ratio of ZPL emission to the total emission, also called the Debye-Waller factor, is about 8–9% [32]. In development of bright indistinguishable single photon sources for application in quantum communications, quantum photonic cavities with embedded Vsi centers have been demonstrated, showing the enhancement of the ZPL intensity by a factor of 75 and 22 for V1′ and V1 ensembles, respectively, [45], and the PL intensity of single V1 emitters by a factor of 120 [46].

In *n*-type commercial SiC materials, the Si vacancy can be in the negative charge state V_{Si}^- in equilibrium and has stable PL emission if its concentration is enough to compensate the N shallow donor [47]. However, it is not possible to obtain stable single V_{Si}^- emitters even in ultrapure *n*-type SiC materials since the residual concentration of the N shallow donor is at least in the low 10^{13} cm^{-3} ranges while the concentration of the Si vacancy and other intrinsic defects created by irradiation is much lower (typically in the 10^9–10^{10} cm^{-3} for isolation of single emitters [11]). The Fermi level is therefore pinned at the N donor level and the Si vacancy will be in the (3–) charge state. The PL observation of V_{Si}^- then requires the negative charge state to be activated by optical excitation or off-resonant excitation. In ensembles, off-resonant excitation or repump laser can activate and keep a part of the total concentration of the Si vacancy in the single negative charge state. However, it has been shown that in pure *n*-type SiC with the residual N donor concentration in the low 10^{14} cm^{-3} ranges, repumping can recover the negative charge state of V_{Si} but the V_{Si}^- emission still decays exponentially with a time scale of 10 ms, resulting in photobleaching of single emitters [48]. Stable single V_{Si}^- emitters can be obtained in *p*-type like semi-insulating materials [11] with help of the C vacancy [47]. However, with the concentration of residual impurities (N and B) and intrinsic defects being in the 10^{15}cm^{-3} range, long coherence times and narrow optical linewidths may not be expected for the Si vacancy qubits. The charge-state instability and emitter bleaching are general problems for color centers in semiconductors, including diamond and SiC.

Combining repumping with electrical charge control using *p-i-n* [49] or Schottky barrier [50] diodes has been shown to be a very effective way for controlling the charge state of the Si vacancy in SiC, allowing fast switching between the bright and dark states. However, this approach is difficult to implement, e.g., for quantum emitters embedded in photonic devices.

It has been shown recently that in ultrapure SiC materials grown by CVD with the concentration of residual impurities in the mid 10^{13} cm^{-3} ranges, repumping is enough for stabilizing the charge state of V_{Si} and achieving stable single V2 emitters with the optical linewidth approaching the lifetime-limited linewidth [44]. Such pure materials allow indistinguishable single photons to be achieved and two-photon interference contrast close to 90% to be demonstrated in Hong-Ou-Mandel type experiments using off-resonant excitation [51]. Fast frequency-modulated optical transitions for spectral engineering of single photon emission have recently been reported for single V_{Si}^- emitters [52]. Stable single V2 emitters in nanophotonic waveguides have even been achieved under resonant excitation for 30 minutes without repumping [53]. The control of single nuclear spins of V1 [44] and V2 [53] and coupling between the electron spin and nuclear spins in isotopically purified 4H-^{28}Si^{12}C have recently been demonstrated.

2.2 The Divacancy

Divacancy is a fundamental defect in compound semiconductors including SiC. In SiC, the neutral divacancy, i.e., an uncharged complex consisting of a C vacancy (V_C) and a nearest Si vacancy ($V_C V_{Si}^0$), and the Si vacancy are the most studied defects so far.

In earlier EPR experiments, two triplet centers, called P6 and P7, with axial and C_{1h} symmetry, respectively, observed in high-temperature heat treatment under optical excitation were suggested to be related to excited states of exchange-coupled vacancy pairs [54]. A later ODMR study observed spin-1 defects with similar parameters of P6/P7 centers in as-grown and electron irradiated 6H-SiC [55].

The P6/P7 centers in 6H-SiC were later assigned to a PL band at a wavelength range 1154–1242 nm in magnetic circular dichroism of absorption (MCDA) and MCDA-detected EPR studies by Lingner and co-workers [56]. Combining the MCDA results with density functional theory (DFT) modeling, these centers were assigned to the double positively charged C antisite-vacancy pairs $C_{Si}V_C^{2+}$ [56]. However, this PL band has been shown later to be related to the negative charge state of the nitrogen-vacancy pair, i.e., a pair between a nitrogen and a nearest Si vacancy in the negative charge state, $N_CV_{Si}^-$ [13].

The observation at low temperatures in darkness of the P6/P7 EPR centers in 6H-SiC has been suggested to be related to the ground state of the neutral divacancy [57]. Combining DFT calculations and EPR studies of the hf structures from the interaction with nearest ^{13}C neighbors, four different configurations of the P6 and P7 centers in 4H-SiC

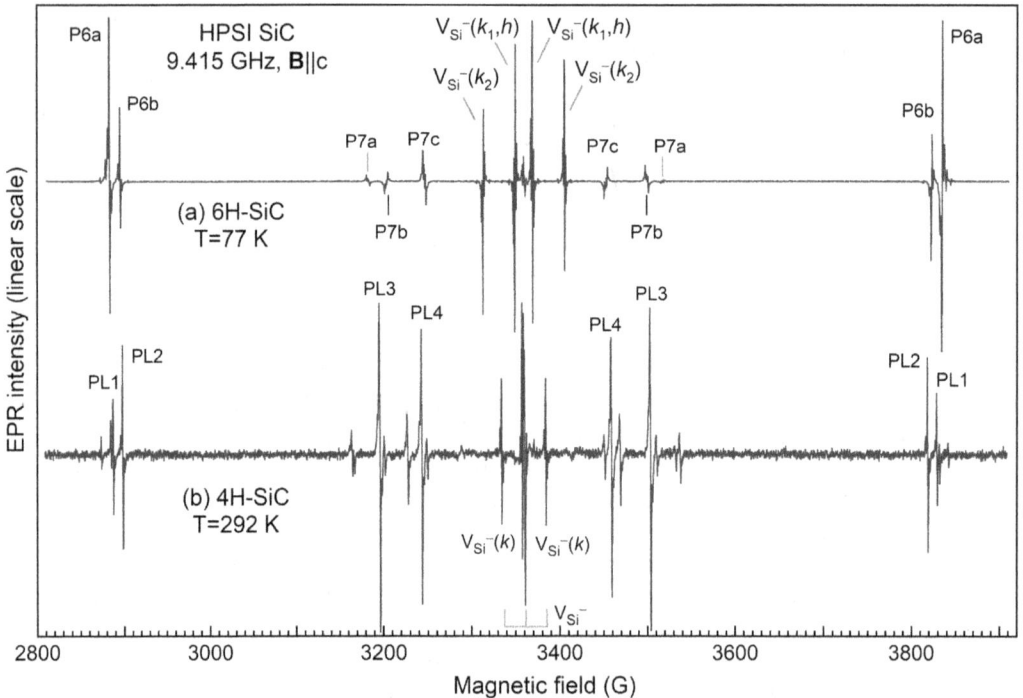

FIGURE 5

EPR spectra measured for $\mathbf{B} \| \mathbf{c}$ showing the neutral divacancy in electron-irradiated HPSI (a) 6H-SiC under illumination at 77 K and (b) 4H-SiC in darkness at 292 K. For 6H-SiC, P6a, P6b, P7a, and P7b are the a, b, c, and d centers observed by ODMR in Ref. [55]. The fine-structure parameters for the P7c center have not been determined and the assignment of P6/P7 EPR centers to ZPLs QL1-QL6 in PL has not been made. For 4H-SiC, the assignment of P6/P7 centers to ZPLs PL1-PL4 is based on comparison of ZFS with the values determined by ODMR in Ref. [4]. In HPSI 4H-SiC, six unidentified spin-1 centers are weakly detected. The V_{Si}^- signals in 6H-SiC are labeled following assignment in Ref. [22]. The ZFSs in 4H-SiC are slightly smaller compared to that in 6H-SiC since the spectrum in (b) was measured at room temperature.

have conclusively been identified to be related to the two axial (*hh*) and (*kk*) and two basal (*hk*) and (*kh*) configurations, respectively, of the neural divacancy $V_CV_{Si}{}^0$ [58,59]. The EPR spectra of $V_CV_{Si}{}^0$ in 4H- and 6H-SiC are shown in Figure 5.

In 4H-SiC, the divacancy introduces in the bandgap a donor level (+ | 0) and two acceptor levels (0 | −) and (− | 2−) at ~ 1.1 eV, ~2.0 eV, and 2.1 eV above the VBM, respectively, as predicted by calculations [60]. The divacancy often has lower concentration compared to the C vacancy even after annealing at high temperatures [58]. With the donor state lying below the donor levels of the C vacancy [60], the divacancy does not therefore play an important role in carrier compensation in *p*-type materials. However, with its acceptor levels close to the mid gap the divacancy is a dominant carrier compensation center in *n*-type materials and has been used for creating HPSI SiC materials [26,27,61].

The PL of the $V_CV_{Si}{}^0$ center has been identified first in 4H-SiC by ODMR [4]. According to Figure 2, in 4H-SiC, C and Si each has two inequivalent lattice sites, hence there are four inequivalent divacancy configurations outlined in Figure 2. Their four ZPLs, PL1-PL4 in the energy (wavelength) range 1.0950 eV (1132.0 nm)–1.1493 eV (1078.5 nm), were assigned to two axial configurations, PL1 (*hh*) and PL2 (*kk*), and two basal configurations, PL3 (*hk*) and PL4 (*kh*) [62] (see Figure 6 and Table 1). These ZPLs are related to optical transitions from the excited state ^3E to the ground state $^3A_2$ of the spin S=1 center, using the notations for the axial configurations with C_{3v} symmetry. With three inequivalent lattice sites in 6H-SiC, a nearest-neighbor paired defect in 6H-SiC has six configurations, three axial (*hh*, k_1k_2, k_2k_1) and three basal (*hk*$_1$, k_2k_2, k_1h). The spectra of the divacancy in the three main polytypes, 6H-, 4H- and 3C-SiC are shown in Figure 6, the ZPL positions and their identifications are listed in Table 1. The corresponding peak positions are listed in Table 1 together with their assignment to specific configurations. The assignment of the six ZPLs of $V_CV_{Si}{}^0$ in 6H-SiC, QL1-QL6 [63], has been suggested based on calculations [22]. 3C-SiC has a single divacancy configuration with ZPL emission at 1106.2 nm (Figure 6).

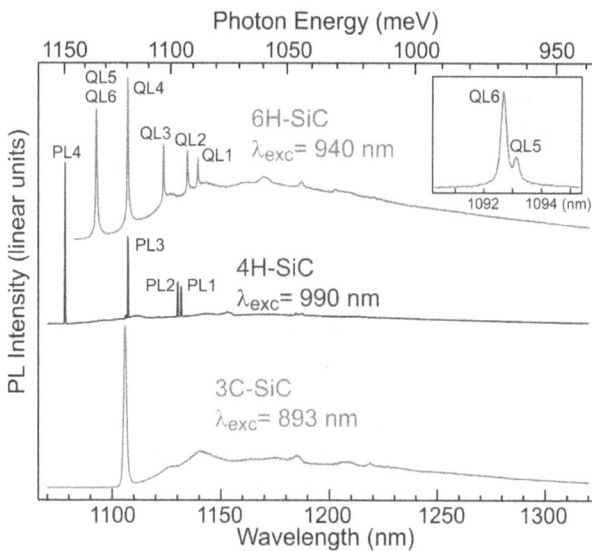

FIGURE 6

Low-temperature (T = 3.5 K) PL spectra of the neutral divacancy in 6H-, 4H-, and 3C-SiC. See Table 1 for the peak positions and their identification. The inset is an enlargement of the QL5 – QL6 lines, which can be resolved using higher-resolution grating.

In 3C-SiC, the ODMR center L3 [64] and EPR center Ky5 [65] have similar magnetic and optical properties to the predicted 3C-SiC neutral divacancy [62] and those observed in the hexagonal polytypes [4], but only tentative assignment as the neutral divacancy has been made.

The first observation of single $V_CV_{Si}^0$ emitters has been reported by Christle et al. [66] in irradiated pure 4H-SiC CVD layers, showing coherent optical control of single spins with coherence times in millisecond ranges. There are favorable optical and spin properties of the divacancy qubits as shown recently [67]. These include the efficient nonradiative decay from the excited state to the singlet $m_S = 0$ ground state that helps to achieve spin polarization of the divacancy in the ground state of ~96% and, hence, high-fidelity spin-to-photon interface. Another favorable feature is the low level of spin mixing in the ^3E excited state (about an order of magnitude weaker compared to the excited state ^3E of the NV center in diamond [68]), that lowers the rate of spin flips and, hence, increases the number of emitted photons per shot. Single $V_CV_{Si}^0$ emitters are therefore rather bright with typical count rates of 40–50 kcts/s without SILs. The enhancement of the Rabi oscillation contrast of single $V_CV_{Si}^0$ emitters from a typical off-resonant value of ~9% to ~94% under resonant excitation has been reported [67].

The neutral divacancy emission is mainly contributed from the phonon side band and only a small amount of ~5% comes from the ZPL. Enhancing the brightness and the emission at the ZPL wavelength of single divacancy emitters embedded in 4H-SiC nanophotonic cavities has recently been reported, demonstrating a Purcell factor of ~50 and an enhancement of the DW factor from ~5% to ~70–75% [69]. However, in these cavities made from commercially available materials, the optical linewidth is about 4–5 GHz, which is well above the lifetime limit of ~11 MHz [67].

The window of the Fermi level for observing the stable neutral charge state of the divacancy is between its donor level $(+|0)$ at ~$E_V+1.1$ eV and acceptor level $(0|-)$ at ~E_V+ 2.0 eV [60]. Within this range, there are acceptor levels $(0|-)$ of the Si vacancy and the N_CV_{Si} pair at ~1.25 eV and ~1.5 eV above the VBM, respectively. Thus, the $V_CV_{Si}^0$ center can be stable in heavily irradiated materials in which the total concentration of the Si vacancy and N_CV_{Si} pair is large enough to compensate the N donor [47]. This condition is rarely satisfied in practice and the divacancy is often in the (2−) charge state. This explains why the EPR observation of $V_CV_{Si}^0$ in *n*-type materials often requires optical excitation to activate the neutral charge state. In HPSI SiC materials, optically induced quenching of $V_CV_{Si}^0$ PL often occurs [70,71,60] and the transform of the charge state from neutral to negative under optical excitation with photon energies less than ~1.3 eV is suggested to be the quenching mechanism [70,60].

Electrical charge-state control of single divacancies in the vicinity of Schottky barriers has been reported, demonstrating fast switching between the negative and neutral charge states [72]. Electrical charge-state modulation has also been demonstrated for single divacancy emitters embedded in the intrinsic *i*-layer of *p-i-n* diodes fabricated from commercially available 4H-SiC [73]. These devices enable deterministic charge-state control and broad Stark-shift tuning exceeding 850 gigahertz. Stable PL emission of single divacancies and their optically detected Rabi oscillations with >98% contrast has been achieved using resonant initialization and readout. Moreover, under reverse bias, depletion of trapped charges in the *i*-layer helps to reduce the fluctuation of the electric environment, resulting in a narrowing of the optical linewidth from ~1 GHz (at zero bias) to ~20 MHz (at 270 V reverse bias), which is about twice the lifetime-limited linewidth of the divacancy. This is a very efficient method for controlling the charge state and mitigating the ubiquitous problem of spectral diffusion in solid-state emitters. However, implementation of *p-i-n*

TABLE 1

Peak positions of the ZPLs of the divacancy emission in 3C-, 4H-, and 6H-SiC. Their association with specific $V_{Si}V_C$ configurations is also given

Lines	Peak position in nm (eV)[a]	Configuration[b,c]
3C-SiC		
$V_{Si}V_C$	1106.2 (1.1205)	*kk*
4H-SiC		
PL1	1132.0 (1.0950)	*hh*
PL2	1130.5 (1.0964)	*kk*
PL3	1107.6 (1.1191)	*hk*
PL4	1078.5 (1.1493)	*kh*
6H-SiC		
QL1	1139.4 (1.0878)	k_1k_2
QL2	1134.7 (1.0924)	*hh*
QL3	1123.6 (1.1031)	k_1h
QL4	1107.1 (1.1196)	k_2k_2
QL5	1093.1 (1.1339)	hk_1
QL6	1092.7 (1.1343)	k_2k_1

[a] According to our data.
[b] According to the inequivalent-site notations presented in Figure 2.
[c] Configurations according to Ref. [62] for 4H-SiC and Ref. [22] for 6H-SiC.

structures in quantum photonic devices, such as nanophotonic cavities, is difficult if not impossible.

Like the case of the Si vacancy in pure SiC with the overall concentration of residual impurities in the mid 10^{13} cm^{-3} ranges, it is possible to stabilize the neutral charge state of single divacancy emitters with using optical repump only. In pure isotopically purified 4H-^{28}Si^{12}C materials, stable PL emission, high fidelity control (99.98%) combined with high level of spin initialization and readout fidelity (>99%) of single divacancies, controlling and entangling of both strongly coupled and weakly coupled single nuclear spins have recently been demonstrated [74]. In such materials, a record long spin coherence time T_2 exceeding 5 seconds has recently been reported [29].

3 Transition Metal Impurities with Near-Infrared Emission

In this section we consider several transition metals in SiC, which have been identified in various SiC polytypes and studied in some detail in the past. The focus of this scope is on the electronic structure and PL properties, but in many cases it is appropriate to consider also EPR results and the location of the ground state in the bandgap determined from electrical measurements, such as deep level transient spectroscopy (DLTS). One transition metal, which is relatively well studied but not included, is titanium, because its PL signature is in the visible part of the spectrum.

3.1 Vanadium

Vanadium (V) is an amphoteric impurity in SiC, meaning that it possesses both donor and acceptor levels deep in the bandgap. V substitutes Si atoms in the host lattice, hence the number of inequivalent centers repeats the number of inequivalent lattice sites in each polytype. Owing to its amphoteric behavior, V can compensate both shallow donors and acceptors. This property has been used to realize semi-insulating (SI) material and room-temperature resistivities in excess of 10^{10} Ω.cm have been achieved by V doping in both 6H- and 4H-SiC [75]. SI epitaxial layers have also been grown to study the V incorporation and the impact of V concentration on the crystalline quality, including nominal off-axis [76,77] and on-axis growth on 4H-SiC [78]. The on-axis growth is especially attractive because the obtained epitaxial layers can be used as (more precisely, instead of) SI substrates for subsequent growth of GaN for various applications [78]. Using deep level transient spectroscopy (DLTS), the solubility limit of V in bulk material grown by physical vapor transport (PVT) has been established to be about 3×10^{17} cm^{-3} [79]. However, in epitaxial growth it has been noticed that degradation of surface morphology (and, consequently, crystalline quality) starts already at concentration of V, [V], in the range of ~0.9–1.5×10^{17} cm^{-3} depending on the growth temperature [77]. Karhu and co-workers [78] also observed the formation of 3C-SiC when the concentration of V, [V] exceeds ~2×10^{16} cm^{-3} and V-related defects (possibly, including V aggregates) when [V] ~1×10^{17} cm^{-3}.

The first report on the optical properties and EPR of V in 4H- and 6H-SiC provides information on the lattice site – V substitutes Si atoms in the host lattice [80]. In EPR, V has been unambiguously identified based on the hf structure consisting of eight hf lines due to the interaction between the electron spin and the nuclear spin of ^{51}V (I= 7/2, 99.75% natural abundance). Both centers associated with V^{4+}(3d^1) (the donor state) and, in *n*-type material with the ionized acceptor state V^{3+}(3d^2), have been observed [80-82]. MCDA absorption of V^{4+} in 6H-SiC has been reported [83-85]. MCDA-detected EPR (MCD-EPR) is site selective, allowing the contributions from the α, β and γ sites in the EPR spectrum to be distinguished and allowing correlation between EPR and PL centers [83].

Vanadium is stable in its neutral charge state V^{4+} in a wide range of Fermi level positions, thus, the number of luminescent and EPR centers related to V replicates the number of inequivalent lattice sites in each polytype, three in 6H-, two in 4H-, and one in 3C-SiC.

The PL spectra associated with vanadium in its neutral charge state V^{4+} have been observed in 4H-, 6H-, and 3C-SiC and are illustrated in Figure 7(a) [80,86]. The peak positions are listed in Table 2. It should be noted that the PL emission from V-doped 3C-SiC was deemed impossible before the report in Ref. [86], because the corresponding excited state of the V^{4+} donor was estimated to be resonant with the CBM [87]. The authors of the latter work measure the photoionization threshold of the EPR spectrum from V^{4+} in *p*-type material by illuminating the sample *in situ* in the microwave cavity. They find a threshold of 1.7 $\pm$ 0.1 eV for the photon energy needed for enhancement of the V^{4+} EPR signal in the process

$$V^{3+} + h\nu = V^{4+}, \tag{2}$$

i.e., for exciting an electron from the VBM to the donor level V^{4+}/V^{3+} (equivalently, exciting the hole from V^{3+} to the VBM). Considering that the bandgap of 3C-SiC is $\approx$ 2.4 eV, an excited state of V^{4+} with an energy of more than 0.7 eV (= 2.4 − 1.7 eV) above the ground state would likely be degenerate with the CBM. Nevertheless, in contemporary work the same authors observe clear absorption peak from V^{4+} in MCDA [88]. That the peak is indeed related to V^{4+} is demonstrated by measuring MCD-EPR with optical detection on this peak; the characteristic eightfold splitting associated with the 7/2 spin of V is manifested in the resulting MCD-EPR spectrum. The quoted absorption peak is at 828 meV [87], which is essentially the same as the low-energy peak for 3C-SiC in Figure 7(a). The

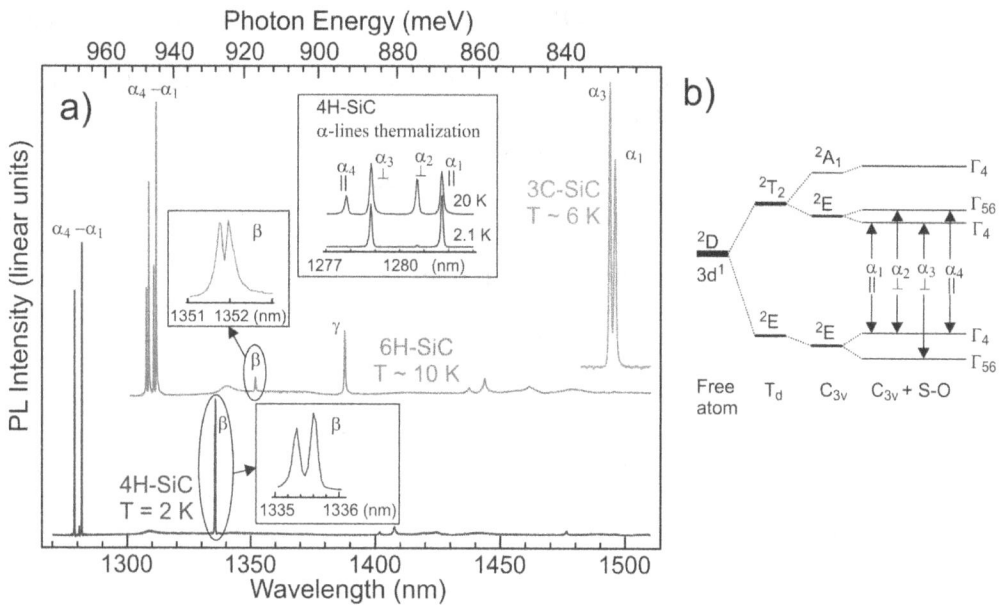

FIGURE 7

(a) Fourier-transform PL of V at low temperature in 3C-, 4H-, and 6H-SiC. The insets next to the β-lines in 4H- and 6H-SiC enlarge the corresponding β-line to illustrate its splitting. The top inset displays the thermalization of the α-lines in 4H-SiC (similar to 6H-SiC), showing that α_2 and α_4 originate from a higher excited state. (b) Qualitative energy-level diagram illustrating the splitting of the five-fold degenerate 3d-level of the vanadium $3d^1$ electron in cubic (T_d), trigonal (C_{3v}) symmetry and in C_{3v} symmetry plus the spin-orbit (S-O) interaction included. The order of the energy levels follows Ref. [89]. See Table 2 for the peak positions.

doublet seen in Figure 7(a) for 3C-SiC is likely not resolved in the MCDA experiments [88]. It remains unclear why the PL was not observed in this early work [87,88], but a possible reason is inferior material quality, which, on one hand, may explain the broadening of the MCDA peak so that the doublet structure is not detected and, on the other hand, may provide competitive nonradiative channels damping the luminescence to an undetectable level. However, the observed PL from V^{4+} in 3C-SiC refutes the suggestion that the excited state is degenerate with the CB. Instead, the excited state is in the bandgap, albeit presumably close to the CBM. With a little stretch this is possible within the given error margins of the photoionization threshold: ~ 1.6 eV as a lower bound for this threshold would provide ~ 0.8 eV for an excited state in the bandgap. So far, the 3C-SiC polytype is not sufficiently investigated and only the low-temperature spectra (near 2 K) are available. The fact that the observed PL of V^{4+} in 3C-SiC is in the telecom S-band close to the minimum absorption of silica fibers makes the V^{4+} center very interesting for long-distance quantum communications and will trigger further detailed investigations.

The luminescence in all three polytypes presented in Figure 7(a) can be understood if we consider the electronic structure of V in SiC [81,84,89,90]. Here we provide a generic level scheme depicted in Figure 7(b). The ^2D state of the $3d^1$ electron of the free vanadium atom, which has five-fold orbital degeneracy in cubic symmetry T_d, splits into orbital triplet $^2T_2(3)$ and doublet $^2E(2)$ states. Spin-orbit coupling will further split the $^2T_2$ level into a $^2A_1$ and ^2E levels, but since now the electron spin is accounted for, the wave functions of the corresponding levels are spinors classified in accord with the irreducible representations of the double group $\bar{T}_d$: $^2A_1 \rightarrow \Gamma_7$, $^2E \rightarrow \Gamma_8$. In this latter case, we overgo from the Mulliken symbols to those of Ref. [91] for irreducible representations of the double group $\bar{T}_d$. In the hexagonal 4H and 6H polytypes, the symmetry reduces further to trigonal (C_{3v}) and the $^2T_2$ state splits into $^2A_1$ and ^2E. Due to spin-orbit interaction further splitting of the ^2E levels into Γ_4 and Γ_{56} states occurs (Γ_{56} denotes the direct sum of the Γ_5 and Γ_6 representations in the double group $\bar{C}_{3v}$ [91] subject to Kramers' degeneracy). The order of the energy levels in Figure 7(b) is taken from Ref. [89] and applies to the α lines in 6H-SiC, albeit it is noted that reversing the order of the Γ_4 and Γ_{56} states in both the ground and the excited state would result in the same polarization selection rules as investigated in detail in Ref. [89]. The dominating polarization of each of the ZPLs α_1–α_4 is denoted below the corresponding transition with the symbols $||$ ($\perp$) for parallel (perpendicular) polarization of the emitted photons with respect to the crystal c-axis.

The α-lines in both 4H- and 6H-SiC are associated in early work with the hexagonal (h) sites of V. The main argument is that the larger splitting for the ground state of these sites is due to stronger crystal field on just the h sites. In addition, 6H-SiC has two inequivalent quasicubic (k) sites and, indeed, two of the ZPLs (β and γ) exhibit much smaller splitting, together with the β ZPL in 4H-SiC, therefore, they should be associated with the k sites in both polytypes. However, recent *ab initio* calculations for 4H-SiC suggest that the cubic site is related to α and the hexagonal – to β [90]. Another prediction of their theoretical model is that at the hexagonal site the excited $^2A_1$ has lower energy than the excited ^2E state. Thus, according to their model, the emission at the k sites (α-lines) comes from transitions of the ^2E manifold (Γ_4 and Γ_{56}) to the manifold of the ground state ^2E, whereas for the h sites (β-lines) the transitions are involving the $^2A_1$ (Γ_4) excited state. This has implications on the fine structure and the selection rules of the different components for the two cases, but corresponding experimental investigation of this concept has not been presented so far. In a later work [92], based on comparison between the 4H and 6H polytypes, the latter theoretical assignment is questioned again, so that the site association of the different lines remains an open question. We notice also that the PL spectrum originating from V^{4+} in 15R-SiC has been reported, too, and five optical centers corresponding to the five inequivalent sites for V$_{Si}$ in this polytype are seen in the spectrum [93].

We notice also that the conventional absorption spectrum measured on several occasions (e.g., [80,83] for 6H-SiC) shows essentially the same lines as observed in PL. However, in this case, additional higher excited states can be observed at low temperature in absorption without the need of being thermally populated. This is the case of the so-called β_2 and β_3 "hot" lines – a closely spaced doublet appearing about 2.5 meV above the β-line, which can be seen also in PL at higher temperatures but is hardly resolved due to temperature-induced line broadening [80].

The position of the vanadium spectrum is beneficial for applications in quantum information – the ZPLs of both 4H- and 6H-SiC lie entirely in the O-telecom band (~1265–1360 nm). This circumstance has motivated new detailed studies of the defect from the point of view of future applications as a single photon source and/or qubit. The above-mentioned Ref. [90] investigates in some detail the optical properties of the ensemble of V defects in 4H-SiC and attempts determination of the Debye-Waller factors for the α and β emissions and their lifetimes. The Debye-Waller (DW) factor of certain defect visible in luminescence is defined as the ratio of the ZPL intensity to the total intensity in the spectrum including the phonon sideband (PSB). The bounds estimated in this work are 31% < DW$_\alpha$ < 60%, and DW$_\beta$ > 10%. The lifetimes measured with off-resonant excitation (at 441 nm) are $\tau_\alpha \approx 160$ ns and $\tau_\beta \approx 40$ ns. A later work, in addition to ensemble investigations, reports also on realization and coherent control of single V centers in both 4H- and 6H-SiC [92].

A very comprehensive study of the optical and magnetic properties (Zeeman effect) of V^{4+} in 6H-SiC is presented in Ref. [89]. Apart from considering in detail the crystal-field model for V in 6H-SiC and investigating the selection rules of the optical transitions (mainly for the α-lines), the authors measure the Zeeman effect (up to ~7 T) and obtain also a full angular dependence of the spectral line positions in magnetic field at two different magnitudes of the field, 4.61 T and 6.91 T. They build a phenomenological Hamiltonian allowing them to fit all the experimental data with very few adjustable parameters and with accuracy better than 0.06 cm^{-1} (~0.007 meV) for all experimental points. This allows determination of the parameters not only for the ground states (as usually for EPR), but also for the excited states of the various transitions.

The energy levels of V in 4H- and 6H-SiC have been intensively studied by transport measurements and photo-EPR. The donor V^{4+}/V^{5+} level has been estimated by photo-EPR to be at $\sim E_V + 1.6$ eV in 6H-SiC [81]. This is supported by later Hall-effect measurements in p-type 6H-SiC doped with V, which suggest the donor level to be at $\sim E_C - 1.35$ eV [94]. In 4H-SiC, temperature-dependent Hall-effect and optical admittance spectroscopy suggest the donor level at $\sim E_C - 1.6$ eV, i.e., close to $\sim E_V + 1.6$ eV [95]. The donor level in 3C-SiC is estimated by photo-EPR to be at $\sim E_V + 1.7$ eV [87]. However, the observation of the ZPLs of V^{4+} in 3C-SiC with energies ~0.8 eV [86] with considering the bandgap energy of ~2.4 eV, the donor level is likely at $\sim E_V + 1.6$ eV and cannot be at $\sim E_V + 1.7$ eV (with adding 0.1 eV for band offset at the 3C-SiC/6H-SiC interface as suggested by Dombrowski et al. [87]. However, for the acceptor V^{4+}/V^{3+} level, the reported data are very scattered. For V in 4H-SiC, the acceptor level has been unambiguously determined by DLTS studies of 4H-SiC implanted with radioactive isotope ^{48}V at $\sim E_C - 0.97$ eV [96,97]. This is also supported by EPR and temperature dependence of resistivity studies of V-doped 4H-SiC, which show an activation energy of ~0.99 eV in the heavily V-doped sample [98]. Other temperature dependence of resistivity and Hall-effect measurements suggest the acceptor level of V at

TABLE 2

Peak positions of the vanadium-related ZPLs in 6H-, 4H-, and 3C-SiC. The two lines in 3C-SiC are given the same notations as the low-temperature lines in 4H- and 6H-SiC, i.e., α_1 and α_3

Line	Peak position in nm (eV)
3C-SiC	
α_1	1495.9 (0.8286)
α_3	1493.7 (0.8298)
4H-SiC	
α_1	1281.7 (0.9670)
α_2	1280.7 (0.9678)
α_3	1278.8 (0.9693)
α_4	1277.8 (0.9700)
β (low-energy component)	1335.6 (0.9280)
β (high-energy component)	1335.3 (0.9282)
6H-SiC	
α_1	1311.6 (0.9450)
α_2	1310.6 (0.9457)
α_3	1308.6 (0.9472)
α_4	1307.6 (0.9479)
β (low-energy component)	1352.0 (0.9168)
β (high-energy component)	1351,7 (0.9169)
γ	1387.8 (0.8931)

~E_C – 0.8 eV [79] or ~E_C – 1.1 eV [99] for 4H-SiC, and ~E_C – 0.66 eV [79] or ~E_C – 0.85 eV [99] for 6H-SiC. The problem in those studies is most likely the presence of other intrinsic defects and V is not the main defect in all materials (only some samples having the concentration of V above 1×10^{17} cm^{-3} [99]). As shown in the study by Son and co-workers [98], all V-doped 4H-SiC samples with the V concentration below 1×10^{17} cm^{-3} show the activation energy ~1.1 eV, which is shown to be related to intrinsic defects, and reduces to ~0.2–0.3 eV (corresponding to the activation energies of shallow Al and B acceptors) after annealing at 1700 °C. Thus, it can be concluded that the donor level is located at ~E_V + 1.6 eV for V in 3C-, 4H- and 6H-SiC, while the assignment of the acceptor level at ~E_C – 0.97 eV for 4H-SiC [96,97] and at ~E_C – 0.85 eV for 6H-SiC [99] appears to be most reliable.

The PL of the V^{3+} center with ZPLs at ~0.62 eV has been reported in the *n*-type V-doped 6H-SiC sample with an activation energy of ~0.86 eV, while it is absent in the V-doped sample with an activation energy of ~1.4 eV [99]. The PL of V^{3+} has been reported for 4H-SiC.

3.2 Chromium

While being a common impurity in III-V compounds and other oxide semiconductors, chromium (Cr) is not present in as-grown SiC. Since Cr-containing precursor gases are not available, doping with Cr in SiC during growth has been carried out by putting a piece of metal Cr inside the reactor [100,101]. Due to lack of samples, only a few studies of Cr in SiC have been reported.

Chromium has four stable isotopes with different natural abundances and nuclear spins: ^{50}Cr (4.35%, I=0), ^{52}Cr (83.79%, I=0), ^{53}Cr (9.50%, I=3/2), and ^{54}Cr (2.37%, I=0). With the ground state electronic configuration 3d^{5}4s^1, Cr is expected to be a paramagnetic center in SiC in different charge states: single positive Cr^{5+} (3d^1, S=1/2), neutral Cr^{4+} (3d^2, S=1), single negative Cr^{3+} (3d^3, S=3/2), and double negative Cr^{2+} (3d^4, S=2).

The negatively charged Cr^{3+} center has been identified by EPR in Cr-doped *n*-type 6H-SiC layers grown by the sublimation "sandwich" method [100] but no PL band related to this center has been reported. In the same sample, another EPR center with spin S=1 has been observed and assigned to the neutral Cr^{4+}. However, a later study showed that this triplet center is not related to Cr but to Mo^{4+} [102]. An EPR center with spin S=2 in Cr-doped 6H-SiC was tentatively assigned to the double negative Cr^{2+} center [102]. This divalent Cr^{2+} center was then suggested to be related to absorption lines in the 9200–9300 cm^{-1} (1.14–1.15 eV) range in Fourier transform infrared absorption (FTIR) and MCDA studies based on an assumption that the double acceptor level (– | 2–) of Cr should be at about ~ E_V+1.8 eV in 6H-SiC to be in line with the value of 1.8 eV inferred from the position of the (– | 2–) level of Cr in GaAs following the so-called Langer-Heinrich rule [103]. No Cr-related EPR spectra have so far been reported for 4H-SiC.

In a deep level transient spectroscopy (DLTS) study of 4H-SiC implanted with radioactive isotope ^{51}Cr, Achtziger and Witthuhn [96] observed three ^{51}Cr-related levels, one located at ~E_C–0.74 eV and two other close to the conduction band edge at 0.15, and 0.18 eV below the CBM. The deeper level is assigned to the (0 | –) charge state while the other two are suggested to be the double acceptor level (– | 2–) of Cr at the hexagonal and quasicubic sites, respectively [97]. The corresponding levels of Cr in 6H-SiC have not been determined. The (– | 2–) level of Cr in the 6H polytype can be at the similar range as in 4H-SiC or even in the conduction band since the bandgap of 6H-SiC is ~0.26 eV lower. Apparently, these DLTS results rule out the possibility that the absorption lines at 9200– 9300 cm^{-1} in 6H-SiC are related to the Cr^{2+} center.

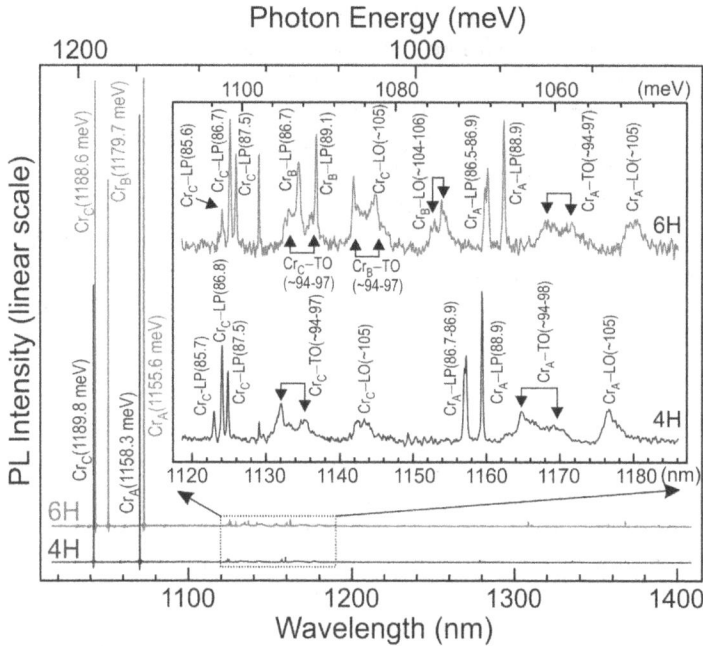

FIGURE 8

Fourier-transform PL spectra of the neutral Cr^{4+} center in 4H- and 6H-SiC layers doped with Cr during HTCVD growth measured at 5 K on a Bomem DA8 FTIR spectrometer with resolution of 0.5 cm^{-1} (~0.06 meV). The inset shows the phonon band in extended intensity scale. The energies of ZPLs and their corresponding lattice (TO and LO) and local (LP) phonons are given in meV in brackets. The multi-line ultraviolet emission (~351.1–363.8 nm) from an Ar ion laser with a power of ~ 20 mW was used for excitation.

The first Cr-related PL centers were observed in Cr-doped 4H- and 6H-SiC layers grown by high-temperature CVD (HTCVD) [101]. The PL spectra show very sharp ZPLs (Cr_a: 1.1583 eV and Cr_c: 1.1898 eV in 4H-SiC, and Cr_a: 1.1556 eV, Cr_b: 1.1797 eV, and Cr_c: 1.1886 eV in 6H-SiC) and very weak phonon sideband as shown in Figure 8. A very high DW factor of ~75% was estimated by Koehl and co-workers [104] from the PL spectra measured at 30 K and confirmed later by Diler et al. [105]. Magneto-optical studies [101] showed that the ground state of these PL centers is a triplet (S=1) with the $g_{||}$ values varying in the range of 2.00–2.04 for 4H-SiC and 1.99–2.02 for 6H-SiC and $g_\perp$=0. The fine-structure parameter D is estimated to be <1.2 GHz for Cr_A in 4H-SiC and Cr_A and Cr_B in 6H-SiC, and ~6 and ~5.4 GHz for Cr_C in 4H- and 6H-SiC, respectively. This axial center is therefore assigned to the isolated Cr in the neutral charge state, Cr^{4+}. The Zeeman data show no fine-structure splitting in the excited state, suggesting that PL emission is related to the transition from the singlet excited state ^1E to the triplet ground state $^3A_2$ of the Cr^{4+} center. The very weak phonon sideband and local vibrational modes observed in PL may be explained by a weak Jahn-Teller coupling expected for the singlet ^1E excited state.

The Cr_C center in both the polytypes was found to have the largest splitting of the excited state under external magnetic field and assigned to Cr^{4+} at the hexagonal lattice site, where the trigonal crystal field is stronger. A later PLE and ODMR study gave a more accurate D-value of 1.06 and 6.7 GHz for the Cr_A and Cr_C centers, respectively, in 4H-SiC [105,104].

The optical decay time of Cr^{4+} under resonant excitation is determined to be ~156 μs (at 30 K). The spin dynamics of its ground state has been studied in the temperature range

15–30 K, showing a long spin-lattice relaxation time T_1 of ~16 s at 15 K, which is limited by spin-orbit interaction [105]. Unlike the cases of the divacancy, the spin initialization of Cr^{4+} is not possible via nonradiative recombination under off-resonant excitation but requires polarizing a subpopulation into a sublevel, e.g., the $m_S = +1$ spin state, by tuning two excitations resonantly with $m_S = 0$ and $m_S = -1$ levels with primary and secondary lasers, respectively. Coherent control has been reported for Cr^{4+} ensembles, demonstrating a high readout contrast of Rabi oscillations (~63% and up to ~79% with reducing the probe time from 50 to 1 μs). This readout contrast places a lower bound on the ensemble spin polarization of at least 77%. The measured spin coherence time T_2 is found to be limited by spin-spin interaction and is comparable to that of the neutral divacancy in ensembles with similar defect densities [105].

3.3 Niobium

Although this transition metal (TM) is not among the commonly observed impurities in SiC, it can be introduced under certain conditions in significant concentrations. Nb has been identified in several SiC polytypes, first by EPR [106] and, later, by PL [107]. Both experimental observations are made on samples grown in a hot-wall CVD reactor with susceptor covered by NbC, hence the appearance of Nb in the samples is anticipated. However, Ref. [107] reports on an intentional doping experiment by placing a Nb flake upstream of the reactor, conducted for the 4H polytype of SiC. This experiment clearly confirms the appearance of the characteristic PL signature (associated with Nb) in the doped sample and its absence in the reference sample; the Nb concentration of 9×10^{16} cm^{-3} in the sample is confirmed also by secondary ion mass spectrometry (SIMS), whereas the reference sample shows concentration below the SIMS detection limit (~ 10^{13} cm^{-3}).

The low-temperature (T = 2 K) PL (LTPL) spectra of Nb in three polytypes, 4H-, 6H-, and 15R-SiC, are displayed in Figure 9. The line positions for 4H- and 6H-SiC are listed in Table 3. It is notable that the number of Nb-associated PL lines is equal to the number of hexagonal lattice sites in each polytype. This feature can be understood in view of the concept that Nb occupies not simply a substitutional site, but a divacancy, more specifically, a divacancy in which both the carbon and silicon vacancies are at hexagonal sites ($V_{Si}V_C$ in *hh* configuration). This concept has been presented in a theoretical work investigating the formation energies of several TMs from the first, second, and third rows of the periodic table [108]. This work shows that for the metals studied from the third row of the periodic table (among which is Nb) the formation energies of the metal at a Si substitutional site (M_{Si}) are comparable to those of the so-called asymmetric split-vacancy (ASV) configuration. The latter can be viewed as the impurity atom (M) occupying a divacancy, but the equilibrium position of M is closer to the silicon vacancy than to the carbon one, hence the asymmetry. In addition, the authors show that among the different nonequivalent configurations of the divacancy the one with both carbon (V_C) and silicon (V_{Si}) vacancies at hexagonal sites has significantly lower formation energy than the others and, therefore, is the preferable one. This is stipulated by the comfortable positioning of the *d*-orbitals' electrons, which form bonds with the unpaired electrons from the ligands of the divacancy [108]. Thus, there is only one preferable configuration of Nb in 4H- and 6H-SiC, which may be denoted by $(Nb_{Si}-V_C)_{hh}$; the subscript '*hh*' denotes the divacancy configuration. Therefore, there is only one zero-phonon line in these two polytypes. In 15R-SiC, there are two inequivalent hexagonal-hexagonal configurations of the divacancy, hence two Nb related lines appear in the LTPL spectrum.

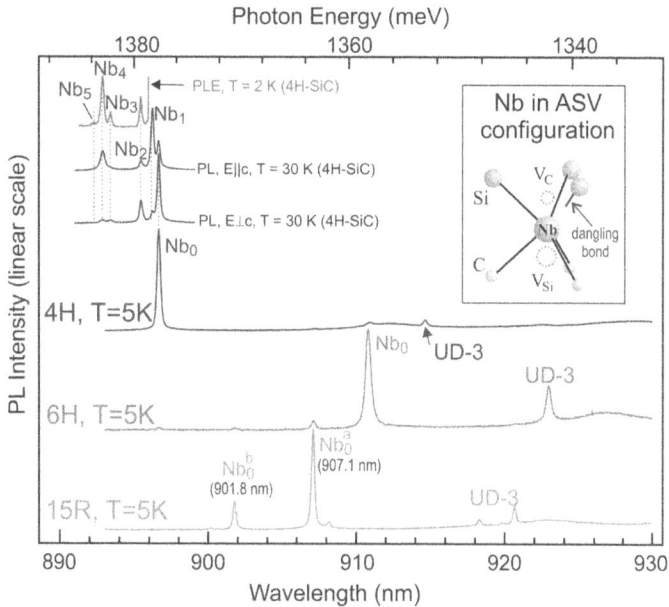

FIGURE 9

LTPL spectra of Nb^0-related emission in 4H-, 6H-, and 15R-SiC. See Table 3 for the peak positions of 4H- and 6H-SiC. The spectra display also weak contribution from the UD-3 defect. The top three curves illustrate the appearance of higher-energy lines at slightly higher temperature (T = 30 K), or in the PLE spectrum at 2 K (top curve), associated with excited states of the Coulombically bound hole. The inset depicts the configuration of Nb in a divacancy with both V_{Si} and V_C at hexagonal sites.

The EPR results [106] also agree with the notion of ASV configuration for the Nb center. The observation of the hf structure due to the interaction between the electron spin S=1/2 and the nuclear spin of ^{93}Nb (I = 9/2, 100% natural abundance) confirms the involvement of one Nb atom in the defect. In 4H-SiC, a considerable spin density of ~37.4% was found on three Si neighbors, suggesting the defect to be a complex between Nb and a nearby carbon vacancy. In both 4H- and 6H-SiC, only one Nb-related EPR spectrum with C_{1h} symmetry has been observed, supporting the ASV model for the neutral center $(Nb_{Si}-V_C)^0_{hh}$. It should be noted that, although a pure substitutional Nb_{Si}^0 also would be subject to Jahn-Teller distortion leading to C_{1h} symmetry, but the calculated hyperfine tensor components with such a model strongly disagree with the experimental ones.

Since the *hh* configuration of both vacancies accommodating the Nb atom is the preferable one, the LTPL spectrum is expected to exhibit the number of ZPLs equal to the number of *hh* divacancy configurations in each polytype, one in 4H- and 6H-SiC, and two in 15R-SiC (cf. Figure 9 and Ref. [107]). Furthermore, the electron configuration of the neutral Nb atom is Nb: [Kr] $4d^4 5s^1$. The five valence electrons will recombine with five of the six dangling bonds of the divacancy (three from the C atoms and three from Si atoms), leaving one of the Si dangling bonds unsaturated. Thus, the neutral Nb center is anticipated to exhibit significant electronegativity, because capturing one more electron (e.g., an electron from an exciton) will complete the bonding and can be seen as analog to filling the incomplete electron shell of a halide atom. This scenario provides an obvious mechanism for binding an exciton: the electron remains strongly localized at the defect forming the negatively charged defect $(Nb_{Si}-V_C)_{hh}^-$, whereas the hole remains weakly bound in the Coulomb field of the defect. The luminescence from the Nb center is therefore due to recombination of the exciton bound at the defect.

According to first-principles calculations, both the positive, the neutral, and the negative charge states of Nb (denoted hereafter shortly as Nb^+, Nb^0, and Nb^-) lie within the bandgap at about 0.2, 0.4–0.6, and 0.9 eV above the valence band edge, respectively. The two levels (0.4 and 0.6 eV) associated with Nb^0 are due to Jahn-Teller splitting of a partly occupied one-electron e level, which splits into a' and a'' levels when the symmetry is lowered from C_{3v} to C_{1h} [108]. An estimate of the exciton recombination energy using the energy position of the Nb^- level is made in [107] and yields the value 1.4 eV, close to the experimental value. This estimate is further corroborated by *ab initio* calculations conducted in the same work. Thus, the Nb center in its excited state consists of a strongly bound electron forming Nb^- and Coulombically bound hole.

When Nb^0 bounds an electron and becomes Nb^-, the reason for Jahn-Teller distortion vanishes and the Nb^- center resumes C_{3v} symmetry. The symmetry of a free hole would be $\bar{C}_{6v}$ (the bar above the group symbol denotes the corresponding double group involved owing to the half-integer spin of the hole). Spin-orbit and crystal-field interactions lead to splitting near the top of the valence band into three bands of Γ_9, Γ_7 and Γ_7 symmetry, using the notations of Ref. [91] for the irreducible representations of $\bar{C}_{6v}$. This splitting is inherited by the weakly bound hole in the Nb-bound exciton; hence the excited state is expected to be a triplet. Indeed, increasing the temperature leads to appearance of three closely spaced lines denoted Nb_0, Nb_1, and Nb_2 (instead of only Nb_0 at 5 K), as illustrated in the spectra at 30 K for 4H-SiC in Figure 9. It should be noted that three more lines appear at slightly higher energy (~ 4–5 meV above the triplet Nb_0–Nb_2). These lines are better resolved in the photoluminescence excitation (PLE) spectrum at 2 K also shown in Figure 9 for 4H-SiC. In Ref. [107], their origin is tentatively associated with the existence of another excited state for the C_{1h} configuration of the center (i.e., when it is excited but not relaxed to C_{3v} symmetry). In this case, not only should the triplet Nb_3–Nb_5 replicate approximately the triplet Nb_0–Nb_2 at slightly higher energy, but it should also exhibit the same selection rules for the composing lines, since the two excited-state configurations differ only by a presumably small perturbation. This is confirmed experimentally, the middle lines in each triplet are polarized parallel to the crystal c-axis (E $||$ c), whereas the rest of the lines do not exhibit preferable polarization.

The selection rules can be easily deduced if one assumes that both the excited and the ground state have C_{3v} symmetry [107]. (The C_{1h} symmetry of the ground state can be seen as a minor perturbation, which does not significantly affect the selection rules derived for C_{3v}.) When the symmetry is reduced from $\bar{C}_{6v}$ (free hole) to $\bar{C}_{3v}$ for the weakly bound hole, the three states inherent to the hole transform according to the compatibility relations: Γ_7 of $\bar{C}_{6v}$ becomes Γ_4 of $\bar{C}_{3v}$, and Γ_9 of $\bar{C}_{6v}$ becomes Γ_{56} of $\bar{C}_{3v}$ (here we use the notation $\Gamma_{56} = \Gamma_5 + \Gamma_6$ for the direct sum of the Kramers-degenerate representations Γ_5 and Γ_6 of $\bar{C}_{3v}$) [91]. Thus, the excited state of the Nb center has the same structure as that of the weakly bound hole, a triplet consisting of two Γ_4 states and one Γ_{56} state. The final state after the exciton recombination is just the Nb^0 center but assumed in C_{3v} symmetry for the sake of deriving the selection rules, as already mentioned above. It can be seen as having its uppermost e orbital occupied by three electrons, or by one hole. The orbital momentum associated with the e orbital couples to the spin of the hole resulting in a spin-orbit splitting of the ground state into Γ_4 and Γ_{56} states. It is argued in Ref. [107] that the splitting in the ground state may well be unresolvable in the optical spectra, so that the observed triplet Nb_0–Nb_2 is associated with transitions from one of the components of the excited state to the Γ_4 or the Γ_{56} counterpart of the ground state. Group-theoretical consideration yields the allowed

TABLE 3

Line positions in nm (photon energy in meV in parentheses) of the six Nb-related transitions in 4H- and 6H-SiC. The polarization of each line is also shown ($||$ – E$||$c, $\perp$ – E$\perp$c)

Polytype	Nb_0 $||$ and $\perp$	Nb_1 $||$	Nb_2 $||$ and $\perp$	Nb_3 $||$ and $\perp$	Nb_4 $||$	Nb_5 $||$ and $\perp$
4H SiC	896.55 (1382.5)	896.14	895.39	893.39	892.88	892.24
6H SiC	910.96	(1383.15)	(1384.3)	(1387.4)	(1388.2)	(1389.2)
	(1360.6)	910.84	909.70	908.79	908.30	906.84
		(1360.8)	(1362.5)	(1363.9)	(1364.6)	(1366.8)

transitions: $\Gamma_{56} \rightarrow \Gamma_{56}$ (with E$||$c polarization) and $\Gamma_4 \rightarrow \Gamma_4$ (allowed with both E$||$c and E$\perp$c polarizations). Transitions $\Gamma_4 \leftrightarrow \Gamma_{56}$ require spin flip and are deemed forbidden.

The experimental data suggest that the order of the states of the weakly bound hole in order of ascending energy is Γ_4–Γ_{56}–Γ_4 with the corresponding splitting represented by the splitting in the Nb_0–Nb_1–Nb_2 lines (with the splitting in the ground state neglected), see Figure 9. Thus, the ordering of the states for the weakly bound hole does not replicate that of a free hole, for which the Γ_9 (compatible with Γ_{56} in $\bar{C}_{3v}$) band has the lowest energy and the two Γ_7 bands have higher energies. The above analysis is equally valid for 6H-SiC, which has the same symmetry as the 4H polytype, but of course the energy positions of the lines and the observed splittings are somewhat different.

Ref. [107] presents also experimental data on the Zeeman effect on the Nb lines, which is analyzed using two models for the spin Hamiltonian, one based on weak to intermediate spin-orbit coupling treating the orbital angular momentum l and the spin of the hole s separately, and another in which l and s are already coupled into total orbital momentum j, which behaves as a single quantity in external magnetic field (strong spin-orbit coupling). It is shown that both models together are needed to provide reasonable description of the Zeeman data, especially the nonlinear dependencies of the high-energy-lines' positions for magnetic field perpendicular to the c-axis.

Finally, we notice that the lifetime of the Nb related emission has been measured [107] to about 180 μs for the Nb_0 line at 2 K. The total emission at 140 K when the contributions from individual lines cannot be discerned exhibits somewhat shorter lifetime, 120 μs. These relatively long lifetimes compared to other color centers in SiC indicate the absence of concurrent nonradiative channels. This is expected since Auger recombination is impossible because the energy stored in the bound exciton (~ 1.4 eV) is insufficient to transfer an electron from the Nb center to the conduction band, which would create Nb$^+$.

3.4 Molybdenum

Molybdenum (Mo) has been observed by EPR in as-grown 4H-, 6H-, and 15R-SiC substrates [109]. EPR of the single donor Mo^{5+} (4d^1) center, with spin S=1/2, has been observed in slightly n-type 6H-SiC. The involvement of Mo is confirmed by the observed well-resolved hf structure due to the interaction between the electron spin and the nuclear spins of isotopes ^{95}Mo (I=5/2, 15.9%) and ^{97}Mo (I=5/2, 9.6%). In all studied polytypes, only one EPR spectrum was detected. It was concluded that the center has a nearly isotropic g-value. However, as shown later from magneto-optical studies, the Mo^{5+} center has g-values for the ground state: $g_{||}$ = 1.87 and $g_{\perp} \approx 0$ for 4H-SiC, and $g_{||}$ = 1.61 and $g_{\perp} \approx 0$ for 6H-SiC [110].

The neutral charge state, Mo^{4+} ($4d^2$), has been observed in 4H-, 6H-, and 15R-SiC [109]. The ground state of the center is a spin-only triplet S=1 state $^3A_2$ with a nearly isotropic g-value. In 6H-SiC, only one EPR Mo^{4+} center was observed [111]. The center shows axial symmetry with nearly isotropic g-value: $g_{||}$ = 1.9787 and $g_\perp$ = 1.9783, and a large fine-structure parameter D = 3.32 GHz. The EPR spectrum of Mo^{4+} could not be detected at temperatures above 77 K. The hf constant was found to be similar for ^{95}Mo and ^{97}Mo and isotropic, $A_{||}$ = $A_\perp$ = 92 MHz. The spin-Hamiltonian parameters for the Mo^{4+} center in 15R- and 6H-SiC are found to be similar [112].

In n-type 6H-SiC, the single acceptor Mo^{3+} ($4d^3$) center has been identified by EPR [111]. The Mo^{3+} center has the effective g-values: $g_{||}$ = 1.945 and $g_\perp$ = 3.939 if analyzing with spin S=1/2. But with the $g_\perp$ value being about double of the g-value of free electron, the true spin of the center should be S=3/2 with ZFS much larger than the microwave frequency (the fine-structure parameter D is larger than 2 cm^{-1} (~60 GHz)). The hf parameter is found to be ~73 MHz for both ^{95}Mo and ^{97}Mo.

Like the case of Nb-related center in SiC, the Mo centers in all charge states, Mo^{5+}, Mo^{4+}, and Mo^{3+}, manifest only one EPR spectrum in 4H- and 6H-SiC [111]. This may suggest that Mo is also in the asymmetric split-vacancy (ASV) configuration and Mo should be in between the C vacancy and Si vacancy at the hexagonal site (Mo_{Si-C}), as suggested by Ivády and co-workers for several TM impurities in SiC [108]. However, no unique opinion on the Mo configuration in SiC has been reached so far in the literature. Early work has not considered the possibility for ASV configuration, but after the publication of Ref. [108] the observation of a single PL center in 4H- and 6H-SiC seems to find its natural explanation in the existence of only one configuration, the ASV at hexagonal site (see Ref. [113] and the PL discussion further below). Both models (ASV configuration and substitutional, Mo_{Si}) have been considered in theory [114] in connection with the PL spectra [113]. Their analysis is based on the search for a suitable charge state with spin S=1 (to agree with the EPR data). As discussed below, it is concluded that the most probable configuration of the Mo center responsible for the observed PL is the double positively charged state of the ASV configuration. However, later work adopts the model of Mo_{Si} in single positively charged state (Mo_{Si}^+) [110,115]. In Ref. [110], the authors conduct all-optical resonant photoluminescence excitation spectroscopy (PLE) in magnetic field with two-laser excitation. The results allow them to conclude that both the ground and the excited states in the PL transition are Kramers-degenerate spin doublets (S = 1/2). Therefore, the authors conclude that Mo_{Si}^+ at the hexagonal site is the center responsible for the PL. The same notion is adopted in a later work by the same group [115], but in both works no reason is provided why only the hexagonal site should be involved in the center.

The PL from Mo in 4H- and 6H-SiC was first reported in 2009 [116], where a defect previously known as I-1 [117] was identified with Mo on the ground of a doping experiment. Similar to EPR, only one optical signature was observed in both polytypes, the corresponding spectra are displayed in Figure 10. Here we must notice that in another paper by Baur et al. [112], two equal intensity EPR spectra of Mo^{4+} are observed in 15R-SiC and two EPR lines with very different intensity have been detected in 6H-SiC. However, the weak line in 6H-SiC has the same D value as one of the two in 15R-SiC. Considering that in the EPR study by Dombrowski and co-workers [111], only one EPR spectrum is observed for the Mo^{4+} center, the additional weak line in 6H-SiC observed by Baur and co-workers [112] may be due to inclusion of the 15R polytype in the 6H-SiC sample. Hence, the number of optical centers in PL and EPR centers corresponds to the number of hexagonal sites in these polytypes. This agrees with the theoretical suggestion for incorporation of Mo in a divacancy with both vacancies at hexagonal sites [108]. In the initial work

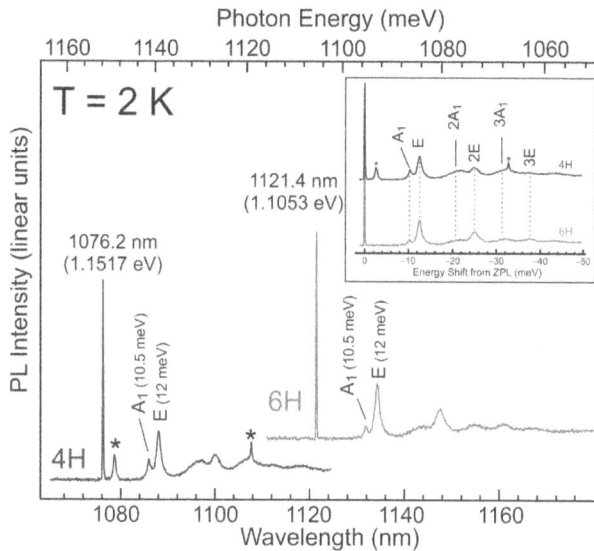

FIGURE 10

Mo-related LTPL spectra in 4H- and 6H-SiC (T = 2 K). Multiline UV lines (351–365 nm) from an Ar ion laser are used as an excitation. The inset is a plot of the same spectra as an energy shift from the corresponding ZPL to emphasize the similarity in both polytypes. A_1 and E denote replicas due to local phonon modes, their overtones are also marked in the inset. The two asterisks in the 4H spectrum denote the divacancy lines PL3 and PL4, which also appear weakly in this spectrum.

[116], Mo was assumed to be at Si substitutional site, but in later work the same group revisited their model and reviewed the experimental data from the point of view of the ASV configuration [113], based on new theoretical results [114].

It should be noticed that if Mo is indeed in ASV configuration instead of mere substitutional (on Si site, Mo_{Si}), then the notations Mo^{3+}, Mo^{4+}, and Mo^{5+} ubiquitous in the EPR literature and used above for the donor, neutral, and acceptor states of Mo, are not rightful notations. If Mo is in the ASV configuration, the EPR data needs new interpretation, which has not been undertaken so far, to the best of our knowledge. This task is beyond the scope of the current script. Mo at the *hh* site in the ASV configuration provides a natural explanation why only one center is observed in 4H- and 6H-SiC, and two centers in 15R-SiC.

The theory [114] finds the charge transition levels $(+|0)$ (i.e., the single donor level) and $(0|-)$ (i.e., the single acceptor level) in the ASV configuration in 4H-SiC at $E_V + 0.68$ eV and $E_V + 2.28$ eV, respectively, by DFT calculations. Here E_V denotes the VBM. The double donor $(2+|+)$ and double acceptor $(-|2-)$ levels are also predicted in the bandgap of 4H-SiC at $E_V + 0.49$ eV and $E_V + 2.9$ eV, respectively. The MCDA measurements in 6H-SiC show several sharp lines in the range 1.10–1.25 eV. ODMR detected on these lines shows different transitions of spin S=1 center, interpreted as belonging to the neutral Mo^{4+} center (in the Mo_{Si} configuration). The proximity of one of these sharp lines in MCDA to the ZPL observed in PL (at 1.1056 eV in PL, and ~ 1.1086 eV in MCDA) inspires the doping experiment proving that indeed the PL spectrum observed in 6H-SiC is related to Mo [116]. Since Dombrowski et al. [111] observe spin-one related ODMR spectrum on this line, the PL spectrum is also associated with a spin S=1 center, although no PL spectra of this center were provided in Ref. [111]. Considering the ASV configuration as the most probable for Mo in SiC, the authors conclude that it is likely that the observed PL is associated with the double-positively charged state of Mo, Mo^{2+}, especially since the PL is strongly observed in p-type or semi-insulating materials. However, as mentioned above, Ref. [110] proves convincingly that the PL center (in 6H and 4H) has ground and excited states, which are Kramers doublets (S=1/2). These apparently contradicting views can be reconciled if one disregards the association of the MCDA line at 1.1086 eV with the PL line at 1.1056 eV and assumes that the PL is due to Mo in the ASV configuration, but in the single-positively charged state, Mo^+, instead of Mo^{2+}. The assumption of the ASV configuration is in contrast with assumption of Ref. [110], but their work does not provide or use information on the exact site positioning or charge state of Mo, as long as the spin of the center is S=1/2. Thus, the assumption of ASV configuration of Mo^+ fits also their results with the sole difference that instead of belonging to unpaired electron the S=1/2 spin belongs to a hole. We notice that the theory has been unable so far to deal with the excited states of Mo in the ASV configuration, which might provide a theoretical estimate of the ZPL energy, because of the complications arising with the need for multi-determinant description [114]. We notice also that if the ASV hh configuration is dominating the incorporation of Mo in SiC, then the EPR data collected in early work needs new interpretation, which is beyond the scope of the present work.

We comment now on the structure of the PL spectrum, which is very alike in 4H- and 6H-SiC, as seen also from the inset of Figure 10. The spectrum in each polytype consists of a sharp ZPL [at 1076.3 nm (1.1519 eV) and 1121.5 nm (1.1055 eV) in the 4H and 6H polytype, respectively] and a phonon sideband. The dominant contribution to the phonon sideband consists of two local phonon modes (labeled A_1 and E in Figure 10) and their overtones. In addition, polarization measurements show that the ZPL, the E-replica and its overtones have parallel to the c-axis polarization (E||c), whereas the A_1 mode and its overtones have polarization E⊥c. Group-theoretical analysis carried out in Ref. [113] shows that if the ground and the excited state are both triplets $^3A_2$ and if the two local phonon modes have A_1 and E symmetries in C_{3v}, then the selection rules are the same as observed experimentally. Phonon modes of A_1 and E symmetry are obtained in a natural way if only the Mo atom in ASV configuration is allowed to move, thus the labeling in Figure 10. This might be a good approximation since the broad phonon sideband usually observed for deep centers is almost absent, indicating weak coupling to lattice phonons. However, if the ground and excited states are Kramers' doublets transforming as the two-dimensional Γ_4 in $\bar{C}_{3v}$ [110], then transitions with both E||c and E⊥c polarization are allowed. In addition, the C_{3v} symmetry assumed here for Mo^+ in the ASV configuration is only an approximation for we neglect the Jahn-Teller distortion inherent to this charge state. Thus, the polarization selection rules in this case (Mo^+ in ASV configuration) cannot be derived on the ground of group theory alone.

In conclusion, more theoretical and experimental work is needed to establish with certainty the microscopic model of Mo in SiC. If the ASV configuration is the true configuration for Mo in SiC, then the EPR data published earlier needs new interpretation.

3.5 Tungsten

Tungsten (W) in SiC has been unambiguously identified by DLTS in combination with radiotracer experiments. In samples implanted with radioactive ^{178}W isotope, Achtziger et al. [118] observe a deep level at 1.43, 1.16, and 1.14 eV below the CBM in 4H-, 6H-, and 15R-SiC, respectively. In addition, a shallow W-related level at 0.17 eV below the CBM is observed in 4H-SiC, but not in 6H- or 15R-SiC. These levels are assigned to deep acceptors in Ref. [118], i.e., the 1.43, 1.16, and 1.14 eV levels in 4H-, 6H-, and 15R-SiC are associated with single deep acceptors corresponding to the $(0 \mid -)$ charge transition levels of W in the three polytypes, whereas the 0.17 eV level observed only in 4H-SiC is associated with a double acceptor $[(- \mid 2-)$ transition]. We notice here, however, that as will be discussed further below, later work suggests that this assignment is incorrect and most likely the observed levels are instead single donor levels $[(+ \mid 0)$ transition levels], with the shallow 0.17 eV level in 4H-SiC being the acceptor level $[(0 \mid -)$ transition]. In Ref. [118] it is observed also that these three deep donor levels have essentially the same offset with respect to the VBM in the three polytypes, which agrees with the Langer–Heinrich rule [103]. This explains why the shallower level is observed only in the 4H polytype, since it should be degenerate with the CB in 6H- and 15R-SiC owing to their smaller bandgaps.

Later work [119] presents DLTS studies on 4H-, 6H-, and 3C-SiC intentionally doped with W. Their results confirm the observation of the same W-related levels in 4H- and 6H-SiC. In addition, they observe a corresponding level in 3C-SiC at an energy 0.47 eV below the CBM, which is thought to be the acceptor state in Ref. [119] but should be associated with the donor state of W, as discussed below. Since the bandgap of 3C-SiC is 2.36 eV vs. 3.26 eV for 4H-SiC at room temperature, the observed energy position of the W level in 3C-SiC is also in accord with the Langer–Heinrich rule. We notice also that the estimated capture cross section for the W donor level is very large, $\sim 1 \cdot 10^{13}$ cm^{-2} in 4H-SiC and 3C-SiC, and $3 \cdot 10^{13}$ cm^{-2} for 6H-SiC [118,119].

In 6H-SiC doped with tantalum (Ta), Irmscher observed two axial EPR spectra under illumination, one with spin S=1 and another with spin S=1/2 [120,121]. The spin S=1/2 defect shows a hf structure, which could be identified as due to the interaction between the electron spin and the nuclear spin of ^{183}W (I=1/2, 14.3%), and was assigned to the positive W^{5+} center. The S=1 spectrum was too weak to detect any accompanying hf structure but is suggested to be related to the neutral W^{4+} center.

The PL of W has been reported in three polytypes, 4H-, 6H-, and 15R-SiC [122]. The spectra in 4H-, 6H-, and 15R-SiC exhibit two, three, and four lines, respectively, as illustrated in Figure 11. The corresponding line positions are listed in Table 4. The centers responsible for the luminescence, seen also in absorption, have been previously denoted as UD-1. However, the association of UD-1 in 4H-SiC with W has been proven by a doping experiment on 4H-SiC [122]. It is shown that the two high-energy lines denoted a and b thermalize in absorption, i.e., the lower-energy counterpart vanishes in absorption at low temperature, as illustrated in Figure 11 for the 6H polytype. No such thermalization is observed in PL, suggesting that the ground state of the center is split, and the observed PL transitions are from a single excited state to the two counterparts of the ground state. It is assumed further that each PL center contributes two closely spaced lines to the spectrum, but in the spectra, they are only resolved for one of the inequivalent sites in each polytype with the highest ZPL energy (corresponding to the a and b lines in each polytype). It has been shown also that the high-energy component b in the a-b doublet is perpendicular to the c-axis polarization (E$\perp$c), whereas the other component has both polarizations (E$\perp$c and E$\mid\mid$c). Assuming $\bar{C}_{3v}$ symmetry for the defect, the line polarizations suggest that the transitions observed in PL are from an excited state of Γ_4 symmetry to a ground state containing both Γ_{56} and Γ_4 counterparts, with the former lying higher in energy (cf. the inset in Figure 11). The double group $\bar{C}_{3v}$ is used in Ref. [122] because the defect is shown to possess half-integer spin (S=1/2) by conducting Zeeman-effect measurements presented in the same work.

The accumulated experimental data (spin S=1/2) suggests that the observed PL centers (single center in 4H) are related to W in either positively (W$^+$) or negatively (W$^-$) charged state because the neutral W atom has an even number of electrons (W: [Xe]6s^{2}4f^{14}5d^4). W$^+$

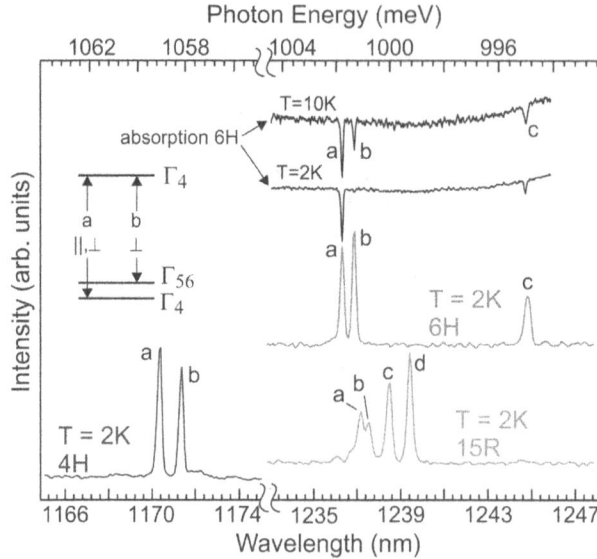

FIGURE 11

LTPL spectra of W in three SiC polytypes, as denoted for each spectrum. See Table 4 for the peak positions. The top two spectra are absorption of W in 6H-SiC at two different temperatures, illustrating the absence of the b-transition at T = 2K and its appearance at 10 K when the higher-energy counterpart of the ground state becomes populated. The inset shows a schematic energy diagram of the ground (Γ_4 and Γ_{56}) and the excited (Γ_4) states. The suggested symmetries are based solely on the observed selection rules for the a and b lines ($||$ − E$||$c, ⊥ − E⊥c), without relation to any underlying model. Notice the horizontal axis brake.

is discussed as the most probable PL center in [122]. Also, the number of observed centers coincides with the number of quasicubic sites in each polytype. The notion of W$^+$ being the luminescence center is corroborated by the mere observation that the corresponding luminescence has been observed either in p-type or in SI material. The results of an older work [123] provide solid evidence in favor of the concept that W$^+$ is the luminescent center and, at the same time, that the deep levels observed by Achtziger et al. and other authors [118,119,121] are donor levels, not acceptor levels. In Ref. [123], the authors investigate the so-called photo-induced absorption of the same W-related center in 4H-SiC that is observed in PL (at that time known as UD-1). They find that the absorption increases abruptly if the sample is illuminated with light of energy larger than ∼ 1.45 eV, which is essentially the same energy determined for the deep level of W in Ref. [118] (1.46 eV). This observation suggests that electrons are pumped from the level to the CB, leaving the W center in positively charged state. Maximum effect is achieved around 1.85 eV for the photon energy of the auxiliary illumination, further increase of the photon energy diminishes the absorption of the center. The observation of the second threshold suggests that the pumping of electrons to the CB starts being counteracted by pumping of electrons from the VB to the center (the two thresholds sum up approximately to the bandgap of 4H-SiC). Thus, for energies above ∼ 1.45 eV photoionization of the deep donor level is observed and the association with acceptor level from Ref. [118] is incorrect, as pointed out also earlier in Ref. [121].

Ref. [122] considers also qualitatively a possible model for the W center, but we will not discuss it here, because it is based on W substituting Si in the host lattice. We believe that

TABLE 4

Line positions of the W-related emission in three SiC polytypes, 4H-, 6H-, and 15R-SiC

Line	Peak position in nm (eV)
4H-SiC	
a	1170.2 (1.0592)
b	1171.2 (1.0583)
6H-SiC	
a	1237.5 (1.0016)
b	1238.1 (1.0011)
c	1246.0 (0.9947)
15R-SiC	
a	1238.4 (1.0009)
b	1238.8 (1.0006)
c	1239.7 (0.9998)
d	1240.6 (0.9991)

the most probable configuration for W in SiC is the so-called asymmetric split vacancy configuration [108] when the W atom occupies a divacancy, slightly displaced towards the Si vacancy V_{Si}. However, the number of observed W centers follows the number of cubic lattice sites in each polytype, so that prevailing *hh* occupation of the ASV in the case of W is unlikely. Other configurations for W have not been considered theoretically in detail so far, to the best of our knowledge.

3.6 Erbium

Erbium (Er) is a rare-earth element with the ground state electronic configuration [Xe]$4f^{12}6s^2$. When incorporated in solids or semiconductors, it exists as 3+ ions. The trivalent erbium (E^{3+}) has 11 electrons in the partly filled 4f shell. The spin-orbit interaction splits the ground state (having angular momentum L=6 and spin S=3/2) into four states $^4I_{15/2}$, $^4I_{13/2}$, $^4I_{11/2}$, and $^4I_{9/2}$, with $^4I_{15/2}$ being the lowest. The intra-4f transition $^4I_{15/2} \rightarrow \, ^4I_{13/2}$ is electric dipole forbidden but becomes partially allowed by crystal-field induced mixing of opposite parity wavefunctions in solids or semiconductors that lack inversion symmetry. This optical transition gives rise to emission at ~1535 nm, which is in the minimum absorption window of silica fibers (the telecommunication wavelength C-band: 1530–1560 nm). So far, the most important application of Er is utilizing the $^4I_{15/2} \rightarrow \, ^4I_{13/2}$ transition in Er-doped fiber amplifiers that are the key devices of current fiber networks, allowing intercontinental communications [124].

The emission of Er^{3+} has been reported for different Er-implanted SiC polytypes: 4H, 6H, 15R, and 3C [125]. For the hexagonal polytypes (4H and 6H) and the rhombohedral polytype (15R), the LTPL (2 K) spectra of Er^{3+} are very similar with a dominating line at ~1534 nm, while for 3C-SiC there are two major lines at 1528 and 1534 nm. It has been found that the integrated PL intensity of Er^{3+} in all studied SiC polytypes varies little in the temperature range of 2–400 K and the emission can still be detected at 525 K. Electroluminescence (EL) of Er^{3+} in Er-implanted 6H-SiC *p-n* junctions has been observed under forward bias conditions. The EL spectrum is identical to the PL spectrum, though the quantum efficiency is very low (7.5×10^{-6}) [126]. A long lifetime of ~2.3 ms has been reported for the Er^{3+} PL center in 3C-SiC polycrystal nanowires [127].

The identification of Er^{3+} in SiC has been reported for the 6H polytype by Baranov and co-workers [128,129]. While the similarity of the PL spectra of Er^{3+} in the hexagonal and rhombohedral polytypes suggest that the dominant factor is the local symmetry with only minor influence from the second and third neighbors [125], several Er-related EPR centers have been observed in Er-implanted 6H-SiC with three having low symmetry C_{1h} and two having axial symmetry [128].

The energy levels of Er in 4H- and 6H-SiC have been determined by DLTS using samples implanted with radioactive isotope ^{120}Er. A donor level located at $\sim E_V +0.75$ eV in 4H-SiC and $\sim E_V +0.78$ eV in 6H-SiC has been observed [130].

4 The Nitrogen-Vacancy Center in SiC

The nitrogen-vacancy center in SiC consists of a nitrogen atom on a C site (N_C) and a nearest-neighbor silicon vacancy, denoted as $N_C V_{Si}$, or shortly NV. In theoretical work, the center in its negative charge state NV^- has been envisaged as an analog of the NV^- center in diamond [9,62,131] well before its experimental identification. In particular, analogous to the NV center in diamond electronic structure is anticipated to lead to analogous properties beneficial for applications in quantum technology, spin S=1 ground state, possibility for optical spin initialization and readout, and long spin coherence times, with the additional benefit of photoluminescence in the infrared at or close to the telecom transmission bands for fibers, instead of in the visible as for the NV center in diamond.

In fact, the NV^- center in 6H-SiC has been detected earlier in 6H-SiC by EPR and magnetic circular dichroism of the absorption [56] and in 4H-SiC in conventional absorption [117], but at that time the observed centers were mistakenly associated with the carbon antisite–carbon vacancy pair, which is also a nearest-neighbors pair defect. Since 4H-SiC has two inequivalent lattice sites for each of the constituents N_C and V_{Si}, of the NV pair, there exist four possible combinations for $N_C V_{Si}$, *hh*, *kk*, *hk*, and *kh*. In the 6H polytype there are three inequivalent sites, one hexagonal (*h*), and two cubic (k_1 and k_2), giving rise to six configurations of $N_C V_{Si}$, *hh*, $k_1 k_2$, $k_2 k_1$, hk_1, $k_2 k_2$, $k_1 h$ (cf. Figure 2). As seen from Figure 2, the first two (three) configurations in 4H- (6H-) SiC are aligned along the crystal c-axis and, therefore, are termed axial, whereas the rest of the configurations are termed basal. It is obvious that, apart from possible distortions of the defect, the axial configurations have higher symmetry (C_{3v}), while the basal configurations have *per se* C_{1h} symmetry, in both 4H- and 6H-SiC. However, the NV^- center in 3C-SiC (cubic crystal) has only one configuration (more precisely, four equivalent configurations oriented along the four equivalent [111] directions of the crystal) with C_{3v} symmetry and this is the closest analog in SiC to the NV^- center in diamond.

The electronic structure of NV^- in SiC has been treated using group theory [132,133] in analogy with that of the divacancy [134]. Quantitative calculations are performed on the ground of the DFT to obtain the positions of the charge transition levels within the bandgap, ZFS, and hf parameters. The ground state of the axial NV^- centers is a spin triplet $^3A_2$, as depicted schematically in the inset of Figure 12. There is also an excited state ^3E, the radiative transitions from which to the ground state give rise to the ZPLs of the different configurations (one in 3C-, four in 4H-, and six in 6H-SiC, in analogy with the divacancy). Theoretical estimates of the ZPL energies have also been made. However, the NV^- centers have also the so-called shelving singlet state (or states), via which nonradiative relaxation

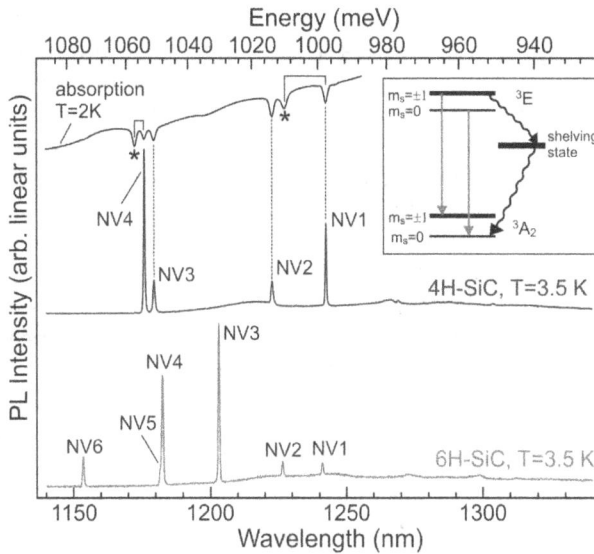

FIGURE 12

LTPL spectra of the N-V pairs in 4H- and 6H-SiC (bottom two curves), excited with 930 and 940 nm laser, respectively. The top curve is an absorption spectrum at T = 2 K illustrating the split excited states of the two basal configurations *hk* and *kh*. The higher-energy excited state is denoted with the asterisk and connected with a bracket to the lower-energy counterpart. The inset shows the mechanism of spin polarization: radiative transitions (straight arrows) are allowed between the same kinds of substates of the ground and excited states (either $m_s = 0$, or $m_s = \pm 1$). Nonradiative decay (wavy arrows) via a shelving state, which decays only to the $m_s = 0$ substate of the ground state ensures spin polarization. See Table 5 for the peak positions.

from the excited state ^{3}E to the $m_s = 0$ component of the ground state is possible, leading to strong spin polarization of the ground state. The singlet states can be built from the one-electron orbitals obtained in a group-theoretical consideration and the singlets ^{1}A$_1$ and ^{1}E are obtained. However, their quantitative investigation requires multideterminant calculation, which has been beyond the scope of present theoretical work [133]. Therefore, in the schematic representation in the inset of Figure 12 simply the presence of shelving state (possibly, more than one) responsible for the spin polarization upon optical excitation is indicated. Thus, like the divacancy, the NV$^-$ center exhibits the possibility of optical spin initialization, beneficial for applications in quantum technology.

The unambiguous identification of the NV$^-$ pair in 4H-SiC was done first by EPR in 4H-SiC [135]. A proton irradiated and annealed *n*-type 4H-SiC sample exhibits one spin S=1 center with a fine-structure parameter $D = 429 \times 10^{-4}$ cm^{-1} (1.286 GHz), close but distinctly smaller than that of the divacancy (440×10^{-4} cm^{-1} or 1.319 GHz). The association of this center with NV$^-$ is due to the three-fold hf splitting of the EPR lines owing to the interaction with the nuclear spin of ^{14}N (I = 1, 100%), thus distinguishing this center from the divacancy. The identification of the center with the axial configurations of NV$^-$ is corroborated by theoretical calculations yielding close to the measured parameters for the ZFS and the hf interactions. The *hh* and *kk* axial configurations were assumed to have too similar parameters to be resolvable in X- and Q-band EPR. In this first work the basal configurations were not observed, but later works from the same group [136,137] show their EPR spectra and determines the relevant ground state parameters. Not only 4H-SiC, but also 3C- and 6H-SiC are included in later studies [136-139,13,140].

The PL spectrum of the NV$^-$ center in 4H-SiC consisting of four lines and the associated phonon sidebands is reported in Ref. [141]. The same spectrum observed previously in absorption [117] and in PL [113] was erroneously assigned in the past to the $C_{Si}V_C$ pair, following assignment of a spectrum in the similar spectral range observed by MCDA in 6H-SiC to $V_C C_{Si}^{2+}$ [56]. (The latter spectrum consisting of six lines corresponding to the six inequivalent NV-pair configurations in 6H-SiC has also been observed in PL later and correctly assigned to the NV$^-$ centers [13].)

The spectra of the NV$^-$ pairs in 4H- and 6H-SiC are displayed in Figure 12. We notice that in the initial work [141], one of the four lines in 4H-SiC is mistakenly assigned to tungsten, but this has been corrected in later works [60, 138]. Also, the authors of Ref. [141] (and in Ref. [138]) report that one (two) of the lines in the 4H-spectrum appear as doublets. Two possible reasons are considered for this doublet structure in Ref. [138]. One possibility is the existence of a second excited state separated by only 0.6–0.7 meV from the lowest excited state, and another is the presence of residual strain due to the proton irradiation used in this work [141, 138]. A second excited state is expected in theory, indeed, for the two basal configurations, due to the splitting of the ^3E state when the symmetry is lowered from C_{3v} (for the axial configurations) to C_{1h} (for the basal ones). However, in this and in a later work [142], the splitting is observed for the axial configurations only, in contrast to the anticipation. In fact, the splitting for the basal configurations in 4H-SiC (PL lines NV1 and NV4) can be seen in low-temperature absorption, as illustrated in Figure 12 (top curve). The higher-energy counterparts of the excited states of the two basal configurations appear as additional lines accompanying the lower energy counterparts (which coincide with the LTPL lines NV1 and NV4). Also, the splitting between the two excited states is significantly larger than that observed in Refs. [141,138,142]. No splitting is seen for the axial configurations. In addition, other groups reporting on the NV$^-$ center in 4H-SiC do not observe such splitting [60,143]. Consequently, the first explanation given in Ref. [138] based on the existence of a second excited state within 0.7 meV from the lowest one can be ruled out, leaving as the most probable reason for the observed splitting residual strain. In fact, no splitting for the axial configurations is observed in neither of both polytypes, but we notice that one of the contributions in the 6H-SiC spectrum appearing as a single peak at ~ 1183 nm actually contains two lines, which are not resolved. The shoulder to the left of the main peak (marked with NV5 in Figure 12) is well visible, thus the two peaks NV4 and NV5 belong to different configurations, see also Table 5.

The possibility for spin polarization of the ground state has proved to be very useful for association (tagging) of the different lines in the PL spectrum and the lines in the EPR spectrum with the different configurations of the NV$^-$ center. Firstly, nonresonant (broadband) excitation of all the configurations facilitates their observation because the spin polarization strongly increases the EPR-signal intensity. The spin polarization is accompanied also with a phase change of the high-field component of the EPR line from absorption to emission. The optically enhanced EPR signal has been used in Ref. [136] to observe the EPR spectra containing contributions from all configurations of NV$^-$ in 3C-, 4H-, and 6H-SiC. Secondly, if resonant laser excitation is used to excite selectively only one of the NV$^-$ configurations via the corresponding $^3A_2 \rightarrow {}^3E$ transition, then the corresponding component in the EPR spectrum will exhibit increased EPR intensity due to the spin polarization. In this manner the ZPLs in the PL spectrum can be associated with the EPR line in the EPR spectrum. We notice here that this has been done for all the three major polytypes of SiC, 3C [140], 4H [138], and 6H [13], and no significant Frank-Condon shift has been found in the corresponding excitation spectra, i.e., the resonant laser excitation coincides in energy with the corresponding ZPL in the common PL spectrum [138,13]. By comparing the calculated parameters of the ground state, such as the ZFS, the hyperfine and superhyperfine tensors, with that determined by EPR, it is possible to distinguish between the configurations of the NV$^-$ center in the hexagonal polytypes [133]. Moreover, using the resonant excitation for tagging the PL lines with certain EPR lines allows also association of the PL lines with the different configurations. This is very useful in providing feedback to theoretical calculations, since often the ZPLs are calculated with insufficient accuracy for comparison with the experiment (~ 0.1 eV), hence, the ordering of the lines (ZPL energy order for the different configurations) cannot be certified from theory alone.

The EPR and PL from the NV$^-$ center in the 3C-SiC polytype has been reported in Ref. [139]. Two ZPLs at 1289 and 1468 nm have been observed in this work and, initially, the 1468 nm line has been associated with the NV$^-$ center in 3C-SiC. A later work using resonant excitation of the center refutes this association and suggests that the 1289 nm (~ 0.961 eV) line is associated with the NV$^-$ center in 3C-SiC [140]. Theoretical estimates of the ZPL energy vary in the range of 0.870–0.965 eV [133,62,140]. Another report claims the observed ZPL at 0.89 eV (about 1390 nm) in 3C-SiC to be related to the NV$^-$ center [144] but without convincing proof. On the other hand, the association of the 1289 nm line with the NV$^-$ center in 3C-SiC is properly corroborated using the resonant optical excitation method with EPR detection, where the EPR spectrum with the observed ^{14}N hf structure confirms the involvement of one N in the center [140]. No PL spectrum is given in Ref. [140], presumably the same sample with the spectrum given in Ref. [139] has been used. The ZPL positions obtained from the PL measurements are summarized in Table 5 where we list the 1289 nm value from Ref. [140] as the ZPL of the NV$^-$ center in 3C-SiC.

The EPR measurements discussed above (including those under optical excitation) were done at a temperature $T = 4$ K. It has been noticed that at this temperature the spin polarization is nearly 100% even if very low optical pumping power in the range of microwatts is used. This observation suggests long spin-lattice relaxation time T_1. Indeed, after switching off the optical excitation the relaxation time of the optically induced spin polarization towards thermal-equilibrium population of the $m_s = 0, \pm 1$ sublevels is determined by the spin-lattice relaxation time T_1, which can be measured by the decay of EPR signal. The time resolution in the EPR measurement of the order of tenths of a second has been sufficient to measure the spin-lattice relaxation times at $T = 4$ K. Measured values are in the range $T_1 \sim 1.5$–4 s for different configurations of the NV$^-$ center in 6H-SiC [13], $T_1 = 17$ s in 3C-SiC [140], and T_1 is of the order of minute in 4H-SiC [138]. The spin coherence time of the NV$^-$ centers is also expected to be long and the first measurements in 4H-SiC indicate $T_2 \approx 30$ μs at $T = 90$ K [137].

The temperature dependence of the fine-structure parameter D has proven useful for local temperature measurements based on the NV$^-$ center in diamond [145]. The temperature dependence $D(T)$ has been measured for the NV$^-$ center in 4H-SiC [135,137] and 3C-SiC [140]. The change of D in the temperature range 50–300 K in 4H-SiC is ~ 15 MHz, twice as large as in diamond. Similar shift in D (by 12 MHz in the range 4–360 K) has been measured in 3C-SiC [140].

Let us now consider the formation of the NV center, which inevitably embarks also on the divacancy formation. Theoretical investigations [9,131,62,133] predict that the negative charge state NV$^-$ in 4H-SiC is stable for the Fermi level in the range ~ 0.3–1.7 eV below the CBM. However, due to the smaller band gap of 3C-SiC the NV$^-$ defect is stable in n-type material, i.e., for Fermi level in the upper half of the band gap, because the (2–/–) charge transition level is degenerate with the CBM in this polytype. These calculations find also rather high formation energies of both divacancies and NV center. However, as noted in Ref. [62] and indicated in Ref. [146], the binding energies of both NV and V_CV_{Si} defects are also high, about 3–4 eV. Therefore, both defects may form during PVT growth at high temperatures (~2200 °C) at which vacancies can be created and highly mobile Si vacancies will find C vacancies and N donors to form divacancies and NV center. The divacancy has been observed in commercial *n*-type 6H-SiC substrates grown by PVT [55] but not in 3C-SiC epilayers grown by CVD on Si substrate at lower temperatures (typically below 1400°C).

It should be noted, however, that divacancies can be observed in as-grown and irradiated material, but NV has only been observed in irradiated/implanted material, to the best of our knowledge. There are two possible reasons for the lack of NV⁻ luminescence from ensemble in as-grown 4H-SiC: (i) the N concentration is too low and even if silicon vacancies are created during growth their concentration is not high enough to combine with N_C and provide sufficient concentration of NV pairs; (ii) if the N concentration is high, the material is *n*-type with Fermi level pinned at the N-donors (ionization energies ~ 61–125 meV), which leads to NV pairs in double-negative charge state NV²⁻ instead of NV⁻. Thus, the observation of the NV⁻ center in 4H- and 6H-SiC requires that the donors are at least partly compensated [136], which is the case in irradiated or implanted material.

The formation of NV centers in 4H-SiC has been studied in Ref. [147]. The authors have irradiated with different ions (H, Si, N, and I) two types of substrates HPSI 4H-SiC with low N doping of 3×10^{15} cm⁻³ (HPSI-4H) and highly N-doped, 9×10^{18} cm⁻³ (N-4H). The implantation doses and ion energies are chosen so that the created vacancies within the first ~ 1.5 μm beneath the sample surface have about the same concentrations. The samples are annealed to allow the vacancies to diffuse and then the efficiency of creation of $V_C V_{Si}{}^0$ and NV⁻ pairs is estimated by the intensity of the corresponding luminescence. The authors present the following findings. (i) The implanted N is less efficient than the N present in the sample in building the NV centers, possibly because of insufficient incorporation of N

TABLE 5

Line positions of the N-V pair emission in 3C-, 4H-, and 6H-SiC. The configuration of $N_C V_{Si}$ for each line is also given, cf. Figure 2. Note that the lines are enumerated in order of increasing energy, which differs from some notations in the literature

Line (configuration)	Peak position in nm (eV)
3C-SiC	
NV (*kk*)	1289 (0.9616)[a]
4H-SiC[b]	
NV1 (*hk*, basal)	1242.8 (0.9973)
NV2 (*kk*, axial)	1223.2 (1013.3)
NV3 (*hh*, axial)	1180.0 (1050.4)
NV4 (*kh*, basal)	1176.4 (1053.6)
6H-SiC[c]	
NV1 (hk_1, basal)	1241.1 (0.9987)
NV2 (k_2k_1, axial)	1226.5 (1.0106)
NV3 (k_2k_2, basal)	1203.0 (1.0303)
NV4 (k_1h, basal)	1182.5 (1.0482)
NV5 (*hh*, axial)	~1181.6 (~1.0490)
NV6 (k_1k_2, axial)	1153.6 (1.0745)

[a] According to Ref. [140].
[b] The configurations are taken from Ref. [138].
[c] The configurations are taken from Ref. [13].

at C sites to form N_C during the annealing. This is evident from the much stronger intensity of the NV⁻ emission in the N-4H than in the HPSI-4H sample for any kind of ions implanted. (ii) It is also evident that the implanted N ions do contribute to creation of NV centers because the N-implanted HPSI-4H samples show much stronger NV⁻ luminescence than those HPSI-4H samples that are implanted with other ions. (iii) In the N-4H samples where the role of the ion irradiation is mainly to create vacancies and mostly the contained N atoms contribute to formation of NV, it is noted that heavier ions, which produce more cascade damage (not only vacancies V_C and V_{Si}, but also more interstitial C atoms) promote higher NV concentration. (iv) The creation of NV is much more efficient than that of $V_C V_{Si}$, since no $V_C V_{Si}$ PL and only NV PL is observed in the implanted and annealed N-4H samples. This observation agrees with the lower formation energy of NV compared to $V_C V_{Si}$ centers [9,131,62]. Thus, this work provides useful information for the development of correct engineering approaches for controllable NV creation and its stabilization in the negative charge state. Similar considerations are expected to be applicable also for the 6H polytype although systematic study is not available.

Controllable creation of single NV⁻ emitters by N-ion implantation has been demonstrated, showing room-temperature photostable single-photon sources [142,143]. By optimizing the implanted conditions, sixfold increase in the concentration of the NV⁻ center ensemble could be achieved [143]. Coherent control of spin ensembles at low temperatures [142] and at room temperature [143] by Rabi and Ramsey oscillations has been demonstrated, showing the coherence time (T_2 ~17 μs [143]) comparable to that of the divacancy and Si vacancy ensembles.

5 Unidentified Defects with Emission in the Near Infrared

In this section we provide a brief account on several defects, which have been observed in the infrared PL spectra of SiC but have not been positively identified yet.

In HPSI 4H-SiC with strong divacancy emission, two additional ZPLs, PL5 (at 1041.9 nm) and PL6 (1037.7 nm) have also been observed [4]. Using ODMR, it is shown in Ref. [4] that PL5 and PL6 are associated with spin-1 centers, similar to the divacancy. In a later theoretical work [148], these lines have been proposed to be due to the neutral divacancy at or in the vicinity of stacking faults. However, further work is needed to confirm or refute this assignment. Ref. [148] proposes also that other lines are also related to neutral divacancies near stacking faults, the so-called PL5′ and PL6′ lines at 1047.3 and 1042.6 nm, respectively [60]. A recent study has observed an ODMR center in 4H-SiC at room temperature with similar properties as the neutral divacancy and labeled as PL7 although the corresponding ZPL has not been identified [149]. In 6H-SiC, three divacancy-like centers, QL7, QL8, and QL9, were also detected by ODMR but their corresponding ZPLs have not been observed either [63].

It should be mentioned that the PL5–PL6 and PL5′–PL6′ lines are much weaker than the bulk divacancy lines (PL1–PL4) and they do not exhibit any quenching for excitation photon energies for which the divacancy lines do quench. Their weaker intensity is interpreted as a sign that they stem from near-surface stacking faults, which have probably been introduced by polishing [148]. Confocal PL scans seem to confirm this idea, and nonspecular X-ray diffraction done on the 4H-SiC samples seems to confirm the presence of 6H-SiC inclusions (one possible type of stacking fault is based on the 6H-SiC stacking sequence). The fact that these lines (PL5–PL6 and PL5′–PL6′) do not quench for laser excitation for which the divacancy quenches is associated with the lowering of the CBM at and near the stacking fault (by up to ~ 0.24 eV). For instance, a laser excitation of, e.g., 1.2 eV is not enough to re-ionize bulk divacancies, which have captured electrons and been transferred to the negative "dark" charge state, but is enough to re-ionize divacancies at stacking faults, thus keeping the emission of neutral divacancies stable.

Another defect with near-infrared sharp emission lines is observed in heavily implanted (with Ge or Sn ions) and annealed 4H-SiC [150]. Three sharp lines, DI_1 at 1002.8 nm, DI_2 at 1004.7 nm, and DI_3 at 1006.1 nm are observed after annealing at 1700–1800°C. Their co-appearance suggests that they are due to the same intrinsic defect (most probably formed of vacancies and antisites [150]) but their microscopic structure remains unknown.

Another family of emission bands peaking in the infrared region between ~ 1100 and 1300 nm is reported in Ref. [151]. The bands remain broad (linewidth between ~ 130 and 200 nm) even at cryogenic temperatures, therefore they are associated with recombination at stacking faults, without a specific microscopic model. However, single photon emission from these defects has been readily observed in this work [151] and the emitters are bright (of the order of several hundred kcounts/s).

Finally, we comment on a defect previously known as UD3 [117] with a single sharp ZPL emission at 914.5 nm. This defect has been commented as due to tantalum (Ta) in Ref. [60], because strong emission of the ZPL of UD-3 has been observed in the 4H-SiC sample intentionally doped with Ta, although no further investigations are carried out. The ZPL is very close to that of V_{Si} (the V2 line, 916.5 nm), but the structure of the phonon sideband (PSB) is completely different: the UD-3 defect has much weaker PSB and, therefore, presumably high DW factor. Its applicability for quantum technology remains to be explored.

References

1. M. Atatüre, D. Englund, N. Vamivakas, S-Y. Lee, and J. Wrachtrup, "Material platforms for spin-based photonic quantum technologies", Nat. Rev. Mater. 3, 38–51 (2018).
2. A. Delteil, Z. Sun, W.-B. Gao, E. Togan, S. Faelt, and A. Imamoglu, "Generation of heralded entanglement between distant hole spins", Nat. Phys. 12, 218–223 (2016).
3. R.J. Warbuton, "Single spins in self-assembled quantum dots", Nat. Mater. 12, 483–493 (2013).
4. W. F. Koehl, B. B. Buckley, F. J. Heremans, G. Calusine, and D. D. Awschalom, "Room temperature coherent control of defect spin qubits in silicon carbide", Nature 479, 84–87 (2011).
5. D.D. Awschalom, R. Hanson, J. Wrachtrup, and B.B. Zhou, "Quantum technologies with optically interfaced solid-state spins", Nat. Photon. 12, 516–527 (2018).
6. B. Hensen, H. Bernien, A. E. Dréau, A. Reiserer, N. Kalb, M. S. Blok, J. Ruitenberg, R. F. L. Vermeulen, R. N. Schouten, C. Abellán, W. Amaya, V. Pruneri, M. W. Mitchell, M. Markham, D. J. Twitchen, D. Elkouss, S. Wehner, T. H. Taminiau, and R. Hanson, "Loophole- free Bell inequality violation using electron spins separated by 1.3 kilometres", Nature 256, 682–686 (2015).
7. A. Tchebotareva, S. L. N. Hermans, P. C. Humphreys, D. Voigt, P. J. Harmsma, L. K. Cheng, A. L. Verlaan, N. Dijkhuizen, W. de Jong, A. Dréau, and R. Hanson, "Entanglement between a Diamond Spin Qubit and a Photonic Time-Bin Qubit at Telecom Wavelength", Phys. Rev. Lett. 123, 064–601 (2019).
8. J.S. Pelc, C. Langrock, Q. Zhang, and M. M. Fejer, " Influence of domain disorder on parametric noise in quasi-phase-matched quantum frequency converters", Opt. Lett. 35, 2804–2806 (2010).

9. J. R. Weber, W. F. Koehl, J. B. Varley, A. Janotti, B. B. Buckley, C. G. Van de Walle, and D. D. Awschalom, "Quantum computing with defects", Proc. Natl. Acad. Sci. USA 107, 8513–8518 (2010).

10. D. J. Christle, A.L. Falk, P. Andrich, P.V. Klimov, J.U. Hassan, N.T. Son, E. Janzén, T. Ohshima, D.D. Awschalom, "Isolated electron spins in silicon carbide with millisecond coherence times", Nat. Mater. 14, 160–163 (2015).

11. M. Widmann, S.-Y. Lee, T. Rendler, N. T. Son, H. Fedder, S. Paik, L.-P. Yang, N. Zhao, S. Yang, I. Booker, A. Denisenko, M. Jamali, S. A. Momenzadeh, I. Gerhardt, T. Ohshima, A. Gali, E. Janzén, and J. Wrachtrup, "Coherent control of single spins in silicon carbide at room temperature", Nat. Mater. 14, 164–168 (2015).

12. N. T. Son, C. P. Anderson, A. Bourassa, K. C. Miao, C. Babin, M. Widmann, M. Niethammer, J. Ul Hassan, N. Morioka, I. G. Ivanov, F. Kaiser, J. Wrachtrup, and D. D. Awschalom, "Developing silicon carbide for quantum spintronics", Appl. Phys. Lett. 116, 190501 (2020).

13. Kh. Khazen, H. J. von Bardeleben, S. A. Zargaleh, J. L. Cantin, M. Zhao, W. Gao, T. Biktagirov, and U. Gerstmann, "High-resolution resonant excitation of NV centers in 6H-SiC: A matrix for quantum technology applications", Phys. Reb B 100, 205202 (2019).

14. H. Itoh, M. Yoshikawa, I. Nashiyama, S. Misawa, H. Okumura, and S. Yoshida, "Radiation induced defects in CVD-grown 3C-SiC" IEEE Trans. Nucl. Sci. 37, 1732–1738 (1990).

15. T. Wimbauer, B. K. Meyer, A. Hofstaetter, A. Scharmann, and H. Overhof, "Negatively charged Si vacancy in 4H SiC: A comparison between theory and experiment", Phys. Rev. B 56, 7384 (1997).

16. H. J. von Bardeleben, J. L. Cantin, and I. Vickridge, "Proton-implantation-induced defects in n-type 6H- and 4H-SiC: An electron paramagnetic resonance study", Phys. Rev. B 62, 10126–10134 (2000).

17. E. Sörman, N. T. Son, W. M. Chen, O. Kordina, C. Hallin, and E. Janzén, "Silicon vacancy related defect in 4H and 6H SiC", Phys. Rev. B 61, 2613–2620 (2000).

18. N. Mizuochi, S. Yamasaki, H. Takizawa, N. Morishita, T. Ohshima, H. Itoh, and J. Isoya, "Continuous-wave and pulsed EPR study of the negatively charged silicon vacancy with S= 3/2 and C_{3v} symmetry in n-type 4H-SiC", Phys. Rev. B 66, 235202 (2002).

19. N. Mizuochi, S. Yamasaki, H. Takizawa, N. Morishita, T. Ohshima, H. Itoh, and J. Isoya, "EPR studies of the isolated negatively charged silicon vacancies in n-type 4H- and 6H-SiC: Identification of C_{3v} symmetry and silicon sites", Phys. Rev. B 68, 165206 (2003).

20. H. Kraus, V. A. Soltamov, D. Riedel, S. Väth, F. Fuchs, A. Sperlich, P. G. Baranov, V. Dyakonov, and G. V. Astakhov, "Room-temperature quantum microwave emitters based on spin defects in silicon carbide", Nat. Phys. 10, 157 (2014).

21. V. Ivády, J. Davidsson, N. T. Son, T. Ohshima, I. A. Abrikosov, and A. Gali, "Identification of Si-vacancy related room-temperature qubits in 4H silicon carbide", Phys. Rev. B 96, 161114(R) (2017).

22. J. Davidsson, V. Ivády, R. Armiento, T. Ohshima, N. T. Son, A. Gali, and I. A. Abrikosov, "Identification of divacancy and silicon vacancy qubits in 6H-SiC", Appl. Phys, Lett. 114, 112107 (2019).

23. N. T. Son, P. Stenberg, V. Jokubavicius, T. Ohshima, J. Ul Hassan, and I. G. Ivanov, "Ligand hyperfine interactions at silicon vacancies in 4H-SiC", J. Phys.: Condens. Matter 31, 195501 (2019).

24. A. Csóré, N.T. Son, and A. Gali, "Towards identification of silicon vacancy-related electron paramagnetic resonance centers in -SiC", Phys. Rev. B 104, 035207 (2021).

25. K. Szász, V. Ivády, I.A. Abrikosov, E. Janzén, M. Bockstedte, and A. Gali, "Spin and photophysics of carbon-antisite vacancy defect in 4H silicon carbide: A potential quantum bit", Phys. Rev. B 91, 121201(R) (2015).

26. A. Ellison, B. Magnusson, N.T. Son, L. Storasta, and E. Janzén, "HTCVD grown semi-insulating SiC substrates", Mater. Sci. Forum 433–436, 33 (2003).

27. N.T. Son, P. Carlsson, J. ul Hassan, B. Magnusson, and E. Janzén, "Defects and carrier compensation in semi-insulating 4H-SiC substrates", Phys. Rev. B 75, 155204 (2007).

28. H. Itoh, A. Kawasuso, T. Ohshima, M. Yoshikawa, I. Nashiyama, S. Tanigawa, S. Misawa, H. Okomura, and S. Yoshida, "Intrinsic defects in cubic silicon carbide", Phys. Stat. Sol. (a) 162, 173 (1997).

29. C. P. Anderson, E. O. Glen, C. Zeledon, A. Bourassa, Y. Jin, Y. Zhu, C. Vorwerk, A. L. Crook, H. Abe, J. Ul-Hassan, T. Ohshima, N. T. Son, G. Galli, and D. D. Awschalom, "Five-second coherence of a single spin with single-shot readout in silicon carbide", Sci. Adv. 8, eabm5912 (2022).

30. E. Janzén, A. Gali, P. Carlsson, A. Gällström, B. Magnusson and N. T. Son, "The silicon vacancy in SiC", Physica B 404, 4354 (2009).

31. R. Nagy, M. Widmann, M. Niethammer, D. B. R. Dasari, I. Gerhardt, Ö. O. Soykal, M. Radulaski, T. Ohshima, J. Vučković, N. T. Son, I. G. Ivanov, S. E. Economou, C. Bonato, S.-Y. Lee, and J. Wrachtrup, "Quantum properties of dichroic silicon vacancies in silicon carbide", Phys. Rev. Applied 9, 034022 (2018).

32. P. Udvarhelyi, G. Thiering, N. Morioka, C. Babin, F. Kaiser, D. Lukin, T. Ohshima, J. Ul-Hassan, N. T. Son, J. Vuckovic, J. Wrachtrup, and A. Gali, "Vibronic States and Their Effect on the Temperature and Strain Dependence of Silicon-Vacancy Qubits in 4H-SiC", Phys. Rev. Applied 13, 054017 (2020).

33. A. Anisimov, D. Simin, V. Soltamov, S. Lebedev, P. Baranov, G. Astakhov, and V. Dyakonov, "Optical thermometry based on level anticrossing in silicon carbide", Sci. Rep. 6, 33301 (2016).

34. P. G. Baranov, A. P. Bundakova, A. A. Soltamova, S.B. Orlinskii, I.V. Borovykh, R. Zondervan, R. Verberk, and J. Schmidt, "Silicon vacancy in SiC as a promising quantum system for single-defect and single-photon spectroscopy", Phys. Rev. B 83, 125203 (2011).

35. V. A. Soltamov, A. A. Soltamova, P. G. Baranov, and I. I. Proskuryakov, "Room Temperature Coherent Spin Alignment of Silicon Vacancies in 4H- and 6H-SiC", Phys. Rev. Lett. 108, 226402 (2012).

36. D. Riedel, F. Fuchs, H. Kraus, S. Väth, A. Sperlich, V. Dyakonov, A. A. Soltamova, P. G. Baranov, V. A. Ilyin, and G.V. Astakhov, "Resonant addressing and manipulation of silicon vacancy qubits in silicon carbide", Phys. Rev. Lett. 109, 226402 (2012).

37. H. Kraus, V. A. Soltamov, F. Fuchs, D. Simin, A. Sperlich, P. G. Baranov, G. V. Astakhov, and V. Dyakonov, "Magnetic field and temperature sensing with atomic-scale spin defects in silicon carbide", Sci. Rep. 4, 5303 (2014).

38. D. Simin, F. Fuchs, H. Kraus, A. Sperlich, P. G. Baranov, G. V. Astakhov, and V. Dyakonov, "High-precision angle-resolved magnetometry with uniaxial quantum centers in silicon carbide", Phys. Rev. Applied 4, 014009 (2015).

39. S.-Y. Lee, M. Niethammer, and J. Wrachtrup, "Vector magnetometry based on S=3/2 electronic spins", Phys. Rev. B 92, 115201 (2015).

40. M. Niethammer, M. Widmann, S.-Y. Lee, P. Stenberg, O. Kordina, T. Ohshima, N. T. Son, E. Janzén, J. Wrachtrup, "Vector Magnetometry Using Silicon Vacancies in -SiC Under Ambient Conditions", Phys. Rev. Applied 6, 034001 (2016).

41. C. J. Cochrane, J. Blacksberg, M. A. Anders, and P. M. Lenahan, "Vectorized magnetometer for space applications using electrical readout of atomic scale defects in silicon carbide", Sci. Rep. 6, 37077 (2016).

42. M. Niethammer, M. Widmann, T. Rendler, N. Morioka, Y.-C. Chen, R. Stöhr, J. Ul Hassan, S. Onoda, T. Ohshima, S.-Y. Lee, A. Mukherjee, J. Isoya, N. T. Son, J. Wrachtrup, "Coherent electrical readout of defect spins in silicon carbide by photo-ionization at ambient conditions", Nat. Commun. 10, 5569 (2019).

43. Ö. O. Soykal, P. Dev, and S. E. Economou, "Silicon vacancy center in 4H-SiC: Electronic structure and spin-photon interfaces", Phys. Rev. B 93, 081207(R) (2016).

44. R. Nagy, M. Niethammer, M. Widmann, Y.-C. Chen, P. Udvarhelyi, C. Bonato, J. Ul Hassan, R. Karhu, I. G. Ivanov, N. T. Son, J. R. Maze, T. Ohshima, Ö. Soykal, Á. Gali, S.-Y. Lee, F. Kaiser, and J. Wrachtrup, "High-fidelity spin and optical control of single silicon-vacancy centres in silicon carbide", Nat. Commun. 10, 1954 (2019).

45. D. O. Brachera, X. Zhang, and E. L. Hu, "Selective Purcell enhancement of two closely linked zero-phonon transitions of a silicon carbide color center", Proc. Natl Acad. Sci. USA 114, 4060 (2017).

46. D. M. Lukin, C. Dory, M. A. Guidry, K. Y. Yang, S. Deb Mishra, R.Trivedi, M. Radulaski, S. Sun, D. Vercruysse, G. H. Ahn, and J. Vučković, "4H-silicon-carbide-on-insulator for integrated quantum and nonlinear photonics", Nat. Photonics 14, 330 (2020).

47. N.T. Son and I.G. Ivanov, "Charge state control of the silicon vacancy and divacancy in silicon carbide", J. Appl. Phys. 129, 215702 (2021).

48. H.B. Banks, Ö.O. Soykal, R.L. Myers-Ward, D.K. Gaskill, T.L. Reinecke, and S.G. Carter, "Resonant Optical Spin Initialization and Readout of Single Silicon Vacancies in 4H-SiC", Phys. Rev. Applied 11, 024013 (2019).

49. M. Widmann, M. Niethammer, D. Yu. Fedyanin, I.A. Khramtsov, I.D. Booker, J. Ul Hassan, S. Lasse, T. Rendler, R. Nagy, N. Morioka, I.G. Ivanov, N.T. Son, T. Ohshima, M. Bockstedte, A. Gali, C. Bonato, S.-Y. Lee, and J. Wrachtrup, "Electrical charge state manipulation of single silicon vacancies in a silicon carbide quantum optoelectronic device", Nano Lett. 19, 7173 (2019).

50. M.E. Bathen, A. Galeckas, J. Müting, H.M. Ayedh, U. Grossner, J. Coutinho, Y.K. Frodason, and L. Vines, "Electrical charge state identification and control for the silicon vacancy in 4H-SiC", npj Quantum Information 5:111 (2019).

51. N. Morioka, C. Babin, R. Nagy, I. Gediz, E. Hesselmeier, D. Liu, M. Joliffe, M. Niethammer, D. Dasari, V. Vorobyov, R. Kolesov, R. Stöhr, J. Ul-Hassan, N.T. Son, T. Ohshima, P. Udvarhelyi, G. Thiering, A. Gali, J. Wrachtrup, and F. Kaiser, "Spin-controlled generation of indistinguishable and distinguishable photons from silicon vacancy centres in silicon carbide", Nat. Commun. 11, 2516 (2020).

52. D.M. Lukin, A.D. White, R. Trivedi, M.A. Guidry, N. Morioka, C. Babin, Öney O. Soykal, J. Ul-Hassan, N.T. Son, T. Ohshima, P.K. Vasireddy, M.H. Nasr, S. Sun, J.-P.W. MacLean, C. Dory, E.A. Nanni, J. Wrachtrup, F. Kaiser and J. Vučković, "Spectrally reconfigurable quantum emitters enabled by optimized fast modulation", npj Quantum Information 6:80 (2020).

53. C. Babin, R. Stöhr, N. Morioka, T. Linkewitz, T. Steidl, R. Wörnle, D. Liu, E. Hesselmeier, V. Vorobyov, A. Denisenko, M. Hentschel, C. Gobert, P. Berwian, G. V. Astakhov, W. Knolle, S. Majety, P. Saha, M. Radulaski, N.T. Son, J. Ul-Hassan, F. Kaiser, and J. Wrachtrup, "Fabrication and nanophotonic waveguide integration of silicon carbide colour centres with preserved spin-optical coherence", Nat. Mater. 21, 67–73 (2022).

54. V.S. Vainer, and V.A. Il'in, "Electron spin resonance of exchange-coupled vacancy pairs in hexagonal silicon carbide", Sov. Phys. Solid State 23, 2126 (1981).

55. N.T. Son, P.N. Hai, Mt. Wagner, W.M. Chen, A. Ellison, C. Hallin, B. Monemar, and E. Janzén, "Optically detected magnetic resonance studies of intrinsic defects in 6H-SiC", Semicond. Sci. Technol. 14, 1141 (1999).

56. Th. Lingner, S. Greulich-Weber, J.-M. Spaeth, U. Gerstmann, E. Rauls, Z. Hajnal, Th. Frauenheim, and H. Overhof, "Structure of the silicon vacancy in 6H-SiC after annealing identified as the carbon vacancy-carbon antisite pair", Phys. Rev. B 64, 245212 (2001).

57. P.G. Baranov, I.V. Il'in, E.N. Mokhov, M.V. Muzafarova, S.B. Orlinskii, and J. Schmidt, "EPR identification of the triplet ground state and photoinduced population inversion for a Si-C divacancy in silicon carbide", JETP Lett. 82, 441 (2005).

58. N.T. Son, P. Carlsson, J. Ul Hassan, E. Janzén, T. Umeda, J. Isoya, A. Gali, M. Bockstedte, N. Morishita, T. Ohshima and H. Itoh, "Divacancy in 4H-SiC", Phys. Rev. Lett. 96, 055501 (2006).

59. N.T. Son, T. Umeda, J. Isoya, A. Gali, M. Bockstedte, B. Magnusson, A. Ellison, N. Morishita, T. Ohshima, H. Itoh, and E. Janzén, "Divacancy model for P6/P7 centers in 4H-and 6H-SiC", Mater. Sci. Forum 527–529, 527 (2006).

60. B. Magnusson, N. T. Son, A. Csore, A. Gällström, T. Ohshima, A. Gali, I.G. Ivanov, "Excitation properties of the divacancy in -SiC", Phys. Rev. B 98, 195202 (2018).

61. J.R. Jenny, D.P. Malta, M.R. Calus, St.G. Müller, A.R. Powell, V.F. Tsvetkov, H. McD. Hobgood, R.C. Glass, and C.H. Carter, Jr., "Development of large diameter high-purity semi-insulating 4H-SiC wafers for microwave devices", Mater. Sci. Forum 457–460, 35 (2004).

62. L. Gordon, A. Janotti, and C.G. Van de Walle, "Defects as qubits in 3C- and 4H-SiC", Phys. Rev. B 92, 045208 (2015).

63. A. Falk, B.B. Buckley, G. Calusine, W.F. Koehl, V. V. Dobrovitski, A. Politi, C.A. Zorman, P.X.-L. Feng, and D.D. Awschalom, "Polytype control of spin qubits in silicon carbide", Nat. Commun. 4, 1819 (2013).

64. N.T. Son, E. Sörman, W. M. Chen, C. Hallin, O. Kordina, B. Monemar, E. Janzén, and J.L. Lindström, "Optically detected magnetic resonance studies of defects in electron-irradiated 3C SiC layers", Phys. Rev. B 55, 2863 (1997).

65. V.Y. Bratus, R.S. Melnik, S.M. Okulov, V.N. Rodionov, B.D. Shanina, and M.I. Smoliy, "A new spin one defect in cubic SiC", Physica A (Amsterdam) 404**B**, 4739 (2009).

66. D. J. Christle, A.L. Falk, P. Andrich, P.V. Klimov, J.U. Hassan, N.T. Son, E. Janzén, T. Ohshima, D.D. Awschalom, "Isolated electron spins in silicon carbide with millisecond coherence times", Nat. Mater. 14, 160 (2015).

67. D.J. Christle, P.V. Klimov, C.F. de las Casas, K. Szász, V. Ivády, V. Jokubavicius, J. Ul Hassan, M. Syväjärvi, W. F. Koehl, T. Ohshima, N. T. Son, E. Janzén, A. Gali, D.D. Awschalom, "Isolated spin qubits in SiC with a high-fidelity infrared spin-to-photon interface", Phys. Rev. X 7, 021046 (2017).

68. A. Batalov, V. Jacques, F. Kaiser, P. Siyushev, P. Neumann, L. J. Rogers, R. L. McMurtrie, N.B. Manson, F. Jelezko, and J. Wrachtrup, "Low Temperature Studies of the Excited-State Structure of Negatively Charged Nitrogen-Vacancy Color Centers in Diamond", Phys. Rev. Lett. 102, 195506 (2009).

69. L. Crook, C.P. Anderson, K.C. Miao, A. Bourassa, H. Lee, S.L. Bayliss, D. O. Bracher, X. Zhang, H. Abe, T. Ohshima, E.L. Hu, and D.D. Awschalom, "Purcell enhancement of a single silicon carbide color center with coherent spin control", Nano Lett. 20, 3427 (2020).

70. G. Wolfowicz, C.P. Anderson, A.L. Yeats, S. J. Whiteley, J. Niklas, O.G. Poluektov, F.J. Heremans, and D.D. Awschalom, "Optical charge state control of spin defects in 4H-SiC", Nat. Commun. 8, 1876 (2017).

71. D.A. Golter and C.W. Lai, "Optical switching of defect charge states in 4H-SiC", Sci. Rep. 7, 13406 (2017).

72. F. de las Casas, D.J. Christle, J. Ul Hassan, T. Ohshima, N.T. Son, and D.D. Awschalom, "Stark tuning and electrical charge state control of single divacancies in silicon carbide", Appl. Phys. Lett. 111, 262403 (2017).

73. P. Anderson, A. Bourassa, K. C. Miao, G. Wolfowicz, P. J. Mintun, A. L. Crook, H. Abe, J. Ul Hassan, N. T. Son, T. Ohshima, D. D. Awschalom, "Electrical and optical control of single spins integrated in scalable semiconductor devices", Science 366, 1225 (2019).

74. A. Bourassa, C.P. Anderson, K.C. Miao, M. Onizhuk, H. Ma, A.L. Crook, H. Abe, J. Ul-Hassan, T. Ohshima, N.T. Son, G. Galli, and D.D. Awschalom, "Entanglement and control of single nuclear spins in isotopically engineered silicon carbide", Nature Materials 19, 1319 (2020).

75. H. McD. Hobgood, R.C. Glass, G. Augustine, R.H. Hopkins, J. Jenny, M. Skowronski, W.C. Mitchel and M. Roth, "Semi-insulating 6H–SiC grown by physical vapor transport", Appl. Phys. Lett. 66, 1364 (1995).

76. Y. Koshka, M. Mazzola, S. Yingquan and C.U. Pittman, Jr., "Vanadium doping of 4H SiC from a solid source: Photoluminescence investigation", J. Electron. Mater. 30, 220 (2001).

77. B. Krishnan, S. Kotamraju, R. Venkatesh, K.G. Thirumalai and Y. Koshka, "Vanadium doping using VCl4 source during the chloro-carbon epitaxial growth of 4H-SiC", J. Cryst. Growth 321, 8 (2011).

78. R. Karhu, E.Ö. Sveinbjörnsson, B. Magnusson, I.G. Ivanov, Ö. Danielsson and J. Ul Hassan, "CVD growth and properties of on-axis vanadium doped semi-insulating 4H-SiC epilayers", J. Appl. Phys. 125, 045702 (2019).

79. J.R. Jenny, J. Skowronski, W.C. Mitchel, H.M. Hobgood, R.C. Glass, G. Augustine and R.H. Hopkins, "Deep level transient spectroscopic and Hall effect investigation of the position of the vanadium acceptor level in 4H and 6H SiC", Appl. Phys. Lett. 68, 1963 (1996).

80. J. Schneider, H.D. Müller, K. Mayer, W. Wilkening, F. Fuchs, A. Dörnen, S. Leibenzeder and R. Stein, "Infrared spectra and electron spin resonance of vanadium deep level impurities in silicon carbide", Appl. Phys. Lett. 56, 1184 (1990).

81. K. Maier, J. Schneider, W. Wilkening, S. Leibenzeder and R. Stein, "Electron spin resonance studies of transition metal deep level impurities in SiC", Mater. Sci. Eng. **B**11, 27 (1992).

82. M.E. Zvanut, V.V. Konovalov, H. Wang, W.C. Mitchel, W.D. Mitchel and G. Landis, "Defect levels and types of point defects in high-purity and vanadium-doped semi-insulating 4H-SiC", J. Appl. Phys. 96, 5484 (2004).

83. M. Kunzer, H.D. Müller and U. Kaufmann, "Magnetic circular dichroism and site-selective optically detected magnetic resonance of the deep amphoteric vanadium impurity in 6H-SiC", Phys. Rev. B 48, 10846 (1993).

84. J. Reinke, H. Weihrich, S. Greulich-Weber, and J.-M. Spaeth, "Magnetic circular dichroism of a vanadium impurity in 6H-silicon carbide", Semicond. Sci. Technol. 8, 1862 (1993).

85. J. Reinke, S. Greulich-Weber, and J.-M. Spaeth, "Optically detected electron paramagnetic resonance of a vanadium impurity in 6H-silicon carbide", Solid State Commun. 85, 1017 (1993).

86. R. Yakimova, I.G. Ivanov, L. Vines, M.K. Linnarsson, A. Gällström, F. Giannazzo, F. Roccaforte, P. Wellmann, M. Syväjärvi and V. Jokubavicius, "Growth, Defects and Doping of 3C-SiC on Hexagonal Polytypes", ECS J. Solid State Sci. Technol. 6, P741 (2017).

87. K.F. Dombrowski, U. Kaufmann, M. Kunzer, K. Maier, J. Schneider, V.B. Shields and M.G. Spencer, "Deep donor state of vanadium in cubic silicon carbide (3C-SiC)", Appl. Phys. Lett. 65, 1811 (1994).

88. K.F. Dombrowski, U. Kaufmann, M. Kunzer, K. Maier, J. Schneider, V.B. Shields and M.G. Spencer, "Identification of the neutral V^{4+} impurity in cubic 3C-SiC by electron-spin resonance and optically detected magnetic resonance", Phys. Rev. B 50, 18034 (1994).

89. B. Kaufmann, A. Dörnen and F.S. Ham, "Crystal-field model of vanadium in 6H silicon carbide", Phys. Rev. B 55, 13009 (1997).

90. L. Spindlberger, A. Csóré, G. Thiering, S. Putz, R. Karhu., J. Ul Hassan, N.T. Son, T. Fromherz, A. Gali and M. Trupke, "Optical Properties of Vanadium in 4H Silicon Carbide for Quantum Technology", Phys. Rev. Appl. 12, 014015 (2019).

91. G. F. Koster, J. O. Dimmock, R. G. Wheeler, and H. Statz, *Properties of the Thirty-Two Point Groups*, MIT Press, Cambridge, MA, 104 pp, 1963.

92. G. Wolfowicz, C.P. Anderson, B. Diler, O.G. Poluektov, F. J. Heremans, D.D. Awschalom, "Vanadium spin qubits as telecom quantum emitters in silicon carbide", Sci. Adv. 6: eaaz1192 (2020).

93. J. Baur, M. Kunzer, and J. Schneider, "Transition metals in SiC polytypes, as studied by magnetic resonance techniques", phys. stat. sol. (a) 162, 153 (1997), and references therein.

94. J. R. Jenny, M. Skowronski, W.C. Mitchel, H.M. Hobgood, R.C. Glass, G. Augustine, and R.H. Hopkins, "On the compensation mechanism in high-resistivity 6H–SiC doped with vanadium", J. Appl. Phys. 78, 3839 (1995).

95. W. C. Mitchel, R. Perrin, J. Goldstein, A. Saxler, M. Roth, S. R. Smith, J. S. Solomon, and A. O. Evwaraye, "Fermi level control and deep levels in semi-insulating 4H–SiC", J. Appl. Phys. 89, 5040 (1999).

96. N. Achtziger and W. Withuhn, "Band gap states of Ti, V, and Cr in 4H–silicon carbide", Appl. Phys. Lett. 71, 110 (1997).

97. N. Achtziger and W. Withuhn, "Band-gap states of Ti, V, and Cr in 4H-SiC: Identification and characterization by elemental transmutation of radioactive isotopes", Phys. Rev. B 57, 12181 (1998).

98. N. T. Son, P. Carlsson, A. Gällström, B. Magnusson, and E. Janzén, "Deep levels and carrier compensation in V-doped semi-insulating 4H-SiC", Appl. Phys. Lett. 91, 202111 (2007).

99. W.C. Mitchel, W.D. Mitchel, G. Landis, H.E. Smith, W. Lee and M.E. Zvanut, "Vanadium donor and acceptor levels in semi-insulating 4H- and 6H-SiC", J. Appl. Phys 101, 013707 (2007)

100. P.G. Baranov, V.A. Khramtsov, and E.N. Mokhov, "Chromium in silicon carbide: electron paramagnetic resonance studies", Semicond. Sci. Technol. 9, 1340 (1994).

101. N.T. Son, A. Ellison, B. Magnusson, M.F. MacMillan, W.M. Chen, B. Monemar, and E. Janzén, "Photoluminescence and Zeeman effect in chromium-doped 4H and 6H SiC", J. Appl. Phys. 86, 4348 (1999).

102. K.F. Dombrowski, M. Kunzer, U. Kaufmann, and J. Schneider, P.G. Baranov, and E.N. Mokhov, "Identification of molybdenum in 6H-SiC by magnetic resonance techniques", Phys. Rev. B 54, 7323 (1996).

103. J.M. Langer and H. Heinrich, "Deep-level impurities: A possible guide to prediction of band-edge discontinuities in semiconductor heterojunctions", Phys. Rev. Lett. 55, 1414 (1985).

104. W.F. Koehl, B. Diler, S. J. Whiteley, A. Bourassa1, N.T. Son, E. Janzén, and D.D. Awschalom, "Resonant optical spectroscopy and coherent control of Cr^{4+} spin ensembles in SiC and GaN", Phys. Rev. B 95, 035207 (2017).

105. B. Diler, S.J. Whiteley, C. P. Anderson, G. Wolfowicz, M.E. Wesson, E.S. Bielejec, F. Joseph Heremans, and D.D. Awschalom, "Coherent control and high-fidelity readout of chromium ions in commercial silicon carbide", npj Quantum Inf. 6:11 (2020).

106. N.T. Son, X.T. Trinh, A. Gällström, S. Leone, O. Kordina, E. Janzén, K. Szász, V. Ivády and A. Gali, "Electron paramagnetic resonance and theoretical studies of Nb in 4H-and 6H-SiC", J. Appl. Phys. 112, 083711 (2012).

107. A. Gällström, B. Magnusson, S. Leone, O. Kordina, N.T. Son, V. Ivády, A. Gali, I.A. Abrikosov, E. Janzén and I.G. Ivanov, "Optical properties and Zeeman spectroscopy of niobium in silicon carbide", Phys. Rev. B 92, 075207 (2015).

108. V. Ivády, A. Gällström, N.T. Son, E. Janzén and A. Gali, "Asymmetric split-vacancy defects in SiC polytypes: a combined theoretical and electron spin resonance study", Phys. Rev. Lett. 107, 195501 (2011).

109. J. Baur, M. Kunzer and J, Schneider, "Transition methods in SiC polytypes, as studied by magnetic resonance techniques", phys. stat. sol. (a) 162, 153–172 (1997).

110. T. Bosma, G.J.J. Lof, C.M. Gilardoni, O.V. Zwier, F. Hendriks, B. Magnusson, A. Ellison, A. Gällström, I.G. Ivanov, N.T. Son, R.W.A. Havenith and C.H. van der Wal, "Identification and tunable optical coherent control of transition-metal spins in silicon carbide", npj Quantum Inf. 4:48 (2018).

111. K.F. Dombrowski, M. Kunzer, U. Kaufmann, and J. Schneider, P.G. Baranov, and E.N. Mokhov, "Identification of molybdenum in 6H-SiC by magnetic resonance techniques", Phys. Rev. B 54, 7323 (1996).

112. J. Baur, M. Kunzer, K.F. Dombrowski, U. Kaufmann, J. Schneider, P.G. Baranov, E.N. Mokhov, "Electrically and optically active molybdenum impurities in commercial SiC substrates", Mater. Sci. Eng. B 46, 313 (1997).

113. Gällström, A. "Optical characterization of deep level defects in SiC", PhD dissertation, No. 1674, ISSN 0345-7524, Linköping University (2015).

114. A. Csóré, A. Gällström, E. Janzén, and A. Gali, "Investigation of Mo defects in 4H-SiC by means of density functional theory", Mater. Sci. Forum 858, 261 (2016).

115. C.M. Gilardoni, T. Bosma, D. van Hien, F. Hendriks, B. Magnusson, A. Ellison, I.G. Ivanov, N.T. Son and C.H. van der Wal, "Spin-relaxation times exceeding seconds for color centers with strong spin–orbit coupling in SiC", New J. Phys. 22, 103051 (2020).

116. A. Gällström, B. Magnusson and E. Janzén, "Optical Identification of Mo Related Deep Level Defect in 4H and 6H-SiC", Mater. Sci. Forum 483–485, 405–408 (2009).

117. B. Magnusson and E. Janzén, "Optical characterization of deep level defects in SiC", Mater. Sci. Forum 483–485, 341–346 (2005).

118. N. Achtziger, G. Pasold, R. Sielemann, C. Hülsen, J. Grillenberger, and W. Witthuhn, "Tungsten in silicon carbide: Band-gap states and their polytype dependence", Phys. Rev. B 62, 12888 (2000).

119. F.C. Beyer, C.G. Hemmingsson, A. Gällström, S. Leone, H. Pedersen, A. Henry and E. Janzén, "Deep levels in tungsten doped n-type 3C-SiC", Appl. Phys. Lett. 98, 152104 (2011).

120. K. Irmscher, I. Pintilie, L. Pintilie, D. Schulz, "Deep level defects in sublimation-grown 6H silicon carbide investigated by DLTS and EPR", Physica B 308–310, 730–733 (2001).

121. K. Irmscher, "Electrical properties of SiC: characterisation of bulk crystals and epilayers", Mat. Sci. Eng. B91–92, 358–366 (2002).

122. A. Gällström, B. Magnusson, F.C. Beyer, A. Gali, N.T. Son, S. Leone, I.G. Ivanov, C.G. Hemmingsson, A. Henry and E. Janzén, "Optical identification and electronic configuration of tungsten in 4H- and 6H-SiC", Physica B 407, 1462–1466 (2012).

123. B. Magnusson, A. Ellison and E. Janzén, "Properties of the UD-1 Deep-Level Center in 4H-SiC", Mater. Sci. Forum 389–393, 505–508 (2002).

124. Desurvire, "The golden age of optical fiber amplifiers", Physics Today 47, 20 (1994), and references therein.

125. W.J. Choyke, R.P. Devaty, L.L. Clemen, M. Yoganathan, G. Pensl, and Ch. Hässler, "Intense erbium-1.54-μm photoluminescence from 2 to 525 K in ion-implanted 4H, 6H, 15R, and 3C SiC", Appl. Phys. Lett. 65, 1668 (1994).

126. M. Yoganathan, W.J. Choyke, R.P. Devaty, G. Pensl, J.A. Edmond, "1.54 μm photoluminescence and electroluminescence in erbium implanted 6H SiC", Mater. Res. Soc. Sym. Proc. 422, 339 (1996).

127. N. Tabassum, V. Nikas, A. E. Kaloyeros, V. Kaushik, E. Crawford, M. Huang, and S. Gallis, "Engineered telecom emission and controlled positioning of Er3+ enabled by SiC nanophotonic structures", Nanophotonics 9, 1425 (2020).

128. P.G. Baranov, I.V. Ilyin and E.N. Mokhov, "Electron paramagnetic resonance of erbium in bulk silicon carbide crystals", Solid Stat. Commun. 103, 291 (1997).

129. P.G. Baranov, I.V. Ilyin, E.N. Mokhov, A.B. Pevtsov, and V.A. Khramtsov, "Erbium in silicon carbide crystals: EPR and high-temperature luminescence", Phys. Solid Stat. 41, 32–34 (1999).

130. G. Pasold, F. Albrecht, J. Grillenberger, U. Grossner, C. Hülsen, W. Witthuhn, and R. Sielemann, "Erbium-related band gap states in 4H–and 6H–silicon carbide", J. Appl. Phys. 93, 2289–2291 (2003).

131. J.R. Weber, W.F. Koehl, J.B. Varley, A. Janotti, B.B. Buckley, C.G. Van de Walle and D.D. Awschalom, "Defects in SiC for quantum computing", J. Appl. Phys. 109, 102417 (2011).

132. A. Csóré and A. Gali, "Density functional theory on NV center in 4H SiC", Mater. Sci. Forum 897, 269–274 (2017).

133. Csóré, H.J. von Bardeleben, J.L. Cantin and A. Gali, "Characterization and formation of NV centers in 3C-, 4H- and 6H-SiC: An *ab initio* study", Phys. Rev. B 96, 085204 (2017).

134. Gali, A. Gällström, N.T. Son and E. Janzén, "Theory of neutral divacancy in SiC: a defect for spintronics", Mater. Sci. Forum 645–648, 395–397 (2010).

135. H.J. von Bardeleben, J.L. Cantin, E. Rauls and U. Gerstmann, "Identification and magneto-optical properties of the NV center in 4H-SiC", Phys. Rev. B 92, 064104 (2015).

136. H.J. von Bardeleben, J.L. Cantin, A. Csóré, A. Gali, E. Rauls and U. Gerstmann, "NV centers in 3C, 4H and 6H silicon carbide: A variable platform for solid-state qubits and nanosensors", Phys. Rev. B 94, 121202(R) (2016).

137. H.J. von Bardeleben, J.L. Cantin, "NV centers in silicon carbide: from theoretical predictions to experimental observation", MRS Commun. 7, 591–594 (2017).

138. S.A. Zargaleh, H.J. von Bardeleben, J.L. Cantin, U. Gerstmann, S. Hameau, B. Eblé and Weibo Gao, "Electron paramagnetic resonance tagged high-resolution excitation spectroscopy of NV-centers in 4H-SiC", Phys. Rev. B 98, 214113 (2018).

139. S.A. Zargaleh, S. Hameau, B. Eblé, F. Margaillan, H.J. von Bardeleben, J.L. Cantin and Weibo Gao, "Nitrogen vacancy center in cubic silicon carbide: A promising qubit in the 1.5 μm spectral range for photonic quantum networks", Phys. Rev. B 98, 165203 (2018).

140. H.J. von Bardeleben, J.L. Cantin, U. Gerstmann, W.G. Schmidt and T. Biktagirov, "Spin Polarization, Electron–Phonon Coupling, and Zero-Phonon Line of the NV Center in 3C-SiC", Nano Lett. 21, 8119–8125 (2021).

141. S.A. Zargaleh, B. Eble, S. Hameau, J.-L. Cantin, L. Legrand, M. Bernard, F. Margaillan, J.-S. Lauret, J.-F. Roch, H. J. von Bardeleben, E. Rauls, U. Gerstmann and F. Treussart, "Evidence for near-infrared photoluminescence of nitrogen vacancy centers in 4H-SiC", Phys. Rev. B 94, 060102(R) (2016).

142. Z. Mu, S.A. Zargaleh, H.J. von Bardeleben, J.E. Fröch, M. Nonahal, H. Cai, X. Yang, X. Li, I. Aharanovich and W. Gao, "Coherent manipulation with resonant excitation and single emitter creation of nitrogen vacancy centers in 4H silicon carbide", Nano Lett. 20, 6142–6147 (2020).

143. J.-F. Wang, F.-F. Yan, Q. Li, Zh.-H. Liu, He Liu, G.-P. Guo, L.-P. Guo, X. Zhou, J.-M. Cui, J. Wang, Z.-Q. Zhou, X.-Y. Xu, J.-S. Xu, Ch.-F. Li and G.-C. Guo, "Coherent control of nitrogen-vacancy center spins in silicon carbide at room temperature", Phys. Rev. Lett. 124, 223601 (2020).

144. G.D. Cheng, Y.P. Wan and S.Y. Yan, "Optical and spin coherence properties of NV center in diamond and 3C-SiC", Comp. Mater. Sci. 154, 60–64 (2018).

145. M. W. Doherty, V. M. Acosta, A. Jarmola, M. S. J. Barson, N. B. Manson, D. Budker, and L. C. L. Hollenberg, "Temperature shifts of the resonances of the NV$^-$ center in diamond", Phys. Rev. B 90, 041201(R) (2014).

146. L. Torpo, T.E.M. Staab and R.M. Nieminen, "Divacancy in 3C and 4H-SiC: An extremely stable defect", Phys. Rev. B 65, 085202 (2002).

147. Sh.-I. Sato, T. Narahara, Y. Abe, Y. Hijikata, T. Umeda and T. Ohshima, "Formation of nitrogen-vacancy centers in 4H-SiC and their near infrared photoluminescence properties", J. Appl. Phys. 126, 083105 (2019).

148. V. Ivády, J. Davidsson, N. Delegan, A.L. Falk, P.V. Klimov, S.J. Whiteley, S.O. Hruszkewycz, M.V. Holt, F.J. Heremans, N.T. Son, D.D. Awschalom, I.A. Abrikosov, and A. Gali, "Stabilization of point-defect spin qubits by quantum wells", Nat. Commun. 10, 5607 (2019).

149. Q. Li, J.-F. Wang, F.-F. Yan, J.-Y. Zhou, H.-F. Wang, H. Liu, L.-P. Guo, X. Zhou, A. Gali, Z.-H. Liu, Z.-Q. Wang, K. Sun, G.-P. Guo, J.-S. Tang, H. Li, L.-X. You, J.-S. Xu, C.-F. Li, and G.-C. Guo, "Room temperature coherent manipulation of single-spin qubits in silicon carbide with a high readout contrast", Natl. Sci. Review, nwab122 (2021). https://doi.org/10.1093/nsr/nwab122

150. T. Kobayashi, M. Rühl, J. Lehmeyer, L.K.S. Zimmermann, M. Krieger and H.B. Weber, "Intrinsic color centers in 4H-silicon carbide formed by heavy ion implantation and annealing", J. Phys. D: Appl. Phys. 55, 105303 (2022).

151. J. Wang, Yu Zhou, Z. Wang, A. Rasmita, J. Yang, X. Li, H.J. von Bardeleben and W. Gao, "Bright room temperature single photon source at telecom range in cubic silicon carbide", Nat. Commun. 9, 4106 (2018).

11

SiC Substrate and its Epitaxial Layers' Analysis by Spectroscopic Ellipsometry

ChangCai Cui

National and Local Joint Engineering Research Center for Intelligent Manufacturing Technology of Brittle Material Products, Huaqiao University, Xiamen, China

HuiHui Li

Institute of Manufacturing Engineering, Huaqiao University, Xiamen, China

1 Introduction

The thickness and optical properties of film layers are important indexes of semiconductor devices. The film layers are significantly thinner (e.g., down to monolayer thickness) for optoelectronic applications. The usual methods for determining thickness (calipers, micrometers, yardsticks, etc.) are ineffective for films thinner than about one micrometer and the commonly used transmission electron microscope (TEM) for semiconductors is destructive. Optical measurement technology may be an alternative option for thickness measurement, which plays an important role in various fields of scientific research and manufacturing as a possible precise and nondestructive method. For the optical properties, optical measurement technology may be the only selection, which includes the commonly used light absorption method, differential reflectance spectroscopy (DRS), reflection electron energy loss spectroscope (REELS) and spectroscopic ellipsometry (SE), and so on [1-2]. Compared with other optical measurement techniques, SE can achieve higher accuracy in a non-destructive way, which is suitable for not only studying the optical properties of materials but also for obtaining geometrical characteristics, such as the thickness of films.

Spectroscopic ellipsometer is used to obtain certain parameters (Psi/Delta or Mueller matrix) by measuring the changes of polarized light, which is reflected upon or transmitted through a sample over a wide spectral range. Then, the optical properties of materials, which are associated with bandgap, free carrier concentrations and phonon modes, etc. can be obtained from the data analysis [1]. The thickness of the film sample can also be deduced. At the same time, SE can be used to obtain more information such as material composition, interface layer properties, surface roughness, etc. [2-3]. In addition, SE can be used to study birefringence, anisotropy, or some optical phenomena caused by strain. Therefore, SE is of great significance for obtaining comprehensive information on optical, electrical, mechanical, structural (multilayer structure), and chemical properties of various materials [4-5].

Silicon carbide (SiC), as the third-generation semiconductor material, has a series of features, such as large band gap, high thermal conductivity, excellent chemical inertia, and

DOI: 10.1201/9780429198540-14

mechanical robustness [6]. There is a variety of crystal types in SiC crystal, among which the polytypes of 4H- and 6H-SiC are of great interest [7]. 4H-SiC has been demonstrated to be a promising nonlinear optical material, which has been successfully used to fabricate tunable mid-infrared lasers by phase-matched difference-frequency generation [8]. Optoelectronic devices like wireless optical sensors, optical waveguide, photonic crystals, UV photodiodes, etc. have been successfully fabricated based on 6H-SiC substrate. The performance of the devices depends on the quality of the substrate and their epilayers generated by homoepitaxy and heteroepitaxy techniques. Therefore, it is very important to precisely characterize their key index such as optical properties and thickness of a film device.

In terms of the advantages of SE, the optical properties of SiC substrates and the film thickness of their epilayers can be precisely obtained in a non-destructive way [9-12]. In particular, the ultra-thin damage layer of SiC substrate caused by machining can be evaluated by SE technology, too.

In this chapter, the basic principle of spectroscopic ellipsometry is introduced first. Then, the SE measurement theory and data analysis procedures for 4H- and 6H-SiC materials are overviewed. Finally, the SE analysis for homoepitaxy and heteroepitaxy layers (SiC epitaxy layers on substrates of SiC, Si, ZrO_2, MgO, etc. or epitaxy layers of GaN, AlN and graphene, etc. on SiC substrate) in normal environment are reviewed together with some cases under extreme temperatures, e.g., GaN layer at 10 K and 295 K.

2 Background of Spectroscopic Ellipsometry

2.1 Basic Theory of Spectroscopic Ellipsometry

Spectroscopic ellipsometry is a very powerful optical technique for investigating the optical functions and the thickness of thin films, etc. simultaneously. The mathematical theory for ellipsometry analysis is based on the Fresnel reflection or transmission equations for polarized light encountering boundaries in planar multi-layered materials. These come from solutions to Maxwell's equations. In general, the ellipsometer measures the change of polarized light upon reflection or transmission on a sample and compares it with an optical stack model [13]. The general schematic of an ellipsometer can be seen in Figure 3.1 of [15]. The general configuration mainly consists of light source, polarization state generator (PSG), polarization state analyzer (PSA), and detector.

As indicated, the incident beam and the normal to the sample surface define a plane, which is called the incident plane. θ_0 is the angle of the incident light with the normal. In the SE measurement, two parameters Psi (Ψ) and Delta (Δ) are revealed from the expression of complex reflectance ratio ρ:

$$\rho \equiv \tan(\Psi)e^{i\Delta} \equiv r_p / r_s \equiv \left(E_{rp}/E_{ip}\right)/\left(E_{rs}/E_{is}\right) \qquad (1)$$

where, Ψ and Δ are the amplitude ratio and phase difference of reflected p- to s-polarized light, respectively; i_p and i_s are the incident coefficients of p- and s-polarized lights; r_p and r_s are the reflection coefficients of p- and s-polarized lights; E_p and E_s are the electric fields for p- and s-polarized lights, respectively.

The field amplitudes are usually calculated using Jones vectors and matrices in properly chosen coordinate systems. In order to describe the interaction of p-waves and s-waves with the sample, a 2×2 Jones matrix is usually used to describe an isotropic sample, or any optical element that may alter the light polarization, there is no cross-polarization between p-waves and s-waves. For anisotropic samples, the off-diagonal elements of Jones matrix are complex numbers. Therefore, the Jones description is limited to polarized light and is not able to describe partially polarized or unpolarized light. As a comparison, an alternative convenient scheme uses Stokes vectors and Mueller matrices [14].

For the Stokes-Mueller description, an isotropic sample will have off-diagonal 2×2 blocks that are zero (because of no cross-polarization between p- and s-waves):

$$\begin{bmatrix} S_0 \\ S_1 \\ S_2 \\ S_3 \end{bmatrix}_{out} = m_{11} \begin{pmatrix} 1 & -N & 0 & 0 \\ -N & 0 & 0 & 0 \\ 0 & 1 & C & S \\ 0 & i & -S & C \end{pmatrix} \begin{bmatrix} S_0 \\ S_1 \\ S_2 \\ S_3 \end{bmatrix}_{in} \tag{2}$$

Where $N = \cos(2\Psi)$, $C = \sin(2\Psi)\cos(\Delta)$, $S = \sin(2\Psi)\sin(\Delta)$. m_{11} is the reflected intensity of unpolarized light.

If the sample does not depolarize the incident light beam, a Mueller-Jones matrix can be defined as

$$M = A(J \otimes J^*)A^{-1} \tag{3}$$

where the matrix A is given by

$$A = \frac{1}{\sqrt{2}} \begin{pmatrix} 1 & 0 & 0 & 1 \\ 1 & 0 & 0 & -1 \\ 0 & 1 & 1 & 0 \\ 0 & i & -i & 0 \end{pmatrix} \tag{4}$$

where the symbol $\otimes$ indicates the Kronecker product of the sample Jones matrix J and its complex conjugate J^*. If the sample is anisotropic, the sample Mueller matrix in the $N/C/S$ representation (Eq. 2) can be found in Brosseau [16] and Tompkins [14]. Moreover, only when one of the optical axis and periodic structure is parallel or perpendicular to the plane of incidence, the off-diagonal Mueller matrix (MM) elements are zero, and the MM measurement cannot provide more additional information about the optical or structural anisotropic samples than conventional ellipsometry.

When the measured Mueller matrices are not Mueller-Jones matrices and show some level of depolarization, there are various methods of quantifying the extent to which a Mueller matrix depolarizes light [17]. To investigate the depolarization features of a medium, it is useful to use the polarization fraction DI of a Mueller matrix (m_{ij} is an element of Mueller matrix) that was introduced by Gil and Bernabeu [18]:

$$DI = \frac{\sqrt{\left(\sum_{ij} m_{ij}^2\right) - m_{11}^2}}{\sqrt{3}m_{11}} \tag{5}$$

A value of $DI = 1$ indicates a non-depolarizing Mueller matrix, whereas a value of $DI = 0$ means completely depolarizing.

2.2 Analysis Strategies of Spectroscopic Ellipsometry

2.2.1 Forward Modeling and Reverse Fitting

The purpose of ellipsometry characterization is to transform the measured quantities into sample properties, such as optical functions and film thickness. For a bare substrate, the transformation is straightforward. The measured values of Ψ and Δ for each wavelength related to the complex reflectance ratio ρ can be directly inverted to give the complex optical functions:

$$[\tilde{\varepsilon}] = [\tilde{n}^2] = \sin^2(\theta_0)\left[1 + \tan^2(\theta_0)\left(\frac{1-\rho}{1+\rho}\right)\right] \tag{6}$$

The optical functions with triangle brackets represent "pseudo" optical functions. The amplitude ratio Ψ is characterized by the refractive index n, while Δ represents light absorption described by the extinction coefficient k as described later. For one film, the Ψ and Δ cannot be directly transformed into sample properties, and a more complex analysis is used, instead of simply plugging values into an equation and solving algebraically.

The typical analysis method is model-based regression using the data measured with a reflection mode. It needs to construct an optical stack model of the representative samples to extract significant physical and optical parameters from raw reflection SE data, which usually accounts for a number of distinct layers with individual optical dispersion. This includes a few steps, as illustrated in Figure 1 of [30].

In general, there are four steps as follows. Step 1: the experimental data Ψ and Δ are measured by an ellipsometer. Step 2: the sample is described by an optical model. Step 3: the ellipsometric experimental spectra are fitted to model calculations by adjusting any unknown sample properties. Step 4: the optical constants and thickness are deduced.

Generally, the optimum wavelength range for reflection SE to determine both thickness and refractive index accurately is from 50 nm to 2 μm. On the one hand, if the layer is ultra-thin (<10 nm), it is difficult to determine the index of refraction. Researchers proposed some methods to solve this problem to a certain extent, for instance, combining reflectivity to provide more information [19], solving the polynomial equation to provide an analytical method [20], using specific wavelength and angle to provide higher data sensitivity [21]. On the other hand, if the films are several tens of microns thick, the spectral features come very close together and would be affected by small nonidealities from either sample properties or measurement. Several methods have been developed for a general formalism describing incoherent reflection or transmission by thick isotropic/anisotropic multilayers or substrates [22-23].

In order to know how well the model fits the data, error-based merit, such as the reduced χ^2 or MSE is given by [4]

$$\chi^2 = \frac{1}{NM - m - 1}\sum_{i=1}^{M}\sum_{j=1}^{N}\frac{\left(\rho_{j.exp}(\lambda_i) - \rho_{j.calc}(\lambda_i)\right)^2}{\delta_{\rho_j}(\lambda_i)^2} \tag{7}$$

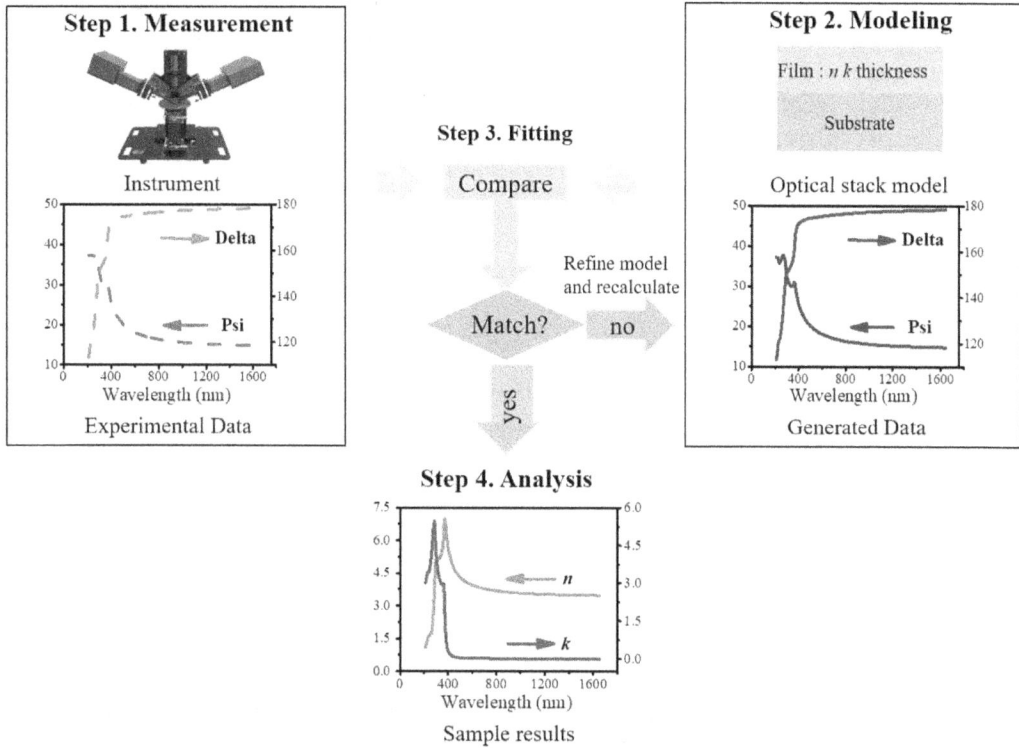

FIGURE 1

The typical flowchart for SE data analysis. From [30] Figure 9, reproduced with permission of Elsevier.

$$MSE = \frac{1}{NM - m - 1} \sum_{i=1}^{M} \sum_{j=1}^{N} \left(\rho_{j.exp} (\lambda_i) - \rho_{j.calc} (\lambda_i) \right)^2 \qquad (8)$$

where, $\rho_{j.exp} (\lambda_i)$, $\rho_{j.calc} (\lambda_i)$, and $\delta_{\rho_j} (\lambda_i)$ represent the experimental, calculated, and error quantities at wavelength λ_i and data set j, N is the total number of data points, m is the number of fitted parameters, and M is the number of data sets. The reduced χ^2 in the simulation is often minimized using a Levenberg-Marquardt regression algorithm.

For Mueller matrix ellipsometry, additional analysis strategies can also be chosen: Mueller matrix decomposition and Mueller matrix transformation methods. Both can extract more information from Mueller matrix than Jones matrix.

2.2.2 Matrix Decomposition

Mueller matrix decomposition including Lu-Chipman [24], differential [25], Cloude [26] methods aims to decompose a matrix into product or sum of multiple sub-matrices. Each sub-matrix represents a purely optical effect. Lu-Chipman, a polar decomposition, assumes that the order of polarization effects dichroism, phase retardation, and generalized depolarization, which is widely utilized in the field of biomedical diagnosis. Differential decomposition does not consider the order of occurrence of the polarization effects in the sample but assumes that the polarization properties of the sample are uniform along the depth of the light. Accordingly, it deals with the polarization parameter changing with the thickness of the sample. Cloude decomposition called spectral decomposition, decomposes Mueller matrix into no more than four non-depolarization matrices, especially applicable to the depolarization sample. In principle, there is not a universal Mueller matrix decomposition method and each method is only applicable to its corresponding optical path model.

2.2.3 Matrix Transformation

Mueller matrix transformation, including rotation, mirror, and reciprocal transformation, aims to extract an invariant value incorporated in the Mueller matrix [27]. The rotation transformation calculates the rotation invariant of the azimuth angle based on matrix transformation theory, eliminating the influence of the selected reference system on measurement results. The azimuth angle of mirror symmetry can be solved by mirror transformation of the Mueller matrix, when only one anisotropy exists in the sample. Reciprocal transformation aiming to handle multiple anisotropic superposition situations, and it can determine the order of optical properties, which is a help to Mueller decomposition.

2.2.4 Combined with Other Methods

If necessary, SE can be combined with other thin film measurement technologies to evaluate the test results of multiple platforms at the same time to ensure the accuracy of the measurement results. Ellipsometry is often used in conjunction with TEM, XRD, AFM, Raman, and other test methods. Figure 2 is a statistical chart of the "cooperation" between ellipsometry and other characterization techniques.

(a) Thin films   (a) Semiconductor films

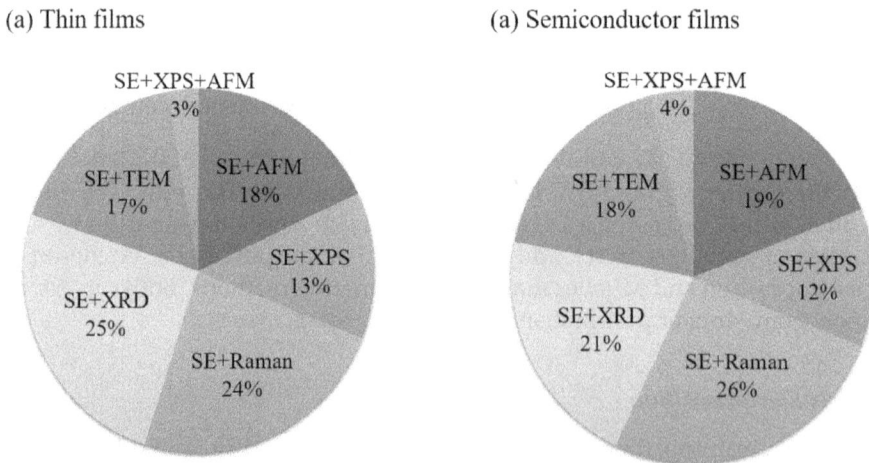

FIGURE 2
Distribution of techniques corroborating spectroscopic ellipsometry for analysis of (a) thin films and (b) nanostructures (AFM: atomic force microscopy; SEM: scanning electron microscopy; TEM: transmission electron microscopy; XPS: X-ray photoelectron spectroscopy; XRD: X-ray diffraction).

2.3 Typical Applications of Spectroscopic Ellipsometry

Ellipsometry has been widely used in physics, chemistry, material science, microelectronics, thin film technology, surface, and interface technology. With the development of ellipsometry technology, more and more applications have been developed in scientific research and industry monitoring. At the same time, the measurement objects of ellipsometry have also been expanded from solid samples to liquid samples, which has promoted its applications in biological sciences and medicine.

The traditional application areas of SE include semiconductor devices, optical coatings, data storage, flat panel display, etc. [12]. In recent years, with the development of nano processing technology, more and more requirements for high device performance, that is, pursuing the flexibility and miniaturization of devices and further evolving to wearable devices, make researchers pay more attention to two-dimensional materials and photonic crystals. Ellipsometric parameters can be used to determine the number of atomic layers of two-dimensional graphene [28] and MoS_2 [29], and further analysis can be used to obtain their optical constants and other photoelectric properties.

In addition, it is important to recognize that many photoelectric semiconductor materials are not unique stable entities, such as perovskite thin films [30]. This requires not only the measurement of the steady-state nanostructures, but also to the measurement of their interaction with the environment, in various atmospheres and conditions, and in real-time during processes [31-32].

2.4 Development of Spectroscopic Ellipsometry

As the diversity and complexity of ellipsometric applications continue to increase, increasing ellipsometric data accuracy is required [33]. Measurement accuracy is generally influenced by the instrument's systematic and random errors [10-11]. Most commercial instruments meet 0.5% accuracy and Aspnes suggests a 0.1% best target for ellipsometric accuracy [35-36]. There are some methods to reduce measurement error, for example, making optimization measurement configuration to improve resistance ability to error [37], evaluating and calibrating systematic error by comparing well-known standard samples like silicon substrate and air [38-39], increasing signal to noise ratio by incorporating a high-quality light source and detector in the system, which can improve random error significantly.

In addition, ellipsometry is an indirect measurement method, and its accuracy also depends on the consistency of the sample optical model established for ellipsometry parameter analysis. Therefore, selecting an appropriate optical dispersion model for the material under the current test band and environmental conditions is also to ensure the accuracy of fitting results. If necessary, SE can be combined with other measurement technologies, such as AFM, TEM, XRD, Raman, etc. to ensure the accuracy of the fitting results. Expanding the wavelength measurement range and combining other measurement techniques to broaden the field of ellipsometry to reveal more characteristics of materials is another development direction of ellipsometry [40].

This part only gives a brief introduction and an outlook about SE. Books on the ellipsometry topic include the classic 1977 work of Azzam and Bashara [41], which has been followed by, among others, a collection of monographs edited by Irene and Tompkins [14], and the work by Tompkins and Hilfiker [15] and Fujiwara [13]. More details can be found in those books.

3 Ellipsometric Analysis of Anisotropic SiC

3.1 Reflection and Transmission of Light by Bulk SiC

Figure 3 illustrates two optical models for an ambient / (anisotropic substrate) / ambient structure with and without backside reflection. The light reflection and transmission at the ambient/film interface are described below.

N_0 and N_2 show the complex refractive indices of the ambient. The anisotropic substrate is assumed to be uniaxial material (4H- or 6H-SiC) with the complex refractive index of N_1 (N_{1x}, N_{1y}, N_{1z}), where $N_{1x} = N_{1y}$. The incident beam and the normal to the surface define a plane called the "plane of incidence", which is perpendicular to the surface of the sample. It is assumed that the principal axes of the uniaxial substrate are parallel to the (x, y, z) coordinates. Figure 3(a) shows a simple structure (semi-infinite substrate), where only considering a single interface between the first medium (ambient) and the second medium (uniaxial substrate). The Fresnel reflection coefficients for p-waves and s-waves are given by

$$r_{01ss} = \frac{N_{1x}N_{1z}\cos\theta_0 - N_0(N_{1z}^2 - N_0^2\sin^2\theta_0)^{1/2}}{N_{1x}N_{1z}\cos\theta_0 + N_0(N_{1z}^2 - N_0^2\sin^2\theta_0)^{1/2}} \tag{9}$$

$$r_{01pp} = \frac{N_0\cos\theta_0 - (N_{1y}^2 - N_0^2\sin^2\theta_0)^{1/2}}{N_0\cos\theta_0 + (N_{1y}^2 - N_0^2\sin^2\theta_0)^{1/2}} \tag{10}$$

For samples with a non-negligible backside reflection, the transmitted waves are separated into p- and s-polarizations, as shown in Figure 3(b). The light transmission at the ambient/interface can be expressed as follows,

$$t_{01pp} = \frac{2N_0\cos\theta_0[N_0^2\sin^2\theta_0\left(N_{1x}^2 - N_{1z}^2\right) + N_z^4)^{1/2}}{N_z[N_zN_x\cos\theta_0 + N_0\left(N_{1z}^2 - N_0^2\sin^2\theta_0)^{1/2}\right]} \tag{11}$$

$$t_{01ss} = \frac{2N_0\cos\theta_0}{N_0\cos\theta_0 + (N_{1y}^2 - N_0^2\sin^2\theta_0)^{1/2}} \tag{12}$$

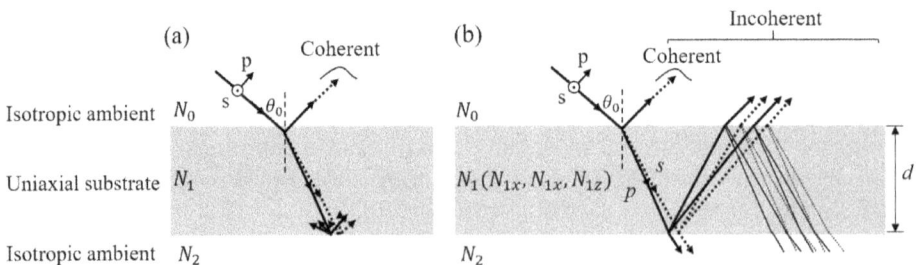

FIGURE 3

Optical model for an ambient/(uniaxial substrate)/ambient structure (a) without backside reflection. (b) With backside reflection. The principal axes of the uniaxial substrate are assumed to be parallel to the (x, y, z) coordinates. From [45] Figure 1, reproduced with permission of AIP.

When the thickness of the specimen approaches the coherence length of light, the super-position of outgoing waves is no longer fully coherent. Therefore, the partially coherent or incoherent description needs to be taken into consideration for the varying phase [22]. Particularly when the anisotropic samples are not oriented in specific directions, the data analysis is more complicated, and a method referred to as the 4 × 4 matrix needs to be used. Several methods have been developed for a general formalism describing incoherent reflection or transmission by thick isotropic/anisotropic multilayers or substrates [23] [42-44]. These methods have applications in transparent or partially transparent systems involving thick or moderately thick layers. In the case of SiC wafers, a partial coherent treatment was considered in Refs. [42] [45], of which the weak absorption was ignored in Ref. [42]. The calculation procedure program of the coherence and partial coherence Mueller matrix method from Nichols can be found in Ref. [46]. The complete derivation of the transfer matrix method, partial wave method, and methods to deal with partial coherence was explained in Ref. [46].

3.2 Determination of Optical Functions from Reflection Ellipsometry

SE is a very useful tool in measuring the optical functions of bulk materials, and it is the best available technique to measure the bulk optical functions of a semiconductor or insulator with photon energies greater than their band edges [14]. The optical function means complex refractive index $\tilde{n} = n + ik$ (n is the refractive index and k is extinction coefficient) or, equivalently, complex dielectric constant $\varepsilon_r = \varepsilon_1 + i\varepsilon_2$. The direct relationship between the complex dielectric function and optical constants, is $\varepsilon_1 = n^2 - k^2$, $\varepsilon_2 = 2nk$ in the linear regime. The dielectric tensors are diagonal as $\varepsilon = diag\left(\varepsilon_x, \varepsilon_y, \varepsilon_z\right) = diag\left(\tilde{n}_x^2, \tilde{n}_y^2, \tilde{n}_z^2\right)$. For uniaxial crystals such as α-SiC, the non-zero matrix elements of the dielectric tensor are $\varepsilon_x = \varepsilon_y = \varepsilon_o$ and $\varepsilon_z = \varepsilon_e$. The two independent quantities, ε_o and ε_e, determine the propagation of linearly polarized light with electric field vector (**E**) perpendicular and parallel to the c-axis, respectively, and are known as the ordinary and extraordinary dielectric functions (DFs).

The refractive index of SiC is easy to fit for photon energies below the band edge by using Sellmeier or Cauchy oscillator model. Most of reported ellipsometric characterizations in SiC wafers have been done in configurations without backside reflection by some pre-treatment, for example, roughening the backside surface. In this case, the analysis can be simplified because the crystal is treated as a semi-infinite substrate and only light reflected from the surface reaches the detector. This allows for a fully coherent treatment, and the resulting optical properties are usually fitted by an oscillator model.

Usually, considering the high absorption above the bandgap, the very weak anisotropic extinction coefficients of SiC below the bandgap due to the doping cannot be detected. The ordinary and the extraordinary refractive index of SiC can be seen in [47] Figure 9.

The absorption caused by doping below the bandgap is often ignored because the spectral dependence of k_o and k_e (extinction coefficients of ordinary and extraordinary component of the dielectric function, respectively) are unknown and they are too weak to fit. Kildemo [48] studied the uniaxial dielectric function below the bandgap of double-sided polished 4H-, 6H-, and 15R-SiC. The ordinary component of the dielectric function is extracted from phase-modulated spectroscopic ellipsometric measurements, including mathematical removal of the overlayer. The extinction coefficient was obtained by spectroscopic ellipsometry in transmission mode. Transmission measurements were performed at a 0° angle of incidence for s-polarized light (T_s) and at a 70° angle of incidence for p-polarized

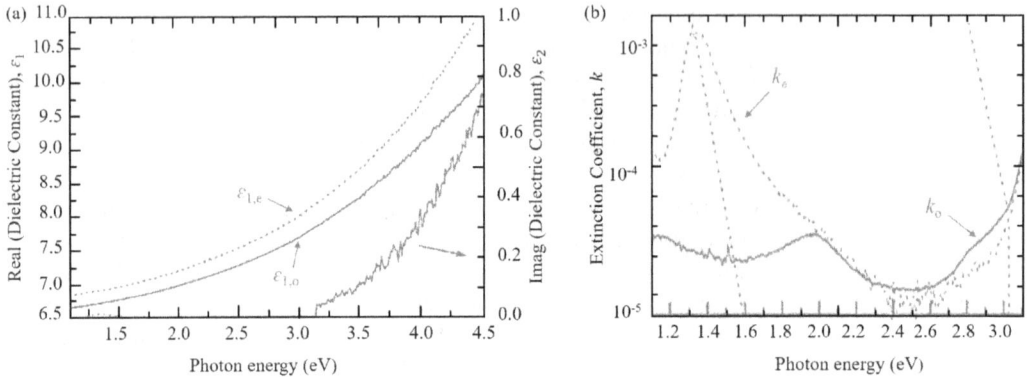

FIGURE 4

(a) The extracted complex dielectric functions for 6H-SiC. (b) The extinction coefficients k_o and k_e from T_s measurements at normal incidence and T_p measurements at 70° incidence. From [48] Figures 3 and 9, reproduced with permission of AIP.

light (T_p). After having determined k_o from the T_s measurements, the extraordinary component (k_e) was extracted from the T_p measurements. Figure 4 shows the complex dielectric functions of 6H-SiC. It should be noted that the calculated birefringence and extinction coefficients for the extraordinary component are approximate results.

Li et al. used the partial wave theory combining the point-by-point method for the data analysis of double-sided polished n-doped SiC substrates [45]. Their optical functions below the bandgap were determined. The optical axis of SiC substrate deviates from the surface normal by $4 \pm 0.5°$ (toward 11$\overline{2}$0). Measurements at different angles of incidence and different azimuth angles were carried out. Figure 5 shows a representative example of a MM ellipsometry measurement on the double-sided polished 6H-SiC wafer. For this sample, the measured MM is "rich" in information, which can be verified by two facts: first, it contains no vanishing elements and second, it is depolarizing (see the depolarization index (DI)), which means that it cannot be reduced to only six parameters (assuming a normalized Mueller matrix) as it can be done for nondepolarizing measurements. As point-by-point fitting tends to involve many parameters, it is important that, even for the initial step of the fitting process, the model should be as close as possible to the real sample to avoid falling in local minima that might deliver totally unrealistic values of the optical functions. Therefore, appropriate fitting parameters and optical models need to be carefully considered in each fitting step.

The method is powerful to determine the ordinary and extraordinary optical constants (as shown in Figure 6) from comparatively few experimental measurements in one measurement mode. This is possible because the measured Mueller matrix, which are depolarizing due to the incoherent superposition of multiple reflected waves in the relatively thick substrate, are "rich" in information as they contain information from both the interfaces and the bulk of the substrate. In addition, the properties of the ultra-thin damage layer on the surface that results from the SiC wafer polishing processes also have been determined. The fitting results indicate that the refractive index of the damage layer is significantly smaller than that of the SiC crystal. The thickness of the damage layer obtained by ellipsometry is also in good agreement with the value measured by TEM.

Moreover, the Mueller matrix spectroscopic ellipsometry is also proposed to evaluate the subsurface damage of SiC wafers in rough grinding, fine grinding, and CMP stages. The

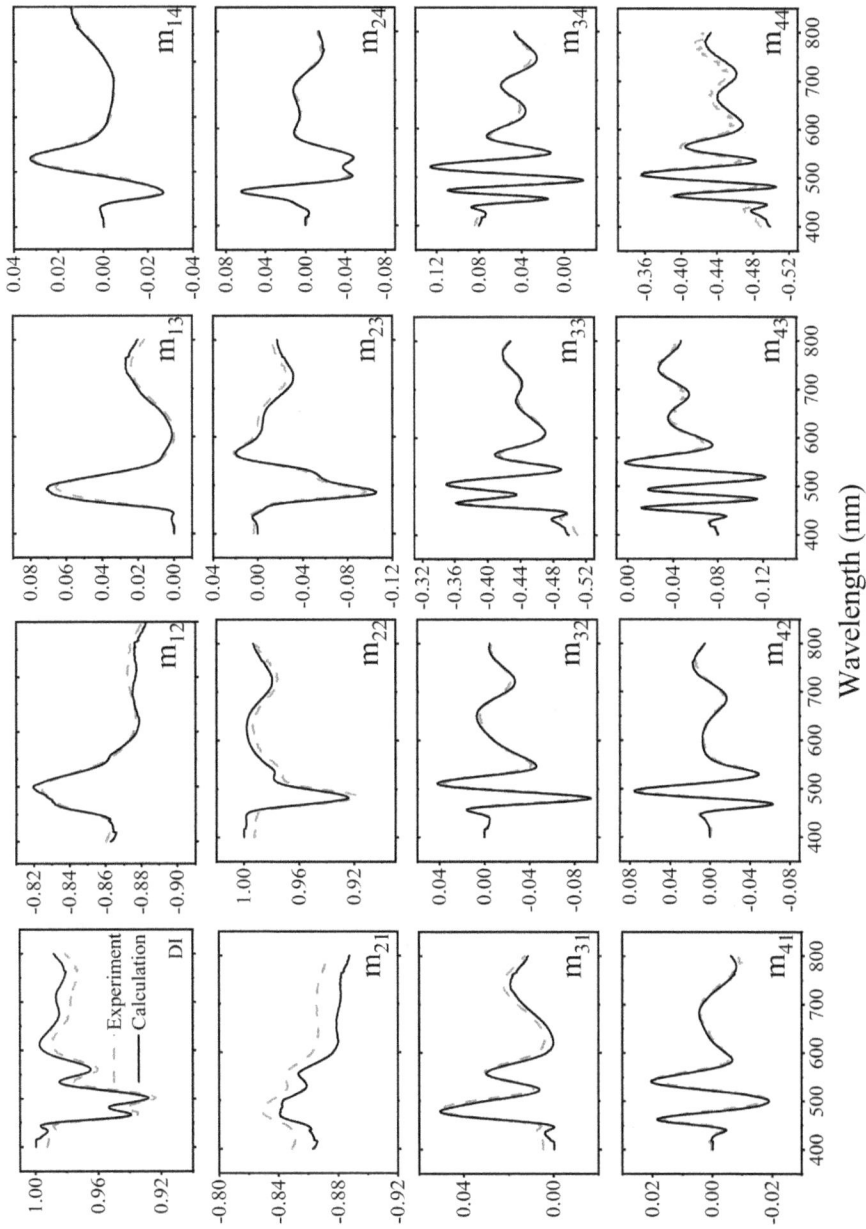

FIGURE 5

Comparison of the experimental (dashed) and calculated (solid) reflection MM of 336 μm thick 6H-SiC wafer in the range of 400 nm (3.1 eV) to 800 nm (1.55 eV) at $\theta_0 = 60°$, $\beta = 3.73°$, and $\gamma = 135°$. MM elements are normalized by m_{11} element, and their values range between -1 and 1. The graph in the top left corner shows the depolarization index (DI), that is 1 in a non-depolarizing situation and lower than 1 when depolarization is present. From [45] Figure 3, reproduced with permission of AIP.

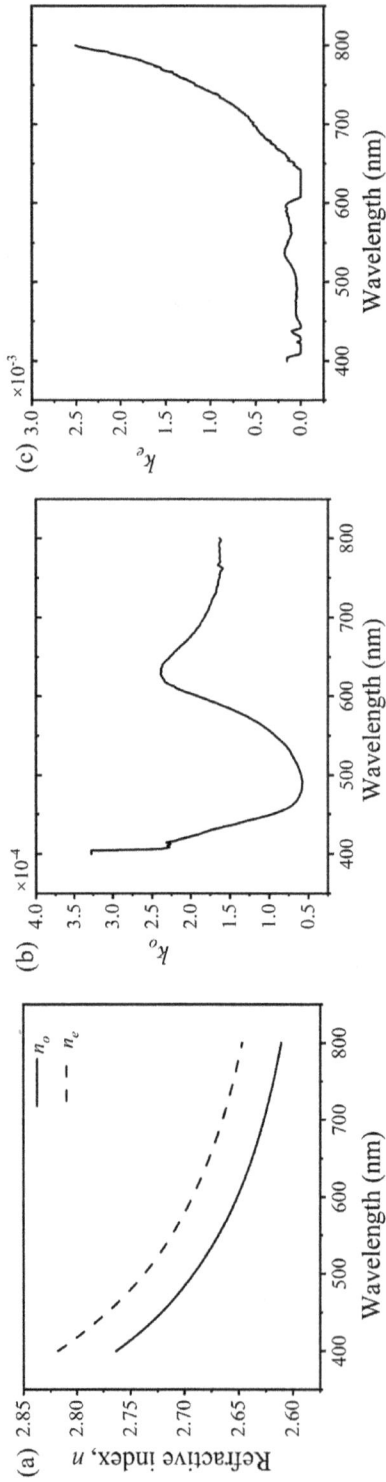

FIGURE 6

(a) The refractive indexes and extinction coefficients (b) k_o (c) k_e of n-doped 6H-SiC wafer. From [45] Figures 4 and 6, reproduced with permission of AIP.

elements of the Muller matrix are sensitive indicators of the damaged layer and surface texture. When the surface texture direction is perpendicular to the incident plane, Mueller matrix can obtain the maximum response from the damage and interface. Particularly, the fitting thickness of the damage layer is consistent with the value from TEM and the refractive index of the damage layer matches the surface elements' analysis result from X-ray photoelectron spectroscopy (XPS). This provides a possible method to achieve rapid quality assessment of SiC wafer in the entire production line [49].

3.3 Characterization of Anisotropy From Transmission Ellipsometry

In uniaxial crystals, the most employed crystal cut for the determination of optical anisotropy is with the optic axis (*c*-axis) lying in the sample surface, as this cut permits a direct measurement of the phase retardation between the ordinary and extraordinary wavefronts passing through the crystal plate. However, the obtained retardation will be much larger than 360° in this configuration. If the light propagates parallel to the *c*-axis for *c*-cut sample, no retardation can be observed, and small values of retardation appear for small off-axis incidence.

The Mueller matrix can be inverted to obtain eight physical parameters in an optical system with a transmission light path measurement, namely, mean absorption (*k*), mean refraction (*η*), circular birefringence (CB), circular dichroism (CD), linear birefringence (LB, LB) and linear dichroism (LD, LD') [50]. For *n*-doped 4H- and 6H-SiC uniaxial crystal, it only has the linear horizontal anisotropy Mueller matrix. In this special case, the MM is [51]

$$\text{MM} = \begin{bmatrix} 1 & m_{12} & 0 & 0 \\ m_{21} & 1 & 0 & 0 \\ 0 & 0 & m_{33} & m_{34} \\ 0 & 0 & m_{43} & m_{44} \end{bmatrix} \tag{13}$$

Where $m_{12} = m_{21}$, $m_{33} = m_{44}$, $m_{34} = -m_{43}$, and $m_{12}^2 + m_{33}^2 + m_{34}^2 = 1$. $(m_{12} + m_{21})$ and $(m_{34} - m_{43})$ correspond to the linear horizontal anisotropy.

The determination of linear birefringence by using off-axis transmission ellipsometry in *c*-cut plates was pioneered by Jellison and Rouleau [12]. The linear dichroism is also considered in Li's work [52]. A closed-form expression for determination of the linear birefringence ($\Delta_n = n_e - n_o$) and linear dichroism ($\Delta_k = k_e - k_o$) of uniaxial crystals utilizing transmission ellipsometry measurements was derived by

$$\Delta = \frac{2\delta n_o - n_o \sin^2\theta_1 (n_o + 2\delta) + (n_o + \delta)\sin\theta_1 \sqrt{n_o^2\sin^2\theta_1 - 2\delta n_o \sin\theta_1 \cos^2\theta_1}}{(n_o + 2\delta)\sin^2\theta_1 - 2\delta} \tag{14}$$

$$\delta = \sqrt{\frac{\varepsilon_o \varepsilon_e}{\varepsilon_o \sin^2\theta_1 + \varepsilon_e \cos^2\theta_1}} - \sqrt{\varepsilon_o} \tag{15}$$

In general, the equation gives the algebraic relation between complex optical anisotropy ($\Delta = \Delta_n + i\Delta_k$) and measured complex retardance ($\delta = LB + iLD$). θ_1 is calculated from Snell's law ($\theta_1 = \arcsin\left(\dfrac{\sin\theta_0}{n_o}\right)$, θ_0 is the incident angle); n_o is the refractive index for the polarization component perpendicular to the incidence plane.

Assuming a *c*-cut uniaxial sample, the measurement configuration with a small off-normal incident angle is presented in Figure 1 of [52]. The slight offset in linear dichroism caused by interface effect of the refractive index must be modified. For a wafer with miscut angle from sample surface, the measurement configuration can be set at normal incidence in Figure 1 of [52], in which there is no interface effects and the optical path length coincides with the thickness of the slab. The important thing is to ensure that the optical axis is parallel to the incident plane, and the off-diagonal elements are zero. In addition, the miscut angle needs to be accurately measured.

The original raw Mueller matrices for *c*-cut 4H-SiC at three considered angles of incidence are shown in Figure 4 of [52]. The linear birefringence and linear dichroism of 4H-SiC are illustrated in Figure 5 of [52].

The absorption band of linear dichroism for *n*-type 4H-SiC crystal is attributed to free electrons. This transmission ellipsometry measurement method, which works under the small angle of incidence assumption, is fully analytic and does not require numerical inversion procedures and/or least-squares fitting of the measured data to an optical model. In addition, it can be applied to *c*-cut crystal substrates as well as to plates slightly deviating from the *c*-cut.

3.4 Temperature-Dependent Optical Properties of Bulk SiC

Due to the excellent optical properties, SiC has great potential application in civil, commercial, military, and aerospace vehicles as a candidate of high-temperature electromagnetic wave absorbing materials [53]. This demands sufficient knowledge of the optical properties of SiC, especially their temperature-dependent optical properties. The temperature is usually the dominating factor that influences the refractive index of materials.

The temperature-dependent experimental measurements of the optical properties of 6H-SiC in the ultraviolet range were obtained by Petalas et al. [54]. The real (ε_1) and imaginary (ε_2) parts of the dielectric function of 6H-SiC in the energy region 5 to 10 eV at temperatures 92 and 536 K are presented in Figure 1 of [54]. The spectra are dominated by a broad feature that extends between 6.2 and 8.0 eV and consists in the case of ε_2 of two peaks at 6.7 and 7.0 eV. The corresponding second-derivative spectrum of the dielectric function ($d^2\varepsilon(\omega)/d\omega^2$) at T = 92 K is shown in Figure 2 of [54]. The contribution of so-called critical points (CPs) to the dielectric function is of the form,

$$\varepsilon(\omega) = C - Ae^{i\varphi}(\omega - E + i\Gamma)^n \tag{16}$$

where E and Γ are the energy location and broadening parameter of the CP, respectively, A is oscillator strength and the phase factor φ allows for the mixture of CPs of different nature. C stands for the contribution of oscillators at higher energies.

For some specific wavelength from visible to infrared light, Xu et al. [55] obtained the refractive indices of 4H- and 6H-SiC single crystals at wavelengths of 404.7, 435.8, 480.0, 546.1, 587.5, 643.8, 706.5, 852.1, 1014.0, 1529.6, and 2325.4 nm. Figure 7 shows that ordinary (n_o) and extraordinary (n_e) refractive indices for 4H- and 6H-SiC crystals increase with increasing temperature for each wavelength.

For a specific wavelength λ_0, the temperature-dependent refractive index (from 293 to 493 K) can be fitted by the following equation:

$$n|_{\lambda=\lambda_0}(T) = n_0|_{\lambda=\lambda_0} + P_1(T-T_0) + P_2(T-T_0)^2 \tag{17}$$

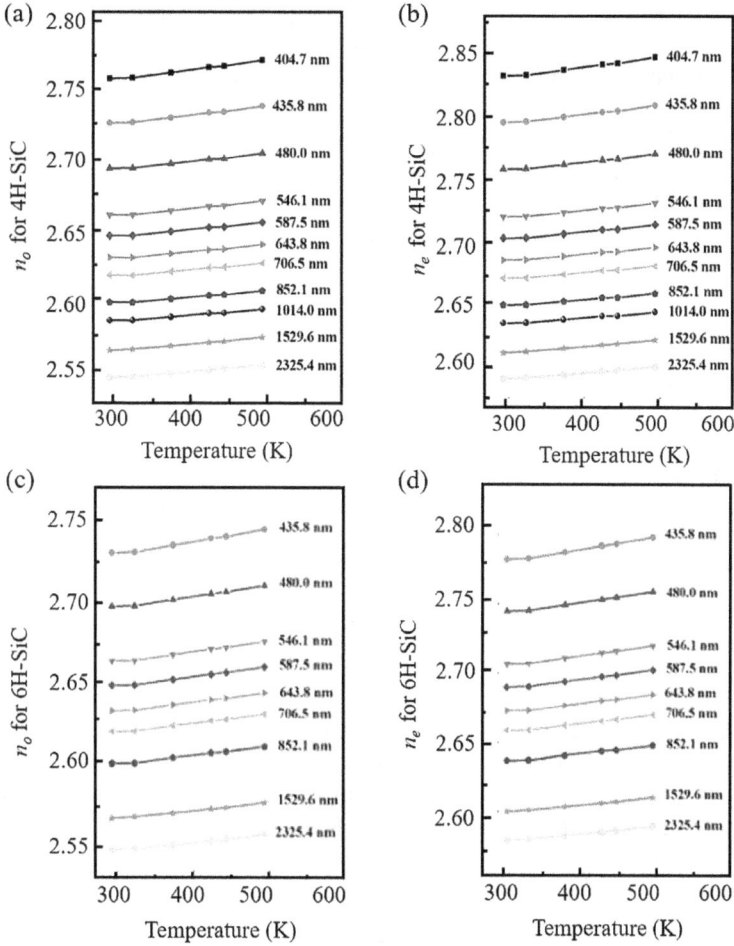

FIGURE 7

The temperature dependence of refractive indices at varying wavelengths (a) n_o, (b) n_e for 4H-SiC and (c) n_o, (d) n_e for 6H-SiC. From [55] Figures 1 and 2, reproduced with permission of AIP.

where n_0 is the refractive index at 293 K (T_0), P_1 and P_2 are constants. The relations of bandgap to temperature for 4H- and 6H-SiC can be described by the following equations, respectively [55]:

$$E_g\big|_{4H-SiC} = E_g(0) - 6.5 \times 10^{-4} \times \frac{T^2}{T+1300}(eV) \tag{18}$$

$$E_g\big|_{6H-SiC} = E_g(0) - 6.0 \times 10^{-4} \times \frac{T^2}{T+1300}(eV) \tag{19}$$

where $E_g(0)$ is the bandgap at 0 K.

For the hexagonal structure SiC, there are a large number of infrared active phonons (six and ten IR active branches for 4H and 6H-SiC, respectively). In these infrared active modes, there are strong infrared modes with large net polarization resulting from all the silicon atoms displaced opposite to carbon atoms, while the others are weak modes, which can be neglected. In order to obtain sufficient information about ε_x, ε_y, and ε_z, the measurements at variable angles of incidence are necessary [56]. Using the generalized Lyddane-Sachs-Teller (LST) relationship and neglecting the mode coupling, the dielectric function of infrared wave can be given by [57]

$$\frac{\varepsilon_j(\omega)}{\varepsilon_j(\infty)} = 1 + \frac{\omega_{j,LO}^2 - \omega_{j,TO}^2}{\omega_{j,TO}^2 - \omega^2 - i\gamma_j\omega}, \quad j = \parallel, \perp \tag{20}$$

where, LO and TO are the frequencies of transverse and longitudinal phonons, $\omega \perp LO$ ($\omega \perp TO$) and $\omega \parallel LO(\omega \parallel TO)$ are the zone-center (approaching to Γ point) frequencies of the longitudinal and transverse infrared active phonon mode that are perpendicular and parallel to c-axis, respectively.

Baugher et al. [58] represented the results of a study of the temperature dependence of the birefringence of 6H-SiC using an interference technique. The results are used to explore the possibilities this material presents as a mid-infrared non-linear optical parametric oscillator. The birefringence data can be fitted with the Sellmeier equation:

$$\Delta_n = A + \frac{B\lambda^2}{\lambda^2(\lambda^2 - C)} + \frac{D\lambda^2}{\lambda^2 - F} \tag{21}$$

In general, there are some studies on the change of SiC optical properties with temperature, as shown in Figure 8. The characteristics of a larger temperature range and a wider spectral range need to be explored. In addition, Yang et al. [59] measured the dielectric properties in the temperature range of 373–773 K at gigahertz range (8.2–12.4 GHz).

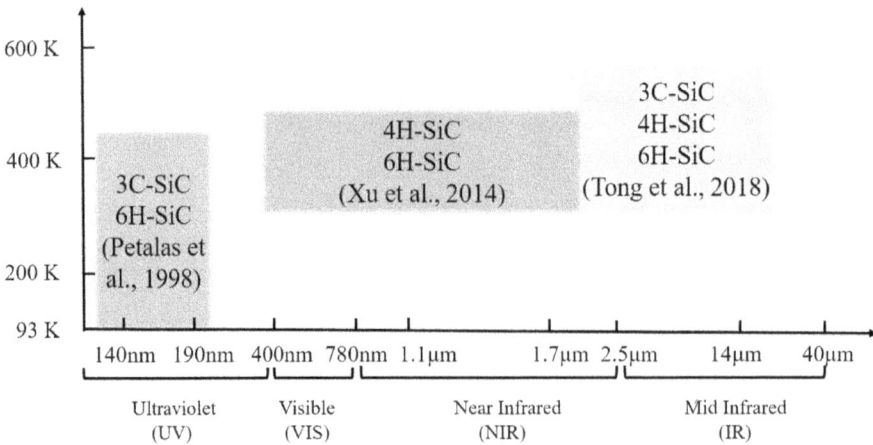

FIGURE 8
The studies about the temperature-dependent refractive index of SiC in different wavelength range and temperature.

4 Ellipsometric Analysis of SiC Epilayers and SiC Substrate-Based Epilayers

4.1 Ellipsometric Analysis of Substrate-Film-Ambient System

An optical model constructed for a thin film formed on a substrate (ambient/thin film/ substrate) can be seen from Figure 2.22 of [13]. N_0, N_1, and N_2 are the complex refractive indices of air, the thin film, and the substrate.

Calculating the Fresnel reflection and transmission coefficients for the thin film and substrate by applying

$$r_{012,p} = \frac{r_{01,p} + r_{12,p}\exp(-i2\beta)}{1 + r_{01,p}r_{12,p}\exp(-i2\beta)}, r_{012,s} = \frac{r_{01,s} + r_{12,s}\exp(-i2\beta)}{1 + r_{01,s}r_{12,s}\exp(-i2\beta)} \qquad (22)$$

$$t_{012,p} = \frac{t_{01,p} + t_{12,p}\exp(-i2\beta)}{1 + t_{01,p}t_{12,p}\exp(-i2\beta)}, t_{012,s} = \frac{t_{01,s} + t_{12,s}\exp(-i2\beta)}{1 + t_{01,s}t_{12,s}\exp(-i2\beta)} \qquad (23)$$

where $\beta = \frac{2\pi d}{\lambda}N_1\cos\theta_1$, which is called the film phase thickness. The amplitude reflection coefficient for the primary beam is r_{01}. The amplitude transmission coefficient for the primary beam is t_{01}.

The optical stack model consists of the structure of all materials, which the light may encounter. Each material should include a proposed thickness and optical functions to allow calculation of the ellipsometric response. A recipe of each layer needs to be defined before fitting the experimental SE data, i.e., a variety of dispersion equations are chosen to describe different film materials. The Sellmeier model and Cauchy equations are designed to work for transparent materials. The Cauchy equation is empirical and developed purely from the observation of the optical constant shapes, without physical meaning for this underlying shape. The Tauc-Lorentz equation is mostly used to describe the semiconductor materials.

Moreover, the surface roughness is usually modeled based on per effective medium approximation (EMA) theory, as a 50:50 model (a mixture of 50 % surface material and 50 % voids) and its dispersive function is referred from the Bruggeman model. Film morphology and surface roughness measured by an AFM can provide a reference of the initial value for the thickness of the roughness layer in the selected model.

4.2 Ellipsometric Analysis of SiC Epilayer on the SiC Substrate and Other Substrates

Besides the SiC substrate, the optical properties of SiC thin films have also attracted more attention in many fields such as photodiodes, phototransistors, photoconductive switches, solar cells, extreme ultraviolet (EUV) reflectors, and astrophysics [60]. The optical dispersion behavior of SiC thin films is necessary to explore in-depth for design and fabrication of such optoelectronic and photonic devices, as the semiconductor materials are characterized by their unique complex dielectric functions.

Mourya et al. investigated the structural, optical, and dispersion parameters of 15R-SiC thin film fabricated on Si/SiC accounting for the surface characteristics (roughness and composition) and interface layer in the optical stack model, as shown in Figures 8–9 [61].

Mourya et al. also indicated that the linear optical properties, i.e., n, k, and E_g of 15R-SiC thin films were strongly substrate-dependent from the SE measurements. By comparing the 15R-SiC films grown on ZrO_2, MgO, SiC, and Si substrates, it was found that the highest optical and dispersion energy parameters were observed for the sample deposited on Si substrate. Their Figure 9 [61] shows the experimental and fitted Ψ and Δ spectra for 1650 nm 15R-SiC thin films deposited on SiC substrate at an incidence angle of 65 degrees. It was also found that different average crystallite sizes and film thicknesses for the sample had been deposited on different substrates. The highest and smallest average crystallite size was found for the samples deposited at Si and MgO, respectively, and the film thickness had also followed the same trend.

The experimental data is fitted with two Tauc-Lorentz dispersion models to extract optical dispersion behavior of 15R-SiC film given by the following equations [62]:

$$\varepsilon_2(E) = \frac{A_L E_0 C\left(E - E_g\right)^2}{\left(E^2 - E_o^2\right) + C^2 E^2} \cdot \frac{1}{E}, \quad E > E_g, \tag{24}$$

$$\varepsilon_2(E) = 0, \quad E < E_g. \tag{25}$$

where E_0 and E_g represent the resonance energy and bandgap energy, E stands for photon energy, and C, A_L are the representation of the broadening coefficient and amplitude of ε_2 peak, respectively.

To describe the dielectric permittivity of crystalline 3C-SiC in the infrared range, considering the contribution of plasmon oscillations, the Lorentz oscillator model is well suited [63]. Particularly, the contribution of plasmon oscillations to the dielectric permittivity can be neglected at a degree of doping smaller than ~10^{18} cm^{-3} [64]. The dielectric permittivity of SiC is determined by [12]

$$\varepsilon_{SiC}(\omega) = \varepsilon_\infty \left(\frac{\omega^2 - \omega_L^2 + i\gamma\omega}{\omega^2 - \omega_T^2 + i\gamma\omega} \right) \tag{26}$$

where ε_∞ is the high-frequency dielectric constant, ω_T and ω_L are the frequencies of the transverse and longitudinal phonons, and γ is the attenuation coefficient. For a single crystal of cubic 3C-SiC, $\varepsilon_\infty = 6.5$, $\omega_T = 796$ cm^{-1}, and $\omega_L = 972$ cm^{-1}.

The free carrier concentration is an important parameter to extract meaningful electrical characteristics of semiconductor devices. The free carrier concentration per unit effective mass (N/m*) can be determined by using the following relationship between n and N/m* as given by

$$n^2 = \varepsilon_\infty - \left(\frac{e^2}{4\pi^2 \varepsilon_0 c^2} \right) \left(\frac{N}{m^*} \right) \lambda^2 \tag{27}$$

where $e = 1.6021 \times 10^{-19}$C is the electronic charge and ε_0 is the permittivity of free space (8.854×10^{-12} F/m). From the plots of n^2 vs. λ^2, the value of N/m* can be calculated with the help of the slope of the linear part of the curve at a higher wavelength, while the extrapolation to λ^2 enables determination of ε_∞.

An increase in the bandgap (5.15–5.59 eV) and a corresponding decrease in the refractive index (2.97–2.77) were noticed with the increase of SiC film thickness from about 20 nm to

450 nm [65]. The thickness-dependent optical properties of SiC thin films can be used for further optimizing the performance of SiC in various applications by tuning the optical properties.

4.3 Ellipsometric Analysis of Graphene on SiC Substrate

Graphene, one atom layer packed in the hexagonal carbon lattice structure, is still challenging as a model system to study the interplay between electron-electron interaction and electron-hole interaction, as well as its optical and electrical properties [66]. The main advantage of graphene grown on SiC substrates is that no transfer is needed for device processing. In experiment, the structural and electronic properties of graphene grown on SiC are strongly affected by its polytype and polarity [67]. Growth of epitaxial graphene was reported on both Si-face (0001) and C-face (000$\bar{1}$) polar surfaces of hexagonal 4H and 6H-SiC, and cubic 3C-SiC substrates.

A stratified layer optical model composed of a substrate, an interface layer between the substrate and the graphene, a graphene layer, and a roughness layer between the graphene and the air (see Figure 7 of [68] and Figure 2 of [69]) is used to analyze the ellipsometric data. The roughness layer is modeled using the effective medium approximation of 50% of graphene and 50% of air. In the optical model, the interface layer accounts for the carbon buffer layer, the roughness of the substrate surface, an effect of the slight off-axis cut of the SiC substrate, and non-uniform sublimation of silicon from the SiC substrate. A linear effective medium approximation comprised of 50% substrate and 50% graphene is used to create a suitable model dielectric function for the combined effect of the buffer layer and surface roughness in a single interface layer. The interface layer accounts for the contributions of (i) surface roughness and a possible slight off-axis cut of the substrate, (ii) step bunching at the SiC substrate surface, (iii) non-uniform sublimation of silicon from the SiC substrate, and (iv) possible existence of an interface/buffer layer with different structural properties from graphene [68].

The dielectric function of graphene is modeled using a sum of Lorentzian and Gaussian oscillators (such as $\varepsilon = 1 + \epsilon_L + \epsilon_G$) in order to describe the critical point (CP) in the DF associated with a Van Hove singularity at ~4.5 eV in the density of states [70]:

$$\epsilon_L(E) = \frac{A_L \gamma_L}{E_L^2 - E^2 - i\gamma_L E}, \tag{28}$$

$$\Im(\epsilon_G(E)) = A_G \left[e^{-\left(\frac{E - E_G}{\sigma}\right)^2} + e^{-\left(\frac{E + E_G}{\sigma}\right)^2} \right], \quad \sigma = \frac{\gamma_G}{2\sqrt{Ln(2)}} \tag{29}$$

where $A_{L,G}$, $E_{L,G}$, and $\gamma_{L,G}$ denote the amplitude, the CP transition energies, and the broadening of the Lorentzian and Gaussian oscillators, respectively [69].

During data analysis, the thicknesses of the surface roughness layer (t_R), epitaxial graphene layer (t_G), and the interface layer (t_I), and the graphene DF parameters are varied until best-match between experimental and model calculated 'Ψ' and 'Δ' spectra is achieved, as shown in Figure 9.

In the ellipsometry analysis, researchers found ellipsometric sensitivity to graphene layer number increases with decreasing layer number [71]. A lower substrate surface roughness resulted in more uniform step bunching and consequently better quality of the grown graphene. Particularly, measurements made on samples of graphene grown on

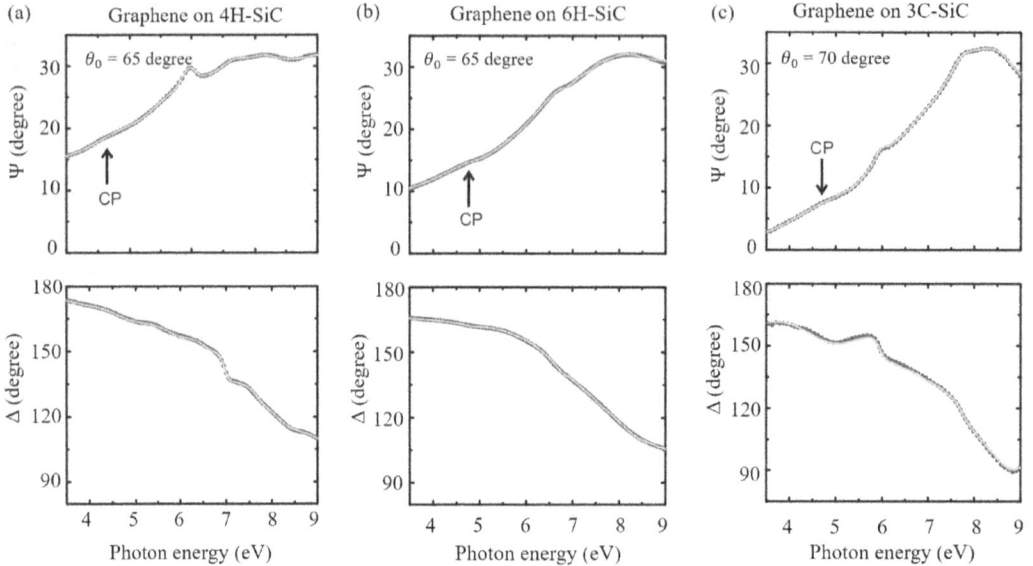

FIGURE 9

SE experimental (points) and best-match calculated (lines) Ψ and Δ spectra for epitaxial graphene (E_g) on the Si-face 4H-, 6H-, and 3C-SiC substrates. From [67] Figure 1, reproduced with permission of Elsevier.

the Si-face (0001) show significantly better uniformity than that on C-face (000-1) [74-75]. Graphene on the C-face has different electronic properties and and a different structure from graphene on the Si-face, as well as in different polytype SiC [72]. Boosalis et al. [69] reported a study of visible to vacuum ultraviolet dielectric functions of epitaxial graphene on 3C- and 4H-SiC polytypes determined by spectroscopic ellipsometry. They showed substantial differences for the dielectric function of graphene grown on the Si- and C-face of 4H-SiC, and on the Si-face of 3C-SiC substrates, in Figure 3 of [69]. The results indicate different polarizabilities and a different interaction with the substrates depending on substrate polytype and polarity [71].

4.4 Ellipsometric Analysis of GaN on 4H-SiC Substrate

GaN with the wurtzite (α) structure is thermodynamically stable. The advantage of 4H-SiC and 6H-SiC substrates is that their lattice mismatch with GaN is reduced ($\sim$3.4%) compared to sapphire ($\sim$16%). The ideal wurtzite structure is a natural optically anisotropic uniaxial crystal [73]. If the z direction is chosen along the optical axis (c-axis), the non-zero matrix elements of the dielectric tensor for such crystals are $\varepsilon_x = \varepsilon_y = \varepsilon_o$ and $\varepsilon_z = \varepsilon_e$ (same as 4H- and 6H-SiC).

In situ spectroscopic ellipsometry has been receiving increasing attention as non-invasive and non-destructive diagnostic tools of metalorganic chemical vapor deposition (MOCVD) growth processes. The growth mechanism for GaN film deposited on 4H-SiC by plasma assisted molecular beam epitaxy [74] or MOCVD [75] has been investigated in real-time by using *in situ* spectroscopic ellipsometry. The complex pseudodielectric function of the GaN film can be used to evaluate the morphological and optical quality of the sample.

Shokhovets [73] showed that standard reflection mode variable angle SE is capable of determining the ordinary and extraordinary DFs in the transparent region of nitride films with the optical axis oriented normal to the film boundaries. The ordinary and extraordinary high-frequency dielectric constants are calculated by

$$\varepsilon(\infty) = 1 + \frac{2}{\pi}\left(A_0 \ln\frac{E_1}{E_0} + \frac{A_1}{E_1} \right) \tag{30}$$

where the parameters of magnitude A_0 and A_1, photon energies E_0 and E_1 can be found in [73]. $\varepsilon_o(\infty) = 5.18 \pm 0.02$ and $\varepsilon_e(\infty) = 5.31 \pm 0.05$ were obtained by SE techniques. The experimental results indicate that α-GaN remains positive birefringent in the infrared spectral region too.

Losurdo et al. [74-75] illustrated the real (ε_1) and imaginary (ε_2) parts of the GaN/SiC sample pseudodoeletric function acquired at the end of the nucleation step for the H_2&N_2 plasma treatment and H_2 plasma treatment. The experimental results emphasized the role of the plasma in the pre-treatment of the SiC surface and subsequently, the impact of the plasma modified SiC surface on the morphology of the GaN nucleation and growth. Furthermore, the impact of the SiC pre-treatment on GaN nucleation and epitaxial growth can be investigated exploiting (Ψ-Δ) trajectories, which can be modeled yielding information on the growth mode. Cleaning and treatments of 4H-SiC are crucial for promoting coalescence of GaN nuclei, which results in a layer-by-layer growth mode of GaN directly on 4H-SiC.

4.5 Ellipsometric Analysis of AlN on SiC Substrate

AlN is an ionic compound semiconductor with hexagonal wurtzite structure. It is an important direct wide bandgap (E_g=6.2 eV) semiconductor with high temperature stability, high resistivity (10^{11}-10^{13} Ω-cm), high thermal conductivity (285 W/mK) and high breakdown field (1.2–1.8$\times10^6$ V/cm). Usually, AlN layers were deposited on Si-terminated surfaces of on-axis 6H-SiC (0001) substrates.

Owing to the negative splitting of the lattice field at the VBM, AlN has strong dichroism near the band edge [76], which makes the measurement of optical constant near the band edge difficult. Therefore, the fit procedure quickly fails at photon energies close to the bandgap where the interference of light does not contribute to optical spectra due to the onset of strong absorption. This means that a different approach is needed for this spectral region [73].

The data of Figure 10 are represented well by

$$\varepsilon_1 = 1 + \frac{2}{\pi}\left(\frac{A_0}{2} \ln\frac{E_1^2 - E^2}{E_0^2 - E^2} + \frac{A_1 E_1}{E_1^2 - E^2} \right) \tag{31}$$

which follows from the Kramers–Kronig relation:

$$\varepsilon_1(E) = 1 + \frac{2}{\pi}\int_0^\infty \frac{\varepsilon_2(E')E'dE'}{(E')^2 - E^2} \tag{32}$$

where ε_1 and ε_2 are the real and imaginary parts of the DF explained in [72]. In this model, the contribution of the fundamental absorption edge due to excitonic and band-to-band transitions is represented by a step-like function with magnitude A_0 for photon energies from E_0 to E_1. The contribution of all high-energy optical transitions is represented by a

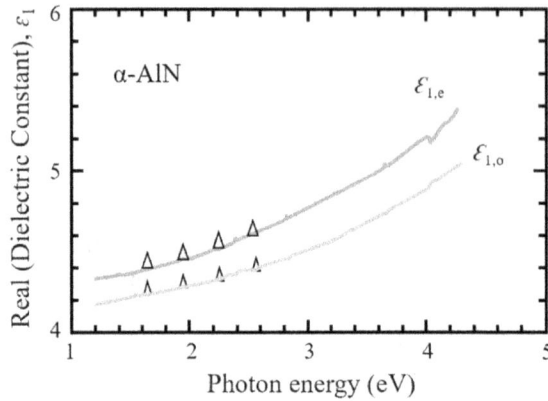

FIGURE 10
Real part of the ordinary and extraordinary dielectric functions for α-AlN in the transparent region. From [73] Figure 4, reproduced with permission of AIP.

delta function at E_1. The quantity E_0 has a meaning of the effective band gap energy. A_0, A_1, E_0, and E_1 are given by [72] too. Figure 10 indicates that α-AlN also remain positive birefringent in the infrared spectral region.

In measurement, standard reflection mode variable angle SE also can determine the ordinary and extraordinary DFs in the transparent region of AlN film. Deciding factors of the SE method are the high resolution and completeness of the spectral and angular ellipsometric scans, which must be performed not only in the near-Brewster region but also at a distance from it on both sides of the Brewster angle.

In fact, surface roughness and non-ideality of the film–substrate boundary are characteristics of epitaxial nitride films when a multilayer model is established. These peculiarities can be considered using the concept of over- and interface layers. Therefore, an optical multilayer model consisted of a semi-infinite 6H-SiC substrate, an interface layer of thickness, an active AlN film, an overlayer, and an ambient from bottom to up is considered.

Specifically, the interface is represented by an effective medium assuming a mixture of 6H-SiC and AlN. The overlayer, which considers the surface roughness, is modeled as a mixture of AlN with voids of equal fraction [76]. Optical anisotropy of the interface layer is introduced by linear interpolation between anisotropic optical constants of 6H-SiC and AlN. All constituents except the isotropic ambient are assumed to be optically uniaxial with the c-axis parallel to the growth direction.

Besides, the infrared spectroscopic ellipsometry (IRSE) spectra of the AlN layers can provide the AlN E_1(TO) and A_1 (LO) phonon modes cause distinct spectral signatures being marked by arrows in Figure 11, which shows the IRSE spectra of the AlN layer grown on the 4H-SiC substrate [77].

Yin et al. [78] studied a 500 nm AlN film prepared by MOCVD grown on 6H-SiC. By using a dual-rotation compensator, the ellipsometric spectra of the wavelength range (195–1600 nm) were measured. According to the structural characteristics of the AlN sample, the physical model of roughness/oxide layer/AlN layer/substrate is used. Figure 12 shows the measured and fitted ellipsometric spectra of AlN film on the 6H-SiC substrate.

In order to obtain the thickness of the AlN epitaxial layer, the Cauchy dispersion model is used to fit the transparent area:

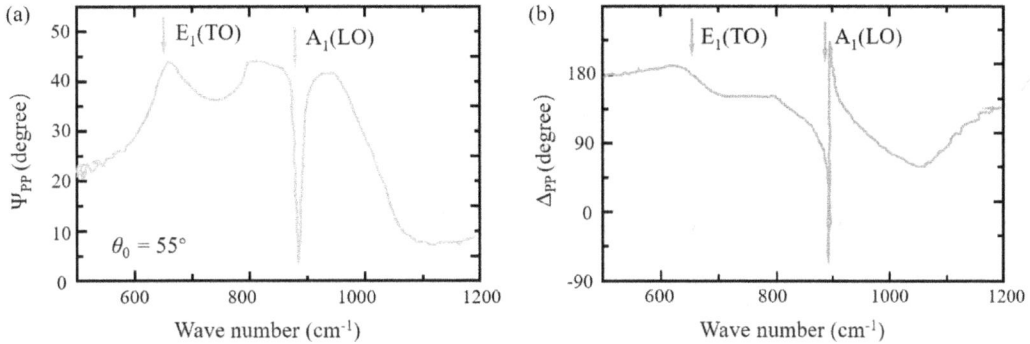

FIGURE 11
Measured and calculated best-fit Ψ and Δ spectra of AlN layer grown on 4H-SiC substrate misoriented towards <11–20> by 8°. The frequencies of the IR-active phonon modes of AlN are indicated by vertical arrows. From [77] Figure 1, reproduced with permission of AIP.

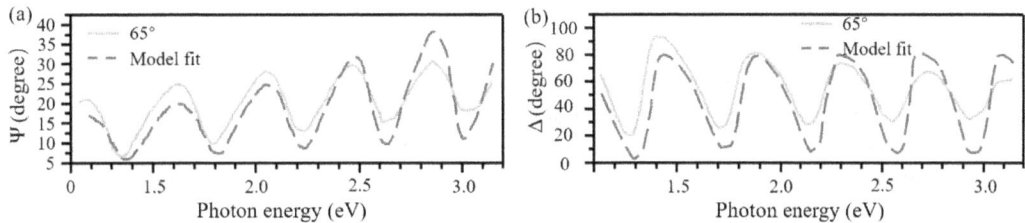

FIGURE 12
SE room temperature (RT) experimental data (solid lines) and model fitting results (dashed lines) for 6H-SiC. From [78] Figure 10, reproduced with permission of Elsevier.

$$n(\lambda) = A + \frac{B}{\lambda^2} + \frac{C}{\lambda^4} \tag{33}$$

$$k(\lambda) = \alpha_0 e^{\beta\left(1.24\mu m\left(\frac{1}{\lambda}-1/\gamma\right)\right)} \tag{34}$$

where n is the refractive index, k is the extinction coefficient, A, B, and C are the fitting parameters, α_0 is the extinction coefficient amplitude, β is the exponential factor, and γ is the band edge. Particularly, when there is an oxide layer on the surface, the Cauchy model can be used to fit the surface oxide layer to get its initial value.

The thickness of the AlN epitaxial layer and the initial value of the surface roughness can be determined in this fitting stage. However, because only the Cauchy model is used to fit the SiC instead of its real optical constants, the fitting results are not particularly good.

From the spectroscopic ellipsometry measurements and analyses, the thickness, roughness, and bandgap thickness of the surface oxide layer of the AlN film grown on substrate can be determined. From [78], the results show that the difference of the aluminum nitride film substrate may affect the roughness of the sample surface and the crystal particle size. The crystal quality of AlN film grown on SiC is better than that grown on sapphire.

Sampieri et al., in the work "A Study of Pt/AlN/6H-SiC MIS Structures for Device Applications Margarita" [79], compared the index of refraction and extinction coefficient of the AlN films grown at 500°C, 640°C, and 800°C via PSMBE. The difference in optical constants between the thinner and thicker films may be related to the higher surface roughness exhibited by the thicker AlN film.

4.6 Temperature-Dependent Optical Properties Analysis of GaN Epilayer on SiC Substrate

Many devices work under extreme temperature conditions so that the temperature-dependent optical properties of epitaxial layer are important for exploiting their performances. Here is a typical example of GaN epitaxial layer on SiC substrate.

Feneberg et al. [80] studied the temperature-dependent optical properties of a high-quality nearly unstrained zincblende GaN (c-GaN) epitaxial layer grown on 3C-SiC. Besides the layer thickness, the interface and surface roughness were also fully considered in a sophisticated multilayer model.

The complete dielectric function of c-GaN is shown in an energy range between 1 and 20 eV measured at 295 K, and the experimental data is compared to theoretical calculations [81]. The imaginary part of the dielectric function ε_2 represents the absorption of c-GaN. The direct bandgap E_0 is found as the onset of ε_2 at ≈ 3.2 eV. Higher critical points are seen as pronounced features in ε_2. The shoulder at around 7.23 eV (E_1) is identified to stem from interband transitions in the region of the L point. E_2 clearly peaking at 7.51 eV is assigned to X-point absorption, where the conduction and valence band levels are more parallel for a larger part of the Brillouin zone, thus leading to a sharp feature in the dielectric function. At even higher energy, at 10.72 eV, the spectrally overlapping E_1' and E_2' transitions was detected. E_1' and E_2' are absorption contributions of photons from the uppermost valence band to the next higher conduction bands at the L and X points, respectively. The same transition at the point of the Brillouin zone is labeled E_0' and detected at 12.87 eV.

The comparison of the imaginary part of the dielectric function ε_2 for both temperatures (10 K and 295 K) was presented [80]. The sharper features at lower temperatures and an overall blueshift of the dielectric function was found. The decreasing linewidth and increasing amplitude, of, e.g., E_2 is an indication of the mainly phononic origin of the broadening instead of sample-quality-dependent inhomogeneous broadening.

The dispersive below-gap region in ε_1 is equal to the square of the refractive index (as long as $\varepsilon_2 = 0$) and thus is of special interest for the design of optoelectronic devices. This dispersion can be modeled by the analytic expression:

$$\varepsilon_1(\hbar\omega) = 1 + \frac{2}{\pi}\left(\frac{A_g}{2} \ln \frac{E_h^2 - (\hbar\omega)^2}{E_g^2 - (\hbar\omega)^2} + \frac{A_h E_h}{E_h^2 - (\hbar\omega)^2} \right) \tag{35}$$

Here, E_g and E_h represent the average energies of the bandgap and the higher energy contributions, respectively. The amplitudes A_g and A_h weight these accordingly. In the case of c-GaN, $E_g = 3.23$ eV, $E_h = 8.65$ eV, $A_g = 1.55$, and $A_h = 45.29$ were obtained [81].

These investigated optical parameters allow one to design and fabricate optoelectronic, photonic, and telecommunication devices for deployment in extreme environments.

5 The Subsurface Damaged Layer of SiC Substrate

The processing flow of SiC substrate mainly includes rough grinding, fine grinding, and chemical mechanical polishing (CMP). Usually, rough grinding will leave large surface texture and a mass of subsurface damage (SSD). Although those damages can be gradually removed by subsequent fine grinding and CMP, it is time-consuming. Therefore, monitoring the depth of damage will provide a useful index for the quality control of SiC wafer production chain and the processing technology optimization, especially for wafers with large diameters, which is a trend with the development of material growth. A nondestructive and precise method for measuring the thickness of SSD layer is indispensable.

As a non-destructive strategy, Yao et al. [82] proposed a quasi-Brewster angle technology to quickly evaluate the polishing quality covering rough- and fine-polishing stages using a variable angle ellipsometer, but the thickness of damage layer was not obtainable. Furthermore, we, Li et al. [49], proposed a method based on the Mueller matrix spectroscopic ellipsometry (MMSE) to detect the nanoscale subsurface damage of 4H-SiC wafers induced by grinding and polishing. In details, 4H-SiC single crystal wafers (*n*-type doped, off-axis cut toward <1120>, 4 inch) after double-sided rough grinding, double-sided fine grinding and double-sided CMP were selected for research. The Mueller matrix of wafers were measured in transmission and reflection modes with a dual-rotation compensator Muller matrix ellipsometer (DRMME, Wuhan Eoptics Technology Co., China) [83]. Measurements were done in the spectral range of 250 nm (4.96 eV) to 1400 nm (0.89 eV). The short axis diameter of the incident beam spot is 3 mm.

5.1 Optical Constants of 4H-SiC

The optical anisotropy of uniaxial 4H-SiC wafer is obtained by analyzing the transmission Mueller matrix (MM), as shown in Figure 13(a). The 4×4 Mueller matrix (**M**) be inverted to obtain physical parameters (**L**) in an optical system by using the differential matrix decomposition, Eq. (36). As the considered uniaxial 4H-SiC crystal, only two polarization properties were non-vanishing, namely, LB, which describes the phase retardation between x and y polarizations and related to m_{43} element ($m_{34} = -m_{43}$), and LD, which describes the diattenuation between x and y polarizations and related to m_{12} element ($m_{12} = m_{21}$).

$$\mathbf{L} = \ln(\mathbf{M}) \tag{36}$$

Based on the measured complex retardance ($\delta' = LB + iLD$), the birefringence and dichroism can be obtained by using Eq. (14). Moreover, the ordinary extinction coefficient $k_o(\lambda)$ of 4H-SiC crystal can be calculated from R and T spectra by using the equation,

$$k_o(\lambda) = \frac{\alpha(\lambda)}{4\pi} \lambda \tag{37}$$

$$\alpha(\lambda) = -\frac{1}{d} \ln \left[\frac{\sqrt{(1-R)^4 + 4T^2 R^2} - (1-R)^2}{2TR^2} \right] \tag{38}$$

where, d is the thickness of wafer along the direction of light propagation. The extraordinary extinction coefficient (k_e) is calculated by the known ordinary extinction coefficient

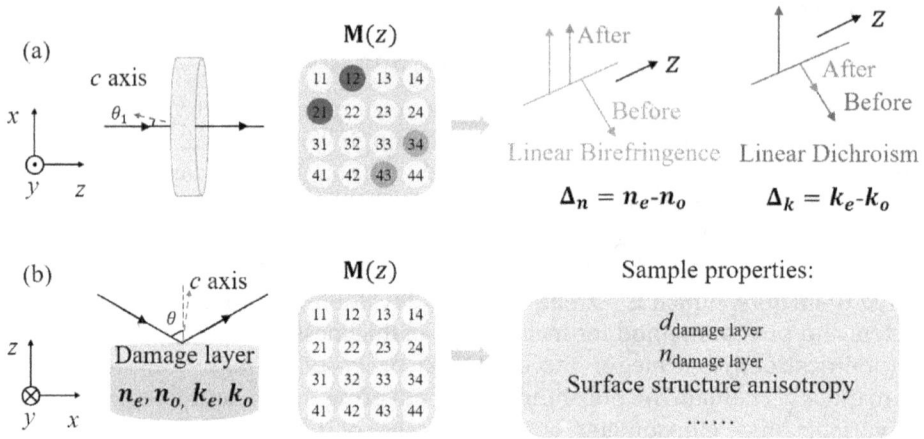

FIGURE 13

(a) The transmission Mueller matrix measurement of off-axis cut uniaxial 4H-SiC wafer. The linear birefringence and linear dichroism can be analyzed. (b) The reflective Mueller matrix measurement of processed 4H-SiC wafer with damage layer. The refractive index and thickness of damage layer and the surface structure anisotropy of the wafers can be analyzed. From [49] Figure 1, an open-access article, Li and Cui as original authors, with reproduction permitted.

(k_o) and linear dichroism $(k_e - k_o)$. In addition, the extraordinary refractive index (n_e) is calculated by using the known ordinary refractive index.

5.2 Optical Stack Model

In MMSE reflection measurement, an appropriate optical stack model is acquired to obtain the accurate parameters of thin layers. The refractive index, thickness of damage layer, and surface structure anisotropy of the wafers can be analyzed, as shown in Figure 13(b).

In general, rough surface, amorphous layer, nonideality boundary (between the damaged zone and pure substrate), and roughed backside are characteristics of the rough grinding wafer surface from top to bottom. These peculiarities are considered for ellipsometric analysis using roughness layer, damage layer, interface layer, and semi-infinite SiC substrate. For fine grinding wafer, there are surface scratches and nonideality damage layer, which are modeled as three layers: roughness layer/ damage layer/SiC substrate. CMP wafer with the sub-nanometer roughness is regarded as specular surface. Therefore, the roughness is omitted, and CMP wafer is modeled as two layers: damage layer/ SiC substrate.

5.3 The Sensitivity of Mueller Matrix

Different from traditional thin film analysis, the sensitivity of the ultra-thin damaged layer needs to be considered in optical model for fine grinding and CMP samples with backside reflection. Besides that, the effect of surface texture caused by rough grinding on the Mueller matrix needs to be considered.

Firstly, in order to figure out the role of surface (Si-face), back (C-face) damage layer and the backside reflection in Mueller matrix, the following simulation is executed in Figure 14. Using partial coherence wave and fully coherence wave theory, four models are used to explore the response of the Mueller matrix elements to ultra-thin layer on Si-face and

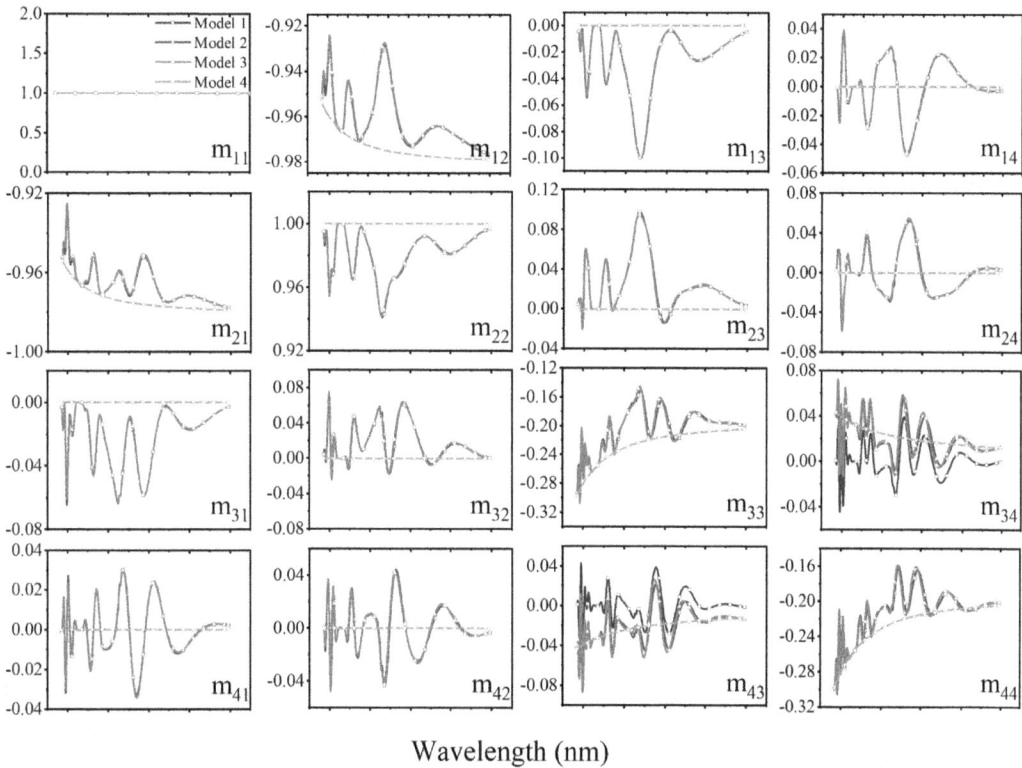

FIGURE 14

The simulation of Mueller matrix. The angle of incidence is 65° and the Euler angle γ is 45°, β is 4°. From [49] Figure 4A, an open-access article, Li and Cui as original authors, with reproduction permitted.

C-face of 4H-SiC substrate. A known SiO_2 layer is used as the material of the ultra-thin film. Model 1, 350 μm 4H-SiC thick layer with backside reflection. Model 2, 2 nm SiO_2 layer on Si-face of 4H-SiC thick layer with backside reflection. Model 3, 2 nm and 5 nm SiO_2 layers on Si-face and C-face of 4H-SiC thick layer with backside reflection. Model 4, 2 nm SiO_2 layers on Si-face of 4H-SiC semi-infinite substrate without backside reflection. The angle of incidence is 65° and Euler angle β is 4°, γ is 45°.

From the simulation results, the diagonal element m_{34} has the highest sensitivity to small changes to the ultra-thin layer compared to other elements. The difference between Model 1 and Model 2 indicates that m_{34} is only sensitive to the thin layer. The difference between Model 2 and Model 3 indicates that the thin layer on C-face of substrate has almost no effect on the values of the Muller matrix. The difference between Model 3 and Model 4 indicates that the fluctuation of MM spectra is related to the backside reflection of substrate. The "position balance" of m_{34} spectra of substrate with backside reflection is the same with that of substrate without backside reflection. Therefore, the backside reflection when only considering the fitting result of the m_{34} element can be omitted.

On the other hand, in order to appropriately describe the influence of the surface texture of rough grinding 4H-SiC wafers on MM spectra, the MM elements are plotted in polar coordinates with wavelength and rotation angle as radial and angular coordinate, respectively. The reflection MM data is measured at 17 Euler rotation angles (γ from 0° to 360° in steps of 22.5°) with the incident angle of $\theta = 65°$.

For demonstration, one point of rough grinding wafer is selected. Figure 15(a) shows the surface morphology measured by 3D optical surface profiler and its initial texture direction is parallel to the plane of incidence. Figure 15(b) shows the schematic diagram of reflection ellipsometry measurement. The anisotropy of rough grinding wafer is reflected in the diagonal and non-diagonal elements of Mueller matrix. The elements of Mueller matrix are also shown as symmetrical. Therefore, only the non-zero off-diagonal elements m_{13}, m_{14}, and diagonal element m_{34} are shown in Figure 15(c), (d), and (e), which can reflect the surface structure anisotropy of rough grinding 4H-SiC wafer.

The measurement data indicate that the off-diagonal elements depend on rotation angles and the Mueller matrix have high sensitivity to structure anisotropy. It can be observed that the off-diagonal elements are zero at 0, 90, 180, and 270° rotation angles. The maximum and minimum values of m_{13} and m_{14} spectra are located at 135, 315° and 45, 225° rotation angles, respectively. Therefore, off-diagonal Mueller matrix elements can be set as an indicator to judge the direction of surface rough texture. In this view, the roughness does not appear as an intrinsic characteristic of the surface, which depends on the wavelength and on the direction of propagation of the incident wave.

5.4 Reflection Mueller Matrix Analysis

Figure 16(a), (b), and (c) show the schematic diagrams of multilayer optical models corresponding to the cross-section characteristics of 4H-SiC wafers after three machining stages. From the simulation results, the fully coherence wave theory can be used for wafers with and without back reflection, and the value of the m_{34} element can be used as an indicator of damage layer. Therefore, only the fitted m_{34} elements are compared with the experimental data of three processed wafers, as shown in Figure 16(d), (e), and (f). The smooth curve of measured m_{34} in Figure 16(d) indicates backside reflection is absent from rough grinding wafer. The fluctuating curves of measured m_{34} in Figure 16(e) and (f) are affected by backside reflection, and their amplitude is related to the absorption of samples. The fluctuations decrease in three higher extinction coefficient ranges, corresponding to 430–480 nm (k_e), 520–650 nm (k_e), and 1100–1400 nm (k_o and k_e).

The thicknesses of roughness layer of rough grinding and fine grinding wafers are initialed by 3D optical surface profiler results, while no roughness layer is set for CMP wafer. Based on the prior knowledge [45], the Cauchy parameters of damage layer are initialed as follows: A is 2, B is 0.05 μm², and the initial thicknesses of damage layers are set to 50 nm, 4 nm, and 2 nm for rough grinding, fine grinding, and CMP wafers according to their processing techniques, respectively.

By fitting the MMSE data, the damaged layer thicknesses of rough grinding, fine grinding, and CMP wafers are obtained as 53.7 ± 0.9 nm, 4.6 ± 0.6 nm, 2.4 ± 0.2 nm, respectively. Besides that, the interface layer under the damage layer of rough grinding wafer is obtained as 49.2 ± 0.6 nm, which reflects the non-ideality of the damage-substrate boundary, residual stress, and other damage types. To verify the reflection Mueller matrix analysis method, TEM experiments are carried out. The damaged layers were 47.8 nm, 4.35 nm (average of 3.6 nm and 5.1 nm), and 2.6 nm thick for rough grinding, fine grinding, and CMP 4H-SiC wafers, as shown in Figure 16(g), (h), and (i). They have a close agreement with those values obtained by MMSE. Because the inhomogeneity of the surface, the results of different positions on wafers show some difference, especially for the rough grinding wafer. Three positions of each sample are measured by TEM for comparison. Their average thickness value and standard deviation of damage layers are 47.7

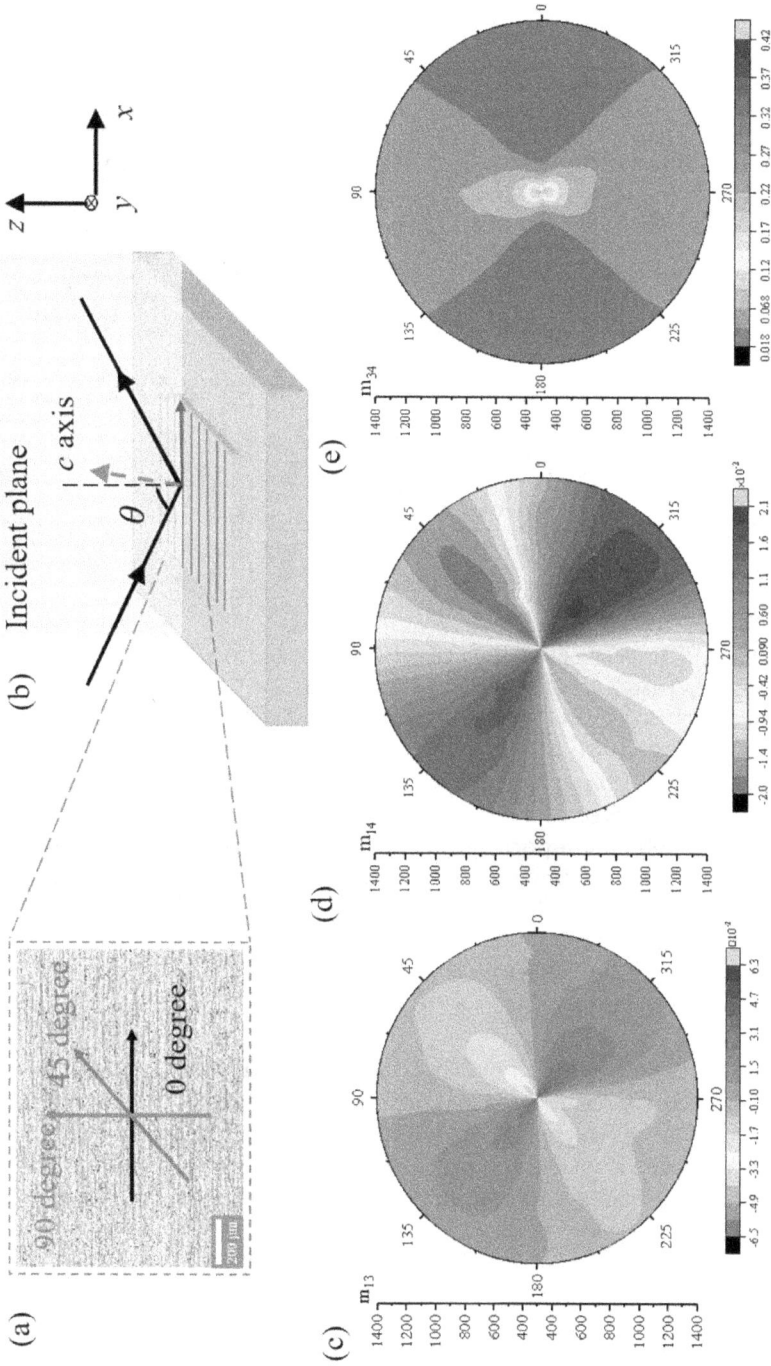

FIGURE 15

(a) The surface morphology of rough grinding wafer and its initial texture direction parallel to the plane of incidence where the direction of rotation angle is anticlockwise. (b) The schematic diagram of reflection ellipsometry measurement. (c), (d), and (e) are the experimental MMSE data of individual m_{13}, m_{14}, and m_{34} elements from 0° to 360° rotation angles in 65° incident angle, respectively. MM intensities are plotted in polar coordinates with wavelength and rotation angle as radial and angular coordinate. From [49] Figure 5, an open-access article, Li and Cui as original authors, with reproduction permitted.

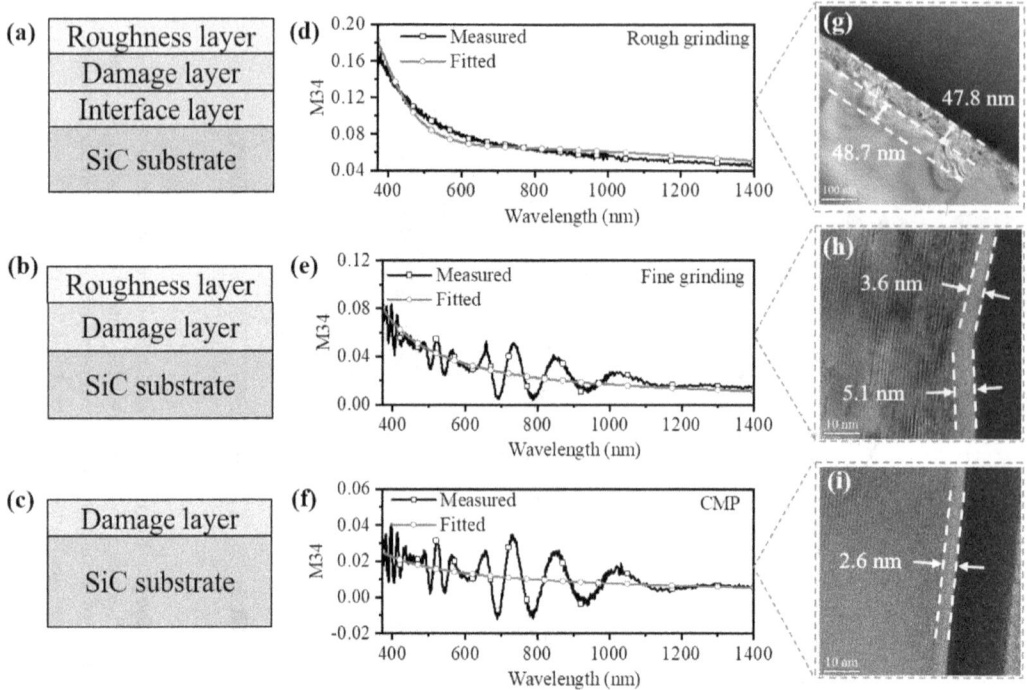

FIGURE 16

(a), (b), and (c) are the schematic diagrams of optical stack models. (d), (e), and (f) are fitting results of M34. (g), (h), and (i) are the typical TEM images of rough grinding, fine grinding, and CMP 4H-SiC wafers. From [49] Figure 7, an open-access article, Li and Cui as original authors, with reproduction permitted.

± 8.9 nm, 4.5 ± 0.7 nm, 2.6 ± 0.1 nm, respectively. The rough surface shows larger damage inhomogeneity than the other two.

The rough grinding also left an interface inhomogeneity (about 48.7 nm thick in one position) under the damage layer, as shown in Figure 16(g). The average thickness is about 53.7 ± 7.5 nm of three positions, which is also close to that result of MMSE. In this view, although these damages are very inhomogeneous distributed in different areas, the interface layer analyzed by MMSE can reflect them to a certain extent. The experiment shows that MMSE can analyze the damage layers of grinding and polishing wafers in a non-destructive way.

As a conclusion, the elements of Mueller matrix are sensitive indictors of damage layer and surface texture. Particularly, the change of m_{34} is significantly induced by the damage layer. When the surface texture direction is perpendicular to the incident plane, the Mueller matrix can obtain maximum response from the damage and interface. According to simulation and experiment, the optical model can even be simplified to ignore the backside reflection.

6 Summary

In summary, this chapter briefly reviews the basic theory, the analysis strategies, and the typical application in semiconductors of SE. Optical anisotropy and optical constants of SiC substrate can be obtained from transmission and reflection mode. The Mueller matrix decomposition method is used for transmission analysis, which is sensitive to the characteristics of bulk materials. Forward modeling and reverse fitting method are used for reflection analysis, which is sensitive to the characteristics of the interface. Besides the SiC epitaxial layer on different substrates, the epitaxial layer on SiC substrate attracts attention such as SiC, graphene, GaN, and AlN films, etc. Each system was introduced from ellipsometry measurement, modeling to analysis of the properties of materials. For silicon carbide and its different epitaxial materials, we introduced the optical stack model and analysis of the optical constants obtained by different researchers. We hope this chapter can provide guidance for those who use ellipsometry to characterize silicon carbide and related materials.

On the basis of available literatures, we have concluded that there are still several key challenges such as polytypism, crystallinity, phase purity, surface characteristics, and broad spectral range pertaining to the investigation of the optical properties of the proposed films.

Acknowledgements

We acknowledge Subiao Bian, Xi Chen, Baorong Zhao, Qimiao Liang, Mingxin Wang, Liyuan Ma, Ziqing Li, and Wanpei Yu and for various help and support on the works in this chapter. Authors would like also thank Editor/Professor Zhe Chuan Feng for his good suggestions and help in the preparation of this chapter.

References

1. O. P. Alexander Lindquist, Hans Arwin, A. Henry and Kenneth Järrendahl, "Infrared Optical Properties of 3C, 4H and 6H Silicon Carbide", Materials Science Forum, 433–436, 329–332 (2003). doi: 10.4028/www.scientific.net/msf.433-436.329.
2. Kaushik Vedam, "Spectroscopic ellipsometry: A historical overview", Thin Solid Films, 313–314, 1–9 (1998). doi: 10.1016/S0040-6090(97)00762-1.
3. Jr. Gerald E. Jellison, "Data Analysis for Spectroscopic Ellipsometry", Thin Solid Films, ch3, 234, pp. 416–422, 1993. doi: 10.1016/B978-081551499-2.50005-8.
4. Jr. Gerald E. Jellison, "Spectroscopic ellipsometry data analysis: measured versus calculated quantities", Thin Solid Films, 313–314, 33–39 (1998). doi: 10.1016/S0040-6090(97)00765-7
5. Débora Gonçalves and Eugene A. Irene, "Fundamentals and applications of spectroscopic ellipsometry", Quimica Nova, 25(5), 794–800 (2002). doi: 10.1590/S0100-40422002000500015.
6. J. B. Casady and Robert Wood Johnson, "Status of silicon carbide (SiC) as a wide-bandgap semiconductor for high-temperature applications: A review", Solid-State Electronics, 39(10), 1409–1422 (1996). doi: 10.1016/0038-1101(96)00045-7.
7. E. Sörman, Nguyen Tien Son, Weimin M. Chen, O. Kordina, C. Hallin and E. Janzén, "Silicon vacancy related defect in 4H and 6H SiC", Physical Review B, 61(4), 2613–2620 (2000). doi: 10.1103/PhysRevB.61.2613.
8. K. Kamitani, M. Grimsditch, J. C. Nipko, C. K. Loong, M. Okada and I. Kimura, "The elastic constants of silicon carbide: A Brillouin-scattering study of 4H and 6H SiC single crystals", Journal of Applied Physics, 82(6), 3152–3154 (1997). doi: 10.1063/1.366100.

9. Stefan Zollner, J. G. Chen, Erika Duda, T. Wetteroth, S. R. Wilson and James N. Hilfiker, "Dielectric functions of bulk 4H and 6H SiC and spectroscopic ellipsometry studies of thin SiC films on Si", Journal of Applied Physics, 85(12), 8353–8361 (1999). doi: 10.1063/1.370682.

10. Christoph Cobet, Klaus, Wilmers, T. Wethkamp, N. V. Edwards, N. Esser and W. Richter, "Optical properties of SiC investigated by spectroscopic ellipsometry from 3.5 to 10 eV", Thin Solid Films, 364(1–2), 111–113 (2000). doi: 10.1016/S0040-6090(99)00893-7.

11. Maria Losurdo, Giovanni Bruno, Tong-Ho Kim, Soojeong Choi and April Brown, "Study of the dielectric function of hexagonal InN: Impact of indium clusters and of native oxide", Applied Physics Letters, 88(12), 86–89 (2006). doi: 10.1063/1.2190461.

12. Jellison J G E and Rouleau C M, "Determination of optical birefringence by using off-axis transmission ellipsometry", Applied Optics, 44(16), 3153–3159 (2005). https://doi.org/10.1364/AO.44.003153

13. Hiroyuki Fujiwara, Spectroscopic Ellipsometry Principles and Applications. Tsukuba: National Institute of Advanced Industrial Science & Technology, 2007.

14. Harland G. Tompkins and Eugene A. Irene, Handbook of ellipsometry. William Andrew, Inc. All rights reserved, 2005.

15. James N. Hilfiker, Harland G. Tompkins, Spectroscopic Ellipsometry: Practical Application to Thin Film Characterization. New York: The United States of America, 2016.

16. Christian Brosseau, Fundamentals of polarized light: A Statistical Optics Approach. 424 pp, 1998.

17. Razvigor Ossikovski, "Alternative depolarization criteria for Mueller matrices", Journal of Optical Society America, 27(4), 808 (2010). doi: 10.1364/josaa.27.000808.

18. José Jorge Gil and Eusebio Bernabeu, "Depolarization and polarization indices of an optical system", Optical Acta: International Journal of Optics, 33(2), 185–189 (1986). doi: 10.1080/713821924.

19. Jiamin Liu, Jianbin Lin, Hao Jiang, Honggang Gu, Xiuguo Chen, Chuanwei Zhang, Guanglan Liao and Shiyuan Liu, "Characterization of dielectric function for metallic thin films based on ellipsometric parameters and reflectivity", Physica Scripta, 94(8), (2019). doi: 10.1088/1402-4896/ab1606.

20. Honggang Gu, Simin Zhu, Baokun Song, Mingsheng Fang, Zhengfeng Guo, Xiuguo Chen, Chuanwei Zhang, Hao Jiang and Shiyuan Liu, "An analytical method to determine the complex refractive index of an ultra-thin film by ellipsometry", Applied Surface Science, 507, 145091 (2020). doi: 10.1016/j.apsusc.2019.145091.

21. James N. Hilfiker, Michale Stadermann, Jianing Sun, Tom Tiwald, Jeffrey S, Hale, Philip E. Miller and Chantel Aracne-Ruddle, "Determining thickness and refractive index from free-standing ultra-thin polymer films with spectroscopic ellipsometry", Applied Surface Science, 421, 508–512 (2017). doi: 10.1016/j.apsusc.2016.08.131.

22. Oleg V. Ivanov, Dmitry I. Sementsov, "Coherent and incoherent reflection and transmission of light in anisotropic layer structures", Crystallography Reports, 45(5), 827–832 (2000). doi: 10.1134/1.1312930

23. Nichols Shane, Martin Alexander, Choi Joshua, and Kahr Bart "Gyration and permittivity of ethylenediammonium sulfate crystal", Chirality, 28(6), 460–465 (2016). doi: 10.1002/chir.22603

24. Shih-Yau Lu, Russell A. Chipman, "Interpretation of Mueller matrices based on polar decomposition", Journal of The Optical Society of America A-Optics Image Science and Vision, 13(5), 1106–1113 (1996). doi: 10.1364/JOSAA.13.001106

25. Noe Ortega-Quijano, Jose Luis Arce-Diego, "Mueller matrix differential decomposition", Optics Letters, 36(10), 1942–1944 (2011). doi: 10.1364/ol.36.001942

26. Hans Arwin, Roger Magnusson. Enric Garcia-Caurel, Clement Fallet, Kenneth Jarrendahl, Martin Foldyna, Antonello De Martino and Razvigor Ossikovski, "Sum decomposition of Mueller-matrix images and spectra of beetle cuticles", Optics Express, 23(3), 1951–1966 (2015). doi: 10.1364/oe.23.001951

27. Chao He, Honghui He, Jintao Chang, Binguo Chen, Hui Ma, and Martin J Booth, "Polarisation optics for biomedical and clinical applications: a review", Light Science Applications, 10(1), (2021). doi: 10.1038/s41377-021-00639-x

28. Faraj Saeed Al-Hazmi, Gary Beall, Ahmed Al-Ghamdi, Ahmed Alshahrie, F.S. Shokr, and Waleed E. Mahmoud, "Raman and ellipsometry spectroscopic analysis of graphene films grown directly on Si substrate via CVD technique for estimating the graphene atomic planes number", Journal of Molecular Structure, 1118, 275–278 (2016). doi: 10.1016/j.molstruc.2016.04.028

29. Bo Zou, Zuoxu Wu, Yu Zhou, Yan Zhou, Jian Wang, Lianghui Zhang, Feng Cao and Huarui Sun, "Spectroscopic ellipsometry investigation of Au-Assisted exfoliated large-area Single-crystalline Monolayer MoS$_2$", Physica Status Solidi-rapid Research Letters, 15(11), (2021). doi: 10.1002/pssr.202100385

30. Huihui Li, ChangCai Cui, Xipeng Xu, Subiao Bian, "A review of characterization of perovskite film in solar cells by spectroscopic ellipsometry", Solar. Energy, 212, 48–61 (2020). doi: 10.1016/j.solener.2020.10.066

31. Erik Langereis, S.B.S. Heil, Harm C.M. Knoops, Wytze Keuning, M.C.M. Van De Sanden and W.M.M. Kessels, "In situ spectroscopic ellipsometry as a versatile tool for studying atomic layer deposition", Journal of Physics D-Applied Physics, 42(7), (2009). doi: 10.1088/0022-3727/42/7/073001

32. M. Losurdo, P. Capezzuto, G. Bruno, "Plasma cleaning and nitridation of sapphire (α-Al$_2$O$_3$) surfaces: New evidence from in situ real time ellipsometry", Journal of Applied Physics, 88(4), 2138–2145 (2000). doi: 10.1063/1.1305926

33. Subiao Bian, ChangCai Cui, Oriol Arteaga, "Mueller matrix ellipsometer based on discrete-angle rotating Fresnel rhomb compensators", Applied Optics, 60(16), 4964–4971 (2021). doi: 10.1364/ao.425899

34. Zhengwei Miao, Yuanyuan Tang, Kai Wei and Yudong Zhang, "Random error analysis of normalized Fourier coefficient in dual-rotating compensator Mueller matrix ellipsometer", Measurement Science Technology, 32(12), (2021). doi: 10.1088/1361-6501/ac1a80

35. Blaine Johs, C.M. Herzinger, "Quantifying the accuracy of ellipsometer systems", Physica Status Solidi Current Topics in Solid State Physics, 5(5), 1031–1035 (2008). doi: 10.1002/pssc.200777755

36. David E. Aspnes, "Expanding horizons: New developments in ellipsometry and polarimetry", Thin Solid Films, 455–456, 3–13, (2004). doi: 10.1016/j.tsf.2003.12.038

37. Dale Gottlieb, Oriol Arteaga, "Optimal elliptical retarder in rotating compensator imaging polarimetry", Optics Letters, 46(13), 3139–3142 (2021). doi: 10.1364/ol.430266

38. Weiqi Li, Chuanwei Zhang, Hao Jiang, Xiuguo Chen and Shiyuan Liu, "Depolarization artifacts in dual rotating-compensator Mueller matrix ellipsometry", Journal of Optics, 18(5), (2016). doi: 10.1088/2040-8978/18/5/055701

39. Shiyuan Liu, Xiuguo Chen and Chuanwei Zhang, "Development of a broadband Mueller matrix ellipsometer as a powerful tool for nanostructure metrology", Thin Solid Films, 584, 176–185 (2015). doi: 10.1016/j.tsf.2015.02.006

40. Xudan Zhu, Rongjun Zhang, Yuxiang Zheng, Songyou Wang and Liangxiao Chen, "Spectroscopic ellipsometry and its applications in the study of thin film materials", Chinese Optics, 12(6), 47–86 (2019). doi: 10.3788/CO.20191206.1195

41. R. M. A. Azzam, N. M. Bashara, "Ellipsometry and polarized light", Physics Today, 31(11), 72 (1977). doi: 10.1063/1.2994821

42. Razvigor Ossikovski, Morten Kildemo, Michel Stchakovsky and Marcus Mooney, "Anisotropic incoherent reflection model for spectroscopic ellipsometry of a thick semi-transparent anisotropic substrate", Applied. Optics, 39(13), 2071–2077 (2000). doi: 10.1364/ao.39.002071

43. Alexander T. Martin, Shane M. Nichols, Melissa Tan and Bart Kahr, "Mueller matrix modeling of thick anisotropic crystals with metallic coatings", Applied Surface Science, 421, 578–584 (2016). doi: 10.1016/j.apsusc.2016.10.124

44. Kamil Postava, Tomuo Yamaguchi and Roman Kantor, "Matrix description of coherent and incoherent light reflection and transmission by anisotropic multilayer structures", Applied Optics, 41(13), 2521 (2002). doi: 10.1364/ao.41.002521.

45. Huihui Li, ChangCai Cui, Subiao Bian, Jing Lu, Xipeng Xu and Oriol Arteaga, "Double-sided and single-sided polished 6H-SiC wafers with subsurface damage layer studied by Mueller matrix ellipsometry", Journal of Applied Physics, 128(23), 235304 (2020). doi: 10.1063/5.0026124.

46. Shane Nichols, Coherence in Polarimetry "smn-thesis", Github, 2017. https://github.com/shane-nichols/smn-thesis.

47. John A. Woollam, James N. Hilfiker, Thomas E. Tiwald, Corey L. Bungay, Ron A. Synowicki, Duane E. Meyer, Craig M. Herzinger, Galen L. Pfeiffer, Gerald T. Cooney, and Steven E. Green, "Variable angle spectroscopic ellipsometry in the vacuum ultraviolet", Optical Metrology, Roadmap for the Semiconductor, Optical, Data Storage Industries, 4099, 197 (2000). doi: 10.1117/12.405820.

48. Morten Kildemo, Fredrik Hansteen and Ola Hunderi, "Details of below band-gap uniaxial dielectric function of SiC polytypes studied by spectroscopic ellipsometry and polarized light transmission spectroscopy", Journal of Applied Physics, 91(9), 5677–5685 (2002). doi: 10.1063/1.1461887.

49. Huihui Li, ChangCai Cui, Jing Lu, Zhongwei Hu, Wuqing Lin, Subiao Bian and Xipeng Xu, "Mueller matrix ellipsometric characterization of nanoscale subsurface damage of 4H-SiC wafers: from Grinding to CMP," Front. Phys., 2022, doi: 10.3389/fphy.2021.820637.

50. Oriol Arteaga and Adolf Canillas, "Analytic inversion of the Mueller–Jones polarization matrices for homogeneous media: erratum", Optics Letters, 35(20), 3525 (2010). doi: 10.1364/ol.35.003525.

51. Oriol Arteaga, Enric Garcia-Caurel and Razvigor Ossikovski, "Anisotropy coefficients of a Mueller matrix", Optics Society of America A, 28(4), 548 (2011). doi: 10.1364/josaa.28.000548.

52. Huihui Li, ChangCai Cui, Subiao Bian, Jing Lu, Xipeng Xu and Oriol Arteaga, "Model-free determination of the birefringence and dichroism in c-cut crystals from transmission ellipsometry measurements", Applied Optics, 59(7), 2192 (2020). doi: 10.1364/ao.386583.

53. Simeon Agathopoulos, "Influence of synthesis process on the dielectric properties of B-doped SiC powders", Ceramics International, 38(4), 3309–3315 (2012). doi: 10.1016/j.ceramint.2011.12.040.

54. J. Petalas, S. Logothetidis, M. Gioti and C. Janowitz, "Optical properties and temperature dependence of the interband transitions of 3C- and 6H-SiC in the energy region 5 to 10 eV", Physica Status Solidi Basic Res, 209(2), 499–521 (1998). doi: 10.1002/(SICI)1521-3951(199810)209:2<499::AID-PSSB499>3.0.CO;2-M.

55. Chunhua Xu, Shunchong Wang, Gang Wang, Jingkui Liang, Shanpeng Wang, Lei Bai, Junwei Yang and Xiaolong Chen, "Temperature dependence of refractive indices for 4H- and 6H-SiC", Journal of Applied Physics, 115(11), 113501 (2014). doi: 10.1063/1.4868576.

56. Zhen Tong, Linhua Liu, Liangsheng Li and Hua Bao, "Temperature-dependent infrared optical properties of 3C-, 4H- and 6H-SiC", Physica B: Condensed Matter, 537, 194–201 (2018). doi: 10.1016/j.physb.2018.02.023.

57. H. Mutschke, Anja Cetti Andersen, Dominique Clément, Thomas Henning and G. Peiter, "Infrared properties of SiC particles", Astron Astrophys, 345(1), 187–202 (1999).

58. Benjamin Baugher and Jonathan Goldstein, "Temperature dependence of the birefringence of SiC", Optics Materials, 23(3–4), 519–528 (2003). doi: 10.1016/S0925-3467(03)00017-X.

59. Huijing Yang Jie Yuan, Yong Li, Zhiling Hou, Haibo Jin, Xiaoyong Fang, Maosheng Cao, "Silicon carbide powders: Temperature-dependent dielectric properties and enhanced microwave absorption at gigahertz range," Solid State Communications, 163, 1–6 (2013). doi: 10.1016/j.ssc.2013.03.004.

60. Jiafa Cai, Xiaping Chen, Rongdun Hong, Weifeng Yang and Zhengyun Wu, "High-performance 4H-SiC-based p-i-n ultraviolet photodiode and investigation of its

capacitance characteristics", Optics Communications, 333, 182–186 (2014). doi: 10.1016/j. optcom.2014.07.071.

61. Satyendra Mourya, Jyoti Jaiswal, Gaurav Malik, Brijesh Kumar and Ramesh Chandra, "Structural and optical characteristics of in-situ sputtered highly oriented 15R-SiC thin films on different substrates", Journal of Applied Physics, 123(2), 023109 (2018). doi: 10.1063/ 1.5006976.

62. Bernhard Von Blanckenhagen, Diana Tonova and Jens Ullmann, "Application of the Tauc-Lorentz formulation to the interband absorption of optical coating materials," Applied Optics, 41(16), 3137–3141 (2002). doi: 10.1364/AO.41.003137.

63. Sergey A. Grudinkin, Sergey A. Kukushkin, Aleksey Viktorovich Osipov and Nikolay A. Feoktistov, "IR spectra of carbon-vacancy clusters in the topochemical transformation of silicon into silicon carbide", Physics of the Solid State, 59, 2430–2435 (2017). doi: 10.1134/ S1063783417120186.

64. Jonathan E. Spanier and Irving P. Herman, "Use of hybrid phenomenological and statistical effective-medium theories of dielectric functions to model the infrared reflectance of porous SiC films", Condensed Matter and Materials Physics, 61(15), 10437–10450 (2000). doi: 10.1103/PhysRevB.61.10437.

65. Aakash Mathur, Dipayan Pal, Ajaib Singh, Rinki Singh, Stefan Zollner and Sudeshna Chattopadhyay, "Dual ion beam grown silicon carbide thin films: Variation of refractive index and bandgap with film thickness", Journal of Vacuum Science & Technology B, 37(4), 041802 (2019). doi: 10.1116/1.5097628.

66. Radhakrishnan Nair, Patricia Blake, Alexander N.Grigorenko, Kostya S. Novoselov, Timothy J. Booth, Tobias Stauber, Numo M.R. Peres, Andre K. Geim, "Fine structure constant defines visual transparency of graphene", Science (New York, N.Y.), 320(5881), 1308 (2008). doi: 10.1126/science.1156965.

67. Rositsa Yakimova, Tihomir Iakimov, G. Reza Yazdi, Chamseddine Bouhafs, Jonas Eriksson and Alexei Zakharov, "Morphological and electronic properties of epitaxial graphene on SiC", Physica B: Condensed Matter, 439, 54–59 (2014). doi: 10.1016/j.physb. 2013.12.048.

68. Chamseddine Bouhafs, Vanya Darakchieva, Inga-Lill Persson, Antoine Tiberj, Per O. Å. Persson, Matthieu Paillet, Ahmed-Azmi Zahab, Perine Landois, Sandrine Juillaguet, Stefan Schöche, Markus Schubert and Rositsa Yakimova, "Structural properties and dielectric function of graphene grown by high-temperature sublimation on 4H-SiC(000-1)", Journal of Applied Physics, 117, 085701 (2015). doi: 10.1063/1.4908216.

69. Alex Boosalis, Thomas Hofmann, Vanya Darakchieva, Rositsa Yakimova and Markus Schubert, "Visible to vacuum ultraviolet dielectric functions of epitaxial graphene on 3C and 4H SiC polytypes determined by spectroscopic ellipsometry", Applied Physics Letters, 101(1), 1–5 (2012). doi: 10.1063/1.4732159.

70. Tobias Stauber, Numo M. R. Peres and Andre K. Geim, "Optical conductivity of graphene in the visible region of the spectrum", Physical Review B – Condensed Matter Mater and Materials Physics, 78(8), 1–8 (2008). doi: 10.1103/PhysRevB.78.085432.

71. Peter E. Gaskell, Helgi S. Skulason, Wlodek Strupinski and Thomas Szkopek, "High spatial resolution ellipsometer for characterization of epitaxial graphene", Optics Letters, 35(20), 3336–3338 (2010). doi: 10.1364/ol.35.003336.

72. Luxmi, Neelabh Srivastava, Guowei He, Randall M. Feenstra and Patricia J. Fisher, "Comparison of graphene formation on C-face and Si-face SiC {0001} surfaces", Physical Review B – Condensed Matter Mater and Materials Physics, 82(23), 1–11 (2010). doi: 10.1103/ PhysRevB.82.235406.

73. Sviatoslav Shokhovets, Ruediger Goldhahn and Gerhard Gobsch, "Determination of the anisotropic dielectric function for wurtzite AlN and GaN by spectroscopic ellipsometry", Journal of Applied Physics, 94(1), 307–312 (2003). doi: 10.1063/1.1582369.

74. Maria Losurdo, Giovanni Bruno, Tae-Hun Kim, Soojeong Choi, April Brown and Akihiro Moto, "Nucleation and growth mode of the molecular beam epitaxy of GaN on 4H – SiC

exploiting real time spectroscopic ellipsometry", Journal of Crystal Growth, 284,156–165 (2005). doi: 10.1016/j.jcrysgro.2005.07.016.

75. Maria Losurdo, Maria M. Giangregorio, Pio Capezzuto, Giovanni Bruno, Tae-Hoon Kim, Sungyeol Choi, Alastair Brown, "Remote plasma assisted MOCVD growth of GaN on 4H-SiC: growth mode characterization exploiting ellipsometry", The European Physical Journal Applied Physic, 31(3), 159–164 (2005). doi: 10.1051/epjap:2005056.

76. Georg Rossbach, Marcus Röppischer, Pascal Schley, Gerhard Gobsch, Cassidy Werner, Christoph Cobet, Nathalie Esser, Armin Dadgar, Matthias Wieneke, Annalena Krost, Ruediger Goldhahn, "Valence-band splitting and optical anisotropy of AlN", Phys. Status Solidi Basic Res., 247(7), 1679–1682 (2010). doi: 10.1002/pssb.200983677.

77. A. Kakanakova-Georgieva, Per O.A. Persson, Amela Kasic, Lars Hultman and E. Janzén, "Superior material properties of AlN on vicinal 4H-SiC", Journal of Applied Physics, 100(3), 19–22 (2006). doi: 10.1063/1.2219380.

78. Junhua Yin, Daihua Chen, Hong Yang, Yao Liu, Devki N. Talwar, Tianlong He, Ian T. Ferguson, Kaiyan He, Lingyu Wan, Zhe Chuan Feng, "Comparative spectroscopic studies of MOCVD grown AlN films on Al$_2$O$_3$ and 6H–SiC", Journal of Alloys and Compounds, 857, 1–13 (2020). doi: 10.1016/j.jallcom.2020.157487.

79. Margarita P. Thompson, Gregory W. Auner, Changhe Huang and James N. Hilfiker, "A Study of Pt/AlN/6H-SiC MIS Structures for Device Applications", MRS Online Proceedings Library, 622, 651(2000). doi: 10.1557/PROC-622-T6.5.1.

80. Martin Feneberg et al., "Optical properties of cubic GaN from 1 to 20 eV", Physical Review B – Condensed Matter Mater and Materials Physics, 85(15), 1–7 (2012). doi: 10.1103/PhysRevB.85.155207.

81. Lorin X. Benedict, T. Wethkamp, Klaus Wilmers, Christoph Cobet, Nathalie Esser, Eric L. Shirley, Wito Richter, Magnolia Cardona, "Dielectric function of wurtzite GaN and AlN thin films", Solid State Communications, 112(3), 129–133 (1999). doi: 10.1016/S0038-1098(99)00323-3.

82. Chengyuan Yao, Shuchun Huo, Wanfu Shen, Zhaoyang Sun, Xiaodong Hu, Xiaotang Hu, Chunguang Hu, "Assessing the quality of polished brittle optical crystal using quasi-Brewster angle technique", Precision Engineering, 72, 184–191, (2021), doi:10.1016/j.precisioneng.2021.04.019.

83. Shiyuan Liu, Xiuguo Chen, Chuanwei Zhang, "Development of a broadband Mueller matrix ellipsometer as a powerful tool for nanostructure metrology", Thin Solid Films, 584, 176–185, (2015), doi:10.1016/j.tsf.2015.02.006.

12

Raman Microscopy and Imaging of Semiconductor Films Grown on SiC Hybrid Substrate Fabricated by the Method of Coordinated Substitution of Atoms on Silicon

Tatiana S. Perova

Department of Electronic and Electrical Engineering, Trinity College Dublin, The University of Dublin, Dublin, Ireland

Sergey A. Kukushkin and Andrey V. Osipov
Institute for Problems in Mechanical Engineering of the Russian Academy of Sciences, Saint Petersburg, Russia

1 Introduction

1.1 Raman Mapping of Various SiC Structures

Structures based on SiC are widely used in many fields of science and technology. Areas of application include microelectronics, power electronics, light emitting devices, MEMS, and biomedical devices [1-4]. There are more than 250 polytypes of SiC [5]. The most common forms used for applications are 4H, 6H (known as hexagonal α-SiC types) and the cubic 3C-SiC type (β-SiC). It is necessary to know which type is formed when making SiC. It is also desirable to investigate the homogeneity of SiC structures. One of the most important methods for studying these properties in SiC structures is Raman spectroscopy – a fast, contact-free technique with easy sample preparation.

Conventional Raman spectroscopy has been used to study SiC structures since the 1970s [6]. This research expanded significantly in 1975, when the Raman microscope, or micro-Raman spectroscopy, was invented [7]. Micro-Raman spectroscopy makes it possible to analyze relatively thin SiC layers (> 2 µm) deposited on various substrates (Si or other semiconductor materials) due to the high-power density characteristics of this method. In addition, this method provides an opportunity not only to determine the specific polytype of the SiC material, but also to perform Raman mapping (or imaging) of samples of various shapes and sizes. With recent developments in micro-Raman instruments, the ability to rapidly scan, or image, the wafer surface is now possible. Raman mapping allows detection of structural inhomogeneities, as well as the presence of various defects, stresses, and strains within SiC structures.

One of the purposes of this chapter is to review recent achievements in Raman mapping of SiC structures of different types and forms, including crystals, films, and fibers, as well

DOI: 10.1201/9780429198540-15

as the application of this technique to mapping of SiC thin films grown by the new method of coordinated substitution of atoms, recently proposed by two authors of this chapter (see Section 2.1).

The Raman spectrum of the most common SiC structures have different active modes in the range from 100 to 1000 cm^{-1}. Typically, the Raman spectrum of SiC of various polytypes [6-10] represented by a number of peaks, appears in the region 150–210 cm^{-1} (folded transverse acoustic modes FTA), between 760 and 800 cm^{-1} (transverse optical phonons E2 (TO)) and in the range 966–982 cm^{-1} (longitudinal optical phonons A1 (LO)), see Figure 1 and Table 1. The relative intensity and position of these peaks enable identification of the SiC polytype, while analysis of the phonon line width gives information about the structural disorder (or defects) inside the structure. The shift in the position of the phonon peak for a certain polytype makes it possible to determine the type and magnitude of stresses in the structure. Thus, measurement of Raman spectra over the entire sample with a typical

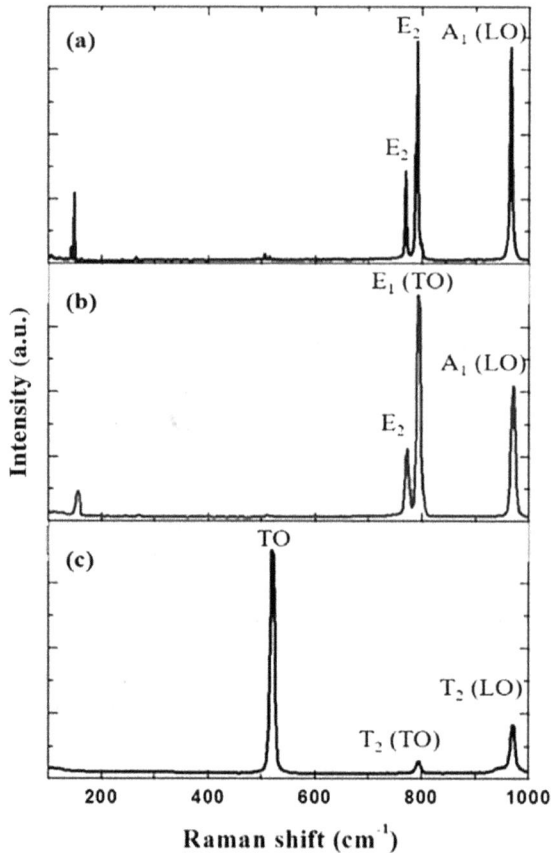

FIGURE 1

Raman signature spectra for the SiC polytypes: (a) 6H-SiC (0001), (b) 4H-SiC (000-1), (c) 3C-SiC (100)/Si (100). The main Raman modes are indicated. The strong peak at 500 cm^{-1} in (c) corresponds to the TO (G) vibrations in the silicon wafer Si (100). Reproduced from [11].

TABLE 1

Raman frequencies of the optical and acoustic phonon modes for bulk 3C, 2H, 4H, and 6H polytypes of SiC obtained by different groups from [8,9]

		Raman frequency, cm⁻¹			
Polytype	$x = q/q_B$	Planar acoustic FTA	Axial acoustic FLA	Planar optic FTO	Axial optic LTO
3C	0	-	-	796 (794.2)*	972 (972.7)*
2H	0	-	-	799	968
	1	264	-	764	-
4H	0	-	-	796 (795.6)	964 (867.3)
	2/4	196, 204	-	776 (775.7)	-
	4/4	266	610		838
6H	0	-	-	797	965 (969.5)
	2/6	145, 150	-	789 (787.9)	-
	4/6	236, 241	504, 514		889
	6/6	266	-	767 (765.6)	-

* Data from [9] are given in brackets.

step size of 0.1 µm provides high resolution information on the quality of the structure under study.

1.1.1 Crystalline Bulk SiC Structures

The Raman mapping technique is becoming more popular in semiconductor physics and microelectronics. This is because it allows control of crystal quality, composition, doping level, and homogeneity of as-grown semiconductors. It also allows stress measurements, both over the entire wafer area and at the level of an individual device [7]. Monitoring of the doping level, the spatial distribution of the dopant, and the associated defects in various types of SiC wafers (namely, 4H-SiC and 6H-SiC) was demonstrated in [9,10,12-14]. A study of the doping distribution in a 6H-SiC crystal at different doping levels over the wafer thickness was reported in [15].

The use of Raman mapping to analyze the residual stress in SiC samples caused by destructive treatment [16,17] or by indentation [18] is important in the field of MEMS and sensor technologies. An example of these studies from [18] is shown in Figures 2 and 3. In this case, both Raman lines, the TO phonon at 889.3 cm⁻¹ and the LO phonon at 971.5 cm⁻¹, were analyzed on a 6H-SiC wafer.

Silicon–silicon carbide (Si–SiC) and C/C-SiC composites have emerged as a new class of materials with properties such as high resistance against oxidation and corrosion, low thermal expansion coefficient, and high mechanical strength. This enables these composites to be successfully used in industrial applications such as combustion chambers, gas turbines, heat exchangers, fusion reactors, seal rings, welding nozzles, valve discs, and ceramic engine parts. The strength of these composites is dependent on their internal granular structure, which affects the local stress distribution in the composite. In Refs. [19-21] the Raman mapping technique was used to monitor the local stress fields in these composites in order to establish the best process conditions for manufacturing these substrates and also for protective layer (such as PyC/SiC) deposition. In the paper by Fu et al. [21], Raman

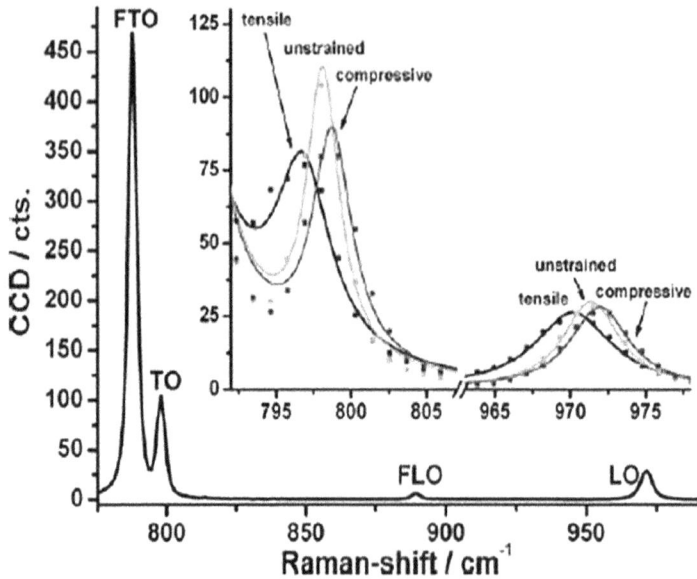

FIGURE 2
Typical Raman spectrum of unstressed 6H-SiC of the most prominent phonon lines at 889.3 (TO-plane) and 971.5 cm^{-1} (LO-axial). The insert shows the TO and LO peaks under tensile (black), neutral (light gray), and compressive (dark gray) loading causing corresponding line-shifts and line-broadening (symbols: experiment; lines Lorentzian fit). (From [18] – Figure 1, reproduced with permission of © The Optical Society).

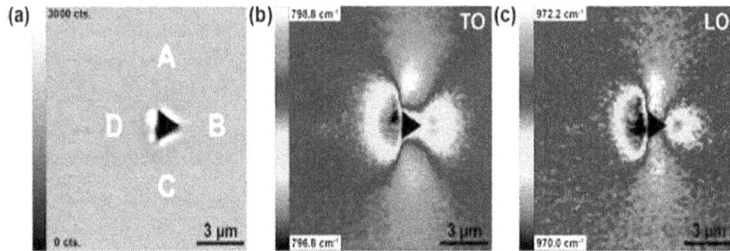

FIGURE 3
Confocal Raman microscopy of a locally stressed SiC sample. (a) Rayleigh intensity map. Four regions of interest are labelled A-D, (b,c) maps of the spectral position of TO and LO phonon lines as obtained by fitting with Lorentzian. Black triangles mark the contour of the indent. (From [18] – Figure 2a, reproduced with permission of © The Optical Society).

mapping was used in order to investigate the effect of laser ablation on SiC ceramics used as a third-generation material for mirrors.

1.1.2 SiC Layers Grown on Different Substrates

It is notable that most Raman mapping studies have been carried out for SiC films with a thickness of over 2 µm [22]. This is because the Si-C phonon bands, which usually appear in the region from 700 to 1000 cm^{-1}, are very weak. This is due to the low efficiency of

Raman scattering for the Si – C bond, since the cross section of the Si – C bond is low [23] compared to Si – Si and C – C bonds, the number of which increases with an excess of Si or C atoms during deposition.

In Ref. [24], Raman mapping was used to study the effect of the local temperature and hydrogen level during selective laser deposition of SiC from tetramethylsilane (TMS). The results obtained confirm that temperature is one of the most important factors in the deposition of pure SiC using a TMS precursor, which is consistent with thermodynamic predictions. In Ref. [25], Raman imaging of both the surface and the cross-section of the film/substrate was performed for better visualization of the stress distribution in the 3C-SiC epitaxial layer grown by chemical vapor deposition (CVD) on Si substrates with various orientations, viz. (100), (110), (111), and (211). Residual tensile (in the epitaxial layer) and compressive (in the substrate) stresses were detected in this study.

An interesting study of thin SiC layers was recently presented in [26]. The authors developed a new approach to analyzing the quality of a homoepitaxial 4H-SiC thin film using Raman mapping of the forbidden E1 (TO) mode at 798.6 cm^{-1} under UV excitation. This mode is usually not allowed while registering the spectrum with a backscattering geometry. It has been demonstrated that forbidden modes are an indication of crystal defects, allowing them to be used as a measure of crystal quality.

1.1.3 SiC Fibers

A large body of Raman research over the past 15 years has been devoted to Raman imaging of SiC fibers. Most natural and synthetic fibers have both radial and structural variations. Since typical fiber diameters vary in the range of tens of microns, investigation of fiber properties can be performed using Raman mapping along the fiber diameter [7]. Some recent works on the study of SiC-based fiber composites are presented in Refs. [27-31]. Figure 4 shows the results of Raman mapping of residual stress, generated by a melt infiltration process around a Sylramic fiber, for SiC and Si phonon lines [29]. The measured Raman band's position shifts were imported into MATLAB® to calculate the residual stress maps in the area sampled. The corresponding reference Raman shifts (without residual stress) were 515.3 cm^{-1} (LO mode) and 796 cm^{-1} (TO mode) for Si and SiC, respectively. The reference for the Raman peak position deviates from the typical Si peak position (at ≈520 cm^{-1}) due to the presence of boron dopant in the silicon.

1.1.4 Gr and III-V Semiconductors Deposited onto SiC

Deposition of thin films of graphene (Gr) and III-V semiconductors (GaN, AlN) onto SiC is required for device fabrication in microelectronics and optoelectronics. Since there is a potential for stress to arise during deposition or growth of these films, the use of Raman imaging to investigate the quality of these layers is extremely important. The most recent work on Gr/SiC is published in [11,32-34]. The first three of these references are reviews, summarizing an extremely large number of papers published in this area. One example of these studies is shown in Figure 5 (a,b) and shows Raman mapping of the ratio of intensities of G (graphitic band at ~1600 cm^{-1}) and 2D (disorder band at ~2700 cm^{-1}) bands for quasi-free-standing monolayer graphene (QFMLG). Note, that mapping the I_D/I_G distribution allows visualization of the spatial homogeneity of structural properties of the hydrogen-intercalated graphene QFMLG.

In [35], Raman mapping was applied to study microcracks in GaN films grown on SiC substrates with an AlN nucleation layer using the metalloorganic CVD technique.

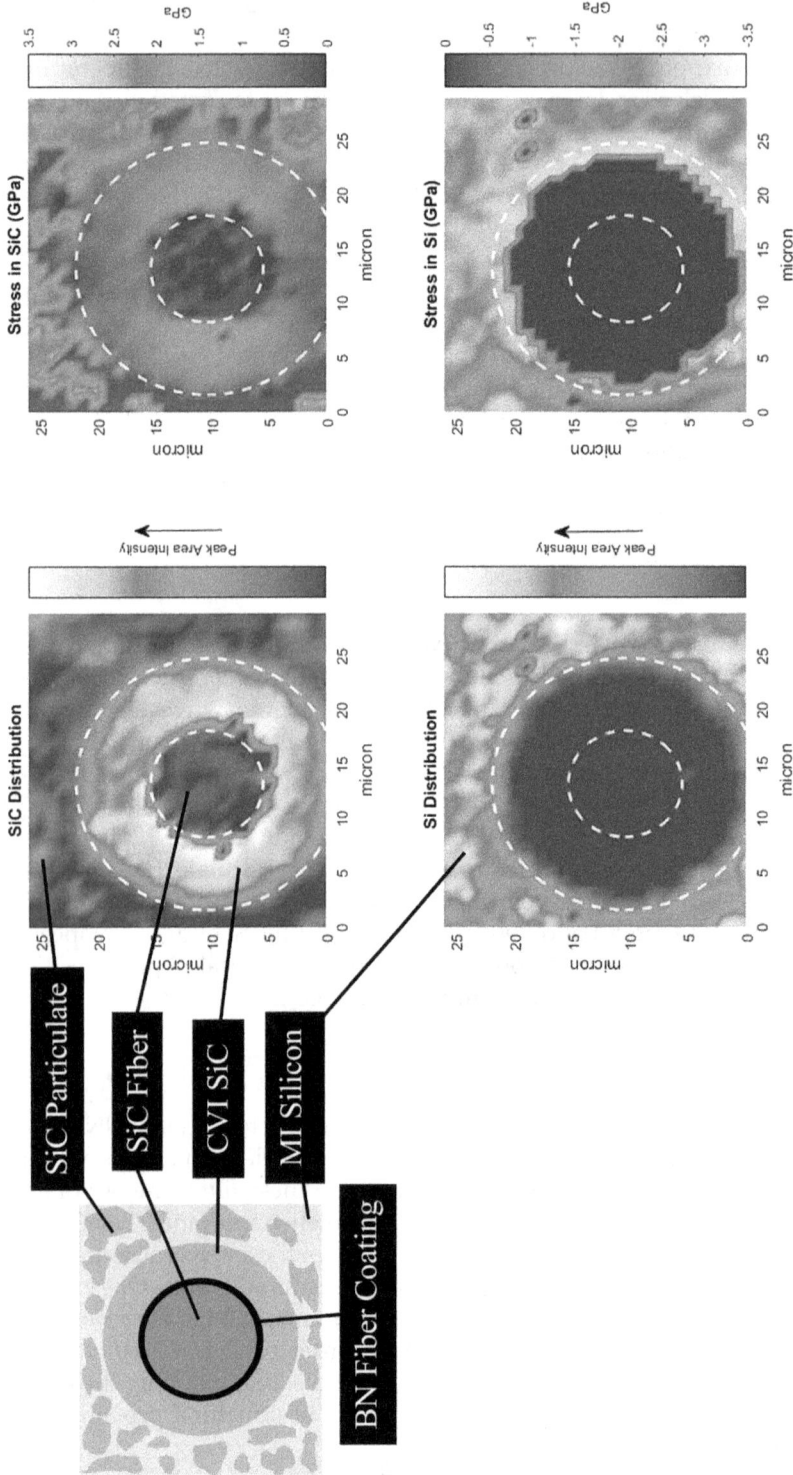

FIGURE 4

Residual stresses of silicon carbide and unreacted silicon around the Sylramic fibers measured by micro-Raman spectroscopy in a 30 μm × 30 μm map. (30 μm × 30 μm area of the machined surfaces was scanned point by point with the individual points spaced 1 μm apart, the spot size was smaller than 1 μm from 514.5 nm Ar+ laser). The following notations are used: CVI: chemical vapor infiltrated, MI: melt-infiltrated, BN: boron nitride. (From [29] – Figure 3, reproduced with permission of Elsevier).

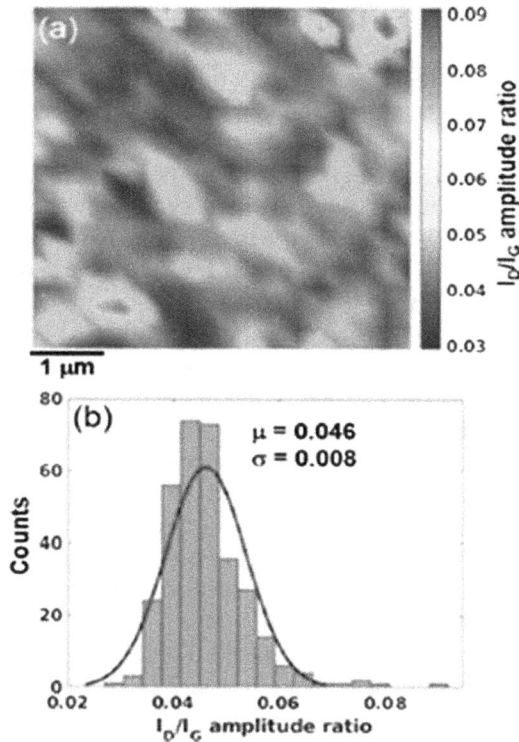

FIGURE 5
(a) Map of the I_D/I_G peaks amplitude ratio for the QFMLG sample obtained from two-dimensional Raman mapping. (b) Histogram of the I_D/I_G amplitude ratio distribution fitted with a Gaussian function. (From [34] – Figure 3, reproduced with permission of IOP Publishing).

Microcracks in GaN films were observed in some of the samples grown using various imaging techniques. Raman imaging results reveal the presence of significant residual tensile stress in the nucleating layer of AlN, leading to tensile stresses in the GaN films and, ultimately, cracking.

1.2 Method of coordinated Substitution of Atoms and its Distinctive Features

In 2008, in [36,37], a new method was developed for the synthesis of epitaxial films of SiC on Si. The method [36,37] is based on the chemical transformation (conversion) of the surface layers of Si into epitaxial layers of SiC due to the chemical interaction of gaseous carbon monoxide (CO) with the surface of the silicon substrate according to reaction

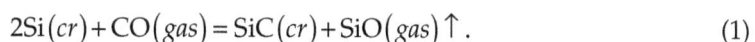

$$2\text{Si}(cr) + \text{CO}(gas) = \text{SiC}(cr) + \text{SiO}(gas)\uparrow. \tag{1}$$

In recent works, this method was called as the method of coordinated substitution of atoms (CSA) [38]. The term "coordinated" means that new chemical bonds are formed simultaneously with the destruction of old bonds. Later, in [38-47], a consistent theory was developed that describes the entire spectrum of interrelated physicochemical processes occurring during the topochemical transformation of Si into SiC. In contrast to the standard growth of SiC by the CVD method, in which the growth of the SiC layer occurs due to the chemical reaction of reagents falling on the surface of the substrate, when using the CSA [36-47], the growth of the layer occurs inside the surface layers of the Si substrate due to the replacement of some of the silicon atoms with carbon atoms. For the sake of fairness, we note that earlier attempts were made to grow SiC by replacing Si atoms with C atoms (see, for example, Refs. [48-53]). The authors of these works used methane and other hydrocarbons. However, the resulting SiC films were of poor quality. These works stimulated the further development of the Si carbonization method. A detailed analysis of the work on carbonization can be found in the review [54]. Note that a consistent quantum-chemical model of the growth of SiC on Si by the CSA method is described in [38].

An amazing feature of the CSA method is that the dependence of the thickness of the grown SiC layer on the pressure of carbon monoxide (CO) is domed; first it increases to a certain maximum value, and then drops to almost zero [55]. The theoretical model of epi-taxy developed in Ref. [55] enables explaination of such a behavior of the layer thickness from CO pressure. It turned out that the gaseous reaction product SiO interferes with the flow of the gaseous reactant CO through the channels of the crystal lattice, reducing their hydraulic diameter. Naturally, the thickness of the SiC films is not large in this case. In particular, at a growth temperature of 1300 °C the highest quality films on Si (111) have a thickness of 20–100 nm and are obtained at CO pressures $P_{CO} \approx$ 200–500 *Pa*. Figure 6(a) shows a typical SEM image of a SiC/Si cross-section, which demonstrates a continuous layer of SiC and pores in Si covered with SiC.

However, for some applications it is necessary to obtain high-quality single-crystal SiC films with a noticeably larger thickness, for example, 200–1000 nm. In [56], a new technique was proposed and implemented, which allows the thickness of the SiC layer to be increased by about an order of magnitude. The technique is based on the theoretical conclusions made in [44] and consists in the fact that, before the growth process, the surface of the silicon substrate is saturated with vacancies by annealing in vacuum at T = 1350 °C for ~ 1–30 min. The annealing time is determined by the need to obtain films of one thickness or another. Thus, at the beginning of the synthesis, silicon vacancies are deliberately formed in the Si substrate prior to the growth of SiC. This allows new pathways to be created for CO diffusion and SiO removal from the substrate. Silicon vacancies penetrate from Si into SiC and provide a high rate of removal of the reaction product SiO from the reaction zone to the outside. Therefore, SiO interferes with the growth of SiC much less; moreover, CO, with the help of vacancies, also penetrates deep into SiC more efficiently, which ultimately leads to significantly thicker SiC layers. This leads to the fact that the mechanism of mass transfer from interstitial to vacancy changes when Si atoms are replaced by C atoms. As a

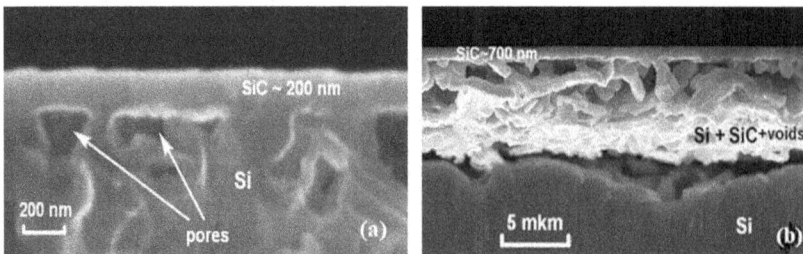

FIGURE 6

Typical SEM images of a cross-section of a SiC layer grown on a Si (111): (a) sample with SiC layer of 200 nm thick grown by standard CSA method; (b) sample of SiC layer of 700 nm thick grown by modified CSA method at a temperature T = 1350 °C and CO pressure p_{CO} = 80 Pa for 10 min with preliminary saturation of Si by vacancies. The SiC layer and pores in Si, covered with SiC, are clearly visible.

result, not only the thickness of the SiC layer increases, but also delamination of SiC from the Si substrate occurs if the thickness of the SiC layer is more than 400 nm. Figure 6(b) shows a SEM image of a thick layer of SiC grown by the method described in [56].

There are two main distinctive features of SiC obtained by the CSA method from SiC crystals and films grown by other methods. These differences are due to the fact that the SiC synthesis reaction (1) proceeds in two stages. At the first stage, complexes between a Si vacancy and an interstitial C atom are formed. In the second stage, C atoms are displaced towards Si vacancies, forming SiC. Activated complexes "Si vacancy – interstitial C atom" are transformed into SiC, and the vacant vacancies merge into pores under the SiC layer. As a result, a SiC film is formed, partially hanging over the pores in Si [36-47]. In this case, the orientation of the film is set by the "old" crystal structure of the initial Si matrix, and not only by the surface of the substrate, as is usually realized in traditional methods of growing films. In this case, at the second and final stage of the transformation of Si into SiC, a process of shrinkage or "collapse" of the material occurs, in which SiC, as a new phase, is separated from the silicon matrix. In the process of "collapse", some of the Si atoms on the (111) and (110) faces at the interface undergo short-term (about 10^{-4} s) compression (impact) with a pressure of about 200 GPa (about 2 ml of atmospheres), which arises in the process of shrinkage of the material. As a result, silicon at the interface transforms into a semi-metal state, which leads to the negative coeffcicient of dielectric permittivity in the ellipsometric spectrum [57,58]. When SiC is grown by a standard technique, Si is in a normal semiconductor state. In the samples of SiC films grown by the CSA method, two quantum effects arise due to the presence of carbon-vacancy in SiC at room temperature in weak magnetic fields [59]. The first of these effects is the formation of a hysteresis of the static magnetic susceptibility (Figure 7a) and the second is the appearance of Aharonov–Bohm oscillations in the field dependences of the static magnetic susceptibility (Figure 7b). The first effect is associated with the Meissner–Ochsenfeld effect, and the second, with the presence in these structures under the SiC layer, in addition to carbon-vacancy structures, microdefects in the form of nanotubes and micropores formed during the synthesis of SiC [59]. For the first time, electroluminescent radiation in the middle and far infrared

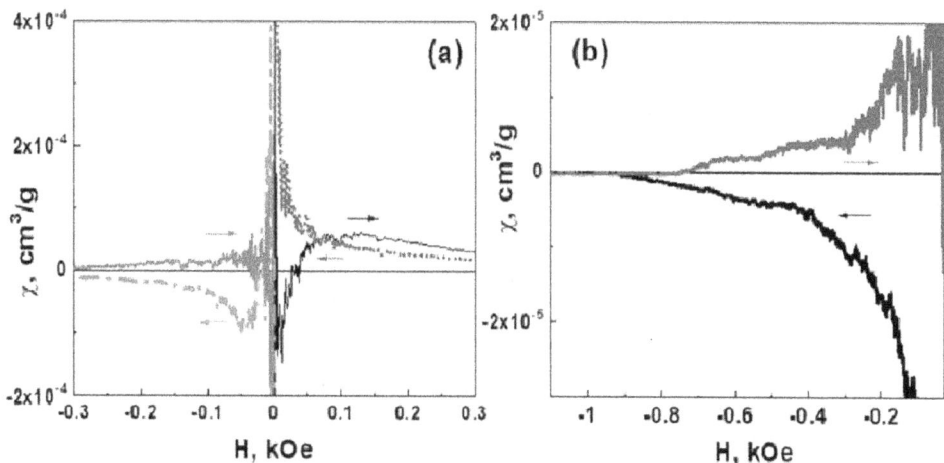

FIGURE 7

(a) Field dependences of static magnetic susceptibility, demonstrating characteristic hysteresis in weak magnetic fields and (b) field dependences of static magnetic susceptibility, demonstrating Aharonov-Bohm oscillations with a large period in SiC-3C (111)/Si (111) samples. The arrows indicate the direction of scanning of the external magnetic field.

ranges, emitted from SiC on Si nanostructures, was discovered. Electroluminescence from the edge channels of nanostructures is induced using a longitudinal source-drain current. The electroluminescence spectra obtained in the terahertz frequency range, 3.4 and 0.12 THz, arise due to the quantum Faraday effect. It has been shown experimentally that the longitudinal current induces a change in the number of magnetic flux quanta in the edge channels, which leads to the formation of a generation current in the edge channel and, accordingly, to terahertz radiation [60,61] (see Figure 7). Nothing of the kind is observed in SiC layers grown by a standard technique. The field dependences of the static magnetic susceptibility are shown in Figure 7, demonstrating characteristic hysteresis in weak magnetic fields and Aharonov–Bohm oscillations with a long period in SiC-3C (111)/Si (111) samples at room temperature.

Depending on the preparation conditions, as a result of the "collapse" process, excess Si vacancies are present in the SiC layers in an amount of 0.5–5%. It is the presence of Si vacancies that is a distinctive feature that determines all the basic properties of SiC formed by conversion from SiC obtained by standard methods [38,57,58]. Ordinary SiC contains only C vacancies [62]. This is due to the fact that the energy of formation of a silicon vacancy in SiC is of the order of ~ 8.7 eV, while for a carbon vacancy ~ 4.6 eV. In SiC grown by the CSA method, in contrast, excess Si vacancies are formed. Using the methods of infrared spectroscopy and spectroscopic ellipsometry, we were able to experimentally prove the existence of such formations [63–65]. Figures 8(a,b) show an image of a C-vacancy cluster in the cubic polytype of SiC (3C-SiC), calculated by the methods of quantum chemistry, and its infrared spectrum.

Figure 8(b) shows the experimentally obtained IR transmission spectra of epitaxial SiC samples grown on boron-doped Si. As a result of our studies in all samples, the IR spectra revealed a vibrational band lying in the region of 960 cm^{-1}, which was not previously observed either in SiC single crystals or in SiC films grown both on silicon substrates and on substrates of other materials. Note that the same band was also found in IR reflection spectra [63–65].

A typical Raman spectrum of SiC samples grown on boron-doped Si, the cross-section of which is shown in Figure 6(b), is demonstrated in Figure 9. Since this sample has already

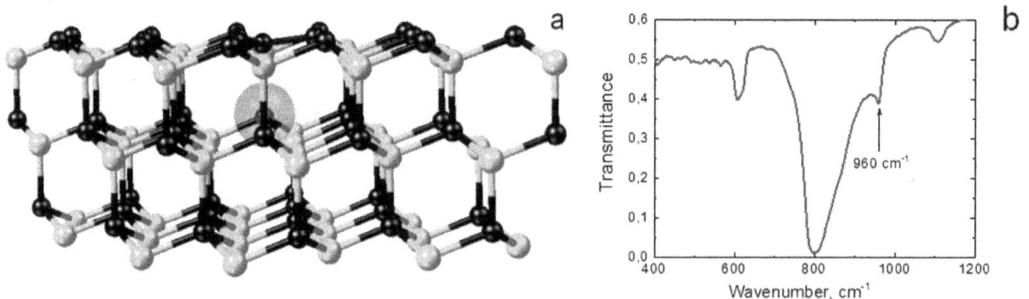

FIGURE 8
(a) Image of a carbon-vacancy structure in 3C-SiC, calculated by quantum chemistry methods. Carbon-vacancy structure, consisting of an almost flat cluster of four carbon atoms and a void with a diameter of 2.1 Å, shown as a semi-transparent sphere; the <111> axis points up; a cluster of four carbon atoms in the (111) plane is formed due to a jump of a carbon atom from below to the place of a silicon vacancy. (b) experimental IR transmission spectrum of epitaxial SiC.

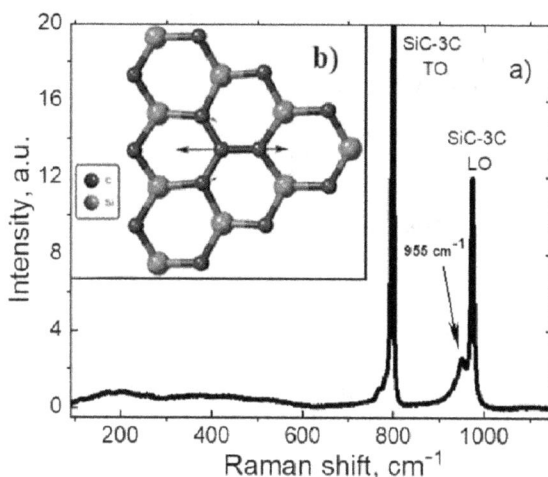

FIGURE 9

(a) Raman spetrum of 3C-SiC/Si (111) sample prepared on boron-doped Si. An additional peak at 955 cm^{-1} is attibuted to carbon-vacancy structures in a stable state. Insert (b) – vibrations of the carbon-vacancy structure corresponding to the theoretical frequency in the Raman and IR spectra at 970 cm^{-1}. The largest movement is performed by one C – C bond with a length of 1.57 Å.

begun to peel off, the silicon line is absent. In addition to the two peaks of 3C-SiC, TO and LO, there is an additional peak at 955 cm^{-1}, which was attributed to carbon vacancy structures that have passed into a stable state [66]. Density functional theory (DFT) modeling showed that it is with this frequency that the C – C bonds, that have arisen in 3C-SiC after a C atom jump to the place of a Si vacancy, vibrate [66]. This line was previously repeatedly observed [63–65] in infrared transmission and reflection spectra of 3C-SiC samples, obtained by the CSA method, but with a significantly lower concentration of vacancies.

The main polytype formed in the process of coordinated substitution of atoms, as shown in our numerous studies, including those carried out using the powerful apparatus of quantum chemistry [38], is the cubic polytype 3C-SiC. However, this does not mean that at certain time stages of the transformation of Si into SiC and on the Si faces with a different orientation from (111), other SiC polytypes cannot be formed. During topochemical transformation, polytypes 4H-SiC, 6H-SiC [36,42,44,67–70] and even rather rare polytypes 2H-SC and 8H-SiC can be formed (see Figure 1 in [71]). In addition in Ref. [72], based on group-theoretical analysis, the formation of a new, previously unknown trigonal (rhombohedral) phase of SiC during the conversion of Si to SiC was theoretically predicted, and in [73] this phase was experimentally discovered. It turned out that if the synthesis time of SiC films does not exceed 5 min, then in the Raman spectra, in addition to the 3C-SiC phonon lines, there is also line at 258 cm^{-1}, which refers to the new intermediate rhombohedral phase of SiC. In the Raman spectra of the samples that were synthesized for more than 5 min, this line was absent. The Raman spectrum of the discussed structure is shown in Figure 10. In this figure, the line at 258 cm^{-1} is clearly visible, which is close to the theoretically calculated line at 266 cm^{-1}, and, in our opinion (within the error of the growth experiment), related to the new intermediate rhombohedral phase of SiC.

Note that in our last published works, the SiC/Si substrate fabricated by the CSA method was called a hybrid SiC/Si substrate.

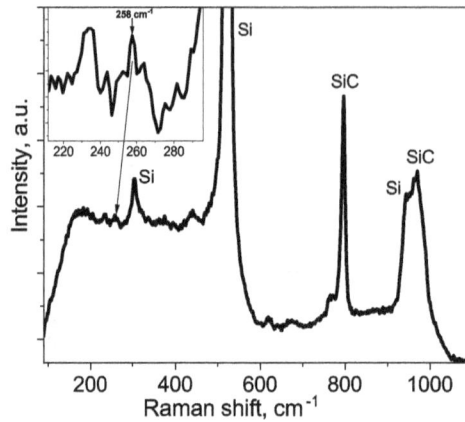

FIGURE 10

Raman spectrum of the intermediate rhombohedral phase of SiC grown by the method of coordinated substitution of atoms on Si.

2 Experimental Details

2.1 Investigation of 3C-SiC Layers Deposited by Method of Coordinated Substitution of Atoms

2.1.1 Fabrication of SiC/Si Hybrid Substrate

This section is concentrated on fabrication and characterization of SiC thin films epitaxially grown on Si substrate. The main attention is paid to the characterization of these films using a Raman mapping technique. A series of SiC films with thickness varying from ~20 nm up to 200 nm and with different polytype structure (3C, 6H, 4H) were grown on Si (111) and 6H-SiC substrates by the described in Section 1.2 CSA method, resulting in the formation of voids in the Si substrate near the SiC/Si interface (Figure 6). The deposition of SiC is achieved by means of a chemical interaction between mono-crystalline Si and CO in gas phase described by equation (1) and schematically shown in Ref. [36] (see Figure 1).

A low-pressure CVD system with a vertical cold-wall reactor made from sapphire, with a diameter of 40 mm and length of 50 mm, with the heated central zone, was used for SiC film deposition. The silicon wafer was placed on a graphite holder, with a thermocouple attached to the end. The sapphire tube was connected to a high vacuum system, consisting of diffusion and turbo-molecular pumps. Initially, the system was pumped down to a pressure of 10^{-5}–10^{-6} Torr. For SiC deposition, a 2-inch (111)-orientated Si wafer with a thickness of 300 μm and a tilt of 4° was used. CO gas was supplied at a rate of 1–10 ncm³/min and a pressure of 0.1–10 Torr. Growth of SiC layers occurred in the temperature range 1100°C to 1350°C and growth durations of 10–60 min were used. Due to the fabrication procedure, the SiC samples obtained are lightly doped with nitrogen at a level of 10^{14} cm⁻³.

2.1.2 Microscopy Characterization Techniques

The following samples were examined using a variety of microscopy imaging techniques: presumably 3C-SiC samples 378, 389, 439, 452, 469, 501 grown on a Si substrate, and sample AZ, definitely consisting of cubic SiC, deposited on a 6H-SiC substrate.

The cross section and surface morphologies of SiC films were examined using a Tescan Mira scanning electron microscope (SEM) and NT-MDT atomic force microscopy (AFM) operating in contact mode. The quality of the crystalline structure was investigated using energy dispersive x-ray spectroscopy (EDX).

Raman spectra were registered in a backscattering geometry using a RENISHAW 1000 micro-Raman system equipped with a CCD camera and a Leica microscope. A 1800 lines/mm grating was used for all measurements, providing a spectral resolution of ~1 cm^{-1}. Two types of measurements were performed: single spot measurements both from a void area and outside the void of the SiC layers, as well as line and area mapping measurements conducted across the voids. As an excitation source for single measurements an Ar^+ laser at wavelength of 457 nm with a power of 10 mW was used, while for line and area mapping the excitation wavelength λ was 633 nm using a HeNe laser with a laser power of 10 mW. The line mapping was performed at the distance X, ranging from 0 to 13 μm with a step size of 0.5 μm, where zero corresponds to the starting point of the measurements. Whilst the area mapping was performed in the area of 8×8 μm^2 with a step size of 0.2 μm. The laser radiation was focused onto the sample using a 100× microscope objective with short focus working distance, providing a spot size of ~ 1 μm. UV Raman spectra were collected using the micro-Raman system HR800 model supplied by Horiba Jobin Yvon. As an excitation source the He-Cd laser at 325 nm with power of 7 mW was used. The laser spot was focused on the sample surface using 40× objectives with short focus working distance. MATLAB software was used to simulate the enhancement of a Raman signal within voids.

As shown in Figures 6(a,b) and 11(a) the growth process suggested in [36,37] is accompanied by the formation of voids in Si substrate under the SiC layer. The voids typically have an inverted pyramid shape when using (111) Si or (100) Si. The formation of voids at the initial stage of SiC film growth (carbonization process) on Si substrate has already been discussed in the literature for different types of growth processes [74–76]. The voids formed during the process described by reaction (1) are often filled with some type of SiC crystallite grown inside the voids and attached to the Si (110) planes (see Figures 6(a,b) and 11(a)). We assume that those SiC crystallites have orientations different than the SiC material from the top layer. Figure 11(a) presents the SEM image of a cross section of the SiC/Si sample with observed triangular voids. The voids were also observed using optical microscopy (see Section 2.1.3). The presence of the voids is confirmed for the first time by micro-Raman spectroscopy, as reflected by the strong enhancement of Raman signal observed at the voids. In addition, micro-Raman measurements were applied for investigation of the voids' influence on properties of the SiC layer as discussed in the next section.

The surface morphology of 3C-SiC layers on Si (111), investigated by SEM and AFM microscopy, is demonstrated in Figures 11(b,c,d). Figures 11(b,c) show the SEM and AFM images of sample 501, which was grown with improved growth conditions. This SiC layer also reveals the mixture of 6H and 3C polytypes, which is discussed in the next paragraph. It can be seen from Figure 11(c) that the surface of the SiC layer is relatively smooth with roughness of ~5 nm rms. A SEM image of cubic SiC layer deposited on the 6H-SiC substrate is shown in Figure 11(d), which demonstrates that the surface of the layer grown on SiC is smoother compared to that grown on Si [82].

2.1.3 Raman Microscopy and Mapping of SiC/Si Hybrid Substrate

Raman spectra of bulk 6H-SiC substrate and 3C-SiC layer grown on 6H-SiC substrate, measured with excitation wavelength of 457 nm, are shown in Figure 12 [82,83]. For bulk 6H-SiC three characteristic features are observed: the LO phonon mode with A_1 symmetry

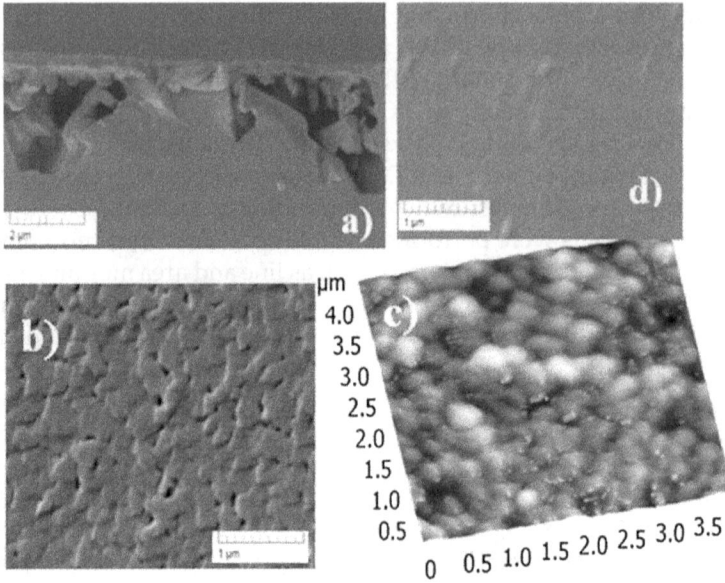

FIGURE 11
(a) SEM images of cross section of sample 439 SiC/(111)Si; (b) SEM and (c) AFM images of the sample 501 with improved growth condition (surface roughness is 5 nm rms), (d) SEM image of the sample AZ (3C-SiC on 6H-SiC substrate).

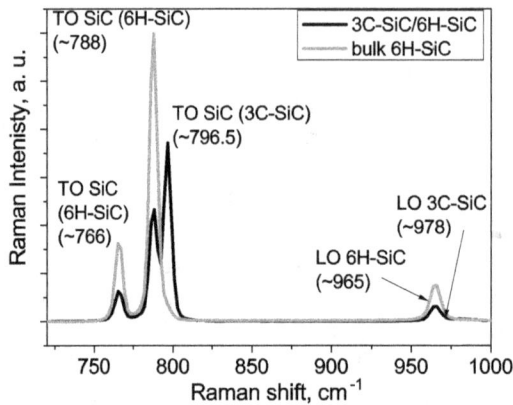

FIGURE 12
Raman spectra of bulk 6H-SiC and sample AZ (3C-SiC layer on 6H-SiC substrate). Reproduced from [83].

at 965 cm^{-1} and two TO phonon modes at about 788 and 766 cm^{-1} with E$_2$ symmetry. This agrees with previously published data [8,9,22]. For the sample AZ with a SiC layer grown on 6H-SiC substrate two additional bands, corresponding to 3C-SiC, were observed: TO phonon mode at ~796 cm^{-1} and low intensity LO phonon mode at ~978 cm^{-1} on the high-frequency side of 6H-SiC LO peak. Table 2 presents the peak position, intensity, and linewidth of TO and LO SiC phonon modes for sample AZ and 6H-SiC substrate.

Figure 13(a) shows Raman spectra of SiC film, grown on Si substrate, measured with 457 nm excitation at the void and outside the void. As seen from Figure 13(a) the cubic

TABLE 2

Peak position, intensity, and linewidth of TO and LO SiC phonon modes for samples AZ and 6H-SiC substrate

Sample name	Phonon mode	Peak position, cm^{-1}	Intensity, a.u.	Linewidth, cm^{-1}
AZ	TO (E_1) SiC (6H-SiC)	766.1	25684.1	5.8
	TO ($2E_2$) SiC (6H-SiC)	787.6	93481.4	5.6
	TO SiC (3C-SiC)	796.5	146079	5.4
	LO (A_1) SiC (6H-SiC)	965	12038.7	7.4
	LO SiC (3C-SiC)	978.6	1640.1	25
	SiC 2nd order	1527.2	3644.4	46.2
Bulk	TO (E_1) SiC (6H-SiC)	766.1	67651.6	5.7
6H-SiC	TO ($2E_2$) SiC (6H-SiC)	787.7	247188	5.6
	LO (A_1) SiC (6H-SiC)	965	30730.4	7.5
	SiC 2nd order	1525.7	4027.1	50.2

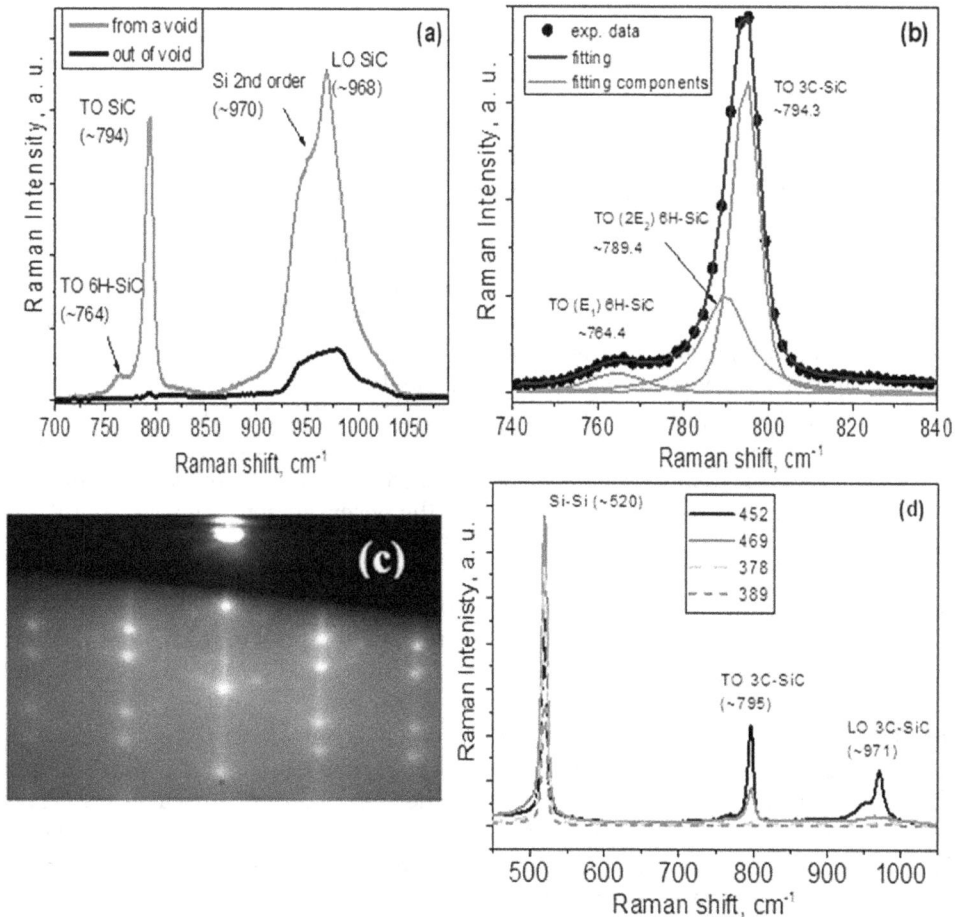

FIGURE 13

(a) Raman spectra of SiC layer grown on Si substrate (sample 452) measured at the void area and from outside the void area. (b) Fitting of TO-band, registered at the void, with three functions (which are the mixture of Lorentzian and Gaussian). c) Electron diffraction pattern of 3C-SiC layer on Si (111). Reproduced from [83]. (d) Raman spectra of samples: 452, 469, 378, 389 measured with UV excitation wavelength of 325 nm.

TABLE 3

Peak position, intensity, and linewidth of TO and LO SiC phonon modes for sample 452

name of sample	mode	TO SiC			SiC 2nd order		
		peak position, cm^{-1}	intensity, a.u.	linewidth, cm^{-1}	peak position, cm^{-1}	intensity, a.u.	linewidth, cm^{-1}
452	TO SiC (3C-SiC)	794.4	19797.5	7.7	1517.6	1736.2	43.6
	TO ($2E_2$) SiC (6H-SiC)	789.7	5031	12			
	TO (E_1) SiC (6H-SiC)	764.4	1057	16.6			

3C-SiC characteristic modes of TO and LO phonons appear at 794 cm^{-1} and at ~ 968 cm^{-1}, respectively. This confirms that the SiC layers, analyzed in this work, consist mainly of cubic polytype structure [8,9,22]. Note that the peak at 794 cm^{-1} may also be attributed to the TO phonon mode for the disordered 6H-SiC structure observed in the backscattering configuration. However, this peak is significantly low intensity compare with another TO peak observed in 6H-SiC at 788 cm^{-1} (see Figure 12). This excludes an assignment of the peak at 794 cm^{-1} to the 6H-SiC polytype. Also, we confirmed the validity of the assignment of the peak at ~794 cm^{-1} to the cubic SiC by measuring the Raman spectra for this sample with a low numerical aperture (NA) objective of ×10 and ×20. The NA of ×100 objective gathers signal over a wide scattering angle. The backscattering geometry is distorted determining the presence of 797 cm^{-1} Raman peak in 6H polytype SiC. This peak is forbidden when the low NA objective is used. The peak at 794 cm^{-1} was observed in the Raman spectra taken with different objectives: ×100, ×50, ×20, and ×10. Therefore, this also excludes assignment of the peak at 794 cm^{-1} to the 6H polytype of SiC. A low intensity shoulder observed at ~ 764 cm^{-1} near the TO band indicates the presence of a small amount of 6H-SiC polytype in this SiC layer. From Figures 13(a) and 13(b), the TO peak at 794 cm^{-1} demonstrates asymmetry from the low-frequency side. At the same time, in accordance with Nakashima [9], structural disorder in the SiC leads to a symmetrical widening of all the TO peaks. We conclude that the observed asymmetry of the TO peak is due to the presence of a 6H-SiC peak at ~789 cm^{-1}, clearly demonstrated by the fitting of the TO band in Figure 13(b).

The wide feature observed in the range 900–1010 cm^{-1} is associated with Si second-order Raman scattering [84]. This peak is overlapped with the LO phonon peak of the SiC layer, which complicates the use of LO peak position for drawing the conclusion on SiC structure. The full width at half maximum (FWHM) of the band at ~794 cm^{-1}, as determined by fitting, is approximately 7.7 cm^{-1} (see Table 3). This is only 2.5 cm^{-1} larger than that for relaxed 3C-SiC on a 6H-SiC substrate presented in Figure 12. This relatively small difference in linewidth leads to the conclusion that the thin SiC layer grown on Si (111) has reasonably good crystalline quality. This was also confirmed by energy dispersive x-ray analysis (see Figure 13(c)). Almost no circles characteristic of the polycrystalline phase are seen on Figure 13c, but there are well-defined Kikuchi lines, which indicate that the crystal structure of the film is of good quality [37]. It was shown by numerous x-ray diffraction and electron diffraction studies and luminescence analysis that either the hexagonal or cubic polytype structure or a mixture of both can be grown using the method of coordinated substitution of atoms introduced in [36,37] (see also Section 1.2). It will also be experimentally demonstrated in Section 2.2. that semiconductor layers grown on a hybrid SiC/Si substrate with a SiC layer, composed of a mixture of cubic and hexagonal polytypes, have excellent structural quality and exhibit some unique electrical, magnetic, and optical properties.

The UV Raman spectra of representative SiC samples are presented in Figure 13(d). The UV light has smaller depth of penetration than visible laser light. Due to this fact, SiC LO phonon mode related to cubic SiC structure is more pronounced as can be seen from Figure 13(d). Weak bands, observed at ~ 765 cm^{-1} and 790 cm^{-1}, are related to TO 6H-SiC peaks for samples 452 and 469. This provides evidence of the presence of a small amount of 6H-SiC in 3C-SiC layers.

A large enhancement of the Raman peak intensity, by up to 30 times for some samples, for both TO and LO modes, is observed at the void area in Figure 13(a). This enhancement enables the acquisition of a reasonably good Raman spectrum from ultra-thin SiC layers (~ 80 nm), as shown in Figure 13(a). Three mechanisms may contribute to the observed enhancement of the Raman signal: (i) multiple reflection of the incident light at the void, (ii) interference and multiple reflection of the Raman signal in the SiC layer above the void and (iii) the presence of additional SiC material grown on the (110) Si ribs of the pyramid inside the voids [37,74]. The first mechanism is also responsible for the moderate enhancement of the Si second order peak, by approximately 4 times, from the Si ribs. A somewhat similar effect was discussed for porous Si and SiC in Refs. [85,86]. For the second mechanism mentioned, the enhancement of the Raman signal in thin films, surrounded by media with low refractive indices, was discussed recently for graphene in Refs. [87,88]. We use a similar approach in the estimation of the effect of multiple reflections of the Raman signal on the peak intensity from the thin film. This is done by using a three-layer model consisting of SiC-air-Si. The application of this model to explain the results of line and area mapping will be discussed further in Section 2.1.4.

Figure 14(a) presents an optical microscopy image of a 3C-SiC/Si sample, where the brighter areas correspond to the voids seen under thin SiC layers. The arrow on Figure 14(a) shows the route of the line-mapping measurements. Figures 14(b), 14(c) and 14(d) show Raman line-maps for the peak position, peak intensity, and linewidth of the SiC TO-peak along the voids for the 3C-SiC/Si sample 469. Since the position of the TO peak is more sensitive to the effect of stress relaxation [85], the TO SiC peak was used to study the relaxation level in 3C-SiC films grown at the top of voids of different thicknesses and sizes. The position of the TO-Raman band of SiC in the relaxed 3C-SiC structure is typically 796 cm^{-1}, but for SiC layers grown on Si, the TO band is shifted to the low-frequency side [89]. We observed the position of the TO-SiC peak at ~794cm^{-1}, which indicates that the SiC layer is under tensile stress. Tensile stress in the SiC layer is observed, since the lattice constant of SiC (a_{SiC}=4.3Å) is less than for Si (a_{Si} = 5.38Å). Figure 14(b) shows the peak position of the TO-SiC band as a function of distance X. As seen in the figure, the peak position changes from 794.5 cm^{-1} in the middle of the void to 793.5 cm^{-1} outside the void. Consequently, a greater tensile stress is observed outside the voids than within the voids, which confirms the partial stress relief in the voids.

Figure 14(d) shows the full width at half maximum (FWHM) of the SiC TO-mode as a function of mapping distance. The linewidth of the TO peak significantly increases at the cavities (by ~ 3 cm^{-1}), probably as a result of the contribution of differently oriented SiC materials inside the void as mentioned earlier. The strong 15-fold increase in the intensity of the Raman peak of the SiC TO mode inside the cavities is confirmed by the line-mapping measurements presented in Figure 14(c). It can be seen that the signal enhancement is significantly greater in the center of the voids, corresponding to a deeper cavity or a thicker air layer.

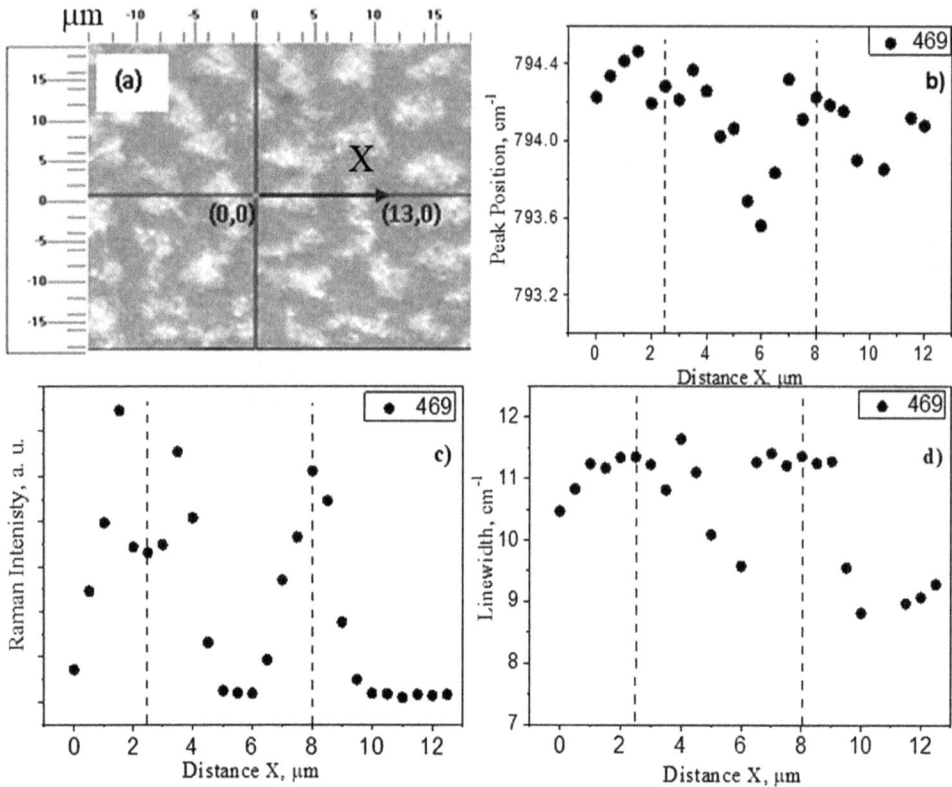

FIGURE 14

(a) Top view of the sample 3C-SiC/Si (111) obtained by optical microscopy, the arrow shows the mapping line. Results of Raman line-mapping for (b) peak position, (c) linewidth, and (d) peak intensity of the TO phonon mode for the 3C-SiC/Si sample 469 (dashed lines correspond to the centres of the voids). Reproduced from [83].

Figure 15(a) demonstrates the evolution of the experimental Raman spectra in the region of the TO SiC peak as a function of the distance extended along four voids for sample 452. At the center of each of the voids, a significant increase in the intensity of phonon mode at ~ 794 cm^{-1} is observed. A Raman area mapping was also made including several voids (see Figure 15(b,c)). Figure 15(c) shows an optical microscopic image of sample 452 with an 8×8 μm^2 mapping area shown in the square. The map of Raman intensity of the TO SiC mode at ~794 cm^{-1}, depending on the position [x, y] of the measured point in the above square is shown in Figure 15(b). As can be seen, the maximum intensity of the TO Si-C mode in Figure 15(b) matches exactly the voids observed in the optical microscopy image in Figure 15(c). The peak intensity increases as the depth of the void increases towards the center of the void.

2.1.4 Mechanism of the Raman Signal Enhancement

The enhancement of the Raman signal was estimated considering two mechanisms: (i) the multilayer interference of incident light and (ii) the multi-reflection of the Raman signal based on Fresnel's equation [88,90]. The MATLAB program was used for this purpose. We consider the incident light from air ($n_0=1$) onto a SiC layer (n_1), air(n_2)/Si(n_3) double-layer

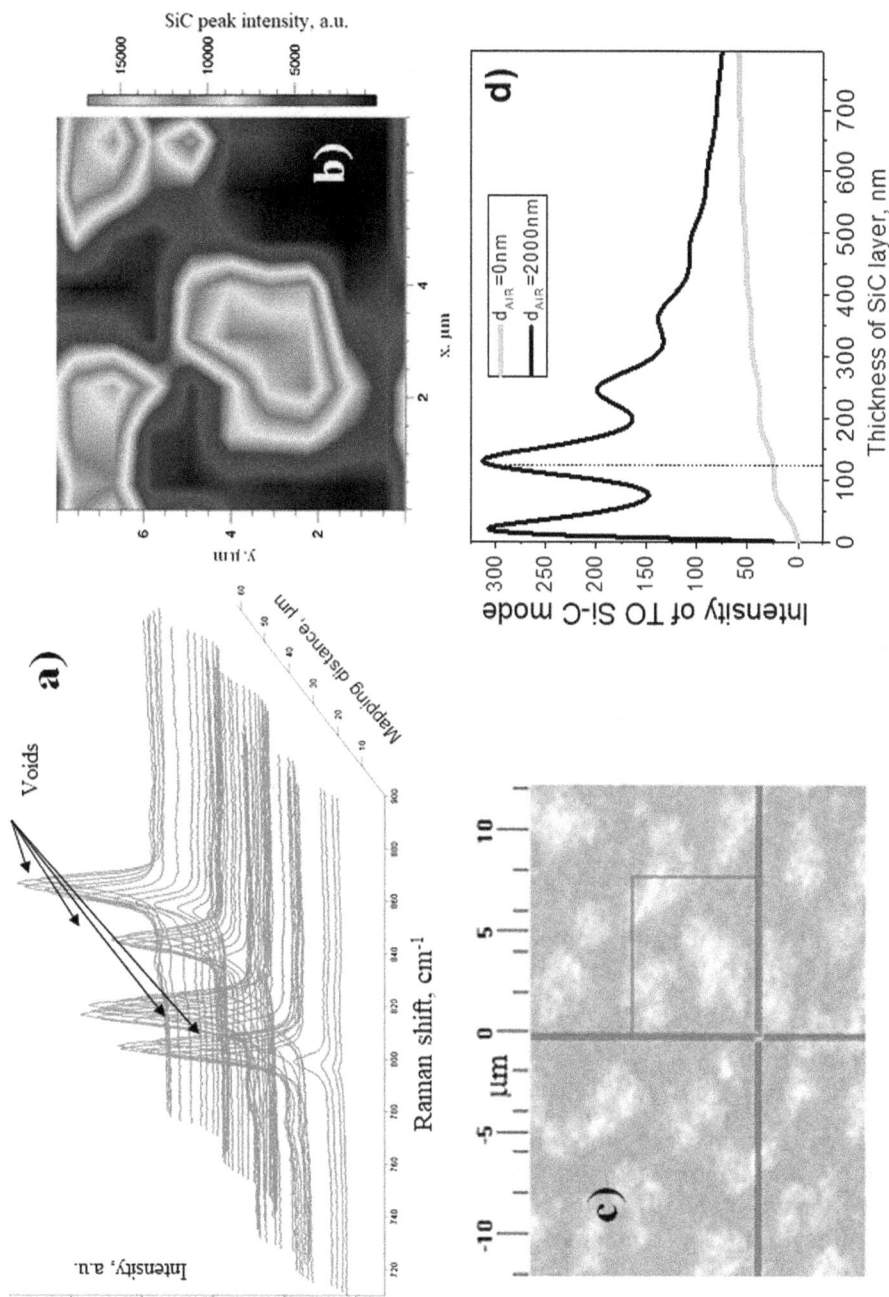

FIGURE 15

(a) Evolution of Si-C peak along four voids for sample 452. (b) Raman area map of the Si-C peak intensity corresponding to (c) the optical microscopy image of mapped area shown in square; (d) results of calculating the Raman intensity of the TO Si-C phonon mode as a function of the thickness of the SiC layer with (black) and without (gray) air void beneath.

system, where $n_1=2.65-0.2i$, $n_2=1.5$, and $n_3=4.15-0.044i$ are refractive indices of SiC, air, and Si at 633 nm, respectively. The extinction coefficient, $k_1=0.2$, was introduced to the complex refractive index of SiC layer due to the observed surface roughness of SiC film. The refractive index of air in the void was taken ~1.5 due to parts of Si and SiC present in the void. d_1 is the thickness of SiC layer, which is considered as a sum of 1 nm thick SiC monolayers. Therefore, the thickness d_1 can be estimated as $d_1=N\Delta d$ and $\Delta d=1$ nm, where N is the number of mono layers, d_2 is the depth of the void filled with air and the Si substrate is considered as semi-infinite. The TO-SiC Raman intensity of SiC layer depends on the electric field distribution, which is a result of interference between all transmitted optical paths in SiC layer. The total amplitude of the electric field at certain depth y in SiC layer is viewed as a sum of the infinite transmitted laser, whose amplitudes are

$$t_1 e^{\beta y} e^{-i(2\pi \tilde{n}_1 y/\lambda)},$$

$$t_1 r' e^{\beta(2d_1-y)} e^{-i(2\pi \tilde{n}_1 (2d_1-y)/\lambda)}, \tag{2}$$

$$-t_1 e^{\beta y} e^{-i(2\pi \tilde{n}_1 y/\lambda)} r_1 \rightleftarrows r' e^{-2if_{i1}} e^{2\beta d_1},$$

$$-t_1 r' e^{\beta(2d_1-y)} e^{-i(2\pi \tilde{n}_1 (2d_1-y)/\lambda)} r_1 \rightleftarrows r' e^{-2if_{i1}} e^{2\beta d_1},$$

where $\beta = -\dfrac{2\pi k_1}{\lambda}$ (λ is the excitation wavelength) emerges as a measure of the absorption in the SiC layer, $t_1 = 2n_0/(n_0+\tilde{n}_1)$ is transmission coefficient at the interface of air/SiC, $r_1 = (n_0-\tilde{n}_1)/(n_0+\tilde{n}_1)$ is the reflection coefficient at the interface of air/SiC, $f_{i1,2} = -\dfrac{2\pi \tilde{n}_{1,2} d_{1,2}}{\lambda}$ are the phase differences when light passes through SiC monolayers and air layers in the void, respectively. Here, $r' = \dfrac{\left(r_2 + r_3 e^{-2if_{i2}}\right)}{\left(1 + r_2 r_3 e^{-2if_{i2}}\right)}$ is the effective reflection coefficient of SiC/(air void in Si) interface, where $r_2 = (\tilde{n}_1 - n_2)/(\tilde{n}_1 + n_2)$ and $r_3 = \dfrac{\left(n_2 - \tilde{n}_3\right)}{\left(n_2 + \tilde{n}_3\right)}$ are individual reflection coefficients at the interface of SiC/air void and air void/Si, respectively. Thus, the total amplitude of the electric field at the depth y is

$$t = \frac{t_1 e^{\beta y} e^{-i(2\pi \tilde{n}_1 y/\lambda)} t_1 r' e^{\beta(2d_1-y)} e^{-i(2\pi \tilde{n}_1 (2d_1-y)/\lambda)}}{1 + r_1 \rightleftarrows r' e^{-2if_{i1}} e^{2\beta d_1}} \tag{3}$$

In addition, further consideration should be applied to the multireflection of scattering Raman light in SiC at the interface of SiC/air and SiC/air void in Si, which contribute to the detected Raman signal. Thus, the detected signal is a result of summation of infinite transmitted light from the interface of SiC/air, which makes the amplitude multiplied by

$$\gamma = \frac{\left(e^{\beta y} + r' e^{\beta(2d_1-y)}\right) t_1'}{1 + r_1 \rightleftarrows r' e^{2\beta d_1}} \tag{4}$$

where $t_1' = (1 - r_1^2)/t_1$ represents the transmission coefficients at the interface of SiC/air. In the equation above, the interference of Raman scattering light is not considered as the phase of spontaneous random Raman lights. Thus, the total Raman signal can be expressed as

$$I = \int_0^{d_1} |t\gamma|^2 \, \Delta y \qquad (5)$$

Figure 15(d) shows the results of calculating the Raman intensity of the TO Si-C band depending on the thickness of the SiC layer. The gray curve shows the calculation result for a SiC layer grown directly on Si without an air void. The black curve shows the results of calculations for the SiC layer grown on top of the air cavity. It can be seen that the Raman intensity of the Si-C peak increases with increasing layer thickness, which is in agreement with the experimental results.

With varying the thickness of the air layer (cavity depth) from 0 to 2000 nm, and the thickness of the SiC layer between 0 and 800 nm, the Raman enhancement at the center of the void was estimated to be approximately 12 times larger than that at the edge of the void for a SiC layer with a thickness of about 120 nm. An increase in the layer thickness to 800 nm reduces the Raman signal enhancement by a factor of ~ 8. This was confirmed experimentally by Raman line-mapping measurements for the sample with an ~800 nm thick SiC layer, where enhancement of the Raman signal by a factor of 1.5 was detected at the void center. These results were also confirmed by the transfer matrix method calculations.

To conclude, it is shown in this section that the SiC layer on Si (111) investigated here is composed of a cubic polytype of SiC with a small amount of 6H-SiC. The presence of the voids has been experimentally confirmed by micro-Raman spectroscopy and scanning electron microscopy. The strong enhancement in the peak intensity of the TO and LO modes is observed for the Raman signal measured in the void area. This enhancement of the Raman signal is advantageous as it allows micro-Raman measurements to be used for the detection of different polytypes in ultra-thin (<100 nm) SiC layers.

2.2 SiC-on-Si – A New, Flexible Template for the Growth of Epitaxial Films and Nanocrystals

It has been experimentally shown [44,47,71,91–130] that a hybrid SiC/Si substrate is very suitable for the growth of continuous films and nanocrystals of II-IV and III-V compounds. Due to the presence of pores along the SiC layer, the Si substrate with the SiC layer is a flexible, elastic system, easily adjusts under the crystal lattices of foreign materials. We noted above that in the course of reaction (1), Si shrinks, followed by its conversion to SiC. The unit cell volume of 3C-SiC is almost half the volume of the Si unit cell. At the moment of shrinkage, the initial coherence is violated and occurs only where every fifth SiC cell coincides with the fourth Si cell. Note that the distance between the C atoms along the (111) plane in projection onto the $(11\bar{2})$ plane in SiC is 3.08 Å [38,44]. A similar distance between Si atoms in silicon is 3.84 Å. Hence, it follows that the distance between the planes of five cells in SiC = 15.40 Å, and the distance between four cells is Si = 15.36 Å. As it was proved in [38,44], it is the conjugation of each fifth and fourth crystal cell of Si and SiC, respectively, that leads to the epitaxial orientation of the SiC film. The rest of the bonds at the interface between SiC and Si, as shown in [44], are broken and pores are formed under the SiC layer in these places (see Figure 6). Thus, coherent conjugation between the film and the substrate occurs only in places where every fifth SiC cell does not completely coincide with the fourth Si cell. The deformation due to the difference between these parameters is insignificant and amounts to 0.3%. It was shown in [131] that if the lattice parameters of the film and the substrate do not exceed 1%, then complete coherence between the film and the substrate can remain and defects in the film will not form, which we observed experimentally with Si – SiC [44–47]. Despite the small value of deformation, it leads to the appearance of sufficiently high elastic stresses. An elementary calculation taking into account the elastic constants given above shows, for example, that the radial component of elastic stresses is σ_{rr} = –0.8GPa [132]. Consequently, it is no longer the SiC film that has a large lattice parameter with respect to the original silicon substrate, but, in contrast, the new ordering substrate (four silicon cells with pores under their surface) has a large lattice parameter. Thus, the substrate will not stretch, but will compress the SiC film.

Moreover, since the SiC layer lies above the pore surface, a part of the boundary of the SiC film surface is not mechanically fixed to the substrate and is in a free-standing state. In general, it is possible to grow SiC layers 90% of the area of which will not be mechanically fixed to the substrate. This will lead to the fact that in the formulas for calculating elastic thermal deformations, it is necessary, instead of the tabular values of the coefficients of linear thermal expansion, to set their modified values taking into account the contact area of the SiC film with the Si substrate. This means that the tabular values of the coefficients per unit area must be multiplied by the proportion of the contacting area. In [133], using the example of the growth of InGaN nanocrystals on a SiC/Si substrate, the elastic stresses caused by the difference between the tabular values of the thermal expansion coefficients of the SiC/Si substrate and the InGaN layer, and their values taking into account the actual contact area of the layers, led practically to a complete absence of thermal deformation in this system.

To date, epitaxial layers of such semiconductors as AlN, GaN, AlGaN, Ga_2O_3, ZnO, ZnS, CdS, CdSe, CdTe have been grown on SiC/Si substrates [91–130,133]. Layers of semipolar structures AlN, GaN [97,98,105–108] were also grown. On the heterostructures AlN/SiC/Si, GaN/AlN/SiC/Si, AlGaN/SiC/Si, device structures were made, namely pyro- and piezo-electric sensors, as well as work began on the fabrication of high-electron mobility transistors [134–139]. For the first time, a semi-industrial technology for creating chips for LEDs on silicon was developed and a working prototype of a white LED on silicon was demonstrated [137]. A detailed description of the routes of most of these studies can be found in the reviews [44–47]. Here we will briefly discuss only the latest data obtained, which were not included or were only partially included in previously published reviews.

2.2.1 Growth of II-VI Compounds on SiC/Si Substrates

2.2.1.1 Selenides, Sulfides, and Tellurides of Cadmium

Using the growth of CdS, CdSe, CdTe, ZnS, and ZnO films as an example, it was proved that the use of SiC as a buffer layer for the growth of II-VI semiconductors on Si significantly improves the growth quality of these films, leading to epitaxy even with a large difference in the lattice parameters of the film and the substrate. This is due to two factors. First, SiC protects Si from its interaction with elements of group VI, which ensures the stoichiometric epitaxial growth of the II-VI semiconductor films. In addition, the voids in the substrate under the SiC layer make the substrate elastic, allowing it to adapt to the growing film and reducing elastic stresses.

(I) CADMIUM SULFIDE   Cadmium sulfide (CdS) is a direct gap semiconductor with a band gap of ~ 2.4 eV, which is used in many microelectronic applications related to solar cells, photovoltaic converters, lasers, and others. When it grows on silicon, the problem arises of the chemical interaction of CdS with Si with the formation of amorphous silicon sulfide (SiS), which greatly impairs the semiconducting properties of growing structures. In Ref. [71], a method was developed for epitaxial growth of films by atomic layer deposition (ALD) on SiC/Si substrates at low temperatures (~ 180 °C), while in [122], another growth method was used, namely the method of evaporation and condensation in a closed volume, which also enables epitaxial CdS layers up to 300 nm thick to be obtained.

Using ALD growth technology, a rare, metastable cubic CdS phase was obtained. As shown in [71], the growth of the cubic phase of CdS is caused by two related processes, namely, due to the growth of CdS at a low temperature (~ 180 °C) and, due to the use of a buffer layer of the cubic 3C-SiC polytype grown as a hybrid substrate on Si by the method of coordinated substitution of atoms. The cubic phase of CdS was identified by x-ray diffraction (XRD) thanks to the fact that the main absorption peak of light in CdS is split in the hexagonal phase into two peaks (4.9 eV and 5.4 eV) and degenerate in the cubic phase (5.1 eV). This layer was also investigated by spectroscopic ellipsometry and Raman spectroscopy. The dependence of the dielectric constant of CdS-c, grown by the ALD method, on the light energy in the range from 0.7 eV to 6.5 eV has been experimentally measured for the first time. Studies have shown that on the SiC/Si surface with orientations (100) and (110), CdS layers, grown under the same conditions as on the SiC/Si surface with orientations (111), were polycrystalline. Figure 16(a) shows a diffraction pattern and SEM image of a CdS-c layer grown on a SiC/Si substrate by the ALD method. Whilst Figure 16(b) shows a typical Raman spectrum of a CdS-c/3C-SiC/Si(111) sample grown by this method. Note that, in addition to the only first-order line of CdS-c 1LO (at 303 cm^{-1}), corresponding to longitudinal optical (LO) vibrations of atoms, the corresponding second-order line 2LO (at 605 cm^{-1}) is clearly visible. The direction of vibrations of sulfur atoms in CdS-c corresponding to the 1LO line, calculated by quantum chemistry methods, is

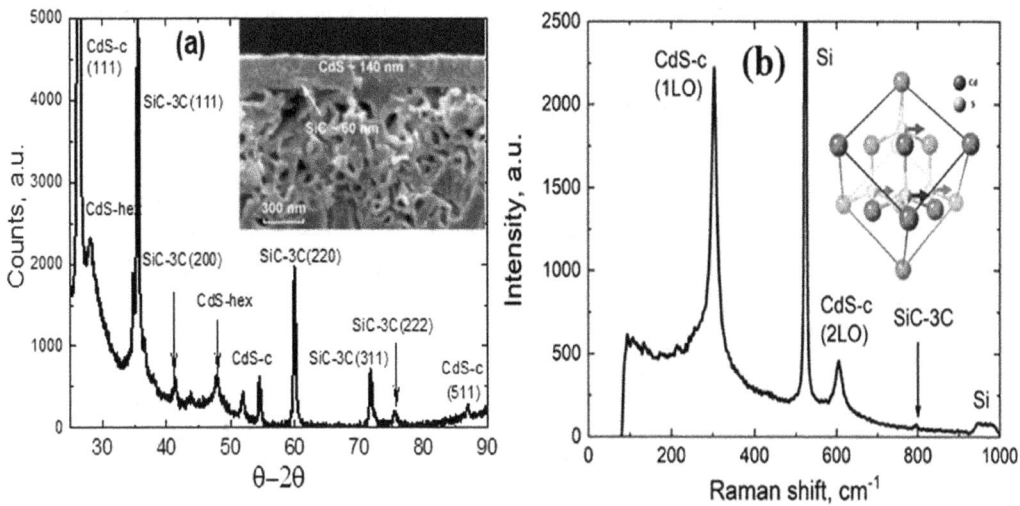

FIGURE 16

(a) X-ray diffractogram of the CdS/SiC/Si (111) sample (the insert shows a SEM image of a cross section of this sample; pores and voids in the bulk of Si, partially filled with SiC, are visible under the SiC layer). (b) Raman spectrum of the CdS-c/SiC (111)/Si (111) sample (the inset shows the vibrations of sulfur atoms in CdS-c corresponding to the 1LO line at 303 cm^{-1}).

shown in the inset in Figure 16(b). Since the width of the direct band gap of CdS-c (2.4 eV) is slightly higher than the laser radiation energy (532 nm ~ 2.33 eV), the Raman spectrum also shows lines of both Si and SiC (Figure 16(b)).

(II) CADMIUM SELENIDE   Cadmium selenide (CdSe) with a slightly smaller band gap than CdS (1.70 eV in cubic CdSe and 1.73 eV in hexagonal CdSe), is a direct-gap semiconductor, and is also of interest in electronic applications. In Refs. [123,124], epitaxial layers of cubic CdSe of ~ 350 nm thick were grown for the first time on SiC/Si substrates by the method of evaporation and condensation in a quasi-closed volume, mentioned earlier. It was found that in this method the optimum substrate temperature is 590 °C, the evaporator temperature is 660 °C, and the growth time is 2 s. The experiment showed [123] that epitaxial CdSe with a cubic sphalerite structure grows on Si substrates with a buffer SiC layer, while only a polycrystalline phase is formed on pure Si substrate. Analysis of the thin film by various characterization techniques [123] showed a high structural perfection of the CdSe layer and the absence of a polycrystalline phase in it.

Figure 17(a) shows the Raman spectrum of the CdSe/SiC/Si sample. Despite the relatively small thickness of CdSe (~ 350 nm), both peaks at ~ 205 cm^{-1}, corresponding to the LO mode of CdSe, and a peak at ~ 411 cm^{-1}, corresponding to the second order optical phonon mode of CdSe (2LO), are clearly seen. In order to determine which polytype of CdSe grew, cubic or hexagonal, and also to compare the measured Raman spectrum of CdSe with the theoretical one, Raman spectra of both cubic and hexagonal CdSe were calculated in [123] using quantum chemistry methods. The DFT calculation was used in the approximation of the gradient functional. Calculations showed [123] that the Raman spectrum of cubic CdSe consists of only one line of the first order at 201 cm^{-1} (note, that the lines of higher orders were not calculated). Due to the symmetry of the cubic CdSe crystal, this frequency corresponds to all three modes of LO vibrations. The Raman spectrum of wurtzite CdSe, obtained in the same approximation, consists of two high-intensity lines at 61 cm^{-1} and 181 cm^{-1} and one low-intensity line at 191 cm^{-1}. The experimental Raman spectrum shows only one line of the first order at 205 cm^{-1}, which is much closer to cubic CdSe.

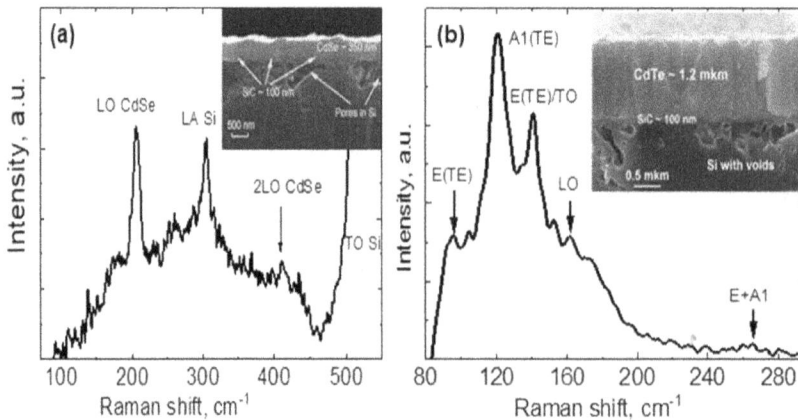

FIGURE 17

(a) Raman spectrum of the CdSe/SiC/Si sample. The CdS peaks correspond to the optical phonon modes of first (1LO) and second (2LO) order. The insert shows a SEM image of a cross-section of a CdSe/SiC/Si (111) sample. (b) Raman spectrum of the CdTe/ SiC/Si sample in the range 80–300 cm^{-1}. The two highest peaks correspond to the Raman shift of 121 и 141 cm^{-1}. Insert shows a SEM image of a cross-section of a CdTe/SiC/Si (111) sample. Pores and voids in the bulk of Si, partially filled with SiC, are visible under the SiC layer.

Insert in Figure 17(a) shows a SEM image of a cross-section of a CdSe/SiC/Si sample. The CdSe layer is about 350 nm thick. Under a SiC layer of ~ 100 nm thick, pores and voids in the bulk of the silicon substrate are clearly visible, which were formed as a result of the topochemical reaction (1). Their formation is directly related to the relaxation of elastic stresses at the SiC/Si interface and the absence of lattice mismatch dislocations on it [44–47]; they do not affect the quality of SiC layers [44–47].

(III) CADMIUM TELLURIDE Cadmium telluride (CdTe) is also a 1.49 eV direct gap semiconductor and finds applications in solar cells, ionizing radiation detectors and photodetectors. In [125], epitaxial films of CdTe of 1–3 μm thick were grown on SiC/Si hybrid substrates. It was found that the optimum substrate temperature is 500 °C at an evaporator temperature of 580 °C, and the growth time is 4 s. Ellipsometric, Raman, x-ray structural and electron diffraction analyses have shown a high structural perfection of the CdTe layer and the absence of a polycrystalline phase in it. Trace element analysis [125] revealed an almost ideal stoichiometry of CdTe grown on SiC/Si. Ellipsometry spectra showed a forward band gap of epitaxial CdTe equal to 1.5 eV. Figure 17(b) demonstrates Raman spectrum registered from sample CdTe/SiC/Si. The presence of very weak LO and E + A1 lines in the spectrum indicates a high structural perfection of CdTe. The inset to Figure 17(b) shows a SEM image of a cross section of a CdSe/SiC/Si sample.

2.2.1.2 Zinc Compounds

(I) ZINC SULFIDE Zinc sulfide (ZnS) – a direct-gap semiconductor – can be represented by many polytype modifications, of which two are considered the main ones: wurtzite (cubic) with a band gap of 3.5 eV and sphalerite (hexagonal) with a band gap of 3.9 eV. In recent years, interest in ZnS has sharply increased due to the huge possibilities of using ZnS in optoelectronics. For many applications, it is required to grow a high-quality transparent ZnS layer on a conductive semiconductor substrate. In work [126] epitaxial films of ZnS on silicon were obtained by the ALD method. In order to avoid the interaction between Si and ZnS on the silicon surface, a high-quality buffer layer of SiC with a thickness of ~ 100 nm was first grown by the CSA method. Fast electron diffraction investigations showed that the ZnS layers are epitaxial. It was proved by spectroscopic ellipsometry methods that the grown ZnS layers are transparent in the range of photon energies up to 3 eV, which is of decisive importance for applications in optoelectronics. Investigation of the Raman spectrum of ZnS/SiC/Si (111) samples revealed the presence of only one (rather weak) phonon line at 351 cm⁻¹, attributed to ZnS (Figure 18). This line corresponds to LO vibrations of a ZnS crystal with cubic symmetry (see inset in Figure 18) and indicates a relatively high quality of ZnS epitaxy, since the thickness of this layer is not large and amounts to ~ 130 nm.

(II) ZINC OXIDE Zinc oxide ZnO is also a direct gap semiconductor with a band gap of 3.4 eV. In recent years, the interest of researchers in ZnO has significantly increased due to the prospect of its use in thin-film transistors, light-emitting diodes, lasers, photodetectors, etc. In [127], ZnO films were grown for the first time on hybrid SiC/Si substrates by the ALD method. The films were grown on Si (100) wafers at a temperature of $T = 250$ °C. The authors of [127] showed that the use of SiC as a buffer layer significantly improves the quality of ZnO, providing stoichiometric and epitaxial growth. Raman spectroscopy data and electron diffraction analysis showed a high structural perfection of the ZnO layer. Figure 19 shows the example of Raman spectrum of the *n*-type ZnO/SiC/ Si (100) sample, while the inset to this figure shows a SEM image of a cross-section of this sample with the ZnO layer of about 200 nm thick. Under a SiC layer of ~ 50 nm thick, pores and voids in

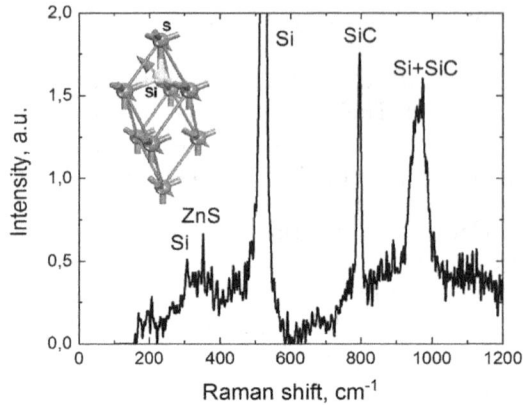

FIGURE 18

Raman spectrum of the ZnS/SiC/Si (111) sample. A single ZnS peak at 351 cm^{-1} corresponds to longitudinal optical (LO) vibrations of a cubic ZnS crystal. The primitive cell of this crystal is shown in the inset, the gray balls are the S atoms, the lighter ball is the Si atom. Arrows show the amplitude and directions of vibrations of atoms in the crystal corresponding to the LO peak at 351 cm^{-1}.

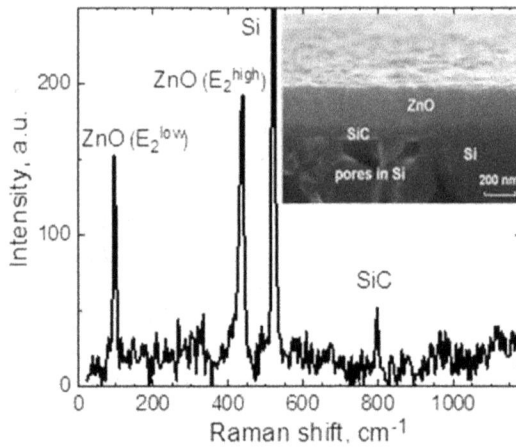

FIGURE 19

Raman spectrum of the *n*-type ZnO/SiC/Si (100) sample with subtracted background. ZnO peaks correspond to vibrations E_2^{low} (at 96 cm^{-1}) and E_2^{high} (at 438 cm^{-1}). Insert shows a SEM image of a cross-section of a ZnO/SiC/Si (100) sample. Pores and voids in the bulk of Si are visible under the SiC layer.

the bulk of the silicon substrate are clearly visible, which were formed as a result of the topochemical reaction (1).

The peaks E_2^{low} $(96\,\text{cm}^{-1})$ and E_2^{high} $(438\,\text{cm}^{-1})$, corresponding to vibrations of Zn and O atoms perpendicular to the symmetry axis of the hexagonal ZnO crystal, are clearly seen in Figure 19. Thus, the hexagonal structure of epitaxial ZnO is manifested. Despite the small thickness of the SiC buffer layer, its main peak at ~ 800 cm^{-1} is clearly visible. The absence of a polycrystalline phase makes this method promising for the growth of ZnO layers on Si with an intermediate SiC layer. It was found in [128] that the mechanism of formation of epitaxial ZnO textures depends on the type of conductivity (*n*- or *p*-type) of the substrate, and a theoretical model was proposed to explain the effect of texture formation and its dependence on the type of conductivity. The effect is associated with the transformation of vicinal Si (100) surfaces into SiC surfaces in the course of its synthesis through the coordinated substitution of atoms. Significant differences were found between the structure and growth mechanisms of ZnO layers on SiC/Si (111) and SiC/Si (100) substrates. The results of experiments on the growth of ZnO by the method of ion-plasma high-frequency magnetron sputtering are presented in Refs. [129,130]. This method also allows epitaxial ZnO films of a sufficiently high crystal quality to be obtained. In [129], the optical constants of these layers were investigated by spectroscopic ellipsometry, and one of the main features found in the samples obtained was the absorption of light in the range of 2.0–3.3 eV. This was explained by elastic stresses in the zinc oxide layer.

Microelement analysis, which was carried out at various points of the ZnO film, both along the surface and in depth, showed a high stoichiometric composition of the film, the excess of oxygen atoms did not exceed 5%. It is important to note that the presence of the SiC buffer layer strongly changes the structure of the ZnO film. The ZnO islands without a buffer layer are ~ 15–20 nm in size and have a strongly elongated cigar-like shape [127,128]. In the presence of a SiC buffer layer, the ZnO islands are larger, ~ 100–150 nm and symmetric, and the film surface is smoother (see Figure 19, insert). Presumably, this is due to the orienting effect of the SiC layer.

2.2.2 Growth of III-V Compounds on SiC/Si Substrates

2.2.2.1 Gallium Oxide (Ga$_2$O$_3$)

Gallium oxide is a promising wide-gap semiconductor ($E_g \approx 4.9$ eV), which is currently poorly understood. This material has a number of physical properties that make it quite competitive with silicon carbide and III-nitrides. First of all, it is transparent in the ultraviolet region of the spectrum and has a high breakdown voltage (8 mV/cm). In addition, Ga$_2$O$_3$ is easily doped, which makes it possible to obtain well-conducting layers of this material. In [120], well-textured layers of gallium oxide (β-Ga$_2$O$_3$) with a thickness of about 1 μm were grown by the method of chloride-hydride epitaxy (HVPE) on SiC/Si substrates. Studies have shown that the films have a texture close to epitaxial and consist of the pure β-phase of Ga$_2$O$_3$ with the ($\overline{2}$01) orientation. β-Ga$_2$O$_3$ films were grown at a temperature of 1050 °C and were studied by XRD, SEM, Raman spectroscopy, and reflection electron diffraction. Figure 20 shows a typical unpolarized Raman spectrum of a β-Ga$_2$O$_3$/SiC/Si sample obtained at room temperature. A$_g$ and B$_g$, respectively, optical and acoustical modes active in the Raman spectra. Modes A$_g$ and B$_g$ in β-Ga$_2$O$_3$ below 200 cm^{-1} are associated with vibrations and displacements of tetrahedral "chains" (ribbons) consisting of gallium and oxygen atoms in the atomic structure of β-Ga$_2$O$_3$, while the modes lying in the range of ~ 480–310 cm^{-1} can be associated with the average bending vibrations of the Ga$_2$O$_6$ octahedral cluster in the β-Ga$_2$O$_3$ cell structure. Vibrations in the frequency range ~ 770–500 cm^{-1} describe stretching and bending vibrations of the GaO$_4$ cluster in the β-Ga$_2$O$_3$ cell structure [120]. The SiC (TO) and SiC (LO) modes, as well as the Si modes, fully correspond to the Raman modes of the SiC film on Si [120]. The inset to this figure shows a SEM image of a cross section of a β-Ga$_2$O$_3$/SiC/Si (111) sample.

Therefore, the SiC buffer layer allows Ga$_2$O$_3$ layers to be grown on silicon substrates, orientates them, and protects the Si substrate from chemical interaction with oxygen, water vapor, and chlorine-containing compounds. This suggests that this method of growing of β-Ga$_2$O$_3$ layers on Si opens up new opportunities for application and new synthesis routes for obtaining such a wide-gap material as Ga$_2$O$_3$ [121].

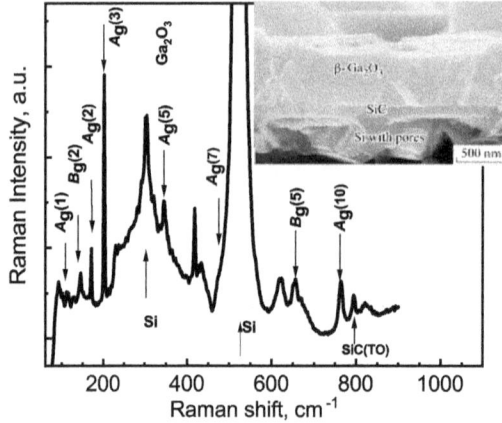

FIGURE 20

Raman spectrum of the β-Ga$_2$O$_3$/SiC/Si sample registered at room temperature. The inset shows a SEM image of a cross-section of the same sample.

2.2.2.2 *Self-organization of the Composition of Al$_x$Ga$_{1-x}$N Epitaxial Layers During Their Growth on SiC Hybrid Substrate*

The phenomenon of the emergence of a spontaneous change in the composition of the layers during the growth of Al$_x$Ga$_{1-x}$N films on SiC hybrid substrate was discovered and described in [140]. It was found that during the growth of Al$_x$Ga$_{1-x}$N films with a low (about 11–24%) Al content, interlayers or domains appear, consisting of stoichiometric AlGaN (Figure 21). In [140], a model was proposed according to which self-organization in composition arises due to the effect of two processes on the growth kinetics of the Al$_x$Ga$_{1-x}$N film. The first process is associated with the competition of two chemical reactions proceeding at different rates. One of these reactions is the AlN formation reaction; the second is the reaction of the formation of GaN. The second process, closely related to the first, is the appearance of elastic compressive and tensile stresses during the growth of Al$_x$Ga$_{1-x}$N films on hybrid SiC/Si (111) substrate. Both processes influence each other, which leads to a complex pattern of aperiodic changes in the composition over the thickness of the film layer. Figures 21(A,B) demonstrate the images of a cross-section of Al$_x$Ga$_{1-x}$N/SiC/Si (111) and Al$_x$Ga$_{1-x}$N/AlN/SiC/Si (111) samples, respectively, obtained using an EDX technique (SEM equipped with an x-ray spectrometer).

The images in Figures 21(A,B) show lines describing the distribution of the atomic fraction of aluminium, gallium, nitrogen, and silicon over the thickness of the layers. It follows from these data that the growth of the Al$_x$Ga$_{1-x}$N layer from the vapor phase containing 50% molar composition with respect to Ga and Al on a SiC/Si (111) substrate occurs as follows. Namely, the very first layers of the film, several nanometers in size, contain AlN, then the predominant growth of Al$_x$Ga$_{1-x}$N begins, in which the atomic fraction of Ga exceeds the atomic fraction of Al. When the layer reaches a thickness of about 1.7 μm, an Al$_x$Ga$_{1-x}$N interlayer begins to grow with an Al content of about 50% and a thickness of about 90–100 nm. Further, only GaN grows with a Ga content close to 85%–95% atomic percent. And, finally, on the surface itself, the Ga content in the layer decreases and the Al content increases again. The distribution of the concentration of Al and Ga over the thickness of the layer of the Al$_x$Ga$_{1-x}$N/AlN/SiC/Si (111) sample is somewhat different. The growth of the layer begins almost with the growth of pure GaN, and AlGaN is not deposited.

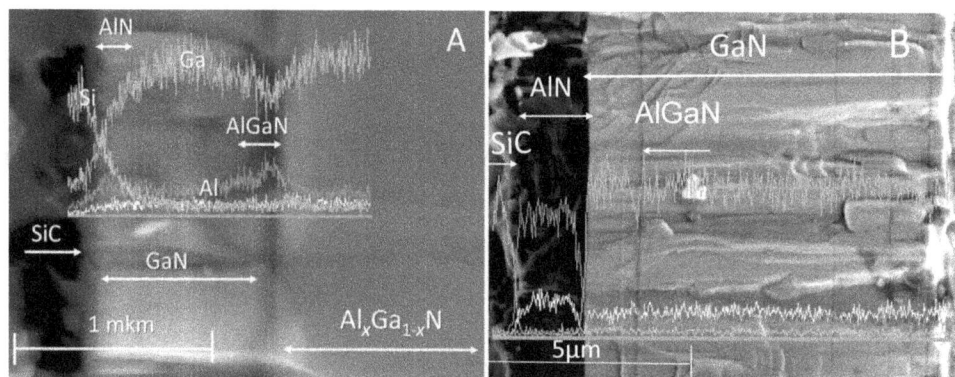

FIGURE 21

SEM images of a cross-section of structures (A) $Al_xGa_{1-x}N/SiC/Si$ (111) and (B) $Al_xGa_{1-x}N/AlN/SiC/Si$ (111). The substrate in the figures is on the left, while the surface of the film layer is on the right. Wavy lines, drawn from the substrate to the film surface, describe the distribution of the atomic fraction of Al, Ga, N, and Si over the thickness of the layers. The darker thin stripes correspond to AlGaN layers with an Al content of about 50%.

Then, when the layer reaches a thickness of about 1.8 μm, the growth of an $Al_xGa_{1-x}N$ interlayer begins, in which approximately 25%–30% of Al and a thickness of about 90–100 nm. When the GaN layer reaches a thickness of about 4.8 μm, an AlGaN interlayer reappears and GaN grows again. Further, this process is repeated aperiodically. Closer to the sample surface, the frequency of the appearance of AlGaN interlayers increases; accordingly, the thicknesses of the layers with an increased Ga content between them decrease. The formation of an interlayer with an increased Al content on the surface of this sample was not detected.

Figures 22(a,b) show Raman spectra of the same samples as discussed above. Figure 22(a) shows the spectrum of the $Al_xGa_{1-x}N/SiC/Si$ (111) sample, and Figure 22(b) shows the spectrum of $Al_xGa_{1-x}N/AlN/SiC/Si$ (111) sample. Spectra were recorded using a Witec Alpha 300R confocal Raman microscope. The vertical dimension of the focused laser beam waste is several micrometers, and thus all layers and the substrate are in focus at the same time. The spectra of both samples show: Si lines (at 525 cm⁻¹, and 300 cm⁻¹); SiC line (at 796 cm⁻¹), AlN line (at 657 cm⁻¹), GaN line (at 144 cm⁻¹) and lines at 579 cm⁻¹ and 575 cm⁻¹ of $Al_xGa_{1-x}N$. Note that the spectrum of pure bulk GaN demonstrates line at 568 cm⁻¹. According to our data, this line for the $Al_xGa_{1-x}N/SiC/Si$ (111) sample is shifted by 11 cm⁻¹ relative to the GaN line, while for the $Al_xGa_{1-x}N/AlN/SiC/Si$ (111) sample it is shifted by only 7 cm⁻¹. The position of the 579 cm⁻¹ line corresponds to GaN containing approximately 50% of AlN (sample $Al_xGa_{1-x}N/SiC/Si$ (111)), and the position of the 575 cm⁻¹ line corresponds to GaN containing approximately 25% of AlN (sample $Al_xGa_{1-x}N/AlN/SiC/Si$ (111)). Thus, Raman spectroscopy confirms with a sufficiently high accuracy the values for composition the $Al_xGa_{1-x}N$, which were found from the ellipsometry method.

2.2.2.3 Piezo- and Pyro-electric Properties of AlGaN Layers Grown on (111) SiC/Si.

In Ref. [138], the piezo- and pyro-electric properties of self-assembled AlGaN layers grown on (111) SiC/Si were investigated. These data are compared with the structure of the SiC/Si (111) interface. To measure the piezo- and pyro-electric properties of AlGaN layers, epitaxial $Al_xGa_{1-x}N$ films were grown and a series of studies of their structural and optical

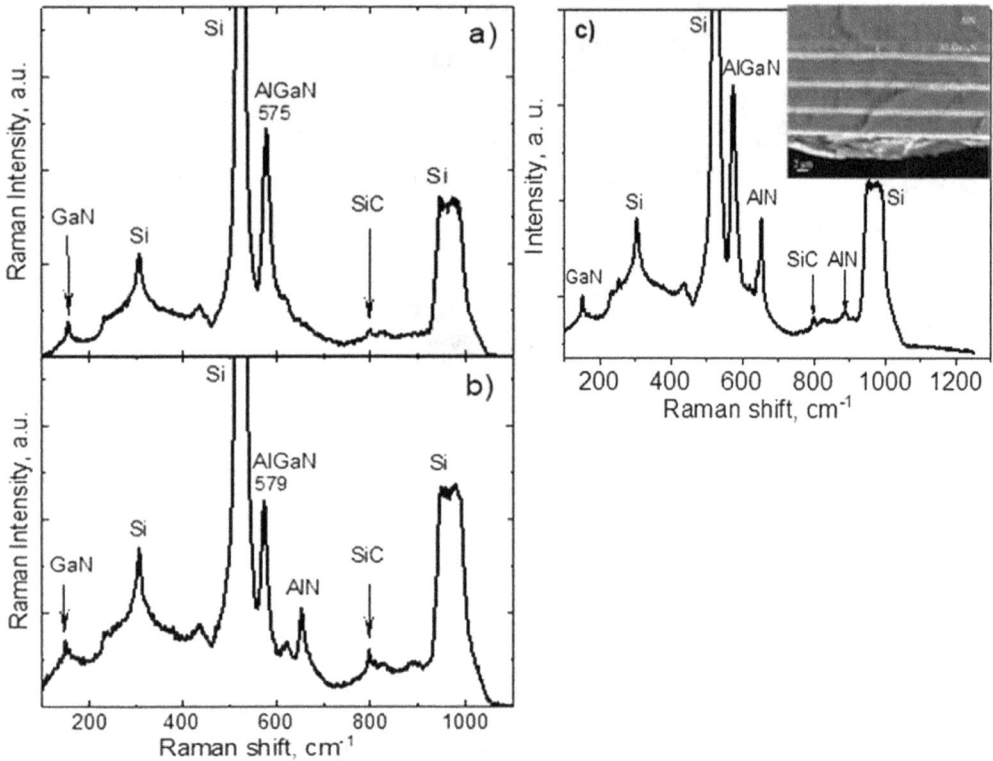

FIGURE 22

Raman spectra of two Al$_x$Ga$_{1-x}$N samples: (a) spectrum of the Al$_x$Ga$_{1-x}$N/SiC/Si (111) sample, (b) spectrum of the Al$_x$Ga$_{1-x}$N/AlN/SiC/Si (111) sample and (c) Raman spectrum of AlN/Al$_x$Ga$_{1-x}$N/SiC/Si (111) sample, insert shows a SEM image of a cross-section of this structure.

properties was carried out. This was done primarily to determine their composition. As in [138], Al$_x$Ga$_{1-x}$N films, with thick AlN layer on the top, were grown on hybrid substrates of two types, namely, on SiC/Si (111) substrates and AlN/SiC/Si (111) substrates by the HVPE method. The Raman spectrum of AlN/Al$_x$Ga$_{1-x}$N/SiC/Si (111) sample is shown in Figure 22(c), and the inset to the figure shows a SEM image of a cross-section of this structure.

In [138], for the first time, studies of the frequency dependences of the permittivity and piezo- and pyro-electric properties of films of Al$_x$Ga$_{1-x}$N solid solutions were carried out. During growth in these solutions the interlayers of Al$_x$Ga$_{1-x}$N of various compositions are formed spontaneously. We note that data on measuring the piezo- and pyro-properties of films of Al$_x$Ga$_{1-x}$N solid solutions are absent in the world literature. Two types of Al$_x$Ga$_{1-x}$N/SiC/Si (111) samples grown by the HVPE method were investigated. Some Al$_x$Ga$_{1-x}$N/SiC/Si (111) samples were grown on SiC/Si (111) hybrid substrates without an AlN sublayer, while other samples were grown on AlN/SiC/Si (111) substrates with an AlN layer. These investigations have shown that in all samples, a significant frequency dispersion of the dielectric constant was observed. The dielectric loss (tanδ) in these samples was an order of magnitude lower than in pure AlN, decreasing to ~ (2–3) × 10^{-3} at frequencies exceeding 100 kHz. A significant difference was also found in the pyroelectric coefficients of the Al$_x$Ga$_{1-x}$N/SiC/Si (111) and Al$_x$Ga$_{1-x}$N/AlN/SiC/Si (111) samples.

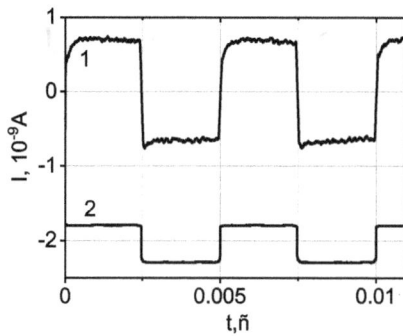

FIGURE 23

Kinetic dependences of the pyroelectric current of the AlN/Al$_x$Ga$_{1-x}$N/SiC/Si (111) sample for a heat flux modulation frequency of 200 Hz (1), heat flux modulation form (2).

Studies have shown [138] that the pyroelectric coefficient of Al$_x$Ga$_{1-x}$N layers grown directly on SiC/Si (111) is 2 times higher than the pyroelectric coefficient of Al$_x$Ga$_{1-x}$N films grown on AlN/SiC/Si (111) substrates and is almost 1.5 times higher than the pyroelectric coefficient of AlN films, which reached theoretically possible values. Typical kinetic dependences of the pyroelectric current of the Al$_x$Ga$_{1-x}$N/SiC/Si (111) sample are shown in Figure 23.

2.2.2.4 LED Devices Based on Heterostructures of III-N Compounds Grown On Hybrid SiC/Si Substrate

In Refs. [135,136], light-emitting III-N heterostructures were grown by the method of gas-phase epitaxy from the metal-organic vapor phase epitaxy (MOVPE) on SiC/Si (111) hybrid substrates, and their optical and structural properties were studied. As a result of these studies, it was found that the pores formed at the SiC/Si interface led to a modification of the initial conditions for the growth of the heterostructure and the formation of blocks with different lateral sizes in the range of 0.1–1 μm. In addition to pores at the SiC/Si interface, pores were revealed in the buffer layer of the heterostructure. These features lead to a high density of dislocations in the buffer layer, the dominant of which are edge dislocations. However, the authors of [136] managed to find such growth conditions for AlInGaN heterostructures under which an increased concentration of pores under the SiC layer has a positive effect on the growth of the heterostructure, significantly reducing the thermal deformations of the layer. The use of an optimized design of the buffer layer enables significant reduction of the dislocation density and formation of an active region with a good structural quality. Figure 24(a) shows a bright-field TEM image of the grown structure obtained in cross section (10-10). It is possible to distinguish the regions of the buffer layer (Figure 24(c)) and the active region (Figure 24(b)) that are fundamentally different in structural quality. Figure 24(c) shows dark-field images obtained in the cross-section (01-10) in the weak-beam mode with diffraction reflections g = 2-1-10 (Figure 24(c) on the left) and g = 0002 (Figure 24(c) on the right). Analysis of the obtained images, taken under these conditions, allows estimation of the density of dislocations of various types (edge, screw, and mixed) in different regions of the heterostructure. In the lower part of the heterostructure, in a sequence of AlGaN/GaN layers of various compositions, the total density of dislocations of all types is ~ 10^{10} cm^{-2}. The use of AlGaN/GaN superlattices leads to a decrease in the dislocation density in the lower part of the

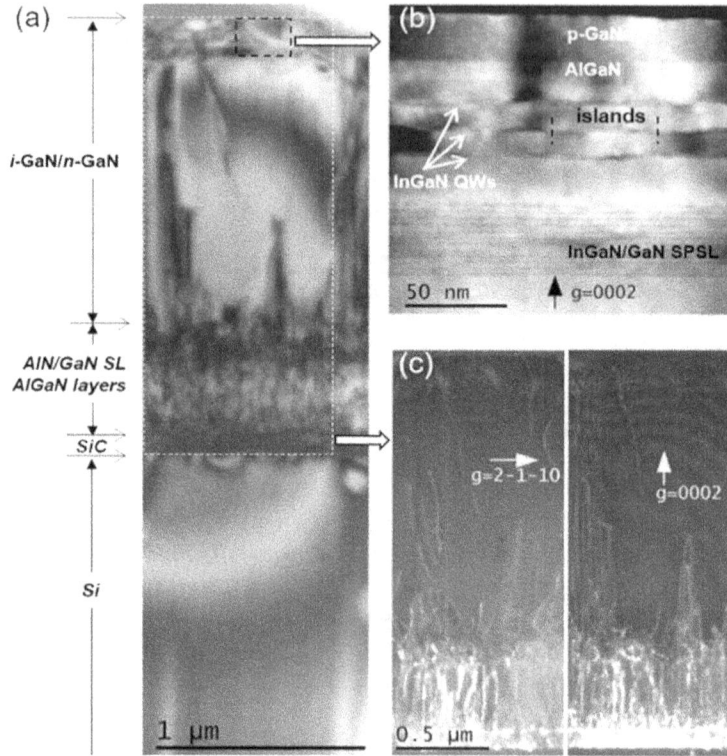

FIGURE 24

TEM images of a cross-section (10-10) of a LED heterostructure grown on a SiC/Si substrate: (a) – bright-field general view; (b) – dark-field image of the active region of the structure obtained with $g = 0002$; (c) general dark-field images obtained in the weak-beam mode with diffraction reflections $g = 2$-1-10 (image on the left) and $g = 0002$ (image on the right).

GaN layer to ~ 3×10^9 cm^{-2} (the density of edge dislocations is ~ 2.2×10^9 cm^{-2}, the density of screw dislocations is ~ 5×10^8 cm^{-2}, and the density dislocations of mixed type is ~ 4×10^8 cm^{-2}). A decrease in the density of threading dislocations occurs as a result of their bending into the (0001) plane, caused by the presence of a stress gradient between the AlGaN layers with a gradually decreasing Al composition. The density of edge dislocations in the active region (Figure 24(b)) is ~ 5×10^8 cm^{-2}, while the density of other types of dislocations remains at the level of the corresponding values in the GaN layer. Thus, the total dislocation density in the active region is typical for structures on sapphire substrates.

The paper [137] describes the method and technology for manufacturing both LED chips and packaged LEDs from AlInGaN/GaN heterostructures grown on SiC/Si hybrid substrates. Their current-voltage characteristics, emission spectra, dependences of the radiation power, and external quantum efficiency on the magnitude of the current were investigated. The studies undertaken in [137] unambiguously showed that the pores, created during the formation of SiC from Si using the CSA method [44,47,71], play a positive role in the fabrication of light-emitting diodes, since they effectively scatter their own radiation. As already noted, from the emission spectrum of LEDs on a SiC/Si substrate shown in Figure 25, it is clearly seen that there are no interference extrema observed in the spectrum of chips on a Si substrate. This means that the pores, formed at the boundary of

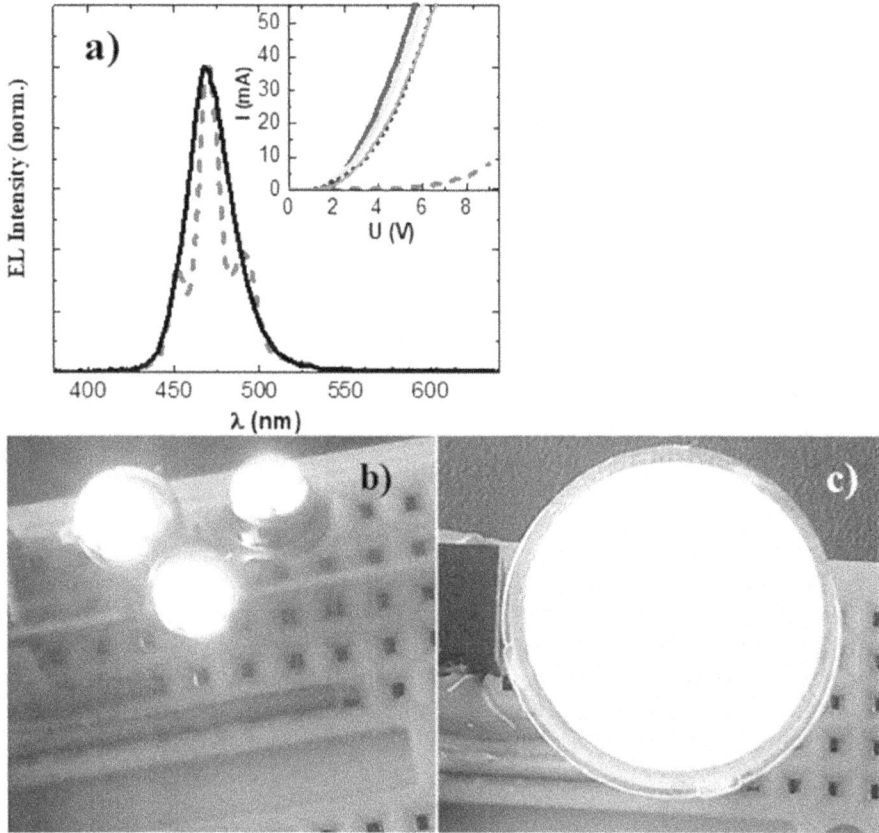

FIGURE 25
LED structures and their characteristics: (a) emission spectrum of LED chips based on heterostructures on SiC/Si (solid line) and Si (dashed line) substrates; the insert shows the $I – V$ characteristics of the corresponding chips. Appearance of the finished LEDs: (b) three LEDs without a phosphor coating and (c) a LED with a polymer lens containing a phosphor.

the Si-SiC layer in the SiC/Si substrate [44,47,71], effectively scatter their own radiation. The presence of such pores at the SiC/Si interface, in addition to strong scattering of light and an increase in its output due to the destruction of the "light output cone", also leads to an increase in light reflection due to the large value of the contrast of the refractive indices of the pore and the SiC material. Consequently, the fraction of light absorbed by the opaque substrate is reduced. In addition, the pores, facilitating the scattering of light at the boundary of the heterostructure, prevent the capture of generated light in the waveguide of the heterostructure with its subsequent absorption upon multiple reflection from its boundaries. To eliminate this effect on sapphire substrates, a special light-scattering relief is first created. In the process of manufacturing LED crystals, the presence of pores in SiC/Si substrates led to a noticeable difference in the appearance of the surface of the samples in comparison with the samples of LED structures on the Si substrate. It can also be noted that LED chips fabricated on a SiC/Si substrate showed a uniform glow over the entire surface of the active region, in contrast to chips on a Si substrate.

Thus, hybrid SiC/Si substrates fabricated by the CSA method can be successfully used to fabricate blue and ultraviolet light-emitting diode chips. In this case, the presence of pore growth at the boundary of the Si-SiC layer of the substrate plays a positive role due to the scattering and increase in the reflection of the LED's own radiation at this boundary.

2.2.3 Hybrid SiC/Si Substrate as an Intermediate Structure for Two-Stage Conversion of Si into a Thin Layer of Diamond-like Graphite

A fundamentally new method was proposed and experimentally demonstrated in [141] for growth of epitaxial layers of nanostructured carbon on Si substrates. Epitaxial growth in this case of seemingly incompatible lattices is achieved by a two-stage conversion of Si crystal using the CSA method. At the first stage of the conversion, the first half of Si atoms are consistently replaced by C atoms thanks to reaction (1), and an epitaxial layer of cubic silicon carbide SiC-3C is formed. At the second stage of conversion, the remaining half of the Si atoms are consistently replaced by C atoms due to the reaction (6) of SiC with gaseous carbon tetrafluoride CF_4:

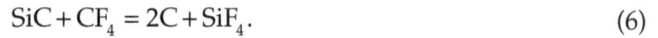

$$SiC + CF_4 = 2C + SiF_4. \tag{6}$$

Depending on the orientation of the silicon surface, reagent gas pressure, temperature and growth time, carbon structures with different properties were fabricated, ranging from nano-diamonds to nanotubes and onion-like carbon. The key feature of this method is that the surface layer of Si substrate orders the resulting structures using the initial chemical bonds between Si atoms. This means that new chemical bonds are formed simultaneously and in conformity with the destruction of old bonds. Characterization of various fabricated nanocarbon structures by scanning electron microscopy as well as by Raman spectroscopy and spectroscopic ellipsometry are presented. Studies performed in [141] suggest that the growth of carbon structures from SiC-3C proceeds via two competing mechanisms. The first mechanism is growth due to the consistent substitution of C atoms for Si atoms in SiC-3C. This mechanism provides the ordered growth of planar carbon structures, possibly with predominant sp^3 hybridization since these structures are very transparent. This mechanism takes place until the end of Si-3C in the thin film obtained in the first stage of conversion. In reality, the reaction may be completed earlier, after the formation of a layer of diamond-like carbon, since the resulting film of diamond-like carbon will "block" the access of CF_4 gas to the lower layers of SiC.

The second growth mechanism is classical; due to elastic stresses, Si atoms intensively enter the surface of the growing carbon structure, where they react with gas CF_4. This mechanism leads to rough amorphous-polycrystalline carbon layers with predominant sp^2 hybridization, since these carbon layers are opaque. Nevertheless, these layers are still several times more transparent than a highly pyrolytic graphite (HOPG). In addition, the D-band (at ~ 1347 cm^{-1} in Figure 26(c)) is absent in the Raman spectrum of graphite, while in the fabricated layers it is quite intense. The second mechanism, apparently, always takes place, but its rate strongly depends on the type of conductivity and the orientation of the surface of the initial Si. In cases where growth by the first mechanism is replaced with the growth by the second mechanism, it is possible to obtain a two-layer carbon structure with an interface between sp^3 and sp^2 hybridization. Figure 26 shows Raman spectra of nanocarbon structures of different type (type I – transparent nanocrystalline carbon, type II – semi-transparent carbon layer, and III – opaque amorphous-polycrystalline carbon layer) fabricated on SiC/Si hybrid substrate [141]. As can be seen from this figure, different peak positions and relative intensities of D, G, D' and 2D bands are observed for fabricated nanostructured carbon of different types. It is

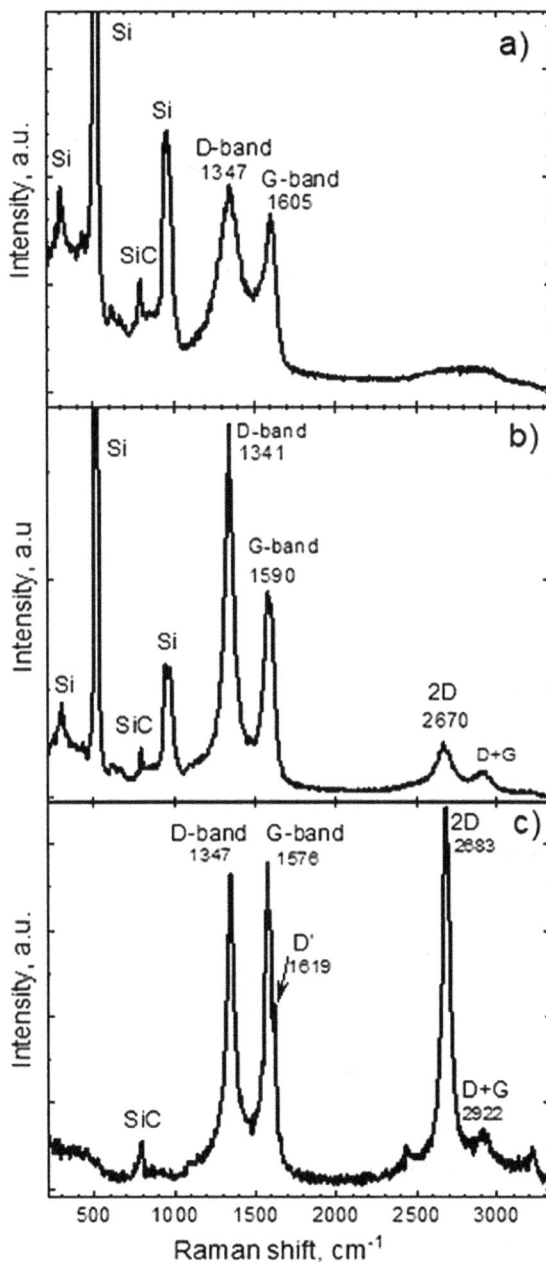

FIGURE 26

Raman spectra of nanostructured carbon layers obtained by conversion of the surface layers of SiC/Si hybrid substrate: (a) Raman spectrum of a type I transparent carbon layer; (b) Raman spectrum of a type II semitransparent carbon layer; (c) Raman spectrum of type III an opaque carbon layer.

also shown in [141] that carbon layers with very high frequency D-band (at ~1350 cm^{-1}) or 2D bands can be realized.

---

3 Conclusions and Summary

The presence of voids during the growth of thin SiC layers on a Si substrate by the method of coordinated substitution of atoms was experimentally confirmed by scanning electron microscopy and Raman imaging. A strong enhancement in the peak intensity of the TO and LO modes of SiC is observed in the Raman signal measured at the voids. This allows structural identification to be performed on very thin SiC films (<100 nm) by Raman measurements, which is typically impossible for layers with thicknesses < 2 μm. The Raman line and area mapping experiments presented in this work confirm that voids formed in the Si substrate under the SiC layer induce relaxation of the elastic stress caused by the lattice mismatch between SiC and Si. Raman data lead to the conclusion that the SiC layers studied consist mainly of the cubic SiC polytype together with small amounts of 6H-SiC. It is shown that various semiconductor films grown on the hybrid SiC/Si substrate described have excellent structural quality and demonstrate unique electrical, magnetic, and optical properties suitable for application to microelectronics and optoelectronics.

To conclude, we would like to answer the following question: why does the standard CVD carbonization method, which uses various hydrocarbon gases, fail to produce high-quality SiC layers on Si? After our analysis described in this chapter, the answer is obvious.

(1) These gases contain more than two atoms. The SiC formation reaction does not take place in two stages.

(2) An ordered ensemble of dilatation [39–41] dipoles is not formed. As a rule, gaseous hydrocarbons, or alcohol vapors containing carbon, are used when growing SiC by CVD. A partial replacement of Si atoms with C atoms would be expected to occur. Naturally, this process proceeds differently and more chaotically than the process using substitution reactions (1). The substitution will be "superimposed" on the main growth process. In our opinion, this is why the SiC-on-Si layer grown using the advanced epitaxy method described in [https://advancedepi.com/sic] does not have a high degree of crystal perfection.

(3) The dilatation dipoles are directed along the <111> direction. When SiC grows by the CSA method on any Si face, viz. (110), (100), (210), etc., the (111) faces are always formed as the main face. Nothing of the kind is observed when SiC is grown on Si using standard procedures. When SiC grows by coating a surface, the orientation of the SiC face is determined by the orientation of the Si face. Growth from polyatomic hydrocarbons forms (100) faces, rather than (111) faces. This is clearly seen from the diffraction patterns given in [142].

(4) When SiC is grown by the CSA method, the SiC layer contains carbon-vacancy structures. When SiC is grown by the standard method, they are absent [63–65]. The presence of a carbon-vacancy structure was demonstrated by the appearance of a peak at 960 cm^{-1} in the IR spectrum. A similar peak is absent in SiC grown by any method except the substitution method [63–65].

(5) During the growth of SiC by the CSA method, the cubic 3C-SiC polytype is mainly formed, but it is also possible to obtain the 4H-SiC, 6H-SiC, 15R-SiC polytypes and, in particular, the rare 2H-SiC and 8H-SiC polytypes [71].

(6) For SiC growth by the CSA method, elastic deformations are either completely absent, or compressive elastic stresses arise. The substrate becomes slightly convex [131]. When SiC grows according to the standard technique, tensile elastic stresses mainly arise, and the substrate becomes concave [53]; except when carbonation is prevented under certain conditions [54].

Acknowledgments

This work was partially supported by the Ministry of Science and Higher Education within the framework of the Russian State Assignment under contract number FFNF-2021-0001.

References

1. W. J. Choyke, H. Matsunami, G. Pensl (eds.), *Silicon Carbide: Recent Major Advances*, Springer-Verlag, Berlin, p.446, 2004.
2. Z. C. Feng (Ed.), *SiC Power Materials – Devices and Applications*, Springer, Berlin, 2004.
3. Peter Friedrichs, Tsunenobu Kimoto, Lothar Ley, Gerhard Pensl (eds.), *Silicon Carbide: Power Devices and Sensors*, Volume 2, Wiley-VCH Verlag GmbH & Co. KGaA, Weinheim, p.500, 2011.
4. Y. K. Sharma (ed.), *Disruptive Wide Bandgap Semiconductors, Related Technologies, and Their Applications*, IntechOpen, London, p.152, 2018.
5. G. R. Fisher and P. Barne, "Towards a Unified View of Polytypism in Silicon Carbide", Philosophical Magazine, B61, 217–236 (1990). https://doi.org/10.1080/13642819008205522
6. D. W. Feldman, J. H. Parker, Jr., W. J. Choyke, and L. Patrick, "Raman Scattering in 6H SiC", Physical Review, 170, 698–704 (1968). https://doi.org/10.1103/PhysRev.170.698.
7. Arnaud Zoubir (ed.), *Raman Imaging: Techniques and Applications*, Springer Series in Optical Science, v 168, Springer-Verlag Berlin Heidelberg, p.383, 2012.
8. H. Nienhaus, T. U. Kampen, W. Monch, "Phonons in 3C-, 4H-, and 6H-SiC", Surface Science, 324, L328-L332 (1995). https://doi.org/10.1016/0039-6028(94)00775-6.
9. S. Nakashima, H. Harima, "Raman Investigation of SiC Polytypes", Physica Status Solidi (a), 162, 39–64 (1997). https://doi.org/10.1002/1521-396X(199707)162:1<39::AID-PSSA39>3.0.CO;2-L.
10. S. Nakashima, H. Harima, N. Ohtani, M. Katsuno, "Raman characterization of local electrical properties and growth process in modulation-doped 6H-SiC crystals", J. Appl. Phys., 95(7), 3547–3552 (2004). https://doi.org/10.1063/1.1655682
11. B. Gouider Trabelsi, F. V. Kusmartsev, A. Kusmartseva, F. H. Alkallas, S. AlFaify and Mohd Shkir, "Raman Spectroscopy Imaging of Exceptional Electronic Properties in Epitaxial Graphene Grown on SiC", Nanomaterials, 10 , 2234/1–37 (2020). doi:10.3390/nano10112234.
12. M. Mermoux, A. Crisci and F. Baillet, "Raman Imaging Characterization of Structural and Electrical Properties in 4H SiC", Materials Science Forum, 457–460, 609–612 (2004). doi:10.4028/www.scientific.net/MSF.457-460.609.

13. Guo Xiao, Liu Xue-Chao, Xin Jun, Yang Jian-Hua, Shi Er-Wei, "Characterization of Polytype Distributions in Nitrogen-doped 6H-SiC Single Crystal by Raman Mapping", J. Inorganic Materials, 27, 609–614 (2012). DOI: 10.3724/SP.J.1077.2012.00609.

14. A. J. Leide, M. J. Lloyd, R. I. Todd, and D. E. J. Armstrong, "Ion Irradiated 6H-SiC: Raman Spectroscopy, Chemical Defects, Strains, Annealing, and Oxidation" arXiv:2004.14335 [cond-mat.mtrl-sci] (2020). https://arxiv.org/abs/2004.14335.

15. A. Flessa, E. Ntemou, M. Kokkoris, E. Liarokapis, M. Gloginjić, S. Petrović, M. Erich, S. Fazinić, M. Karlušić, K. Tomić, "Raman Mapping of 4-MeV C and Si Channelling Implantation of 6H-SiC", Journal of Raman Spectroscopy, 50, 1186–1196 (2019). DOI: 10.1002/jrs.5629.

16. Shin-Ichi Nakashima, Takeshi Mitani, Masaru Tomobe, Tomohisa Kato, and Hajime Okumura, "Raman Characterisation of Damaged Layers of 4H-SiC Induced by Scratching", AIP Advances 6, 015207/1–8 (2016). https://doi.org/10.1063/1.4939985.

17. E. del Corro, J. G. Izquierdo, J. González, M. Taravilloa and V. G. Baonzaa, "3D Raman Mapping of Uniaxially Loaded 6H-SiC Crystals", Journal of Raman Spectroscopy, 44, 758–762 (2013). DOI:10.1002/jrs.4252.

18. A. M. Gigler, A. J. Huber, M. Bauer, A. Ziegler, R. Hillenbrand, R. W. Stark, "Nanoscale Residual Stress-Field Mapping Around Nanoindents in SiC by IR s-SNOM and Confocal Raman Microscopy", Optics Express, 17, 22351–22357 (2009). doi:10.1364/OE.17.022351.

19. M.S. Amer, L. Durgam, M. M. El-Ashry, "Raman Mapping of Local Phases and Local Stress Fields in Silicon–Silicon Carbide Composites", Materials Chemistry and Physics, 98, 410–414 (2006). doi:10.1016/j.matchemphys.2005.09.066.

20. T. S. Perova, R. A. Moore, K. Berreth, K. Maile and A. Lyutovich, "MicroRaman Spectros-Copy of Protective Coatings Deposited onto C/C–SiC Composites", Materials Science and Technology, 23, 1300–1304 (2007). DOI:10.1179/174328407X168810.

21. Chaoli Fu, Yong Yang, Zhengren Huang, Guiling Liu, Hui Zhang, Fang Jiang, Yuquan Wei, Zheng Jiao, "Investigation on the Laser Ablation of SiC Ceramics Using Micro-Raman Mapping Technique", Journal of Advanced Ceramics, 5, 253–261 (2016). DOI: 10.1007/s40145-016-0197-x.

22. Z. C. Feng, A. J. Mascarenhas, W. J. Choyke, and J. A. Powell, "Raman Scattering Studies of Chemical Vapor Deposited Cubic Sic Films of (100) Si", Journal of Applied Physics, 64, 3176–3186 (1988). https://doi.org/10.1063/1.341533.

23. B. Racine, A. C. Ferrari, N. A. Morrison, I. Hutchings, W. I. Milne, and J. Robertson, "Properties of Amorphous Carbon–Silicon Alloys Deposited by a High Plasma Density Source", Journal of Applied Physics, 90, 5002–5012 (2001). DOI: 10.1063/1.1406966.

24. L. Sun, J. E. Crocker, L. L. Shaw and. H. L. Marcus, "Effect of Hydrogen on Silicon Carbide Deposition from Tetramethylsilane-Raman Scattering Studies", Proceedings of Solid Freeform Fabrication 479–486 (1999). http://utw10945.utweb.utexas.edu/Manuscripts/1999/1999-055-Sun.pdf

25. W. L. Zhu, J. L. Zhu, S. Nishino, G. Pezzotti, "Spatially Resolved Raman Spectroscopy Evaluation of Residual Stresses in 3C-SiC Layer Deposited on Si Substrates with Different Crystallographic Orientations", Applied Surface Science, 252, 2346–2354 (2006). doi:10.1016/j.apsusc.2005.04.020.

26. Lingyu Wan, Dishu Zhao, Fangze Wang, Gu Xu, Tao Lin, Chin-Che Tin, Zhaochi Feng, and Zhe Chuan Feng, "Quality Evaluation of Homoepitaxial 4H-SiC Thin Films by a Raman Scattering Study of Forbidden Modes", Optical Materials Express, 8, 313919/119–127 (2018). https://doi.org/10.1364/OME.8.000119.

27. M. Havel, Ph. Colomban, "Raman and Rayleigh Mapping of Corrosion and Mechanical Aging in SiC Fibres", Composites Science and Technology, 65, 353–358 (2005). doi:10.1016/j.compscitech.2004.09.017.

28. O. Y. Goue, B. Raghothamachar, Y. Yang, J. Guo, M. Dudley, K. Kisslinger, A. J. Trunek, P. G. Neudeck, D. J. Spry, A. A. Woodworth, "Study of Defect Structures in 6H-SiC a/M-Plane Pseudofiber Crystals Grown by Hot Wall CVD Epitaxy", Journal of Electronic Materials, 45, 2078–2086 (2016). DOI: 10.1007/s11664-015-4185-7.

29. K. Kollins, C. Przybyla, M. S. Amer, "Residual Stress Measurements in Melt Infiltrated SiC/SiC Ceramic Matrix Composites Using Raman Spectroscopy", Journal of European Ceramic Society, 38, 2784–2791 (2018). https://doi.org/10.1016/j.jeurceramsoc.2018.02.013

30. Xixi Niua, Haoqiang Zhanga, Zhiliang Pei, Nanlin Shi, Chao Sun, Jun Gonga, "Measurement of Interfacial Residual Stress in SiC Fiber Reinforced Ni-Cr-Al Alloy Composites by Raman Spectroscopy", Journal of Materials Science & Technology 35, 88–93 (2019). https://doi.org/10.1016/j.jmst.2018.09.023

31. James Nance, Ghatu Subhash, Bhavani Sankar, Rafael Haftka, Nam Ho Kim, Christian Deckc, Sarah Oswald, "Measurement of Residual Stress in Silicon Carbide Fibers of Tubular Composites Using Raman Spectroscopy", Acta Materialia, 217, 117164/1–8 (2021). https://doi.org/10.1016/j.actamat.2021.117164

32. N. Gogneau, A. B. Gouider Trabelsi, M. G. Silly, M. Ridene, M. Portail, A. Michon, M. Oueslati, R. Belkhou, F. Sirotti, A. Ouerghi, "Investigation of Structural and Electronic Properties of Epitaxial Graphene on 3C–SiC(100)/Si(100) Substrates", Nanotechnology, Science and Applications, 7, 85–95 (2014). http://dx.doi.org/10.2147/NSA.S60324

33. G. Reza Yazdi, T. Iakimov and R. Yakimova, "Epitaxial Graphene on SiC: A Review of Growth and Characterization", Crystals, 6, 53/1–45 (2016). doi:10.3390/cryst6050053.

34. F. Giannazzo, I. Shtepliuk, I. G. Ivanov, T. Iakimov, A. Kakanakova-Georgieva, E. Schilirò, P. Fiorenza and R. Yakimova, "Probing the Uniformity of Hydrogen Intercalation in Quasi-Free-Standing Epitaxial Graphene on SiC by Micro-Raman Mapping and Conductive Atomic Force Microscopy", Nanotechnology, 30, 284003/1–9 (2019). https://doi.org/10.1088/1361-6528/ab134e.

35. S. Saha, D. Kumar, C. K. Sharma, V. K. Singh, S. Channagiri and D. V. Sridhara Rao, "Microstructural Characterization of GaN Grown on SiC", Microscopy and Microanalysis, 1–11 (2019). doi:10.1017/S1431927619014739.

36. S. A. Kukushkin and A. V. Osipov, "New Method for Growing Silicon Carbide on Silicon by Solid-Phase Epitaxy: Model and Experiment", Phys. Solid State, 50(7), 1238–1245 (2008). https://doi.org/10.1134/S1063783408070081

37. S. A. Kukushkin and A. V. Osipov and N. A. Feoktistov, "A Method of Manufacturing a Product Containing a Silicon Substrate with a Silicon Carbide Film on its Surface" RF Patent Appl. No. № 2363067. Registered July 27, 2009 (Priority from January 22. 2008).

38. S. A. Kukushkin and A. V. Osipov, "Quantum Mechanical Theory of Epitaxial Transformation of Silicon to Silicon Carbide", Journal of Physics D: Applied Physics, 50(46), 464006/1–7 (2017). doi:10.1088/1361-6463/aa8f69.

39. S. A. Kukushkin and A. V. Osipov, "Thin Film Heteroepitaxy by the Formation of the Dilatation Dipole Ensemble", Doklady Physics, 57(5), 217–220 (2012). https://doi.org/10.1134/S1028335812050072

40. S. A. Kukushkin and A. V. Osipov, "A New Mechanism of Elastic Energy Relaxation in Heteroepitaxy of Monocrystalline Films: Interaction of Point Defects and Dilatation Dipoles" Mechanics of Solids, 48(2), 216–227 (2013). https://doi.org/10.3103/S0025654413020143

41. S. A. Kukushkin and A. V. Osipov, "A New Method for the Synthesis of Epitaxial Layers of Silicon Carbide on Silicon Owing to Formation of Dilatation Dipoles", Journal of Applied Physics, 113(2), 4909/1–7 (2013). https://doi.org/10.1063/1.4773343.

42. S. A. Kukushkin and A. V. Osipov, "Anisotropy of the Solid State Epitaxy of Silicon Carbide in Silicon", Semiconductors, 47(12), 1551–1555 (2013). https://doi.org/10.1134/S1063782613120129.

43. S. A. Kukushkin and A. V. Osipov, "First Order Phase Transition through an Intermediate State", Physics of the Solid State, 56(4), 792–800 (2014). https://doi.org/10.1134/S1063783414040143.

44. S. A. Kukushkin and A. V. Osipov, "Topical Review. Theory and Practice of SiC Growth on Si and Its Applications to Wide-Gap Semiconductor Films", Journal of Physics D: Applied Physics, 45(31), 313001/1–41 (2014). https://doi.org/10.1088/0022-3727/47/31/313001

45. S. A. Kukushkin, A. V. Osipov and N. A. Feoktistov, "Synthesis of Epitaxial Silicon Carbide Films through the Substitution of Atoms in the Silicon Crystal Lattice: A Review", Physics of the Solid State, 56(8), 1507–1535 (2014). https://doi.org/10.1134/S1063783414080137.

46. S. A. Kukushkin, A. V. Osipov "A New Method of Replacement Atoms for the Synthesis of Epitaxial Layers of SiC on Si: From Theory to Practice", Journal of Physics: Conference Series, 541, 012003/1–9 (2014). https://iopscience.iop.org/article/10.1088/1742-6596/541/1/012003/pdf.

47. S. A. Kukushkin and A. V. Osipov, "Nanoscale Single-Crystal Silicon Carbide on Silicon and Unique Properties of This Material", Inorganic Materials, 57(13), 1–21 (2021).

48. W. G. Spitzer, D. A. Kleinman, and C. J. Frosch, "Infrared Properties of Cubic Silicon Carbide Films", Physical Review, 11(1), 133–136 (1959). https://doi.org/10.1103/PhysRev.113.133.

49. I. H. Khan and R. N. Summergrad, "The Growth of Single-Crystal Films of Cubic Silicon Carbide on Silicon", Applied Physics Letters, 11(1), 12–14 (1967). doi: 10.1063/1.1754939.

50. I. H. Khan and A. J. Learn, "Formation of Epitaxial β-SiC Films on Sapphire", Applied Physics Letters, 15(12), 410–414 (1969). doi: 10.1063/1.1652881.

51. K. E. Haq, and I. H. Khan, "Surface Characteristics and Electrical Conduction of β-SiC Films Formed by Chemical Conversion", J. Vacuum Science & Technology, 7(4), 490–493 (1970). https://doi.org/10.1116/1.1315373.

52. J. Graul and E. Wagner, "Growth Mechanism of Polycrystalline β-SiC Layers on Silicon Substrate", Applied Physics Letters, 21(2), 67–69 (1972). https://doi.org/10.1063/1.1654282.

53. A. Severino, C. Locke, R. Anzalone, M. Camarda, N. Piluso, A. La Magna, S. E. Saddow, G. Abbondanza, G. D'Arrigo, F. La Via, "3C-SiC Film Growth on Si Substrates", ECS Transactions, 35(6), 99–116 (2011). doi:10.1149/1.3570851.

54. G. Ferro, "3C-SiC Heteroepitaxial Growth on Silicon: The Quest for Holy Grail", Critical Reviews in Solid State and Materials Sciences, 40(1), 56–76 (2015). https://doi.org/10.1080/10408436.2014.940440.

55. S. A. Kukushkin, A. V. Osipov, "Drift Mechanism of Mass Transfer on Heterogeneous Reaction in Crystalline Silicon Substrate", Physica B, 512(1) 26–31 (2017). https://doi.org/10.1016/j.physb.2017.02.018.

56. A. S. Grashchenko, S. A. Kukushkin, A. V. Osipov, A. V. Redkov, "Vacancy Growth of Monocrystalline SiC from Si by the Method of self-Consistent Substitution of Atoms", Catalysis Today (2021) (in press).

57. S. A. Kukushkin, A. V. Osipov, "The Optical Properties, Energy Band Structure, and Interfacial Conductance of a 3C-SiC(111)/Si(111) Heterostructure Grown by the Method of Atomic Substitution", Technical Physics Letters, 46(11), 1103–1106 (2020). doi: 10.1134/S1063785020110243.

58. S. A. Kukushkin, A. V. Osipov, "Anomalous Properties of the Dislocation-Free Interface between Si(111) Substrate and 3C-SiC(111) Epitaxial Layer", Materials 14(78), 1–12 (2021). doi: 10.3390/ma14010078.

59. N. T. Bagraev, S. A. Kukushkin, A. V. Osipov, V. V. Romanov, L. E. Klyachkin, A. M. Malyarenko, and V. S. Khromov, "Magnetic Properties of Thin Epitaxial SiC Layers Grown by the Atom-Substitution Method on Single-Crystal Silicon Surfaces", Semiconductors, 55, 137–145 (2021). https://doi.org/10.1134/S106378262102007X.

60. N. T. Bagraev, S. A. Kukushkin, A. V. Osipov, L. E. Klyachkin, A. M. Malyarenko, and V. S. Khromov, "Registration of Terahertz Radiation Using Silicon Carbide Nanostructures", Semiconductors, 55(12) (2021) (in press).

61. N. T. Bagraev, S. A. Kukushkin, A. V. Osipov, L. E. Klyachkin, A. M. Malyarenko, and V. S. Khromov, "Terahertz Radiation from Silicon Carbide Nanostructures", Semiconductors, 55(11), 1027–1033 (2021).

62. S. Yu. Davydov, and A. A. Lebedev, "Vacancy Kinetics in Heteropolytype Epitaxy of SiC", Semiconductors, 41(6), 621–624 (2007). doi: 10.1134/S1063782607060012.

63. S. A. Grudinkin, V. G. Golubev, A. V. Osipov, N. A. Feoktistov, and S. A. Kukushkin, "Infrared Spectroscopy of Silicon Carbide Layers Synthesized by the Substitution of Atoms

on the Surface of Single-Crystal Silicon", Physics of the Solid State, 57(12), 2543–2549 (2015). https://doi.org/10.1134/S1063783415120136

64. S. A. Grudinkin, S. A. Kukushkin, A. V. Osipov, and N. A. Feoktistov, "IR Spectra of Carbon-Vacancy Clusters in the Topochemical Transformation of Silicon into Silicon Carbide", Physics of the Solid State, 59(12), 2430–2435 (2017). https://doi.org/10.1134/S1063783417120186

65. S. A. Kukushkin, K. Kh. Nussupov A. V. Osipov, N. B. Beisenkhanov, and D. I. Bakranova, "Structural Properties and Parameters of Epitaxial Silicon Carbide Films, Grown by Atomic Substitution on the High Resistance (111) Oriented Silicon", Superlattices and Microstructures 111, 899–911 (2017). https://doi.org/10.1016/j.spmi.2017.07.050.

66. S. A. Kukushkin, A. V. Osipov, "Spin Polarization and Magnetic Moment in Silicon Carbide Grown by the Method of Coordinated Substitution of Atoms", Materials 14(55), 5579–5592 (2021). https://doi.org/10.3390/ma14195579

67. S. A. Kukushkin, A. V. Osipov, "Mechanism of Formation of Carbon–Vacancy Structures in Silicon Carbide during Its Growth by Atomic Substitution" Physics of the Solid State, 60(9), 1891–1896 (2018). https://doi.org/10.1134/S1063783418090184.

68. V. K. Egorov, E. V. Egorov, S. A. Kukushkin, A. V. Osipov, "Structural Heteroepitaxy during Topochemical Transformation of Silicon to Silicon Carbide" Physics of the Solid State, 59(4), 773–779 (2017). https://doi.org/10.1134/S1063783417040072.

69. L. M. Sorokin, N. V. Veselov, M. P. Shcheglov, A. E. Kalmykov, A. A. Sitnikova, N. A. Feoktistov, S. A. Kukushkin, A.V. Osipov, "Electron-Microscopic Investigation of a SiC/Si(111) Structure Obtained by Solid Phase Epitaxy", Technical Physics Letters. 34(11), 992–994 (2008). https://doi.org/10.1134/S1063785008110278.

70. S. A. Kukushkin, A. V. Osipov, "Determining Polytype Composition of Silicon Carbide Films by UV Ellipsometry", Technical Physics Letters, 42(2), 175–178 (2016). https://doi.org/10.1134/S1063785016020280.

71. S. A. Kukushkin, A. V. Osipov, A. I. Romanychev, I. A. Kasatkin, and A. S. Loshachenko," Low-Temperature Growth of the CdS Cubic Phase by Atomic-Layer Deposition on SiC/Si Hybrid Substrates", Technical Physics Letters, 46(11), 1049–1052 (2020). http://link.springer.com/article/10.1134/S1063785020110085.

72. Yu. E. Kitaev, S. A. Kukushkin, and A. V. Osipov, "Evolution of the Symmetry of Intermediate Phases and Their Phonon Spectra during the Topochemical Conversion of Silicon into Silicon Carbide", Physics of the Solid State, 59(1), 28–33 (2017). https://doi.org/10.1134/S1063783417010164.

73. Yu. E. Kitaev, S. A. Kukushkin, and A. V. Osipov, and A. V. Redkov, "A New Trigonal (Rhombohedral) SiC Phase: Ab Initio Calculations, a Symmetry Analysis and the Raman Spectra", Physics of the Solid State, 60(10), 2066–2071 (2018). https://doi.org/10.1134/S1063783418100116.

74. R. Scholz, U. Gösele, E. Niemann, F. Wischmeyer, "Micropipes and Voids at β-SiC/Si(100) Interfaces: an Electron Microscopy Study" Applied Physics, A64, 115–125 (1997). https://doi.org/10.1007/s003390050452.

75. A. Severino, G. D'Arrigo, C. Bongiorno, S. Scalese, and F. La Via, G. Foti, "Thin Crystalline 3C-SiC Layer Growth Through Carbonization of Differently Oriented Si Substrates", Journal of Applied Physics, 102, 023518/1–10 (2007). https://doi.org/10.1063/1.2756620.

76. M. Kitabatake, "Simulations and Experiments of 3C-SiC/Si Heteroepitaxial Growth", Physica Status Solidi, (b)202, 405-420 (1997). https://doi.org/10.1002/1521-3951(199707)202:1<405::AID-PSSB405>3.0.CO;2-5

77. J. P. Li, A. J. Steckl, "Nucleation and Void Formation Mechanisms in SiC Thin Film Growth on Si by Carbonization", J. Electrochemical Society, 142, 634–641 (1995). https://doi.org/10.1149/1.2044113.

78. W. Attenberger, J. Lindner, V. Cimalla, J. Pezoldt, "Structural and Morphological Investigations of the Initial Stages in Solid Source Molecular Beam Epitaxy of SiC on (111)Si", Materials Science and Engineering, B61/62, 544–548 (1999). DOI:10.1016/s0921-5107(98)00470-x.

79. R. Anzalone, A. Severino, G. D'Arrigo, C. Bongiorno, G. Abbondanza, G. Foti, S. Saddow, F. La Via, "Heteroepitaxy of 3C-SiC on Different On-Axis Oriented Silicon Substrates", Journal of Applied Physics, 105, 084910/1–7 (2009). https://doi.org/10.1063/1.3095462.

80. M. Portail, T. Chassagne, S. Roy, C. Moisson, M. Zielinski, "Thermally Induced Surface Reorganization of 3C-SiC (111) Epilayers Grown on Silicon Substrate", Material Science Forum, 645–648, 155–158 (2010). DOI: 10.4028/www.scientific.net/MSF.645-648.155.

81. J. Nishizawa, M. Kimura, "Layer Growth in GaAs Epitaxy", Journal of Crystal Growth 74, 331–337 (1986). https://doi.org/10.1016/0022-0248(86)90122-3

82. J. Wasyluk, T. S. Perova, S. A. Kukushkin, A. V. Osipov, N. A. Feoktistov, S. A. Grudinkin, "Raman Investigation of Different Polytypes in SiC Thin Films Grown by Solid-Gas Phase Epitaxy on Si (111) and 6H-SiC Substrates", Materials Science Forum, 645–648, 359–362 (2010). DOI: 10.4028/www.scientific.net/MSF.645-648.359

83. T. S. Perova, J. Wasyluk, S. A. Kukushkin, A. V. Osipov, N. A. Feoktistov, S. A. Grudinkin, "Micro-Raman Mapping of 3C-SiC Thin Films Grown by Solid-Gas Phase Epitaxy on Si(111)", Nanoscale Research Letters, 5, 1507–1511 (2010). DOI: 10.1007/s11671-010-9670-6.

84. P. A. Temple, C. E. Hathaway, "Multiphonon Raman Spectrum of Silicon", Physical Review, **B**7, 3685–3697 (1973). https://doi.org/10.1103/PhysRevB.7.3685.

85. V. Lysenko, D. Barbier, B. Champagnon, "Stress Relaxation Effect in Porous 3C-SiC/Si Heterostructure by Micro-Raman Spectroscopy", Applied Physics Letters, 79, 2366–2368 (2001). DOI: 10.1063/1.1409278.

86. S. A. Dyakov, T. S. Perova, K. A. Gonchar, G. K. Mussabek, K. K. Dikhanbayev, V. Yu. Timoshenko, "Resonance Enhancement of Raman Scattering from One-Dimensional Periodical Structures of Porous Silicon", Journal of Nanoelectronics and Optoelectronics, 7 (N6), 591–595 (2012). https://doi.org/10.1166/jno.2012.1398.

87. Y. Y. Wang, Z. H. Ni, Z. X. Shen, H. M. Wang, Y. H. Wu, "Interference Enhancement of Raman Signal of Graphene", Applied Physics Letters, 92, 043121/1–3 (2008). https://doi.org/10.1063/1.2838745.

88. S. Dyakov, T. Perova, C. Miao, Y.-H. Xie, S. Cherevkov, A. Baranov, "Influence of the Buffer Layer Properties on the Intensity of Raman Scattering of Graphene", Journal of Raman Spectroscopy, 44, 803–809 (2013). DOI 10.1002/jrs.4294.

89. Z. C. Feng, W. J. Choyke and J. A. Powell, "Raman Determination of Layer Stresses and Strains for Heterostructures and Its Application to the Cubic SiC/Si System", Journal of Applied Physis, 64, 6827–6835 (1988). https://doi.org/10.1063/1.341997.

90. Joanna Wasyluk, *Micro-Raman investigation of Si, Ge and Carbon Related Nanostructures*, Thesis, Trinity College Dublin, Dublin (2012).

91. S. A. Kukushkin, A. V. Osipov, V. N. Bessolov, B. K. Medvedev, V. K. Nevolin, and K. A. Tcarik, "Substrates for Epitaxy of Gallium Nitride: New Materials and Techniques", Reviews on Advanced Materials Science, 17, 1–32 (2008). www.ipme.ru/e-journals/RAMS/no_11 708/kukushkin.pdf.

92. S. A. Kukushkin, A. V. Osipov, and I. P. Soshnikov, "Growth of Epitaxial SiC Layer on Si (100) Surface of n- and p-type of Conductivity by the Atoms Substitution Method", Advanced Material Science, 52, 9–42 (2017) www.ipme.ru/e-journals/RAMS/no_15217/05_15217_ku kushkin.pdf.

93. S. A. Kukushkin, A. V. Osipov, M. M. Rozhavskaya, A. V. Myasoedov, S. I. Troshkov, V. V. Lundin, L. M. Sorokin, A. F. Tsatsul'nikov," Growth and Structure of GaN Layers on Silicon Carbide Synthesized on a Si Substrate by the Substitution of Atoms: A Model of the Formation of V-Defects during the Growth of GaN", Physics of the Solid State, 57(9), 1899–1907 (2015). https://doi.org/10.1134/S1063783415090218.

94. M. M. Rozhavskaya, S. A. Kukushkin, A. V. Osipov, A. V. Myasoedov, S. I. Troshkov, L. M. Sorokin, P. N. Brunkov, A. V. Baklanov, R. S. Telyatnik, R. R. Juluri, K. B. Pedersen, V. N. Popok, "Metal Organic Vapor Phase Epitaxy Growth of (Al)GaN Heterostructures on SiC/Si(111) Templates Synthesized by Topochemical Method of Atoms Substitution", Physica Status Solidi, (a)214(10), 1700190/1–7 (2017). DOI 10.1002/pssa.201700190.

95. A. S. Grashchenko, S. A. Kukushkin, A. V. Osipov, A. V. Redkov. "Nanoindentation of GaN/SiC Thin Films on Silicon Substrate", Journal of Physics and Chemistry of Solids, 102, 151–156 (2017). http://dx.doi.org/10.1016/j.jpcs.2016.11.004.

96. S. A. Kukushkin, A. V. Osipov, V. N. Bessolov, E. V. Konenkova, V. N. Panteleev, "Misfit Dislocation Locking and Rotation During Gallium Nitride Growth on SiC/Si Substrates", Physics of the Solid State, 59(4), 36–43 (2017). https://doi.org/10.1134/S1063783417040114.

97. V. Bessolov, A. Kalmykov, E. Konenkova, S. Kukushkin, A. Myasoedov, N. Poletaev, S. Rodin. "Semipolar AlN and GaN on Si(100): HVPE Technology and Layer properties" Journal of Crystal Growth, 457, 202–206 (2017). https://doi.org/10.1016/j.jcrys gro.2016.05.025

98. V. Bessolov, A. Kalmykov, S. Konenkov, E. Konenkova, S. Kukushkin, A. Myasoedov, A. Osipov, V. Panteleev, "Semipolar AlN on Si(100): Technology and Properties", Microelectronic Engineering, 178, 34–37 (2017). http://dx.doi.org/10.1016/j.mee.2017.04.047.

99. S. A. Kukushkin, A. M. Mizerov, A. V. Osipov, A. V. Redkov, S. N. Timoshnev, "Plasma Assisted Molecular Beam Epitaxy of Thin GaN Films on Si(111) and SiC/Si(111) substrates: Effect of SiC and Polarity Issues", Thin Solid Films, 646(1), 158–162 (2018). https://doi.org/10.1016/j.tsf.2017.11.037.

100. S. A. Kukushkin, Sh. Sh. Sharofidinov, A. V. Osipov, A. V. Redkov, V. V. Kidalov, A. S. Grashchenko, I. P. Soshnikov, and A. F. Dydenchuk, "The Mechanism of Growth of GaN Films by the HVPE Method on SiC Synthesized by the Substitution of Atoms on Porous Si Substrates", ECS Journal of Solid State Science and Technology, 7(9), P480-P486 (2018). https://doi.org/10.1149/2.0191809jss

101. K. Yu. Shugurov, R. R. Reznik, A. M. Mozharov, K. P. Kotlyar, O. Yu. Koval, A. V. Osipov, V. V. Fedorov, I. V. Shtrom, A. D. Bolshakov, S. A. Kukushkin I. S. Mukhin, G. E. Cirlin, "Study of SiC Buffer Layer Thickness Influence on Photovoltaic Properties of n-GaN NWs/ SiC/p-Si Heterostructure", Materials Science in Semiconductor Processing, 90, 20–25 (2019). https://doi.org/10.1016/j.mssp.2018.09.024.

102. S. A. Kukushkin, A. V. Osipov, and A. V. Red'kov, "Separation of III–N/SiC Epitaxial Heterostructure from a Si Substrate and their Transfer to other Substrate Types" Semiconductors, 51(3), 396–401 (2017). http://link.springer.com/article/10.1134/S10637 82617030149.

103. V. N. Bessolov, D. V. Karpov, E. V. Konenkova, A. A. Lipovskii, A. V. Osipov, A. V. Redkov, I. P. Soshnikov, S. A. Kukushkin, "Pendeo-Epitaxy of Stress-Free AlN Layer on a Profiled SiC/Si Substrate", Thin Solid Films, 606, 74–79, (2016), https://doi.org/10.1016/ j.tsf.2016.03.034.

104. V. V. Ratnikov, A. E. Kalmykov, A. V. Myasoedov, S. A. Kukushkin, A. V. Osipov, L. M. Sorokin, "Sequential Structural Characterization of Layers in the GaN/AlN/SiC/Si(111) System by X-ray Diffraction Upon Every Growth Stage", Technical Physics Letters, 39(11), 994–997 (2013). https://doi.org/10.1134/S1063785013110230.

105. V. Bessolov, E. Konenkova, M. Shcheglov, S. Sharofidinov, S. Kukushkin, A. Osipov, V. Nikolaev, "HVPE Growth of GaN in the Semipolar Direction on Planar Si(210)", Physica Status Solidi, (c) 10(3), 433–436. (2013). https://doi.org/10.1002/pssc.201200566.

106. V. N. Bessolov, E. V. Konenkova, S. A. Kukushkin, A. V. Myasoedov, A. V. Osipov, S. N. Rodin, M. P. Shcheglov, N. A. Feoktistov, "Epitaxy of Semipolar GaN on a Si(001) Substrate with a SiC Buffer Layer", Technical Physics Letters, 40(5), 386–388 (2014). https://doi.org/ 10.1134/S1063785014050046.

107. V. N. Bessolov, E. V. Konenkova, S. A. Kukushkin, A. V. Myasoedov, A. V. Osipov, S. N. Rodin, "Semipolar Gallium Nitride on Silicon: Technology and Properties", Review of Advanced Materials Science, 38, 75–93 (2014). https://www.ipme.ru/e-journals/RAMS/no_13814/08_13814_ kukushkin.pdf

108. V. N. Bessolov, A. S. Grashchenko, E. V. Konenkova, A. V. Myasoedov, A. V. Osipov, A. V. Red'kov, S. N. Rodin, V. P. Rubets, S. A. Kukushkin, "Effect of the n and p-type Si(100)

Substrates with a SiC Buffer Layer on the Growth Mechanism and Structure of Epitaxial Layers of Semipolar AlN and GaN", Physics of the Solid State, 57(10),1966–1971 (2015). https://doi.org/10.1134/S1063783415100042.

109. V. G. Talalaev, J. W. Tomm, S. A. Kukushkin, A. V. Osipov, I. V. Shtrom, K. P. Kotlyar, F. Mahler, J. Schilling, R. R. Reznik, G. E. Cirlin, "Ascending Si Diffusion into Growing GaN Nanowires from the SiC/Si Substrate: up to the Solubility Limit and Beyond", Nanotechnology, 31(29), 294003/1–8 (2020). https://doi.org/10.1088/1361-6528/ab83b6

110. P. V. Seredin, D. L. Goloshchapov, D. S. Zolotukhin, A. S. Lenshin, A. M. Mizerov, S. N. Timoshnev, E. V. Nikitina, I. N. Arsentiev, S. A. Kukushkin, "Optical Properties of GaN/SiC/por-Si/Si(111) Hybrid Heterostructures", Semiconductors, 54(4), 417–425 (2020). https://doi.org/10.1134/S1063782620040168.

111. P. V. Seredin, D. L. Goloshchapov, D. S. Zolotukhin, A. S. Lenshin, Yu.Yu. Khudyakov, A. M. Mizerov, S. N. Timoshnev, I. N. Arsentyev, A. N. Beltyukov, H. Leiste, S. A. Kukushkin, "Influence of a Nanoporous Silicon Layer on the Practical Implementation and Specific Features of the Epitaxial Growth of GaN Layers on SiC/por-Si/c-Si Templates", Semiconductors, 54(5), 596–608 (2020). https://doi.org/10.1134/S1063782620050115.

112. I. G. Aksyanov, V. N. Bessolov, Yu. V. Zhilyaev, M. E. Kompan, E. V. Konenkova, S. A. Kukushkin, A. V. Osipov, S. N. Rodin, N. A. Feoktistov, Sh. Sharofidinov, M. P. Shcheglov," Chloride Vapor-Phase Epitaxy of Gallium Nitride on Silicon: Effect of a Silicon Carbide Interlayer", Technical Physics Letters, 34(6), 479–482 (2008). https://doi.org/10.1134/S1063785008060084.

113. L. M. Sorokin, A. E. Kalmykov, V. N. Bessolov, N. A. Feoktistov, A. V. Osipov, S. A. Kukushkin, N. V. Veselov, "Structural Characterization of GaN Epilayers on Silicon: Effect of Buffer Layers", Technical Physics Letters, 37(4), 326–329 (2011). https://doi.org/10.1134/S1063785011040158.

114. V. N. Bessolov, Yu. V. Zhilyaev, E. V. Konenkova, L. M. Sorokin, N. A. Feoktistov, Sh. Sharofidinov, M. P. Shcheglov, "Aluminum and Gallium Nitrides on a Silicon Substrate with an Intermediate Silicon Carbide Nanolayer for Ultraviolet Devices", Journal of Optical Technology, 78(7), 435–439 (2011). www.osapublishing.org/jot/abstract.cfm?URI=jot-78-7-435.

115. Sh. Sh. Sharofidinov, S. A. Kukushkin, A. V. Red'kov, A. S. Grashchenko, A. V. Osipov, "Growing III–V Semiconductor Heterostructures on SiC/Si Substrates", Technical Physics Letters, 45(7), 711–713 (2019). https://doi.org/10.1134/S1063785019070277.

116. S. A. Kukushkin, and Sh. Sh. Sharofidinov, "A New Method of Growing AlN, GaN, and AlGaN Bulk Crystals Using Hybrid SiC/Si Substrates", Physics of the Solid State, 61(12), 2342–2347 (2019). https://doi.org/10.1134/S1063783419120254.

117. A. M. Mizerov, S. A. Kukushkin, Sh. Sh. Sharofidinov, A. V. Osipov, S. N. Timoshnev, K. Yu. Shubina, T. N. Berezovskaya, D. V. Mokhov, A. D. Buravlev, "Method for Controlling the Polarity of Gallium Nitride Layers in Epitaxial Synthesis of GaN/AlN Heterostructures on Hybrid SiC/Si Substrates", Physics of the Solid State. 61(12), 2277–2281 (2019). https://doi.org/10.1134/S106378341912031X.

118. A. A. Koryakin, Yu. A. Eremeev, A. V. Osipov, S. A. Kukushkin, "The Influence of the Porosity of the Silicon Layer on the Elastic Properties of Hybrid SiC/Si Substrates" Technical Physics Letters, 47(2), 126–129 (2021). https://doi.org/10.1134/S1063785021020085.

119. A. S. Grashchenko, S. A. Kukushkin, A. V. Osipov, "Elastic Properties of GaN and AlN Films Formed on SiC/Si Hybrid Substrate, a Porous Basis", Mechanics of Solids 55(2), 157–161 (2020). https://link.springer.com/article/10.3103/S0025654420020107.

120. S. A. Kukushkin, V. I. Nikolaev, A. V. Osipov, E. V. Osipova, A. I. Pechnikov, N. A. Feoktistov, "Epitaxial Gallium Oxide on a SiC/Si Substrate", Physics of the Solid State, 58(9), 1876–1881 (2016). https://doi.org/10.1134/S1063783416090201.

121. A. V. Osipov, A. S. Grashchenko, S. A. Kukushkin, V. I. Nikolaev, E. V. Osipova, A. I. Pechnikov, I. P. Soshnikov, "Structural and Elastoplastic Properties of β-Ga_2O_3 Films Grown on Hybrid SiC/Si Substrates", Continuum Mechanics and Thermodynamics, 30, 1059–1068 (2018). https://doi.org/10.1007/s00161-018-0662-6.

122. V. V. Antipov, S. A. Kukushkin, A. V. Osipov, "Epitaxial Growth of Cadmium Sulfide Films on Silicon", Physics of the Solid State, 58(3), 629–632 (2016). https://doi.org/10.1134/S1063783416030033.

123. V. V. Antipov, S. A. Kukushkin, A. V. Osipov, V. P. Rubets, "Epitaxial Growth of Cadmium Selenide Films on Silicon with a Silicon Carbide Buffer Layer", Physics of the Solid State, 60(3), 504–509 (2018). https://doi.org/10.1134/S1063783418030022.

124. A. A. Koryakin, S. A. Kukushkin, A. V. Redkov, "Nucleation of CdSe Thin Films: The Kinetic Model", Journal of Physics: Conference Series 1124, 022044/1–6 (2018). http://iopscience.iop.org/article/10.1088/1742-6596/1124/2/022044/pdf.

125. V. V. Antipov, S. A. Kukushkin, A. V. Osipov, "Epitaxial Growth of Cadmium Telluride Films on Silicon with a Buffer Silicon Carbide Layer", Physics of the Solid State, 59(2), 399–402 (2017). https://doi.org/10.1134/S1063783417020020.

126. V. V. Antipov, S. A. Kukushkin, A. V. Osipov. "Epitaxial Growth of Zinc Sulfide by Atomic Layer Deposition on SiC/Si Hybrid Substrates", Technical Physics Letters, 45(11), 1075–1077 (2019). https://doi.org/10.1134/S1063785019110026.

127. S. A. Kukushkin, A. V. Osipov, A. I. Romanychev, "Epitaxial Growth of Zinc Oxide by the Method of Atomic Layer Deposition on SiC/Si Substrates", Physics of the Solid State, 58(7), 1448–1452 (2016). https://doi.org/10.1134/S1063783416070246.

128. S. A. Kukushkin, A. V. Osipov, I. A. Kasatkin, V. Y. Mikhailovskii, A. I. Romanychev, "Formation of Ordered ZnO Structures Grown by the ALD Method on Hybrid SiC/Si (100) Substrates", Materials Physics and Mechanics, 42(1), 30–39 (2019). DOI: 10.18720/MPM.4212019_4.

129. S. A. Kukushkin, A. V. Osipov, E. V. Osipova, S. V. Razumov, A. V. Kandakov, "The Optical Constants of Zinc Oxide Epitaxial Films Grown on Silicon with a Buffer Nanolayer of Silicon Carbide", Journal of Optical Technology, 78(7), 440–443 (2011). www.osapublishing.org/jot/abstract.cfm?URI=jot-78-7-440.

130. A. V. Osipov, S. A. Kukushkin, N. A. Feoktistov, E. V. Osipova, N. Venugopalb, G. D. Vermab, B. K. Guptac, A. Mitra, "Structural and Optical Properties of High Quality ZnO Thin Film on Si with SiC Buffer Layer", Thin Solid Films, 520(23), 6836–6840 (2012). doi: 10.1016/j.tsf.2012.07.094.

131. J. W. Christian, *The Theory of Transformations in Metals and Alloys*, Pergamon, Amsterdam, p.1200, 2002.

132. S. A. Kukushkin, A. V. Osipov, "The Gorsky Effect in the Synthesis of Silicon-Carbide Films from Silicon by Topochemical Substitution of Atoms", Technical Physics Letters, 43(7), 631–634 (2017). https://doi.org/10.1134/S1063785017070094.

133. V. O. Gridchin, R. R. Reznik, K. P. Kotlyar, A. S. Dragunova, N. V. Kryzhanovskaya, A. Yu. Serov, S. A. Kukushkin, G. E. Tsyrlin, "Molecular Beam Epitaxy of InAs Filamentous Nanocrystals on SiC/Si(111) and Si(111) Substrates: Comparative Analysis", Technical Physics Letters, 47(21) (2021) (in press).

134. I. P. Pronin, E.Yu. Kaptelov, S. V. Senkevich, V. A. Klimov, N. A. Feoktistov, A. V. Osipov, and S. A. Kukushkin, "Thin-Film PZT/SiC Structure on Silicon Substrate: Formation, Structural Features, and Dielectric Properties", Technical Physics Letters, 34(10), 838–840 (2008). https://doi.org/10.1134/S1063785008100088.

135. S. A. Kukushkin, A. V. Osipov, S. G. Zhukov, E. E. Zavarin, W. V. Lundin, M. A. Sinitsyn, M. M. Rozhavskaya, A. F. Tsatsulnikov, S. I. Troshkov, N. A. Feoktistov, "Group-III-Nitride-Based Light-Emitting Diode on Silicon Substrate with Epitaxial Nanolayer of Silicon Carbide", Technical Physics Letters, 38(3), 297–299 (2012). https://doi.org/10.1134/S106378501012030261.

136. N. A. Cherkashin, A. V. Sakharov, A. E. Nikolaev, V. V. Lundin, S. O. Usov, V. M. Ustinov, A. S. Gerashchenko, S. A. Kukushkin, A. V. Osipov, A. F. Tsatsulnikov, "Features of Epitaxial Growth of III-N Light Emitting Heterostructures on SiC/Si Substrates", Technical Physics Letters, 47(15), (2021) (in press).

137. L. K. Markov, S. A. Kukushkin, I. P. Smirnova, A. S. Pavlyuchenko, A. S. Gerashchenko, A. V. Osipov, G. V. Svyatets, A. E. Nikolaev, A. V. Sakharov, V. V. Lundin, A. F.

Tsatsulnikov," LED Based on AlInGaN-Heterostructures Grown on SIC/Si Substrates and Its Manufacturing Technology", Technical Physics Letters. 47(18), (2021) (in press).

138. A. V. Solnyshkin, O. N. Sergeeva, O. A. Shustova, Sh. Sh. Sharofidinov, M. V. Staritsyn, E.Yu. Kaptelov, S. A. Kukushkin, I. P. Pronin, "Dielectric and Pyroelectric Properties of Composites Based on Aluminum and Gallium Nitrides Grown by Chloride-Hydride Epitaxy on a Silicon Carbide-on-Silicon Substrate", Technical Physics Letters, 47(5), (2021) (in press).

139. O. N. Sergeeva, A. V. Solnyshkin, S. A. Kukushkin, A. V. Osipov, Sh. Sharofidinov, E.Yu. Kaptelov, S. V. Senkevich, I. P. Pronin. "New Semipolar Aluminium Nitride Thin Films: Growth Mechanisms, Structure, Dielectric And Pyroelectric Properties" Ferroelectrics, 544, 33–37 (2019). https://doi.org/10.1080/00150193.2019.1598181.

140. S. A. Kukushkin, Sh. Sh. Sharofidinov, A. V. Osipov, A. S. Grashchenko, A. V. Kandakov, E. V. Osipova, K. P. Kotlyar, E. V. Ubyivovk, "Self-Organization of the Composition of $Al_xGa_{1-x}N$ Films Grown on Hybrid SiC/Si Substrates", Physics of the Solid State, 63(3), 442–448 (2021). https://doi.org/10.1134/S1063783421030100.

141. S. A. Kukushkin, A. V. Osipov, N. A. Feoktistov, "Two-Stage Conversion of Silicon to Nanostructured Carbon by the Method of Coordinated Atomic Substitution", Physics of the Solid State, 61, 456–463 (2019). DOI: 10.1134/S1063783419030193.

142. S. A. Kukushkin, A. V. Osipov, E. V. Osipova, V. M. Stozarov "Investigation of the Stages of Transformation of Silicon into Silicon Carbide in the Process of Atomic Substitution by Methods of total External Reflection of X-rays and X-ray Diffractometry", Physics of the Solid State, 64, (2022). (In press).

Part IV

SiC Devices and Developments

13

4H-SiC-Based Photodiodes for Ultraviolet Light Detection

Weifeng Yang

Department of Microelectronics and Integrated Circuit, School of Electronic Science and Engineering (National Model Microelectronics College), Xiamen University, Xiamen, PR China

Zhengyun Wu

Department of Physics, School of Physical Science and Engineering, Xiamen University, Xiamen, PR China

1 Introduction

Ultraviolet (UV) "light" is a type of electromagnetic radiation with wavelengths ranging from 10 to 400 nm [1], which come from either natural or artificial sources. UV is generally classified into three bands according to WHO: UVA (315~400 nm), UVB (280~315 nm), and UVC (100~280 nm). When sunlight passes through the atmosphere, all UVC and most UVB is absorbed by ozone, water vapor, oxygen, and carbon dioxide. UVA is less affected by the atmosphere. Therefore, around 95% of the UV radiation reaching the Earth's surface is UVA radiation. UVA radiation is less hazardous than UVB radiation in general. Excessive exposure to UVB radiation can cause skin cancer, eye damage, and suppression of the body's immune system. Furthermore, due to human-created pollution, reductions in the ozone layer have increased the amount of UV radiation reaching the Earth's surface in recent years, which is harmful to human health. UVC radiation normally comes from some artificial sources, such as deuterium lamps, UV lasers, etc.

Over the last few decades, research on UV radiation has received a lot of attention. The detection of UV radiation is utilized in biotechnology, material chemistry, ecology, astronomy, and ultraviolet radiation measurement [2]. Solar-blind UV detection is one of the most appealing research areas in UV radiation research. Solar radiation with wavelengths longer than 280 nm can reach the Earth's surface. In this scenario, background radiation from sunlight has no effect on the detection of UV radiation in the wavelength range 240~280 nm. As a result, this spectrum is also known as the solar-blind region, which is of great importance in flame monitoring, missile plume detection, and other applications [3,4]. UV radiation has a significant influence on both organic and inorganic materials and serves as an information carrier, resulting in a huge market for UV sensors and related application products [5].

A UV photodiode is a UV-sensitive semiconductor device that converts UV signals to electrical signals via the photoelectric effect. UV photodiodes can be utilized for both military and civilian applications, such as missile guidance, missile early warning, UV

DOI: 10.1201/9780429198540-17

communication, UV radiation monitoring, flame detection, etc. A high-performance photodetector should meet the 5S requirements of high sensitivity, high signal-to-noise ratio, high spectral selectivity, high speed, and high stability [6]. Particularly in military applications, UV photodetectors must be reliable and capable of operating in harsh environments such as extreme temperature and space radiation.

In recent years, UV photodetectors based on wide bandgap semiconductors such as SiC [7], GaN [8], AlN [9], ZnSe [10], ZnO [11], TiO_2 [12], and others have made significant progress. Among them, 4H-SiC has proven its promise for high-power and high-frequency applications because of its properties such as wide bandgap (3.26 eV), high breakdown electric fields, good thermal conductivity, high electron saturation rates, and radiation resistance. Hence, 4H-SiC is suitable for developing high-frequency, high-power, high-temperature, and radiation-resistant semiconductor devices. 4H-SiC UV photodiodes also have the following advantages [13]: (1) Due to the wide bandgap, 4H-SiC-based photodiodes are only sensitive to UV radiation and barely respond to visible light. In this circumstance, they can achieve high response and low noise. (2) 4H-SiC-based photodiodes with planar structures are simple to fabricate, and large-area two-dimensional arrays can be prepared. (3) 4H-SiC-based photodiodes have good reliability and stability. They can work in severe conditions such as extreme temperature and high radiation. (4) 4H-SiC-based photodiodes have a very low leakage current due to the low intrinsic carrier concentration of 4H-SiC. (5) Homogeneous epitaxial growth with the matched lattice of 4H-SiC can be realized because of the comparatively mature crystal growth technology of 4H-SiC [14]. Crystal growth or ion implantation can be used to achieve *p*-type and *n*-type doping [15,16].

In recent decades, 4H-SiC-based UV photodetectors have been greatly improved with various types of photodiode structure including Schottky diodes, metal-semiconductor-metal (MSM), *p-n* or *p-i-n* photodiode and avalanche photodiode (APD). Table 1 presents the comparison of UV photodetectors with these four structures. A Schottky diode has a simple structure that requires fewer fabrication steps, and it has a low sensitivity and a high leakage current. MSM UV photodiodes have a basic structure and can be used to prepare devices with a large photosensitive area. Photoconductive gain is also presented in MSM devices. However, MSM photodiodes demonstrate low sensitivity, a relatively high dark current and low quantum efficiency. The structure of *p-i-n* photodiodes is relatively complicated, yet they have a fast response time and low dark current. APD has attracted extensive attention because of its high sensitivity, fast response speed, and large response bandwidth, which is primarily employed in single photon detection. In particular, 4H-SiC is an excellent material for UV avalanche photodetectors with low multiplication noise due to its low electron and hole ionization coefficient ratio ($k<0.1$).

TABLE 1

The comparison of UV photodetectors with four classic structures

Structure	MSM	Schottky barrier	*p-i-n*	APD
Fabrication Process	Simple	Simple	moderate	complicated
Bias Voltage	Low	Low	low	high
Responsivity	High	Low	high	high
Multiplicative Noise	No	No	no	yes
Sensitivity	low	low	high	very high
Internal Gain	No	No	no	yes

2 Basic Theory of UV Photodetection

2.1 Photoelectric Effect

The photoelectric effect, which is the emission of electrons under the influence of light radiation, is the foundation of UV photodetectors. External and internal photoelectric effects are two types of photoelectric effects [17].

2.1.1 External Photoelectric Effect

The phenomenon of electrons escaping off the surface of the irradiated material is referred to as the external photoelectric effect, also known as the photoelectric emission effect. When the photons are incident on the material, electrons in the material absorb the energy of the photons and transit from the ground state to the excited state. The excited electrons collide with the lattice during travel to the surface, resulting in energy loss. Finally, excited electrons are ejected from the surface with enough energy to surpass the surface barrier. Einstein's photoelectric equation is as follows:

$$E_k = h\nu - E_\varphi \tag{1}$$

where, E_k represents the maximum kinetic energy of emitted electrons from the surface, E_φ is the minimum energy required for the electrons to escape from the surface, which is known as the material's work function, h is Planck's constant, and ν is the light frequency.

2.1.2 Internal Photoelectric Effect

The internal photoelectric effect is divided into photoconductive and photovoltaic types. In contrast to metals, semiconductors generate electrons and holes at temperatures above absolute zero due to thermal excitation, and these electrons and holes also recombine in pairs during thermal motion until a state of dynamic equilibrium is attained. When photons strike a semiconductor material, electrons absorb the energy of the photons and transit from the valence band to the conduction band, while holes are formed in the valence band, which is known as the generation of photogenerated electron-hole pairs. As a result, an increase in carrier concentration leads to an increase in semiconductor conductivity. This phenomenon is called the photoconductive effect.

The photovoltaic effect refers to the phenomenon that a voltage is generated when a semiconductor material with a *p-n* junction or a metal-semiconductor junction is exposed to light. In a *p-n* junction, the depletion region creates a built-in electric field. When the wavelength of the incident light is shorter than the cut-off wavelength, photogenerated electron-hole pairs are formed in the depletion region under zero bias. The depletion region and the built-in electric field can be enhanced by applying a reverse-bias voltage. In this case, the electrons in or near the depletion region accumulate in the *n*-type region, while holes accumulate in the *p*-type region, resulting in a potential difference between the two ends of the *p-n* junction. An open circuit voltage is created across the *p-n* junction. Here the process of converting light into electric current is called photoelectric conversion.

2.2 Key Parameters of UV Photodiodes

2.2.1 Quantum Efficiency and Responsivity

Every incident UV photon tries to produce an electron-hole pair. However, the number of created electron-hole pairs is usually smaller than the number of photons. The ratio of the number of charge carriers to the number of incident photons is known as external quantum efficiency (EQE).

$$EQE = \left(\frac{I_p}{q}\right) \cdot \left(\frac{P_o}{hv}\right)^{-1} \qquad (2)$$

where I_p is the generated photocurrent, P_o is the incident light power, and q is the electron charge. In order to obtain higher quantum efficiency, we should optimize the device structure to increase the number of photons reaching the depletion region and reduce the surface reflection of incident light.

Responsivity is the ratio of photocurrent created from incident light to the incident light power.

$$R = \frac{I_p}{P_o} = \frac{EQE \cdot \lambda}{1240} \qquad (3)$$

The capacitance of the *p-n* junction determines the photodiode's response (speed/time). It is the amount of time needed for charge carriers to pass the *p-n* junction, which is affected by the width of the depletion region.

2.2.2 Cut-off Frequency and Cut-off Wavelength

The cut-off frequency (v_C) of the UV photodetector is the minimum frequency that it can detect. In the meanwhile, the cut-off wavelength (λ_C) is the maximum wavelength that the UV photodetector can detect. The cut-off frequency and cut-off wavelength are calculated using the following formulas, respectively:

$$v \geq \frac{E_\varphi}{h} = v_C \qquad (4)$$

$$\lambda \leq \frac{hc}{E_\varphi} = \lambda_C \qquad (5)$$

where c is the speed of light. According to Eq. 5, the cut-off wavelength of 4H-SiC is calculated to be about 380 nm, making it particularly suitable for visible-blind UV photodetectors.

2.2.3 Photocurrent and Dark Current

The current created in the external circuit is referred to as photocurrent. It may be tested with various biases. When there is no incident light, the current in the photodiode is called dark current. This could be one of the most common sources of photodiode noise. Diffusion current, generation-recombination current, tunneling current, and surface leakage current are all part of the dark current, which is closely related to the material.

2.2.4 Breakdown Voltage

Breakdown voltage is the highest reverse voltage that may be supplied to a photodiode before leakage current or dark current exponentially increases. Photodiodes should be operated at a lower reverse bias than this.

---

3 Classical 4H-SiC-Based Photodetectors

3.1 Schottky Barrier Diodes

The Schottky barrier photodiode (SBD) includes a substrate, a buffer layer adjacent to a substrate, and an epitaxial layer adjacent to the buffer layer, as shown in Figure 1.

The following factors are taken into account when designing the SBD structure: (1) A mesa structure is formed by etching the epitaxial layer for isolation between devices. (2) Metals such as Pt and Au are commonly used in active areas because of their high optical transmittance. (3) The metal should be thin in order to increase optical transmittance and reduce both reflection and absorption loss. (4) The metal should have excellent ohmic contact properties.

High-efficiency photodetectors can be made using Schottky photodiodes. The metal film should be coated with an antireflection film to prevent reflection and absorption loss. Schottky photodiodes can operate in a variety of modes depending on photon energy and bias conditions [18].

(1) As shown in Figure 2 (a), $E_g > h\nu > q\$_{Bn}$, $V < V_B$. Here $\$_{Bn}$ is the Schottky contact potential difference and V_B is the breakdown voltage. Excited electrons in metals can pass across the potential barrier and be collected by semiconductors. This physical process is often used to calculate the Schottky barrier height $q\$_{Bn}$ and also to comprehend hot electron transport in metals.
(2) As shown in Figure 2 (b), $h\nu > E_g$, $V < V_B$. Electron-hole pairs are generated in semiconductors. Schottky photodiodes have similar characteristics to *p-i-n* photodiodes under the conditions of $h\nu > E_g$ and $V < V_B$.
(3) As shown in Figure 2 (c), $h\nu > E_g$, $V \cong V_B$. The Schottky photodiode can operate like an avalanche photodiode in this situation.

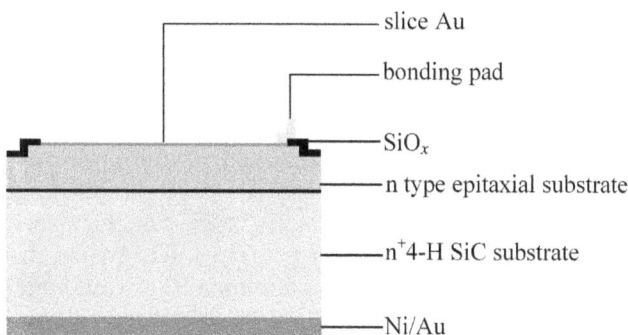

FIGURE 1
Schematic diagram of a 4H-SiC Schottky photodiode.

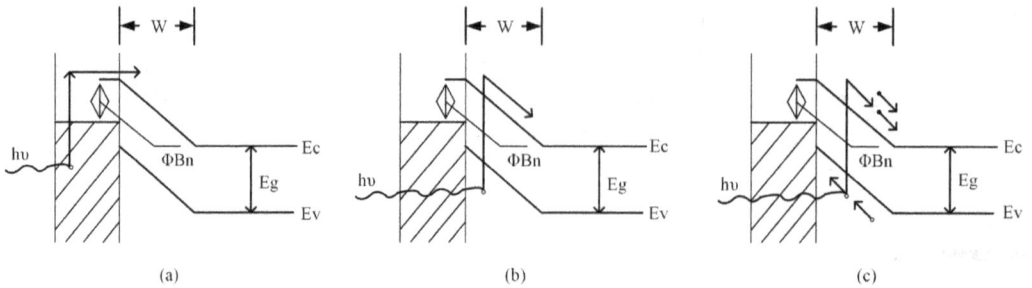

FIGURE 2

Energy band diagram of Schottky photodiodes in different operating modes.

Schottky photodiodes work on the basis of the photovoltaic effect. A Schottky photo-diode separates and collects photogenerated charge carriers using a metal-semiconductor junction. A photocurrent is formed by the photogenerated charge carriers.

In recent decades, researchers have improved the performance of 4H-SiC Schottky photodetectors through enhancing the material quality, choosing proper metals and opti-mizing the structure. Yan *et al.* at Rutgers University developed the first large-area 4H-SiC Schottky photodetector in 2004 [19]. A 7.5 nm semitransparent Pt was deposited on the n^- side to form Schottky contact and reduce UV absorption loss. To reduce dark current, ther-mally oxidized SiO_2, PECVD SiO_2, and PECVD Si_3N_4 were utilized as passivation layers. The dark current is around 0.1 pA at bias of 1 V. From 240 to 300 nm, the highest quantum efficiency is roughly 37% and is practically constant. The peak of the specific detectivity (D^*) at zero bias is 3.6×10^{15} cm•$Hz^{1/2}$/W at 300 nm, which is two orders of magnitude higher than the D^* of Si photodiodes and three orders of magnitude higher than the D^* of Si charge coupled device.

Most of the incident photons are absorbed in the surface layer due to the short penetra-tion depth of extreme ultraviolet (EUV) photons, and the photogenerated charge carriers generated in this region are trapped in high-density surface defects before reaching the depletion region, resulting in low quantum efficiency of the device. As a result, the fabrica-tion of EUV photodetectors is very challenging, which requires a large active area and low dark current. In 2005, Xin *et al.* fabricated the first 4H-SiC EUV detectors with a large detec-tion area for EUV and UV detection [20]. The peak QE in the UV range appears between 260 and 280 nm, with a value of 40% for Ni and 45% for Pt. The rejection ratio of UV to visible light is larger than 1000. The QE decreases for shorter wavelength in the EUV range between 120 and 200 nm. The devices show reasonable QE of 15~18% for Ni and 20% for Pt at 200 nm. At 120 nm, the QE is 4% for Ni and 9% for Pt. For the wavelength below 77.5 nm, the measured QE increases for shorter wavelengths, finally reaching 176% for Ni and 147% for Pt at 21.5 nm, which indicates that multiple electron-hole pairs are generated by one photon.

As demonstrated in Figure 3 (a), Hu *et al.* improved the performance of the 4H-SiC Schottky photodiodes in 2006 [21]. At –4V bias, the photodiodes have a dark current of less than 0.1 pA, as shown in Figure 3 (b). From 230 to 295 nm, the measured QE at 0 V bias is greater than 50%, with a peak of 65% at 275 nm. The QE is greater than 14% in the EUV range, approaches 100 percent at wavelengths less than 50 nm, and eventually exceeds 30 electrons/photon at 3 nm.

Wang *et al.* at University of Science and Technology of China reported Au/ 4H-SiC and Ni/ 4H-SiC Schottky UV photodetectors by using *n*-type 4H-SiC and metal Au (or Ni) to

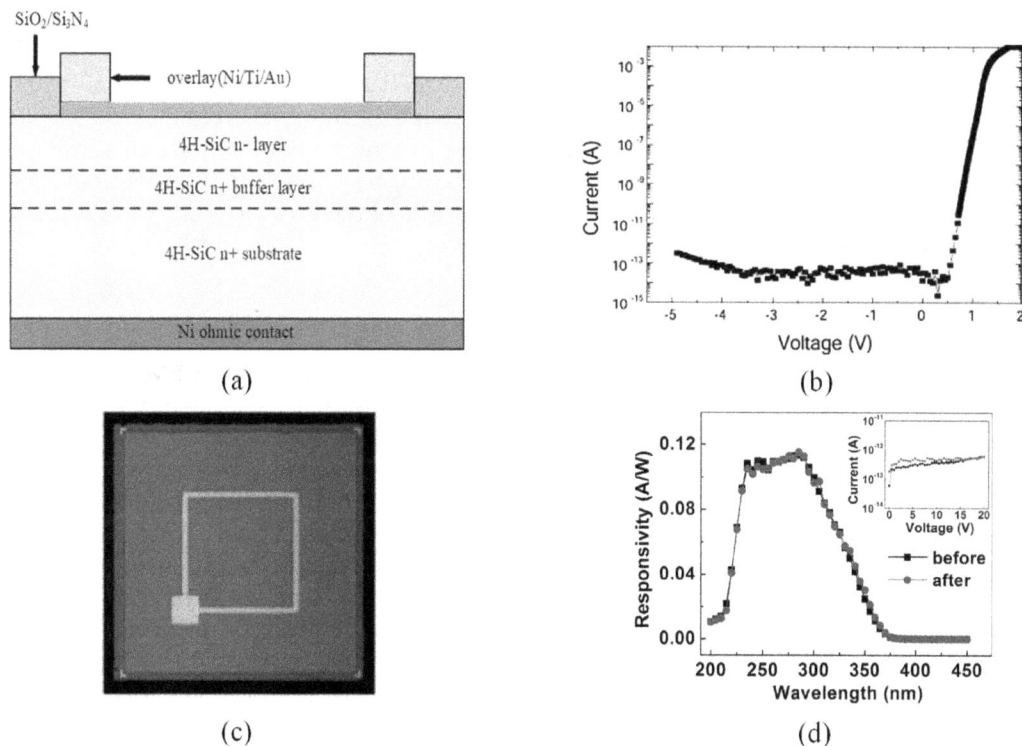

FIGURE 3

(a) The cross-sectional view of the Ni/4H-SiC Schottky photodiode. (b) *I-V* characteristics of the Ni/4H-SiC Schottky photodiode. From [21] Figure 1, reproduced with permission © 2006 Optical Society of America. (c) The top view of 4H-SiC SBD. (d) Room temperature spectral response characteristics of the device measured before and after thermal storage at 200 C in air for 100 h. The inset shows the dark current of the PD before and after the thermal storage treatment. From [25] Figure 1 & Figure 6, reproduced with permission © 2015 Journal of Vacuum Science & Technology B.

form Schottky contact in 2004 [22]. Both leakage currents are smaller than 1×10^{-10} A. At 7 V bias, the maximal spectrum responsivity is 86.72 mA/W for Au/ 4H-SiC and 45.84 mA/W for Ni/ 4H-SiC, respectively, and the maximal quantum efficiency is 37.15% for Au/ 4H-SiC and 18.98% for Ni/ 4H-SiC. The temperature characteristics of Au/ 4H-SiC Schottky UV photodetectors were also investigated [23]. The dark current grows from 0.65 nA to 2.7 nA at 10 V bias as the temperature rises from room temperature to 200 °C, whereas the maximum quantum efficiency drops from 29 percent to around 4% at zero bias. In 2005, Huang *et al.* improved the performance of the Au/ 4H-SiC Schottky UV photodiodes [24]. At 10 V bias, the photodiodes show a dark current of 10 pA. The devices' spectrum response is also dependent on the bias voltage. When the bias voltage is 150 V, a gain of 3.8×10^{4} is attained due to avalanche multiplication. Additionally, the photodiodes maintain good *I-V* characteristics and UV response characteristics at 260 °C.

In 2015, Xu *et al.* [25] at Nanjing University fabricated a vertical type 4H-SiC Schottky-barrier photodiode with ultrathin Ni semitransparent Schottky metal contact, as shown in Figure 3 (c), and evaluated the high-temperature reliability characteristics. In the temperature range of 25 to 200 °C, the photodiode has a very low dark current and a high quantum efficiency. The photodiode can operate at 200 °C with good reliability as shown in Figure 3 (d), implying that the 4H-SiC Schottky-barrier PDs are suitable for applications in medium-high-temperature harsh environment.

In 2020, Wang *et al.* at Nanjing University fabricated a 2.5 × 2.5 mm² large-area 4H-SiC EUV Schottky barrier photodiode with a grid-shaped semitransparent metal electrode [26]. Under 20 V reverse bias at normal temperature, the photodiode has an ultra-low leakage current of 3 pA. A UV-to-visible (285/400 nm) rejection ratio of more than 2000 is obtained, indicating good visible blindness. The QEs of the grid-electrode device at 0 V and −15 V are 739% and 780% at 13.5 nm, respectively.

3.2 Metal-Semiconductor-Metal (MSM) Diodes

The structure of a typical MSM photodiode is shown in Figure 4 (a). The MSM photodiode consists of a single-crystal substrate, a buffer layer adjacent to a substrate to ensure good crystal quality, and an absorption layer or active layer adjacent to the buffer layer that can absorb more than 90% of the incident light with a certain thickness. A thin barrier enhancement layer is applied to the active layer to improve the Schottky barrier height between the metal electrode and the active layer. Interdigitated finger electrodes and an insulating coating deposited on the electrode for passivation and protection are also included in the MSM photodiode. Figure 4 (c) shows the interdigital metal electrode from the top. This type of photodetector has a simple structure, which makes it ideal for producing devices with a large photosensitive area.

The MSM PD consists of back-to-back connected metal-semiconductor Schottky junctions, as shown in Figure 4 (b). When a bias voltage is applied across the MSM photodetector, one diode becomes forward biased and the other diode becomes reverse biased.

Figure 5 depicts the operating principle of MSM photodiodes. Figure 5 (a) shows the one-dimensional device structure of MSM photodetectors. A metal-semiconductor contact is formed on both sides, and the electrode distance is L. Figure 5 (b) shows the energy band diagram at zero bias.

As illustrated in Figure 5 (b), Φ_{n1}, Φ_{n2} is the height of Schottky barrier formed by metal-semiconductor contact on both sides, while V_{D1} and V_{D2} are built-in potentials, respectively. For the same metal on both sides, $\Phi_{n1} = \Phi_{n2}$ and $V_{D1} = V_{D2}$. And Φ_p is the barrier height of the hole. When a bias is applied, junction 1 is reverse biased and junction 2 is forward biased, with the depletion layer widths being W_1 and W_2, respectively. When the applied voltage increases, W_1 increases and W_2 drops, while the total depletion layer width

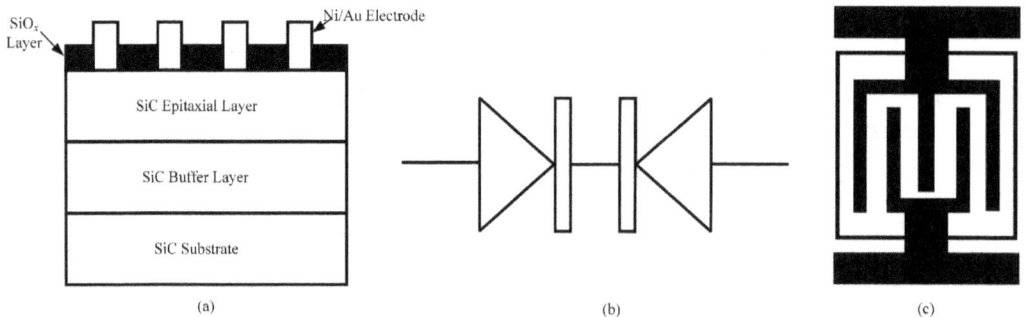

FIGURE 4

(a) Cross-section structure of 4H-SiC MSM photodiode; (b) back-to-back connected metal-semiconductor Schottky junctions; (c) top view of 4H-SiC MSM photodiode.

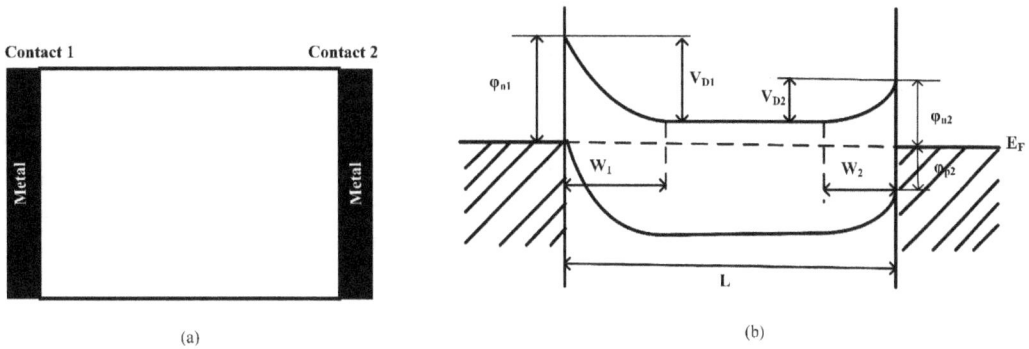

FIGURE 5
(a) One dimensional device structure and (b) energy band diagram of MSM-PD.

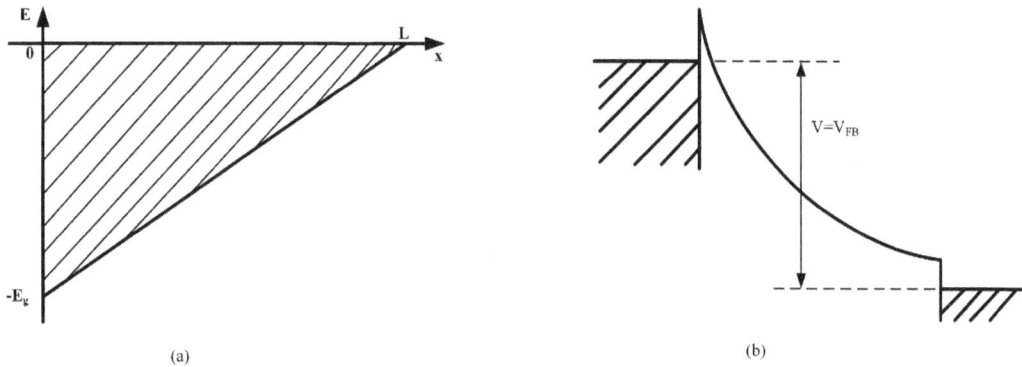

FIGURE 6
(a) Electric field and (b) energy band diagram of the device under flat-band voltage.

steadily increases. When the forward-biased contact meets the reverse-biased contact at the depletion region, the applied potential is called "reach-through voltage" (V_{RT}). Under this circumstance, $W_1 + W_2 = L$.

When the voltage is increased to the point where the forward-biased contact's energy band becomes a flat band and the electric field is zero, the device is completely depleted. In this scenario, the biased voltage is referred to as "flat-band voltage" (V_{FB}). Figure 6 depicts the device's electric field and energy band diagram under flat-band voltage.

The electron current is still relatively small at the flat-band voltage. However, because the potential carrier of the hole is decreasing, hole injection begins from the forward bias. The internal electric field increases as the voltage exceeds the flat-band voltage. Until the avalanche breakdown occurs, which makes the current surge. At this time, the maximum electric field at junction 1 is the breakdown electric field E_B. Usually, the device operates between the flat-band and breakdown voltages.

In 2002, Su *et al.* at National Cheng Kung University fabricated the first 4H-SiC-based MSM ultraviolet (UV) photodetectors with indium–tin-oxide (ITO), Ni/ ITO, and Ni as the semitransparent finger electrodes [27]. Ni was inserted in between ITO and 4H-SiC so as to increase Schottky barrier height and hence reduce leakage current. The contact electrode fingers were 10 μm wide and 200 μm long with a spacing of 10 μm. The device's total active area was 200 × 200 μm^2. The photocurrent of 4H-SiC MSM UV photodetectors with 10 nm Ni/90 nm ITO contact electrodes was significantly higher than that of a 4H-SiC MSM UV photodetector with 100 nm Ni contact electrodes, which can be explained by the fact that the optical transmittance of the 10 nm Ni/90 nm ITO contact electrodes was much larger than that of the 100 nm Ni contact electrodes. With a 5 V applied bias, the photo-current to dark current contrast for the SiC MSM UV photodetector with 10 nm Ni/90 nm ITO contact electrodes was found to be more than three orders of magnitude. The photo-current to dark current contrast might practically reach four orders of magnitude with a 40 V applied bias. The peak of spectrum responsivity of 4H-SiC MSM photodetectors was 0.07 A/W at a 5 V bias, corresponding to an internal quantum efficiency of 33% [28].

Since then, the absorption of ultraviolet light by metal electrodes as well as the effects of metal type, thickness, and spacing of interdigital electrodes were taken into consideration and extensive research was conducted by researchers.

In 2003, Wu *et al.* at Xiamen University demonstrated the first 4H-SiC MSM UV photodetector [29]. The *n*-type SiC photodetectors show a low dark current less than 10 nA at −15V bias while the *p*-type ones show a lower dark current of 0.3 nA at −25V. The higher dark current in *n*-type MSM detectors is likely due to dry-etching-induced surface degradation. The device structure and preparation method have been continuously improved since then. In 2004, Wu *et al.* demonstrated a 4H-SiC MSM UV photodetector with lower dark current, as illustrated in Figure 7 (a) [30]. At the bias about 15 V, the density of dark current is about 70 nA/cm^2. At −45V, the responsivity reaches a high value of about 50A/W for MSM UV detectors in *n*-type 4H-SiC, as shown in Figure 7 (b). The photocurrent is roughly two orders of magnitude higher than the dark current. The UV/visible ratio is larger than 1000, indicating that the devices have significantly enhanced visible-blind performance.

In 2006, Sciuto *et al.* at CNR-IMM Italy demonstrated a vertical 4H-SiC MSM photodetector with self-aligned Ni$_2$Si interdigit semitransparent contacts [31]. A vertical Schottky barrier structure was paired with a planar MSM structure in the suggested structure. The device shows a barrier height of 1.66 ± 0.02 eV and an ideality factor *n* of 1.04 ± 0.01. The dark current was about 200 pA at −50 V. A current increase of more than two orders of magnitude is recorded under a 256 nm UV illumination. The photocurrent saturates at 53 nA at −20 V due to the pinch-off. The maximal detector responsivity at 256 nm is 160 mA/W, with an internal QE of 78%. The optical response of the vertical photodiodes is about a factor of 1.8 higher than that of the conventional MSM structure, which can be attributed to the more efficient depletion of the interdigit structures. And the optically active area in the vertical device is nearly two times higher than in the MSM structure.

In 2008, Yang *et al.* at Xiamen University took the lead in successfully fabricating 40-pixel 4H-SiC MSM photodiode linear arrays [32]. The metal Au was deposited on the SiO$_x$ passivation layer. The metal Au is 40 nm/40 nm thick. The fingers are 500 μm long and 3 μm wide, with a 6 μm spacing. The optically sensitive area of the photodetector is 200 μm × 500 μm. A 1×40 linear array is made up of 40 units. Among them, the detector shows a peak responsivity of about 0.09 A/W, a dark current of less than 5 pA at 20V and a UV/visible ratio is more than 5000. The linear arrays perform consistently. Yang *et al.* also developed high-responsivity 4H-SiC-based MSM UV photodiodes in 2008, as illustrated in Figure 7 (c) [33]. At 20 V, the fabricated 4H-SiC MSM photodetectors with 4 μm-spacing have the highest typical responsivity of 0.103 A/W, as shown in Figure 7 (d). A peak response wavelength of 290 nm and a very low dark current of about 10^{-12} A were obtained. The DUV to visible rejection ratio of the device was larger than 10^3.

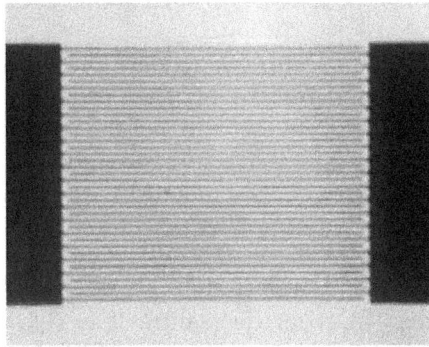

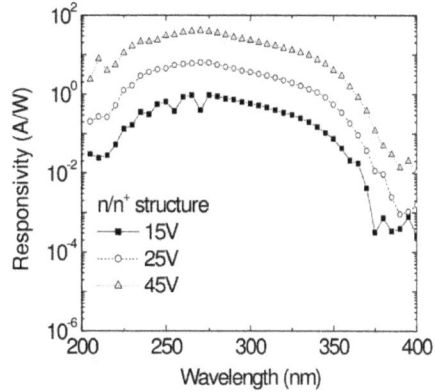

(a)

(b)

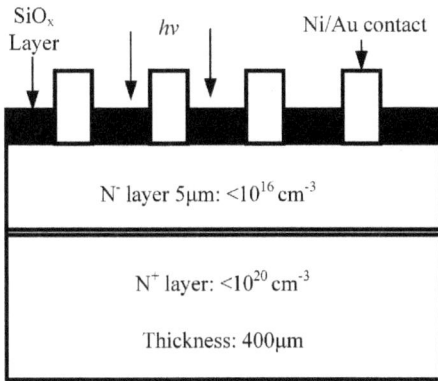

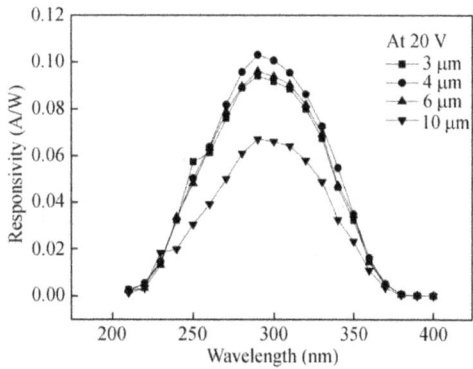

(c)

(d)

FIGURE 7

(a) Cross section of a 4H-SiC MSM PD. (b) Spectral response comparison of the 4H-SiC MSM photodiodes with 3 μm fingers and diverse spacings. From [30] Figure 1 & Figure 5, reproduced with permission © 2004 Materials Science Forum. (c) Schematic top view of 4H-SiC MSM photodiode with AL nanoparticles. (d) Spectral response of PD-0 and PD-1 at 10V bias under deuterium lamp at room temperature. From [33] Figure 1 & Figure 5, reproduced with permission © 2008 Science in China Series G: Physics, Mechanics & Astronomy.

In 2009, Mazzillo *et al.* demonstrated a 4H-SiC large area vertical MSM UV photodiode [34]. The leakage current is about 8 pA at –5 V and 25 °C and increases slowly as the temperature rises. Due to the poor doping of the epilayer, which allows for broad surface depletion zones around each Ni_2Si strip even at low reverse bias, the C-V characteristics showed surface pinch-off at low reverse voltage. At 290 nm, the device had a photoresponsivity peak of roughly 0.106 A/W. The photoresponsivity is independent of the bias, indicating that the active region is almost completely depleted at 0 V.

4H-SiC has high thermal conductivity, which enables 4H-SiC PDs to operate at the high-temperature environment. In 2012, Lien *et al.* at University of California, Berkeley, demonstrated 4H-SiC MSM UV photodetectors in the operation of 450 °C [35]. The operation speed of the MSM PDs is temperature-tolerant; as the temperature rises from 25 °C to 400 °C, the rise and fall time of the PD increase from 594 to 684 μs and from 699 to 786 μs, respectively. The SiC MSM PDs had a photocurrent-to-dark-current ratio of 1.3×10^5

at 25 °C and 0.62 at 450 °C. These results support the use of 4H–SiC PDs in extremely high temperature applications.

In 2012, Chen *et al.* at Xidian University analyzed temperature-dependent properties of a 4H-SiC metal-semiconductor-metal ultraviolet photodetector [36]. They discovered that as the temperature rises, the dark current and photocurrent increase. For the range of 500–800 K, the dark current increases by nearly a factor of 3.5 for every 150 K higher than that of photocurrent, resulting in a drop in the photodetector current ratio (PDCR). At 800 K, the maximum responsivity of 120.3 mA/W and peak quantum efficiency of 51.82% can be obtained at 302 and 280 nm, respectively.

In 2017, Liu *et al.* at Shanghai Normal University demonstrated 4H-SiC metal-semiconductor-metal (MSM) ultraviolet photodiodes (PDs) for deep ultraviolet (DUV) detection by introducing the coupling of localized surface plasmon resonance (LSPR) from Al nanoparticles [37]. When a 10 V bias is applied, the peak responsivity of 165 mA/W at 220 nm is achieved. The quantum efficiency is 93%.

3.3 *p-n* and *p-i-n* Photodiodes

For *p-n* or *p-i-n* photodiodes, the *p-n* junction is of great importance. In a *p-n* junction, some of the free electrons in the *n*-region diffuse across the junction and combine with holes in the *p*-region, leaving the positive ions at the donor impurity sites. Meanwhile, holes in the *p*-region diffuse into the *n*-region, leaving negative ions in the acceptor impurity sites. Because of the depletion of carriers in the region, this region is called the depletion region, also known as the space charge region, as seen in Figure 8. This creates an electric field across the depletion region that provides a force opposing the charge diffusion, which is called the built-in electric field. The direction of the electric field is from *n*-region to *p*-region.

The structure of the *p-n* photodiode is shown in Figure 9 (a). The photovoltaic effect is the fundamental principle of a *p-n* photodiode. When photons with energy greater than or equal to the bandgap of the material strike the depletion region, photogenerated electron-hole pairs are generated. Under the built-in electric field, electrons move towards the *n*-region, while holes flow towards the *p*-region. When an external circuit is connected, electrons pass across the circuit and generate the photocurrent. Since the photogenerated

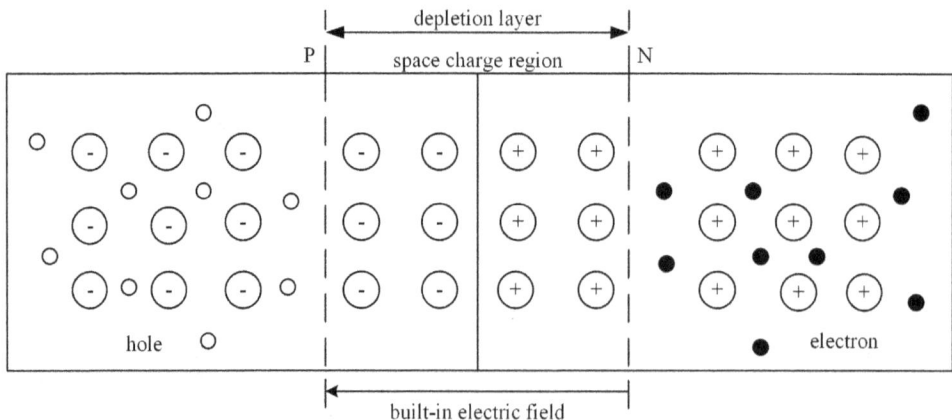

FIGURE 8
Diagram of the space charge region of *p-n* junctions.

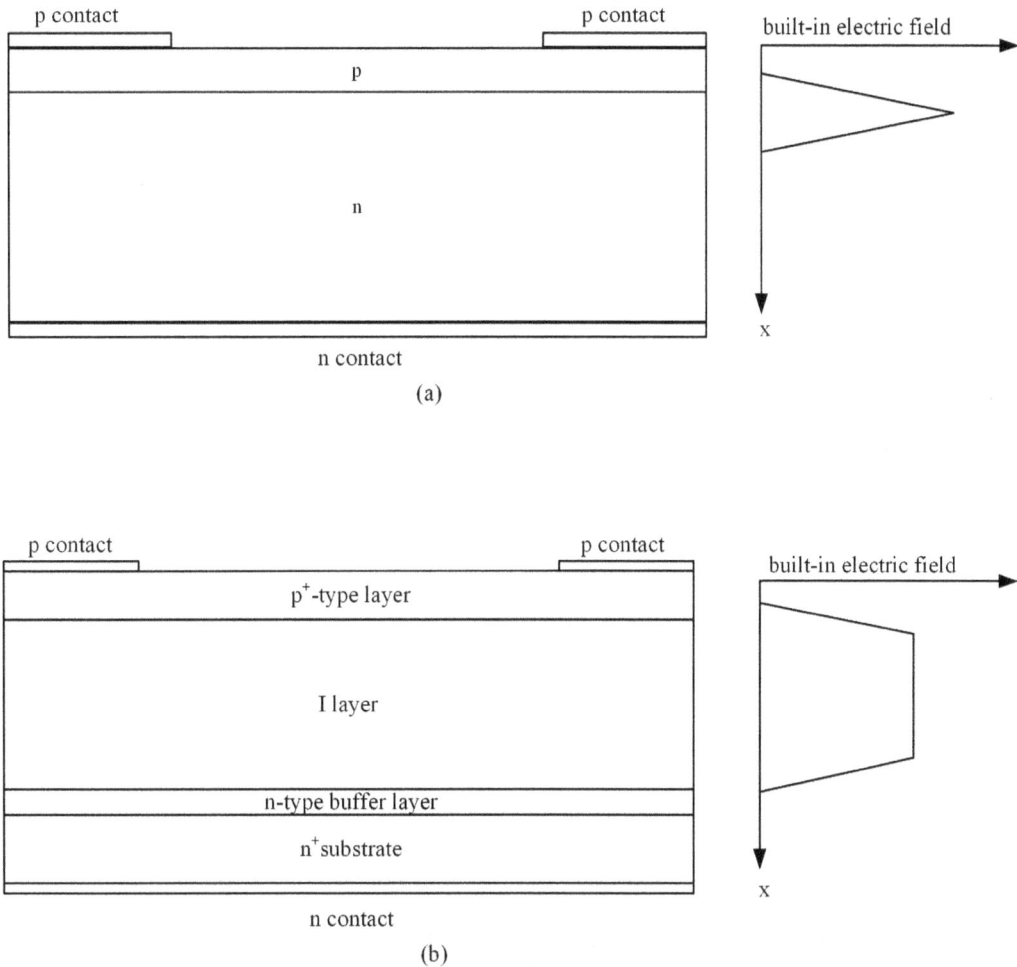

FIGURE 9

(a) Structure of *p-n* photodiode and its electric field intensity distribution; (b) structure of *p-i-n* photodiode and the electric field intensity distribution.

electron-hole pairs are separated by the built-in electric field, the thickness of the depletion layer can be increased to maximize light absorption efficiency. To solve the problem, an intrinsic layer can be sandwiched between a *p* layer and an *n* layer.

The structure of the *p-i-n* photodetector is shown in Figure 9 (b). A *p-i-n* photodiode includes a substrate, an *n*-type buffer layer, an *I* layer, and a p^+-type layer. The thick intrinsic layer has a high resistance and increases the electric field strength, which greatly improves the collection efficiency of photogenerated charge carriers.

The principle of operation of *p-i-n* UV photodiode is shown in Figure 10 [38]. When photons with energy greater than or equal to the bandgap of the material strike the depletion region, electrons in the *p*-, *I*-, and *n*-region are excited, resulting in photogenerated electron-hole pairs, as shown in Figure 11 (b). Since the intrinsic layer has a high resistivity, the potential is mainly over it, which creates a depletion region with width *W*. The intrinsic layer is entirely depleted at modest levels of reverse bias because of the low

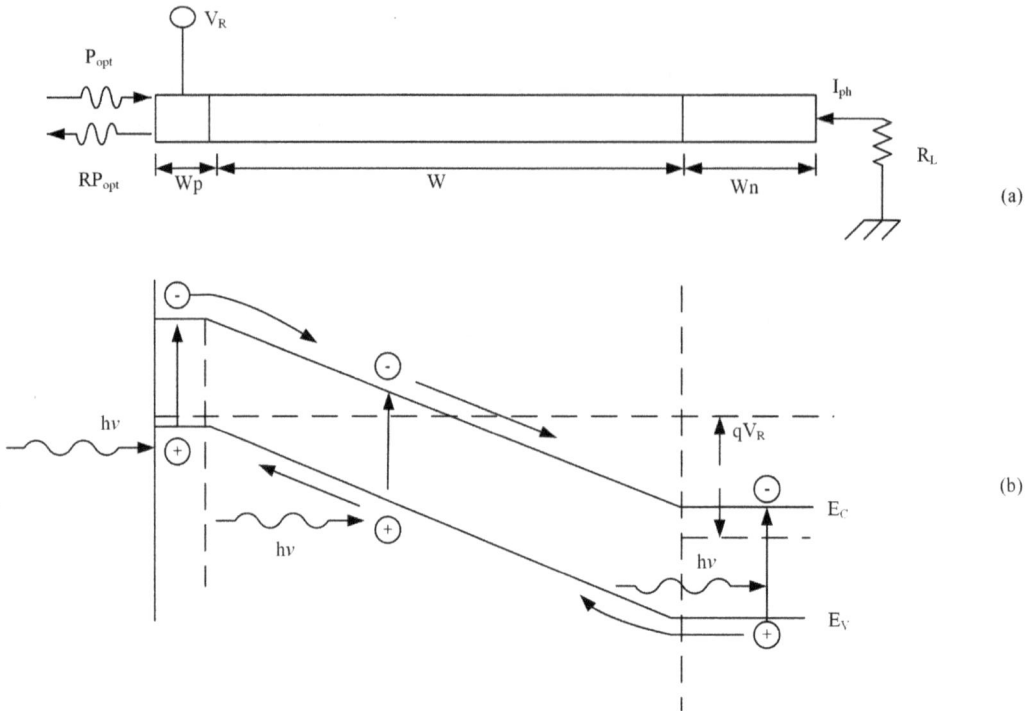

FIGURE 10
The principle of operation of *p-i-n* UV photodiode. From [38], reproduced with permission by John Wiley & Sons, Inc.

doping concentration. Therefore, the depletion region's width is approximately equivalent to the thickness of the intrinsic layer, as shown in Figure 11 (a).

p-i-n photodetectors also have the following advantages over *p-n* photodetectors: (1) The intrinsic layer reduces the dark current of the device. (2) At 0 V or lower reverse bias, the intrinsic layer becomes fully depleted. As the reverse-bias voltage rises, *p-i-n* photodiodes exhibit good stability. (3) Since the built-in electric field extends over the entire intrinsic layer, increasing the thickness of the intrinsic layer can improve the responsivity of the device by enhancing the collection efficiency of photogenerated charge carriers. (4) The intrinsic layer increases the depletion layer and decreases the capacitance of the junction, which presents an excellent frequency response. (5) The response peak wavelength can be adjusted and higher responsivity can be achieved by adjusting the thickness of the p^+-type layer and intrinsic layer.

It should be emphasized that the *p*-layer in a *p-i-n* photodiode should be thin to improve sensitivity to short-wavelength UV photons. The diffusion and drift times of good *p-i-n* photodiodes are in the order of ps, and the junction capacitance can be regulated in the order of several pF.

The *p-i-n* photodetectors based on 4H-SiC have low noise and high-speed response. In 2006, Liu *et al.* at the Institute of Semiconductors of the Chinese Academy of Sciences demonstrated 4H-SiC UV photodetectors with P^+ / π / N^- / N^+ structure [39]. The passivation layer is a 500 nm oxide grown by PECVD, while the optical window is a Pt layer. The size of the active area of the photodetectors was 300×300 mm². At room temperature, the

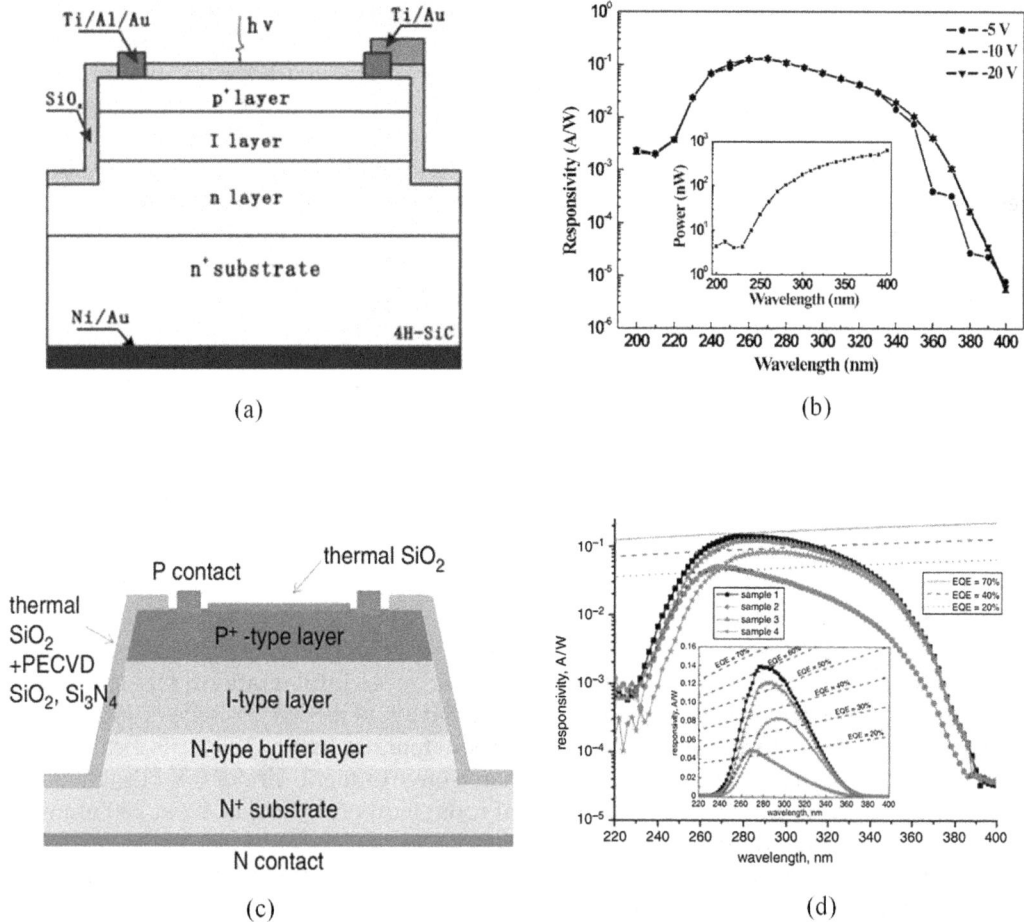

(a)

(b)

(c)

(d)

FIGURE 11

(a) Cross-sectional view of 4H-SiC ultraviolet *p-i-n* photodetector. (b) The spectral responsivities of the 4H-SiC ultraviolet *p-i-n* photodetector under different reverse biases. From [41] Figure 1 & Figure 4, reproduced with permission © 2007 American Institute of Physics. (c) The structure of 4H-SiC *p-i-n* photodiode. (d) Spectral response characteristics of fabricated 4H-SiC *p-i-n* UV PDs under −5 V bias. Inset shows spectral response characteristics plotted in linear scale. From [44] Figure 1 & Figure 3, reproduced with permission © 2019 John Wiley and Sons.

dark and illuminated *I-V* characteristics were measured at reverse biases ranging from 0 to 20 V, with the illuminated current being at least two orders of magnitude greater than that of dark current below 13 V bias. At varied reverse biases, the peak value zones of the photo response were situated at 280~310 nm, and the peak value located at 300 nm was 100 times greater than the cut-off response value in 380 nm at a bias of 10 V, indicating that the device had a good visible-blind performance. However, the dark current is high. The high dark current could be due to the fact that the quality of the PECVD-generated oxide layer is not as good as that of the thermally grown oxide layer. In addition, the quantum efficiency and responsivity of the device are not satisfactory due to the absorption of Pt.

In the same year, Chen *et al.* at Xiamen University reported a 4H-SiC *p-i-n* UV photodetector [40]. SiO$_2$ by thermal oxidation is used as the anti-reflective film and passivation layer. The peak wavelength and cut-off wavelength of the device are 275 nm and 375 nm, respectively, with a response ratio of 100. The photocurrent of the detector was at least two orders of magnitude higher than the dark current. Then, by lowering the intrinsic layer doping concentration and enhancing the passivation process, a *p-i-n* UV photodetector with a low dark current and a high UV-to-visible suppression ratio was prepared. In 2007, they further fabricated a high-performance 4H-SiC *p-i-n* photodetector for visible-blind ultraviolet UV applications [41], as shown in Figure 11 (a). The *I-V* characteristic results revealed that the detector suffered from the significant dark current of 2.5 pA/mm² at a low reverse bias of 5 V. The photodetector exhibited a broad spectral response, with wavelengths ranging from 240 nm to 310 nm, as shown in Figure 11 (b). The peak responsivity of the detector was 0.13 A/W at a wavelength of 270 nm, corresponding to a maximum external quantum efficiency of 61%. The ratio of detector responsivity at 270 nm to that at 380 nm was larger than 10^3.

In 2016, Cai *et al.* at Xiamen University analyzed the capacitance-voltage (*C-V*) characteristics of 4H-SiC *p-i-n* UV photodetector with temperature and bias voltage [42]. Results show that the high-frequency *C-V* characteristics almost do not change with reverse bias due to the fact that the *i*-layer of the detector is in a depletion state under near zero bias. The quantity of thermally ionized free carriers increases as the temperature rises, increasing the high-frequency (1 MHz) junction capacitances. The voltage and temperature dependency of low-frequency (100 kHz) junction capacitances is stronger than that of high-frequency junction capacitances.

In the same year, Yang *et al.* at Nanjing University prepared a high-performance 4H-SiC *p-i-n* ultraviolet photodiode with a *p* layer formed by Al implantation [43]. The device maintained a low dark current density of 1 nA/cm² at a reverse-bias voltage of 100 V when the operating temperature was increased from room temperature to 175 °C, indicating that it can detect UV signals in high-temperature harsh environment. Under 0 V bias, the maximum external quantum efficiency of the PD at room temperature is 44.4% at 270 nm with a UV/visible rejection ratio larger than 10^4.

In 2019, Hou *et al.* at Xiamen University fabricated 4H-SiC ultraviolet *p-i-n* photodiodes with four different epitaxial structures [44], as shown in Figure 11 (c). The results show that for a high-performance UV *p-i-n* photodiode, both a thin P^+-type ohmic contact layer and a large intrinsic layer are required. At 278 nm incident wavelength, a responsivity of 0.139 A/W was achieved, as shown in Figure 11 (d). Within a certain wavelength range, the peak response wavelength of an ultraviolet *p-i-n* photodiode can be modulated by properly adjusting the thicknesses of the P^+-type layer and the intrinsic layer.

3.4 Avalanche Photodiodes

The reverse breakdown of the *PN* junction is mainly divided into avalanche breakdown and Zener breakdown. Avalanche breakdown is when the *PN* junction reverse voltage increases to a certain level, the carrier multiplication is like an avalanche, resulting in a significant increase in the current flow across the *p-n* junction.

Under a high reverse-bias voltage, a high-strength electric field is created across the junction region. When the electric field is strong enough, the photogenerated charge carriers in the depletion layer may be accelerated to collide with lattice atoms, causing electrons excited from the valence band to the conduction band, and producing new electron-hole pairs in the depletion region. The created electron-hole pair acquire sufficient kinetic energies from the field and create additional electron-hole pairs. With the progress of this cascade process, more electron-hole pairs are created. This process is referred to as avalanche carrier multiplication, as shown in Figure 12. Avalanche photodiode is a *p-n* junction type photodiode that amplifies the photoelectric signal

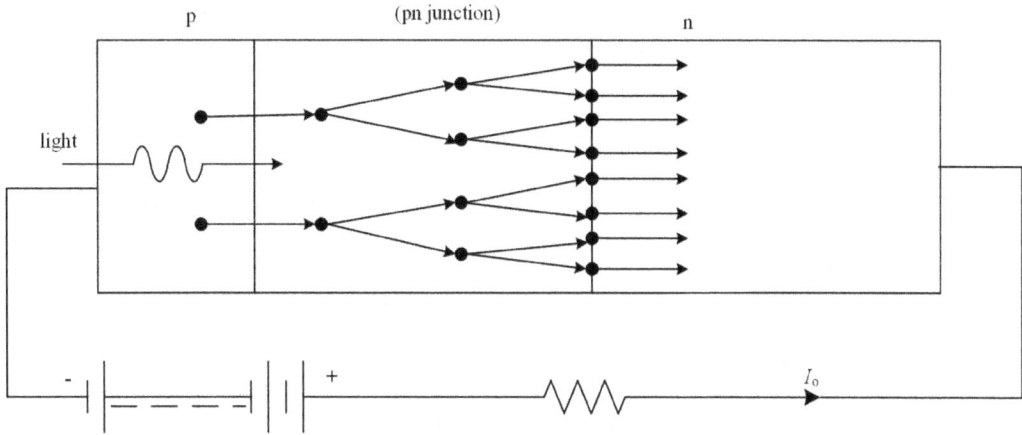

FIGURE 12
Avalanche multiplication effect of APD.

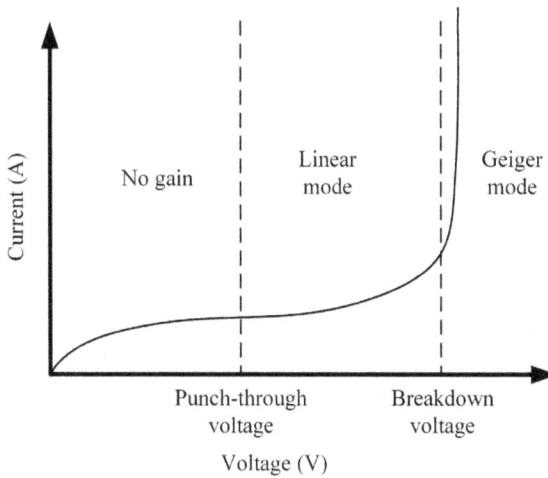

FIGURE 13
Relationship between APD working mode and reverse bias voltage.

to improve detection sensitivity by leveraging the avalanche multiplication action of carriers, which demonstrates an internal current gain effect due to the avalanche phenomenon.

APD has two modes of operation: linear mode and Geiger mode. When the APD's reverse-bias voltage is less than its avalanche voltage, the incident photons are linearly amplified, which is known as the linear mode. The higher the reverse voltage, the greater the gain in linear mode. The APD gain increases rapidly when the reverse-bias voltage is larger than the avalanche voltage, and single photon absorption can saturate the detector output current, which is known as the Geiger mode of APD. As demonstrated in Figure 13, APD's operation mode is linked to its reverse bias.

4H-SiC APD structures mainly include *p-n* structure, reach-through structure, *p-i-n* structure, and absorption layer and separate absorption and multiplication (SAM) structure.

In 1997, Konstantinov *et al.* at Linköping University in Sweden fabricated the first 4H-SiC diodes with uniform avalanche multiplication and breakdown [45]. For diode fabrication, two epitaxial processes were used: sublimation epitaxy and vapor phase epitaxy (VPE). The photocurrent grows by three to four orders of magnitude in the voltage range between 1/2 breakdown voltage and breakdown voltage, owing to avalanche multiplication. And the hole to electron ionization coefficient ratio can reach 50.

In 1999, Yan *et al.* at Rutgers University reported the first 4H-SiC UV avalanche photodiode for visible-blind UV applications with a reach-through structure [46]. The reach-through structure ensures that the depletion region of the device extends from the p^+ layer to the substrate layer before reaching the breakdown voltage. The device shows a positive temperature coefficient for the breakdown voltage, which increases from 93 to 97 V as the temperature increases from 27 to 257°C. In addition, the maximum responsivity is 106 A/W, which is > 600 times higher than that of non-avalanche 6H-SiC *pn* photodiodes and the corresponding optical gain is > 480. The photocurrent is about one order higher than the dark current. The performance of the device is not very ideal, which is largely due to the poor epitaxy quality and immature processes at that time.

In order to reduce the dark current of APD, Yan *et al.* fabricated 4H-SiC UV avalanche photodiodes with edge terminated by 2° positive bevel in 2002 [47]. Low leakage current density at 95% of breakdown and high avalanche gain up to 10^4 have been achieved. The peak of the response spectra is around 320 nm. The ratio between the responsivity at 320 and 400 nm is 40. The effective k of the excess noise factors is found to be 0.1, which makes the 4H-SiC an ideal candidate for visible-blind UV APDs.

In 2003, Yan *et al.* fabricated the first 4H-SiC visible-blind APD linear array and demonstrated a 40-pixel linear array with only one bad pixel [48]. The yield of APDs with breakdown voltage higher than 120 V is found to be 91%. Most 4H-SiC APD pixels show the positive temperature dependence of breakdown voltage. A responsivity higher than 10^5 A/W and a visible rejection ratio higher than 100 have been observed in a very broad range from 250 and 330 nm. However, the dark current is still high since the fabrication of SiC is not very mature at that time.

Also in 2003, Guo *et al.* at the University of Texas prepared low dark current 4H-SiC avalanche photodiodes [49]. The *I-V* characteristics show that the photocurrent gain is greater than 10^3, and the dark current is less than 2 nA. This low dark current was achieved by special attention to the sidewall treatment, including cleaning and passivation.

In 2004, Beck *et al.* at the University of Texas reported suppression of edge breakdown in SiC APDs by employing a 10° sidewall bevel angle [50]. These devices exhibit low dark currents, <10 pA at the onset of avalanche gain.

In 2005, both Zhao's research group at Rutgers University and Campbell's research group at the University of Texas reported single photon counting in 4H-SiC avalanche photodetectors. Xin *et al.* at Rutgers University fabricated the first 4H-SiC UV single photon counting avalanche photodiode using a passive quenching circuit [51]. At −78.0 V, the probability of photons being counted is 75%. The corresponding photon counting efficiency is determined to be 2.6% at 353 nm. The dark count rate at room temperature is 650 kHz, while the photon count rate is 4.5 MHz. The dark count rate is more than an order of magnitude lower than that of InP/InGaAsP and GaN SPADs operating in Geiger mode. Beck *et al.* at the University of Texas reported Geiger mode operation of 4H-SiC UV avalanche photodiodes using the gated quenching circuit [52]. At 325 nm, the unity-gain external quantum efficiency was 10% and the single photon detection efficiency was 2.9%. This result represents around 30% of the maximum attainable detection efficiency.

In 2006, Guo *et al.* reported 4H-SiC APD with a separate absorption and multiplication (SAM) structure [53]. The separate absorption and multiplication (SAM) APDs have the advantage of providing single carrier injection, high quantum efficiency, and lower capacitance, which enables high speed. They have a high external quantum efficiency of 83% (187 mA/W) at 278 nm after reach-through. Gain higher than 1000 was demonstrated without edge breakdown.

In 2008, Cha *et al.* fabricated 4H-SiC separate absorption multiplication region avalanche photodiodes (SAM-APDs) for UV detection in harsh environment applications [54]. The gain of 2500 and quantum efficiency of 45% at room temperature were achieved at the wavelength of 290–300 nm for a packaged device with an active area of 1×1 mm^2. The dark current of this device remains low and increases abruptly near the breakdown voltage.

Xiamen University has made many explorations in 4H-SiC APD. In 2009, Zhu *et al.* at Xiamen University reported a separate absorption and multiplication (SAM) 4H–SiC ultraviolet (UV) avalanche photodetectors (APDs) [55], with a gain higher than 1.8×10^4 at around 55 V, as shown in Figure 14 (a) and Figure 14 (b). At 0 V, the peak absolute responsivity was estimated to be larger than 0.078 A/W at 270 nm, corresponding to a peak external quantum efficiency of over 35.8%. The long-wavelength cutoff was about 380 nm. In addition, the UV-to-visible rejection ratio of around three orders of magnitude was extracted from the spectra response. At the reverse bias of 42 V, the peak responsivity increased to 0.203 A/W at 270 nm, corresponding to a maximum external quantum efficiency of 93%. Furthermore, the ideality factor around 1.65 and the spectral detectivity about 3.1×10^{13} cm Hz$^{1/2}$/W were estimated. However, the dark current of the device is relatively higher.

In 2011, Hong *et al.* at Xiamen University proposed a 4H-SiC nano-pillar-based avalanche photodiode (NAPD) with a separate absorption region and a multiplication region [56]. The nano-pillar for multiplication is designed as small as 0.3 μm in diameter so that the yield of a defect-free array element can be maximized for APD array application. The band diagram, avalanche breakdown voltage and photoresponse of the NAPD are strongly dependent on the device geometry, incident UV wavelength as well as incident UV power density.

In 2013, Sampath *et al.* at the University of Texas reported a deep ultraviolet 4H-SiC APD with high quantum efficiency of around 40% between 200 and 235 nm [57]. This improvement is attributed to the improved collection of carriers generated by deep ultraviolet photons through absorption in the depletion region of the detector.

In 2014, Zhou *et al.* at Nanjing University fabricated a 4H-SiC avalanche photodiode working in Geiger mode and studied the high-temperature performance [58]. At unity gain bias, the maximum quantum efficiency of the APD increases from 53.4% at 290 nm to 63.3% at 295 nm as the temperature rises from room temperature to 150 °C. Meanwhile, the dark current of the APD before breakdown increases by more than two to three orders of magnitude. At a fixed gain of 1.3×10^6, the single photon counting efficiency at 280 nm only slightly drops from 6.17% to 6% in the same temperature range, whereas the dark count rate increases from 22 to 80 kHz. It indicates that SiC APDs have the potential to work in a high-temperature harsh environment with single photon counting capability.

In 2015, Zhong *et al.* at Xiamen University proposed 4H-SiC avalanche photodiodes with a large absorption region and a small multiplication region [59], as shown in Figure 14 (c). The band structure, avalanche breakdown voltage, responsivity and response time of the diode are strongly dependent on the width (D) of the multiplication region as shown in Figure 14 (d). With the adjustment of D, a 4H-SiC ultraviolet APD with the requirement of certain responsivity and response time could be achieved.

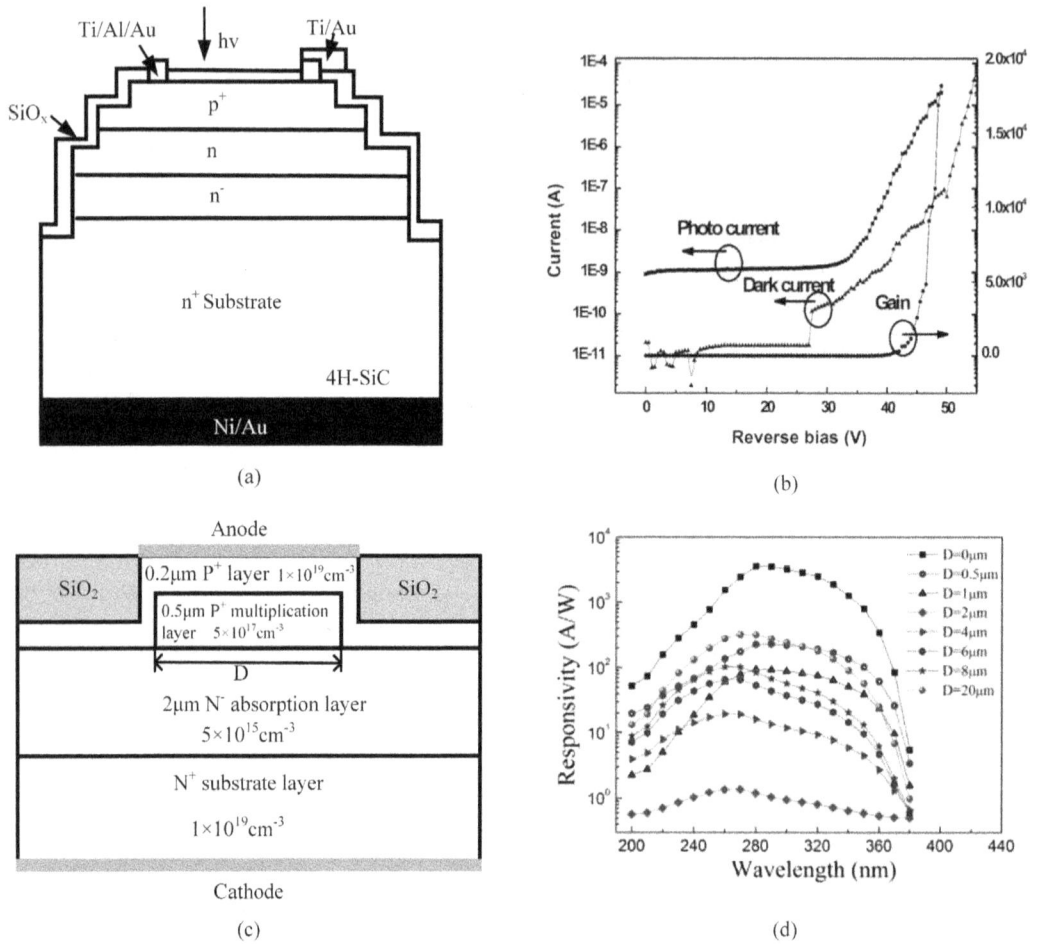

(a)

(b)

(c)

(d)

FIGURE 14

(a) Schematic cross-section of the 4H-SiC SAM-APD. (b) *I-V* characteristics and gain of 4H-SiC APD. From [55] Figure 1 & Figure 2, reproduced with permission © 2008 Elsevier Ltd. (c) Structure of 4H-SiC APD with a large bulk absorption region and a small multiplication region. (d) Spectral responsivities of the 4H-SiC APDs with different *D* at avalanche breakdown voltage. From [59] Figure 1 & Figure 4, reproduced with permission © 2015 The Japan Society of Applied Physics.

In 2018, Cai *et al.* at Nanjing University [60] fabricated vertical 4H-SiC *n-i-p-n* avalanche photodiodes. When a partial trench isolation scheme is applied, the device shows a high fill factor of 78.3%. The dark current curve remains at ~pA level before avalanche break-down occurs at about −250 V. The avalanche gain could reach >1×10⁴ upon 3 V overbias. The peak responsivity of the *n-i-p-n* APD is 0.149 A/W at 280 nm under −10 V bias, corresponding to a quantum efficiency of 66%.

And in 2019, Wu *et al.* at Xiamen University proposed a 4H-SiC separated absorption and multiplication avalanche photodiode with an *n*-type charge layer [61]. A low break-down voltage of 77.6 V and a significant gain of more than 10⁵ were obtained. The peak responsivity for 270 nm illumination of the photodiode biased at −40 V was 83 mA W⁻¹, corresponding to an external quantum efficiency over 38.2%.

In 2020, Zhou *et al.* at Hebei Semiconductor Research Institute reported high-uniformity 1 × 64 linear arrays of 4H-SiC avalanche photodiode for ultraviolet detection [62]. A high pixel yield of 100% and a high-uniformity breakdown voltage with a fluctuation of smaller than 0.5 V are achieved. The dark currents at 95% of breakdown voltage are below 1 nA for all the 64 pixels. Besides, the pixels in the array show a multiplication gain of larger than 106 and a peak responsivity of 0.12 A/W at 285 nm (corresponding to a maximum quantum efficiency of 52%) at room temperature.

4 Novel 4H-SiC-based UV Photodetectors

4.1 Graphene/4H-SiC UV Photodetectors

Graphene is a single layer of carbon atoms arranged in a honeycomb lattice [61]. It exhibits many remarkable optoelectronic properties, which makes it a prominent candidate for future micro- and nano-devices. The carrier mobility of graphene at room temperature is about 15000 cm^2/ (V-s), at low temperatures, this data can even reach 250000 cm^2/ (V-s). Graphene has very good optical properties, and its transmittance is more than 90% in the ultraviolet spectrum of 200–400 nm [63]. Graphene is considered to be used as the transparent conductive electrode for next-generation UV photodetectors. In recent years, the research on graphene and 4H-SiC has attracted much attention.

In 2015, Anderson *et al.* at Naval Research Laboratory fabricated an ultraviolet photodetector based on graphene/SiC heterojunction [64], as shown in Figure 15 (a). The device shows a quantum efficiency above 80% at 4 eV with a dark current <1 nA/cm^2, as shown in Figure 15 (b).

Also in 2015, Kusdemir *et al.* at Izmir Institute of Technology fabricated a graphene-semiconductor-graphene ultraviolet photodetector based on the Schottky junction at the interface between epitaxial graphene and SiC [65]. For a bias voltage of 0 V, the photocurrent of the device is three orders of magnitude larger than its dark current, which was measured as 0.7 pA without UV illumination. The light on/off characteristics showed that epitaxial graphene/SiC heterojunction can be used for UV photodetection with fast response speed. The sample exhibited higher responsivity for an illumination wavelength of 254 nm. The fabricated GSG samples with graphene electrodes have early saturated maximum responsivity so that they can be used for UV sensor systems operating at relatively low voltages.

In 2017, Guo *et al.* at Xidian University reported an ultraviolet photodetector based on SiC-graphene heterojunction with *p*-doped graphene and *n*-doped 4H-SiC [66]. Both the low dark current owing to the *p-n* heterojunctions and the befitting magnitude of responsivity of the conventional *p-i-n* UV photodetectors are obtained. The dark current of the device is at the magnitude of 10^{-14}A ~ 10^{-12}A. The responsivity peak lies near 270 nm, and the maximum response is 0.032 A/W. Under 5 V reverse bias from the illumination of 270 nm UV irradiation, the current reached 1×10^{-8} A magnitude.

In 2020, Bencherif *et al.* at University of Biskra optimized the performance of an interdigitated graphene electrode/4H-SiC MSM photodetector operating in a wide range of temperatures [67]. A responsivity of 238 A/W was obtained under 325 nm illumination. A photocurrent-to-dark-current ratio (PDCR) of 5.75 × 10^5 at 300 K and 270 at 500 K was distinguished. The response time was found to be around 14 s at 300 K and 54.5 s at 500

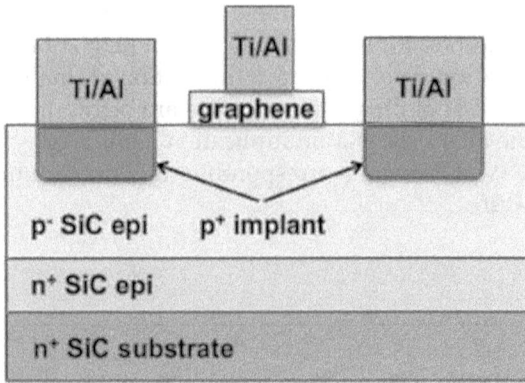

(a)

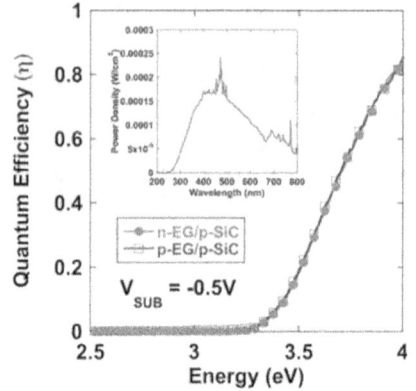

(b)

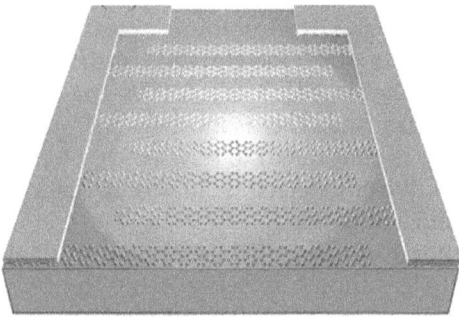

(c)

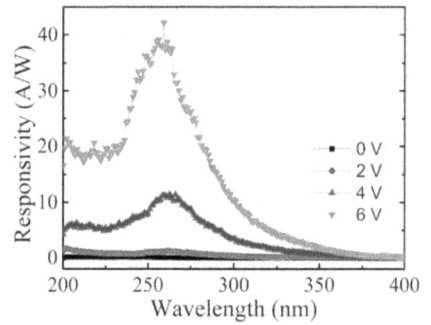

(d)

FIGURE 15

(a) Cross-section schematic of the ultraviolet detector based on graphene/SiC heterojunction. (b) Quantum efficiency of the ultraviolet detector based on graphene/SiC heterojunction. From [64] Figure 3 & Figure 4, reproduced with permission © 2015 The Japan Society of Applied Physics. (c) Schematic of GSG structure devices. (d) The responsivity of the GSG photodetector is measured in the UV illumination wavelength range from 200 to 400 nm under 0, 2, 4, 6 V bias voltages. From [68] Figure 2 & Figure 3, reproduced with permission © 2020 Applied Physics Letters.

K. An ultra-sensitive, high-speed SiC optoelectronic device for extremely high temperature applications can be realized using the proposed method.

In 2020, Sun *et al.* at Xiamen University fabricated and characterized graphene/4H-SiC/graphene photodetectors based on the epitaxial graphene on 4H-SiC as transparent electrodes [68], as shown in Figure 15 (c). High-quality graphene has been grown on an n^--doped 4H-SiC substrate along with a 900 °C hydrogenation process. The bias-dependent Schottky barrier height (varying from 0.43 eV to 0.41 eV) was found and could result mainly from the electrical doping and Fermi level shifting in graphene. The photodetectors showed a peak responsivity of 40 A/W at 270 nm, as shown in Figure 15 (d), an external quantum efficiency of 1.38×10^{4}%, and a detectivity of 9×10^{11} Jones, which are larger than those of previously reported similar devices based on graphene/SiO_2 or graphene/SiC.

4.2 β-Ga₂O₃/4H-SiC UV Photodetectors

Monoclinic gallium oxide (β-Ga$_2$O$_3$), with excellent material properties including a very suitable bandgap value (~4.9 eV) for UV detection, high thermal and chemical stability, very high breakdown field (8 MV/cm), is a promising material for ultraviolet (UV) photodetectors. The heterojunction photodetectors based on β-Ga$_2$O$_3$/4H-SiC have been reported.

In 2016, An *et al.* at Beijing University of Posts and Telecommunications fabricated deep UV *n-n* junction photodiodes of β-Ga$_2$O$_3$/SiC heterojunction [69]. A large rectification ratio of 1900, a high 254 nm ultraviolet photosensitivity of 6308% and a zero response of 365 nm ultraviolet have been achieved by reducing oxygen vacancies.

In 2018, Nakagomi *et al.* at Ishinomaki Senshu University fabricated *p-n* heterojunction diodes based on β-Ga$_2$O$_3$/*p*-type 4H-SiC structures [70], which exhibit good rectification properties and stability under high temperatures. The rectification ratios exceed 1000 even at 500 °C. And they also fabricated deep-ultraviolet (deep-UV) photodiodes are fabricated on the basis of heterojunctions having various β-Ga$_2$O$_3$ thicknesses, which present the maximum responsivity at 250~260 nm and respond to UV pulses as short as about 30 μs in real time.

In 2020, Yu *et al.* at Xidian University fabricated self-powered and fast response photodetectors based on β-Ga$_2$O$_3$/4H-SiC *p-n* heterojunction [71]. The detectors exhibit an ultrahigh current I_{on}/I_{off} ratio more than 10^3 at the light intensity of 91 μW/cm^2 and a fast photoresponse speed (a rise time of 11 ms and a decay time of 19 ms) under zero-bias voltage. In addition, the detectors also show a responsivity of 10.35 mA/W with a high detectivity of 8.8×10^9 Jones, and the maximum linear dynamic range reached 64.38 dB.

5 Conclusions and Outlook

4H-SiC UV photodetectors exhibit excellent performance in harsh environments, such as extreme temperature and space radiation, which is promising in military and civilian applications. This chapter offers a summarized review of 4H-SiC UV photodetectors with different structures, including Schottky diodes, metal-semiconductor-metal (MSM), *p-n* or *p-i-n* photodiodes, avalanche photodiodes (APDs) and some novel-structure photodiodes. Although 4H-SiC UV photodetectors have shown broad application prospects, there are still some challenges to be overcome, such as material growth quality, high cost, and so on. In addition, due to the constraints of material quality and immature preparation process, the yield of large-area devices is not so satisfactory. It is believed that, with the decreasing cost and optimized fabrication process, 4H-SiC UV photodetectors will be widely used in the very near future.

References

1. B. Wang, "Lecture on ultraviolet radiation – definition and classification of ultraviolet radiation", Solar Energy, 4, 6–8 (2003).

2. L. Liu, G. Ni, S. Zhong, Q. Fang and Y. Wang, "Application and detection of ultraviolet and their new development", Optical Technology, 2, 88–91 (1998).
3. G. A. Shaw, A. M. Siegel, J. Model and D. Greisokh, "Recent progress in short-range ultraviolet communication", Proceedings of SPIE, 5796, 214–225 (2005).
4. F. P. Neele and R. M. Schleijpen, "Electro-optical missile plume detection", Proceedings of SPIE, 5075, 270–280 (2003).
5. D. Zhou, H. Lu, D. Chen, F. Ren, R. Zhang and Y. Zheng, "New generation of wide bandgap semiconductor ultraviolet photodetectors", Lamps and Lighting, 4, 25–26 (2016).
6. L. Sang, M. Liao and M. Sumiya, "A comprehensive review of semiconductor ultraviolet photodetectors: from thin film to one-dimensional nanostructures", Sensors, 13, 10482–10518 (2013).
7. Y. G. Zhang, A. Z. Li and A. G. Milnes, "Metal-semiconductor-metal ultraviolet photodetectors using 6H-SiC", IEEE Photonics Technology Letters, 9(3), 363–364 (1997).
8. D. Walker, E. Monroy, P. Kung, J. Wu, M. Hamilton, F. J. Sanchez, J. Diaz and M. Razeghi, "High-speed, low-noise metal-semiconductor-metal ultraviolet photodetectors based on GaN", Applied Physics Letters, 74(5), 762–764 (1999).
9. J. Li, Z.Y. Fan, R. Dahal, M. L Nakarmi, J. Y. Lin and H. X. Jiang, "200 nm deep ultraviolet photodetectors based on AlN", Applied Physics Letters, 89(21), 213510 (2006).
10. H. Ishikura, T. Abe, N. Fukuda, H. Kasada and K. Ando, "Stable avalanche-photodiode operation of ZnSe-based p+-n structure blue-ultraviolet photodetectors", Applied Physics Letters, 76(8):1069–1071 (2000).
11. N. W. Emanetoglu, J. Zhu, Y. Chen, J. Zhong, Y. Chen and Y. Lu, "Surface acoustic wave ultraviolet photodetectors using epitaxial ZnO multilayers grown on r-plane sapphire", Applied Physics Letters, 85(17), (2004).
12. H. Xue, X. Kong, Z. Liu, C. Liu, J. Zhou and W. Chen, "TiO_2 based metal-semiconductor-metal ultraviolet photodetectors", Applied Physics Letters, 90(20):201118 (2007).
13. E. Monroy, F. Omnes and F. Calle, "Wide-bandgap semiconductor ultraviolet photodetectors", Semiconductor Science and Technology, 18(4), R33-R51 (2003).
14. H. Ou, Y. Ou, A. Argyraki, S. Schimmel, M. Kaiser, P. Wellmann, M. K. Linnarsson, V. Jokubavicius, J. Sun, R. Liljedahl and M. Syvajarvi, "Advances in wide bandgap SiC for optoelectronics", European Physical Journal B, 87, 58 (2014).
15. M. Laube, G. Pensl and H. Itoh, "Suppressed diffusion of implanted boron in 4H-SiC" Applied Physics Letters, 74(16):2292–2295 (1999).
16. M. Shur, S. Rumyantsev, M. Levinshtein (ed.), "*SiC Materials and Devices*", World Scientific Publishing Co. Pte. Ltd., Singapore, 2–10pp, 2006.
17. S. Shi and Y. Guo (ed.), "Optoelectronic Technology and its Application", University of Electronic Science and Technology Press, Chengdu, 2001.
18. G. Liu (ed.), "Semiconductor devices – power, sensitive, photonic, microwave devices", Publishing House of Electronics Industry, Beijing, 2000.
19. F. Yan, X. Xin, S. Aslam, Y. Zhao, D. Franz, J. H. Zhao and M. Weiner, "4H-SiC UV photodetectors with large area and very high specific detectivity", IEEE Journal of Quantum Electronics, 40(9), 1315–1320 (2004).
20. X. Xin, F. Yan, T. W. Koeth, C. Joseph, J. Hu, J. Wu and J. H. Zhao, "Demonstration of 4H-SiC visible-blind EUV and UV detector with large detection area", Electronics Letters, 41(21), 1192–1193 (2005).
21. J. Hu, X. Xin, J. H. Zhao, F. Yan, B. Guan, J. Seely and B. Kjornrattanawanich, "Highly sensitive visible-blind extreme ultraviolet Ni/4H-SiC Schottky photodiodes with large detection area", Optics Letters, 31(11), 1591–1593 (2006).
22. L. Wang, J. Xie and W. Liu, "Study on silicon carbide (SiC) Schottky ultraviolet photodetectors", Semiconductor Optoelectronics, 25(1), 25–28 (2004).
23. J. Liang, J. Xie, L. Huang and T. Sun, "Temperature characteristics of Au/n-4H-SiC Schottky UV photodiode", Chinese Journal of Quantum Electronics, 22(6), 932–934 (2005).

24. L. Huang, J. Xie, J. Liang and T. Sun, "A study on the gain performance of SiC ultraviolet photodetectors at high reverse biased voltage", Microelectronics, 35(4), 357–359 (2005).

25. Y. Xu, D. Zhou, H. Lu, D. Chen, F. Ren, R. Zhang and Y. Zheng, "High-temperature and reliability performance of 4H-SiC Schottky-barrier photodiodes for UV detection", Journal of Vacuum Science & Technology, 33(4), 040602 (2015).

26. Z. Wang, D. Zhou, W. Xu, D. Pan, F. Ren, D. Chen, R. Zhang, Y. Zheng, and H. Lu, "High-performance 4H-SiC Schottky photodiode with semitransparent grid-electrode for EUV detection", IEEE Photonics Technology Letters, 32(13), 791–794 (2020).

27. Y. Su, Y. Chiou, C. Chang, S. Chang, Y. Lin and J. F. Chen, "4H-SiC metal-semiconductor-metal ultraviolet photodetectors with Ni/ITO electrodes", Solid-State Electronics, 46(12), 2237–2240 (2002).

28. Y. Chiou, "DC and noise characteristics of 4H-SiC metal-semiconductor-metal ultraviolet photodetectors", Japanese Journal of Applied Physics, 43(5A), 2432–2434 (2004).

29. Z. Wu, X. Xin, F. Yan and J. Zhao, "Demonstration of the first 4H-SiC metal-semiconductor-metal ultraviolet photodetector", Material Science Forum, 457–460, 1491–1494 (2004).

30. Z. Wu, X. Xin, F. Yan and J. Zhao, "Fabrication of MSM structure UV photodetector on 4H-SiC", Chinese Journal of Quantum Electronics, 21(2), 269–272 (2004).

31. A. Sciuto, F. Roccaforte, S. D. Franco and V. Raineri, "High responsivity 4H-SiC Schottky UV photodiodes based on the pinch-off surface effect", Applied Physics Letters, 89(8), (2006).

32. W. Yang, J. Cai, F. Zhang, Z. Liu, Y. LÜ and Z. Wu, "Fabrication of 4H-SiC MSM photodiode linear arrays", Journal of Semiconductors, 29(3), 570–573 (2008).

33. W. Yang, F. Zhang, Z. Liu, Y. LÜ and Z. Wu, "High responsivity 4H-SiC based metal-semiconductor-metal ultraviolet photodetectors", Science in China Series G: Physics, Mechanics & Astronomy, 51(11), 1–5 (2008).

34. M. Mazzillo, G. Condorelli, M. E. Castagna, G. Catania, A. Sciuto, F. Roccaforte and V. Raineri, "High efficient low reverse biased 4H-SiC Schottky photodiodes for UV-light detection", IEEE Photonics Technology Letters, 21(23), 782–1784 (2009).

35. W. Lien, D. Tsai, D. Lien, D. G. Senesky, J. He and A.t P. Pisano, "4H-SiC metal-semiconductor-metal ultraviolet photodetectors in operation of 450°C", IEEE Electron Device Letters, 33(11), 1586–1588 (2012).

36. B. Chen, Y. Yang, X. Xie, N. Wang, Z. Ma, K. Song and X. Zhang, "Analysis of temperature-dependent characteristics of a 4H-SiC metal-semiconductor-metal ultraviolet photodetector", Applied Physics, 57(34), 4427–4433 (2012).

37. S. Liu, T. Wang and Z. Chen, "High-performance of Al nanoparticle enhanced 4H-SiC MSM photodiodes for deep ultraviolet detection", IEEE Electron Device Letters, 38(10), 1405–1408 (2017).

38. H. Melchior, "Demodulation and Photodetection Techniques", a book chapter in *Laser Handbook: Volume 1*, North Holland/American Elsevier, New York, Ch. 7, pp725–835, 1972.

39. X. F. Liu, G. S. Sun, J. M. Li, J. Ning, M. C. Luo, L. Wang, W. S. Zhao and Y. P. Zeng, "Visible blind p+-π-n-n+ ultraviolet photodetectors based on 4H-SiC homoepilayers", Microelectronics Journal, 37(11), 1396–1398 (2006).

40. X. Chen, W. Yang and Z. Wu, "Visible blind p-i-n ultraviolet photodetector fabricated on 4H-SiC", Microelectronic Engineering, 83, 104–106(2006).

41. X. Chen, H. Zhu, J. Cai and Z. Wu, "High-performance 4H-SiC-based ultraviolet p-i-n photodetector", Journal of Applied Physics, 102(2), 024505 (2007).

42. J. Cai, X. Chen, S. Wu and Z. Wu, "Capacitance-voltage characteristics of 4H-SiC p-i-n ultraviolet photodetectors", Chinese Journal of Quantum Electronics, 33(6), 770–774 (2016).

43. S. Yang, D. Zhou, H. Lu, D. Chen, F. Ren, R. Zhang, and Y. Zheng, "High-performance 4H-SiC p-i-n ultraviolet photodiode with p layer formed by Al implantation", IEEE Photonics Technology Letters, 28(11), 1189–1192 (2016).

44. Y. Hou, C. Sun, J. Wu, R. Hong, J. Cai, X. Chen, D. Lin and Z. Wu, "Effect of epitaxial layer's thickness on spectral response of 4H-SiC p-i-n ultraviolet photodiodes", Electronic Letters, 25(4), 216–218 (2019).

45. A. O. Konstantinov, Q. Wahab, N. Nordell, and U. Lindefelt, "Ionization rates and critical fields in 4H silicon carbide", Applied Physics Letters, 71(1), 90–92 (1997).
46. F. Yan, Y. Luo, J.H. Zhao and G.H. Olsen, "4H-SiC visible blind UV avalanche photodiode", Electronic Letters, 35 (11), 929–930 (1999).
47. F. Yan, C. Qin, J.H. Zhao, M. Weiner, B.K. Ng, J.P.R. David and R.C. Tozer, "Low-noise visible-blind UV avalanche photodiodes with edge terminated by 2° positive bevel", Electronic Letters, 38 (7), 335–336 (2002).
48. F. Yan, C. Qin, J.H. Zhao, M. Bush, G. Olsen, B.K. Ng, J.P.R. David, R.C. Tozer, and M. Weiner, "Demonstration of 4H-SiC avalanche photodiodes linear array", Solid-State Electronics, 47, 241–245 (2003).
49. X. Guo, A. L. Beck, B. Yang, and J. C. Campbell, "Low dark current 4H-SiC avalanche photodiodes", Electronic Letters, 39, 1673–1674 (2003).
50. A. L. Beck, B. Yang, X. Guo, and J. C. Campbell, "Edge Breakdown in 4H-SiC Avalanche Photodiodes", IEEE Journal of Quantum Electronics, 40 (3), 321–324 (2004).
51. X. Xin, F. Yan, X. Sun, P. Alexandrove, C.M. Stahle, J. Hu, M. Matsumura, X. Li, M. Weiner and J. H. Zhao, "Demonstration of 4H-SiC UV single photon counting avalanche photo-diode", Electronic Letters, 41 (4), 212–214 (2005).
52. A. L. Beck, G. Karve, S. Wang, J. Ming, X. Guo, and J. C. Campbell, "Geiger Mode Operation of Ultraviolet 4H-SiC Avalanche Photodiodes", IEEE Photonics Technology Letters, 17 (7), 1507–1509 (2005).
53. X. Guo, L. B. Rowland, G. T. Dunne, J. A. Fronheiser, P. M. Sandvik, A. L. Beck, and J. C. Campbell, "Demonstration of Ultraviolet Separate Absorption and Multiplication 4H-SiC Avalanche Photodiodes", IEEE Photonics Technology Letters, 18 (1), 136–138 (2006).
54. H. Y. Cha, S. Soloviev, S. Zelakiewicz, P. Waldrab, and P. M. Sandvik, "Temperature dependent characteristics of nonreach-through 4H-SiC separate absorption and multiplication APDs for UV detection", IEEE Sensors Journal, 8 (3), 233–237 (2008).
55. H. L. Zhu, X. P. Chen, J. F. Cai, and Z. Y. Wu, "4H–SiC ultraviolet avalanche photodetectors with low breakdown voltage and high gain", Solid-State Electronics, 53, 7–10 (2009).
56. R. D. Hong, Y. Zhou, K. L.Wang, and Z. Y. Wu, "4H-SiC nano-pillar avalanche photodiode with illumination-dependent characteristics", IEEE Photonics Technology Letters, 23(12), 816–818 (2011).
57. A.V. Sampath, L.E. Rodak, Y. Chen, Q. Zhou, J.C. Campbell, H. Shen and M. Wraback, "High quantum efficiency deep ultraviolet 4H-SiC photodetectors", Electronic Letters, 49 (25), 1629–1630 (2013).
58. D. Zhou, F. Liu, H. Lu, D. J. Chen, F. F. Ren, R Zhang, and Y. D. Zheng, "High-temperature single photon detection performance of 4H-SiC avalanche photodiodes", IEEE Photonics Technology Letters, 26 (11), 1136–1138 (2014).
59. J. X. Zhong, Z. F. Zhang, Z. Y. Wu, R. D. Hong, and W. F. Yang, "Separated-absorption-multiplication 4H-SiC avalanche photodiodes with adjustable responsivity and response time", Japanese Journal of Applied Physics, 54, 070303 (2015).
60. X. Cai, L. Li, H. Lu, D. Zhou, W. Xu, D. Chen, F. Ren, R. Zhang, Y. Zheng, and G. Li, "Vertical 4H-SiC n-i-p-n APDs with partial trench isolation", IEEE Photonics Technology Letters, 30(9), 805–808 (2018).
61. J. K. Wu, M. K. Zhang, Z. Fu, R. D. Hong, F. Zhang, J. F. Cai, and Z. Y. Wu, "Charge layer optimized 4H-SiC SACM avalanche photodiode with low breakdown voltage and high gain", Japanese Journal of Applied Physics, 58, 100913 (2019).
62. X. Zhou, X. Tan, Y. Lv, J. Li, S. Liang, Z. Feng, and S. Cai, "High-uniformity 1 × 64 linear arrays of silicon carbide avalanche photodiode", Electronic Letters, 56(17), 895–897 (2020).
63. A. H. Castro Neto, F. Guinea, N. M. R. Peres, K. S. Novoselov, and A. K. Geim, "The electronic properties of graphene", Reviews of Modern Physics, 81, 109 (2009).
64. T. J. Anderson, K. D. Hobart, J. D. Greenlee, D. I. Shahin, A. D. Koehler, M. J. Tadjer, E. A. Imhoff, R. L. Myers-Ward, A. Christou, and F. J. Kub, "Ultraviolet detector based on graphene/SiC heterojunction", Applied Physics Express, 8, 041301 (2015).

65. E. Kusdemir, D. Özkendir, V. Fırat and C. Çelebi, "Epitaxial graphene contact electrode for silicon carbide based ultraviolet photodetector", Journal of Physics D: Applied Physics, 48, 095104 (2015).

66. H. Guo, B. Liu, B. Huang and H. Chen, "SiC-graphene heterojunction ultraviolet detector", 2017 14th China International Forum on Solid State Lighting: International Forum on Wide Bandgap Semiconductors China (SSLChina: IFWS), IEEE, (2017). DOI: 10.1109/IFWS.2017.8246020

67. H. Bencherif, L. Dehimi, G. Messina, P. Vincent, F. Pezzimenti, and F.G. Della Corte, "An optimized Graphene/4H-SiC/Graphene MSM UV-photodetector operating in a wide range of temperature", Sensors and Actuators A, 307, 112007 (2020).

68. C. Z. Sun, X. F. Chen, R. D. Hong, X. M. Li, X. G. Xu, X. P. Chen, J. F. Cai, X. A. Zhang, W. Cai, Z. Y. Wu, and F. Zhang, "Enhancing the photoelectrical performance of graphene/4H-SiC/graphene detector by tuning a Schottky barrier by bias", Applied Physics Letters, 117, 071102 (2020).

69. Y. H. An, D. Y. Guo, S. Y. Li, Z. P. Wu, Y. Q. Huang, P. G. Li, L. H. Li, and W. H. Tang, "Influence of oxygen vacancies on the photoresponse of β-Ga2O3/SiC n–n type heterojunctions", Journal of Physics D: Applied Physics. 49, 285111 (2016).

70. S. Nakagomi, T. Sakai, K. Kikuchi, and Y.o Kokubun, "β-Ga$_2$O$_3$/p-type 4H-SiC heterojunction diodes and applications to deep-UV photodiodes", Physica Status Solidi A, 1700796 (2018).

71. J. Yu, L. Dong, B. Peng, L. Yuan, Y. Huang, L. Zhang, Y. Zhang, R. Jia, "Self-powered photodetectors based on β-Ga$_2$O$_3$/4H-SiC heterojunction with ultrahigh current on/off ratio and fast response", Journal of Alloys and Compounds, 821, 153532 (2020).

14

SiC Radiation Detector Based on Metal-Insulator-Semiconductor Structures

Chong Chen, Yuping Jia, Xiaojuan Sun, and Dabing Li
State Key Laboratory of Luminescence and Applications,
Changchun Institute of Optics, Fine Mechanics and Physics, Chinese Academy of Sciences, P. R. China

1 Introduction

SiC-based materials have been recognized as one of the most promising materials for high-resolution semiconductor radiation detectors due to its small *e-h* pair energy of about 7.78eV, which ensures the large number of *e-h* pairs be produced under radiation. Also, its unique properties such as high strength, corrosion resistance, chemical inertness, high thermal conductivity, and low thermal expansion coefficient make it an appealing candidate for extreme environment detection such as high temperature and hard irradiation [1,2]. Now, SiC-based materials' radiation detectors with SBD structures or PIN structures have been used for uncharged or charged particles' detection and especially when it is used for alpha particle detection, it shows high-energy resolution [3-5]. Apart from charged particle detection, SiC has also been reported to be highly sensitive to ion beam, soft X-ray and high-energy resolution to gamma ray [6-12]. Nevertheless, there are still numerous problems to be solved for the practical application of SiC radiation detectors. In terms of material quality, the epitaxial layer thickness of the SiC radiation detector in the experiment is typically 20–30 μm, which is not enough to efficiently detect the high-energy ray [13-16] and the SiC materials' background carrier density is 10^{15}–10^{16} cm^{-3}, which is below the standard of 10^{14} cm^{-3} [17]. For device structure, the method for reducing the reverse leakage current is a crucial problem for the SiC radiation detectors' performance [18]. It is hard to achieve the SiC layer with low carrier density and large thickness thus improving the structure of the detector is more important. As previously reported, the passivation process can effectively reduce the density of trap defects on the semiconductor surface [19,20], therefore, some scholars have designed SiC radiation detectors with MIS structure and achieved good results [21,22]. In this chapter, particular emphasis is given to the SiC radiation detectors on MIS structures with different insulating layers and varying thicknesses.

DOI: 10.1201/9780429198540-18

2 SiC Material Properties

In nature, silicon carbide (SiC), also known as moissanite is only found as a mineral in extraterrestrial rocks and the SiC crystal used in the experiment is artificial. SiC is well known for its occurrence in various stable crystal polytypes, the crystal structures vary in the different stacking sequences of the Si-C double layers in which each Si is surrounded by four C atoms and vice versa. SiC with different crystal structure can be formed under different growth environment conditions [23,24]. At present, more than 200 kinds of SiC crystal structures have been found. The most common structures are 3C SiC with cubic structure, also known as β-SiC, and the 4H-SiC and 6H-SiC with hexagonal wurtzite structure, are called α-SiC. The band gap width of SiC materials with different crystal structures varies from 2.30 eV to 3.28 eV [25]. Due to the larger band gap of 4H-SiC, 4H-SiC materials are more widely used in the research and preparation of radiation detectors [26,27].

Apart from the wide bandgap, SiC is also the semiconductor material with high thermal conductivity (4.9 Wcm^{-1} K^{-1}), high-breakdown field strength (3MV/cm), high electron saturated drift velocity (2×10^7 cm/s), and high threshold displacement energy [28,29]. These excellent properties make SiC, GaN, Ga$_2$O$_3$, and diamond known as third-generation semiconductor materials, which are used in extreme conditions such as high-temperature (573–873 K), high-power, high-frequency, high radiation background environments.

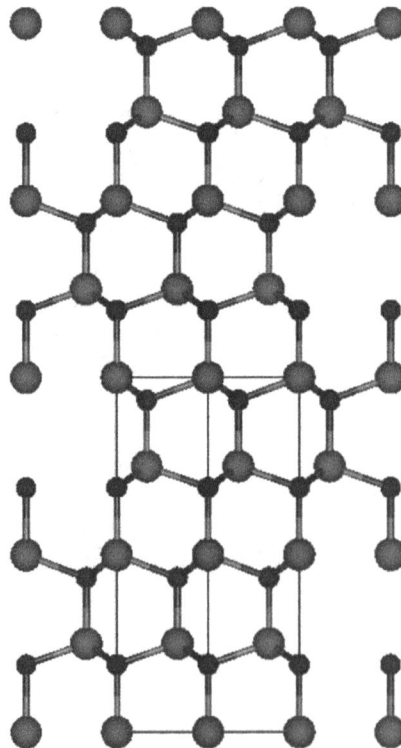

FIGURE 1
Lattice structure of 4H-SiC.

Property	Si	GaAs	4H-SiC	GaN
Bandgap (eV)	1.12	1.43	3.25	3.4
Breakdown field (MV cm^{-1})	0.25	0.3	~3	~3
Saturated electron velocity (10^7 cm s^{-1})	1	2.0 (peak)	2.0	2.5 (peak)
		1.2 (sat)		1.5 (sat)
Electron mobility $N_d \sim 10^{16}$ cm^{-3} (cm^2 V^{-1} s^{-1})	1200	6500	800	900
Thermal conductivity (W cm^{-1} K^{-1})	1.5	0.5	4.9	1.3 (on sapphire)
Dielectric constant	11.8	12.8	9.7	9
Normalized Johnson figure of merit	1	7	360	560

FIGURE 2

Physical properties of wide bandgap semiconductors (a table). From [29] Figure 1, reproduced with permission by Elsevier.

3 SiC Radiation Detector with MIS Structures

SiC radiation detectors with MIS structures have been focused on alpha particles' detection at present. According to the structures with different insulating layer materials or different thicknesses, four SiC radiation detectors are introduced in this chapter.

3.1 Vertical Structure with Thin Al_2O_3 as an Insulator

The SiC radiation detectors with MIS structure, which use Al_2O_3 as the insulating layer, were reported by Kaufmann et al in 2016 [30]. In this study, the SiC Schottky structures were fabricated on 350 μm 4H-SiC (nitrogen doped with $N_D=1\times10^{18}$ cm^{-3}) commercial wafers, 8° off-axis on the Si face with epitaxial layer doped with nitrogen $N_D=1\times10^{15}$ cm^{-3} and 6 μm thick. A Ni layer with a thickness of 150 nm was deposited by sputtering on the SiC wafers and the backside ohmic contacts were formed in a RTA lamp system in an argon flow at 1223 K for 5 min. Al_2O_3 with a thickness of 1 nm was deposited by ALD. The Schottky contact was fabricated by sputtering of Ni through a mechanical mask, forming an electrode with a diameter of 6.0 mm and thickness of 10 nm. A second RTA was performed in argon at 673 K for 5 min for Schottky contacts' improvement. The cross-section view of the fabricated SiC structure with Al_2O_3 insulating layer for alpha particles detection in the RBS experiments was showed in Figure 3.

For the SiC radiation detectors, the reverse currents should be as low as possible to make the charge pulses generated from the particles get a higher signal-to-noise ratio. In this study, a current density of around 20 nA/cm^2 at reverse bias of 40 V was inferred from the fabricated SiC MIS structure. In fact, the 6 μm epitaxial layer depth of SiC was fully depleted at reverse bias of 40 V. In order to compare the device performance, the function of a commercial Si detector was tested and it presented higher reverse current density in this range of the reverse bias in Figure 4. At 40 V, the reverse current density was around 55 nA/cm^2, which is more than twice the current density of the fabricated SiC detector. The active area of the Si detector was 0.25 cm^2 and 0.28 cm^2 for the SiC detector.

Before starting the RBS experiment, it was important to estimate that the alpha collection occurs within the epitaxial layer depth. Combined with the epitaxial layer thickness of the SiC radiation and the simulation results of the SRIM software, the RBS experiment was performed using alpha energies of 1, 1.5, and 2 MeV. This setup ensured that all alpha particles deposited the total energy within the epitaxial layer and the results are shown in Figure 5. The energy resolution for the fabricated detector was estimated to be 76 keV. This energy resolution value was in accordance to other SiC detectors presented in the literature and was poorer than the common Si RBS detectors. However, its major advantage relied on the fact that SiC structures can be used in high-radiation doses or high-temperature ambient without varying its physical and chemical properties.

3.2 Vertical Structure with Thin HfO_2 as the Insulator

In fact, except using Al_2O_3 as the insulator, some work focusing on the selection of the insulating layer material and thickness has been carried out. Through preliminary experiments, it was found that the SiC materials always have a thin insulating film of native silicon oxycarbide (SiC_xO_y) between metal and SiC, which had a measured thickness of about 0.2 nm [31,32]. So when the reverse bias voltage was applied, the trapped charge from the SiC_xO_y would assist the electrons to form a reverse current. This problem has been

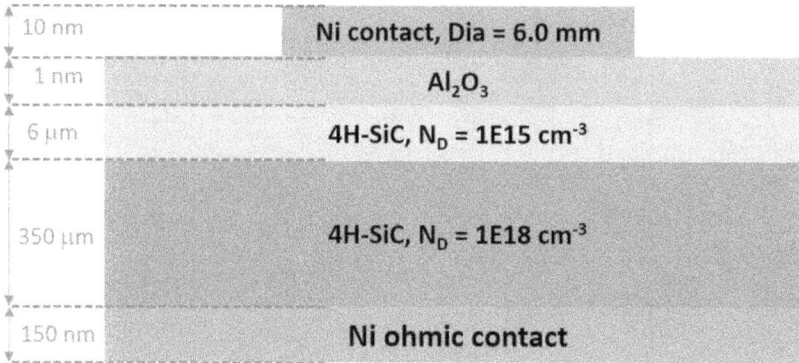

10 nm	Ni contact, Dia = 6.0 mm
1 nm	Al$_2$O$_3$
6 μm	4H-SiC, N$_D$ = 1E15 cm^{-3}
350 μm	4H-SiC, N$_D$ = 1E18 cm^{-3}
150 nm	Ni ohmic contact

FIGURE 3
Cross-section view of the fabricated SiC structure with Al$_2$O$_3$ insulating layer. From [30] Figure 1, reproduced with permission by IOP.

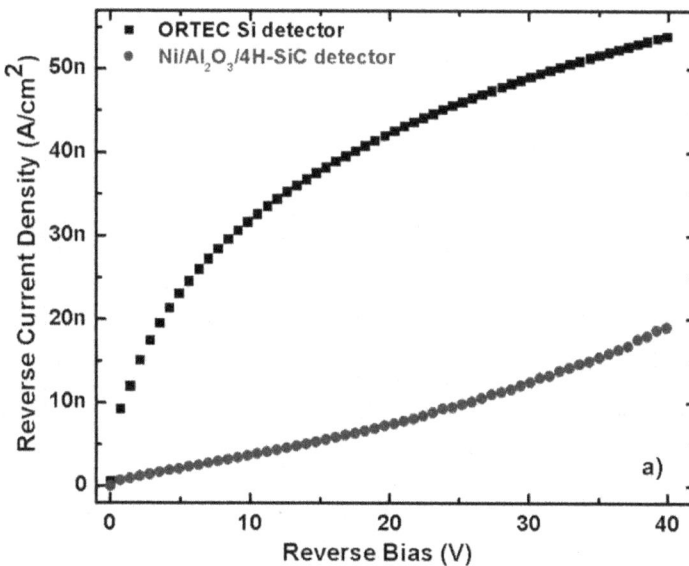

FIGURE 4
Reverse current of the commercial Si detector and the 4H-SiC fabricated detector. From [30] Figure 3, reproduced with permission by IOP.

alleviated with the insertion of a thin interfacial layer. Therefore, HfO$_2$ and TiO$_2$ with thin thickness deposited on SiC in particle detectors were fabricated in this way and the current characteristics were shown in Figure 6.

From Figure 6, all reverse currents of the samples were less than 100 pA for a maximum reverse bias of 40 V and the sample with 1 nm HfO$_2$ presented the lowest leakage current, which was around 5 pA. Based on this result, the SiC radiation detectors with MIS structure using thin HfO$_2$ as the insulating layer were reported by Kaufmann in 2018 [33]. The SiC Schottky detector in this study was also fabricated on 350 μm 4H-SiC (nitrogen doped with N$_D$=10^{18} cm^{-3}) commercial wafers with the 6 μm thick epitaxial layer doped with nitrogen (N$_D$=10^{15} cm^{-3}). The ohmic contact was deposited by Ni sputtering onto SiC

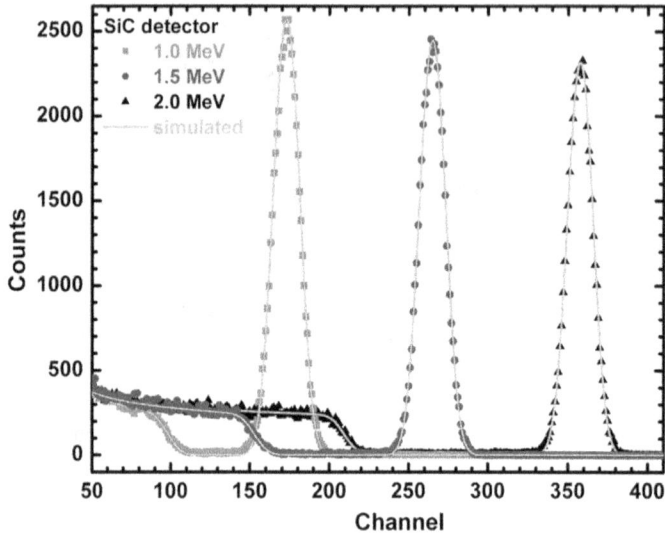

FIGURE 5

RBS spectra collected by the SiC detector for different alpha particles beam energies. From [30] Figure 5, reproduced with permission from IOP.

backside with a thickness of 100 nm. The back side ohmic contact was thermally annealed in an RTA lamp system. HfO_2 with thickness of 1 nm was deposited by the ALD on the front side of the detector sample. After the dielectric deposition, the Ni Schottky contact was deposited by sputtering through a mechanical mask, forming a circular electrode with diameter of 5.4 mm and thickness of 10 nm. A second RTA was performed for Schottky contact improvement.

Figure 7 shows a cross-section view of the fabricated SiC structure for use as an alpha particles' detector in RBS experiments this time. Obviously, the influence of the SiC_xO_y was considered in the SiC radiation detector with MIS structure.

Before the detector was encapsuled and tested in the RBS experiment, the MIS structure was electrically characterized by *I-V* measurements. By testing the forward *I-V* characteristics of the SiC radiation detector and calculating the variable temperature of the ideal factor,it was found that the apparent SBH increased with the temperature. This effect was attributed to the existence of a thin dielectric layer between metal and SiC. In order to judge the electrical performance of the SiC radiation detector with MIS structure, mature commercial Si radiation detector was used for comparison. The reverse leakage current results of the two devices were shown in Figure 8.

Both detectors presented similar reverse currents. For biases lower than 35 V, the SiC detector fabricated presented lower reverse current compared to the commercial Si detector. Both detectors need to be reverse biased with at least 40 V for complete depletion of the epitaxial layer from the SRIM simulation result. When completely depleted, the detectors showed reverse current densities of 62 and 55 nA/cm^2 for the SiC fabricated and commercial Si detectors, respectively. Meanwhile, the active area of the commercial Si detector was 0.25 cm^2, compared to 0.23 cm^2 for the SiC detector. After the *I-V* test, the RBS test results were shown in Figure 9.

In addition to the test at the applied bias voltage of 40 V, the change of SiC energy resolution with the applied voltage and each alpha particle was also measured in this RBS experiments. In general, the greater the applied voltage, the better the energy resolution,

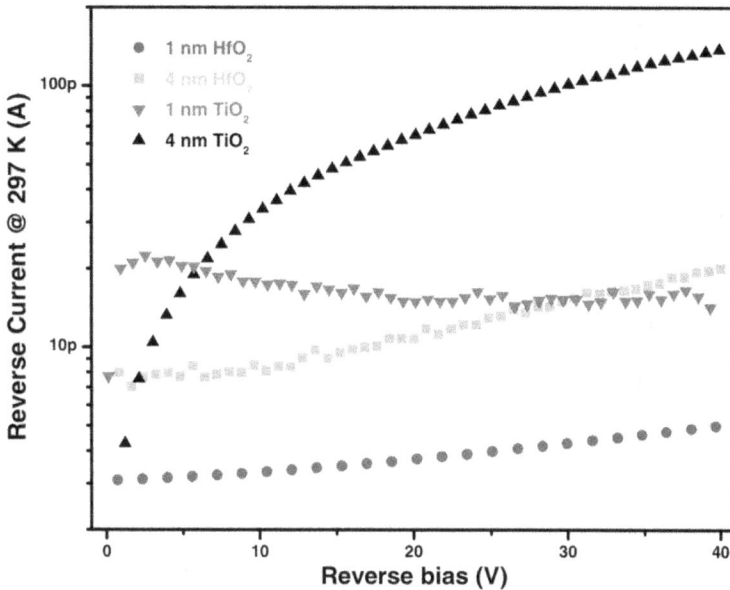

FIGURE 6
Reverse current for the MIS Schottky diodes with 1 and 4 nm of HfO_2 and TiO_2. From [31] Figure 5, reproduced with permission by Elsevier.

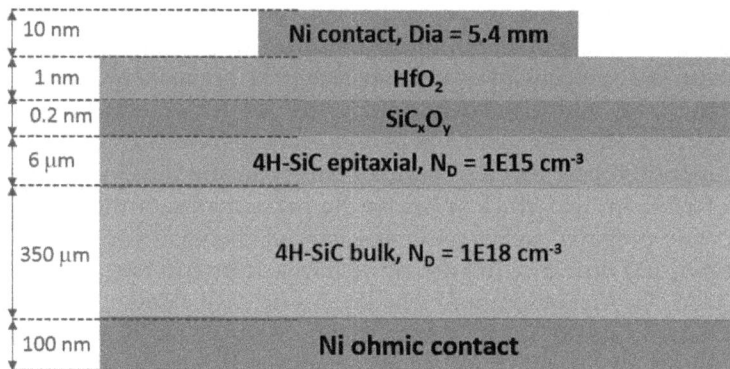

FIGURE 7
Cross-section view of the fabricated $Ni/HfO_2/SiC$ Schottky structure. From [33] Figure 1, reproduced with permission by IOP.

but the overall performance was not as good as that of the Si detector under normal conditions. Nevertheless, the research on the SiC radiation detector with MIS structure should focus on the extreme environment.

3.3 Vertical Structure with Thick SiO_2 as the Insulator

SiC-based material radiation detectors with MIS structure, which achieved the energy resolution of 0.55% under the voltage of 40 V at 5.48 MeV alpha particle irradiation experiment was reported by Yuping Jia et al from the State Key Laboratory of Luminescence and

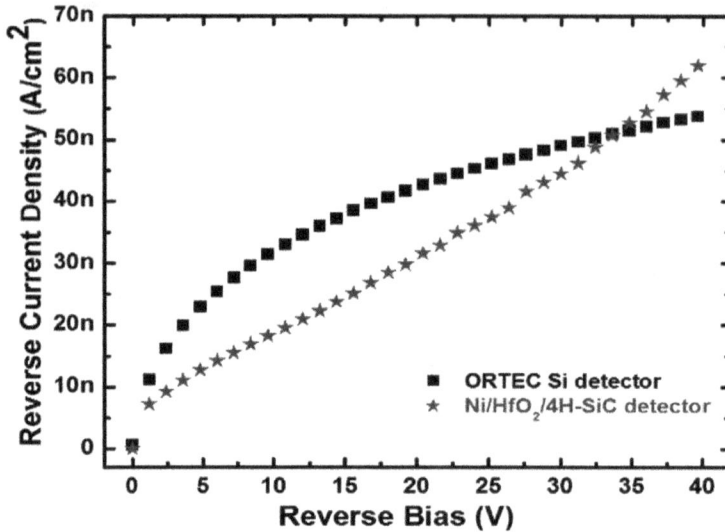

FIGURE 8
Reverse current as a function of the reverse bias for the SiC and Si detectors. From [33] Figure 5, reproduced with permission by IOP.

Applications in 2021. In this study, a SiC radiation detector with a 10×10 mm^2 area was fabricated. An illustration of the detector's structure is presented in Figure 10. A 15 μm epi-taxial SiC layer with doping density of 3.5×10^{14} cm^{-3} was deposited on an *n*-doped 4H-SiC substrate Si-face with off-cut of 4°. A 1 μm buffer layer with a doping density of 10^{18} cm^{-3} was used between the epitaxial layer and substrate. The epitaxial SiC layer was the alpha particle-absorbing layer and the *n*-doped SiC substrate was used as a conductive layer connected to an ohmic electrode, which was important to the device [34-36]. A SiO$_2$ dielec-tric layer was deposited between the Schottky electrode and SiC. In order to understand the influence of different SiO$_2$ thicknesses on the performance of the radiation detectors. Four samples were prepared using plasma-enhanced chemical vapor deposition, which was 5 nm, 50 nm, 100 nm, and 200 nm in thickness, respectively. Then an ohmic elec-trode with a Ti/Al/Ti/Au compound metal layer with each element 30 nm thickness was produced on the backside of the SiC substrate. The Schottky electrode Ni/Au compound metal layer with each element 30 nm thick was deposited on top of the epitaxial SiC.

After device completion, the dark current curves as a function of the bias in the samples with different dielectric thicknesses were measured by the Keithley system. The reverse voltage applied to the Schottky electrode at the largest voltage was 100 V.

From Figure 11, the sample without a SiO$_2$ layer inserted exhibited the highest dark current. The dark currents of the 5 nm, 50 nm, 100 nm, and 200 nm thick samples at a 100 V bias were 7.70×10^{-7} A, 3.73×10^{-7} A, 8.35×10^{-8} A, and 2.14×10^{-8} A. Of note, the dark currents of the 100 nm and 200 nm samples maintained a magnitude of 10^{-8}, suggesting that the optimal effect of the dielectric layer was saturated. ^{239}Pu and ^{241}Am were used as the radi-ation source to provide alpha particles in this study. The results showed that when the thickness of SiO$_2$ is 0 nm, 5 nm, 100 nm, and 200 nm, the energy resolution is 2.52%, 1.15%, 0.55%, and 1.16% in Figure 10.

With the increase of SiO$_2$ thickness, the energy resolution increased firstly and then decreased. In particular, when the dielectric layer's thickness increasing to 100 nm, there

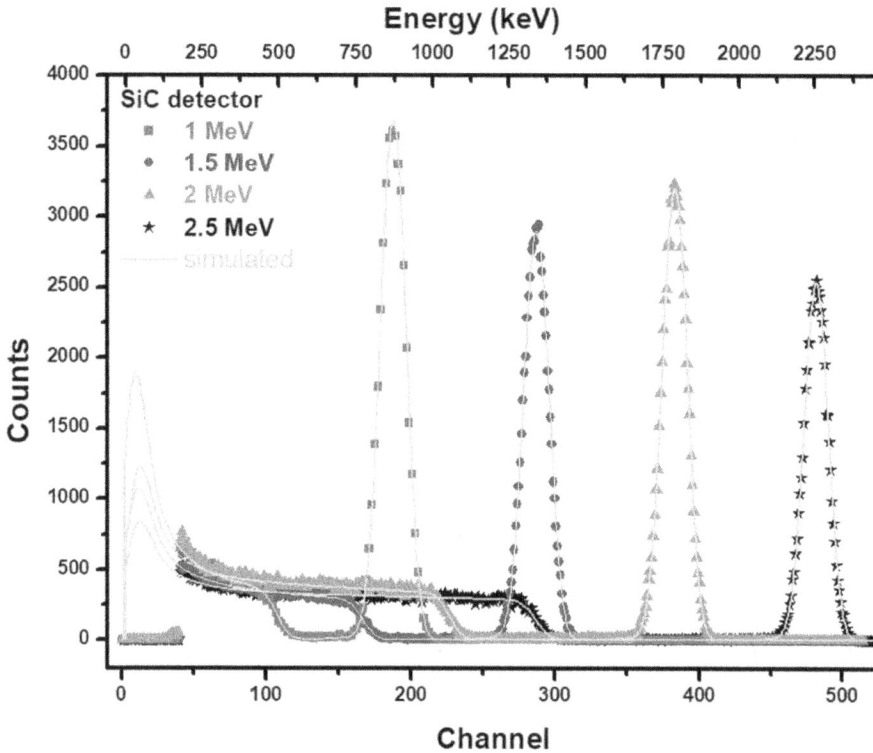

FIGURE 9
RBS spectra collected for SiC detectors, for the four alpha particle energies. From [33] Figure 7, reproduced with permission by IOP.

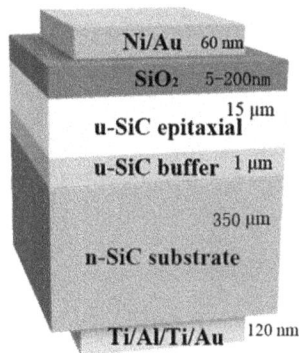

FIGURE 10
Illustration of the SiC detector's structure. From [22] Figure 1.b, reproduced with permission by Elsevier.

were two peaks in the energy spectrum. It showed an obvious detection and distinction between the two signals' source from ^{239}Pu and ^{241}Am. Although the dark current decreased with the increasing of the thickness, the 200 nm device's energy resolution was more inferior than that for 100 nm. The decreasing detection for the 200 nm SiO$_2$ sample might be due to the excessive deposition of the dielectric layer. Although the dark current decreased, the thick dielectric layer impeded the carrier collection, which was due to the scattering effect of the localized charges such as particle boundaries and deep carrier traps. The thick dielectric layer affected the deposition of high-energy particles in the detector's space charge region, and some of the particles were deposited on the metal and dielectric layers.

3.4 MIS Structures with Graphene Insertion in Ohmic Contact Electrode

The SiC radiation detectors based on MIS structures with graphene insertion were also reported by Yuping Jia et al from the State Key Laboratory of Luminescence and Applications in 2021. Based on the MIS structure, the ohmic contact of the SiC radiation detector was effectively optimized by graphene insertion. The graphene inserted device annealed at 400°C achieved a 3.9% energy resolution under the voltage of 40 V and the device even without annealing achieved a 4.4% energy resolution under the voltage of 40 V for ^{239}Pu alpha source, which were both better than that of traditional devices without graphene layer annealed at 880°C got 6% energy resolution under the voltage of 40 V. In this study, the SiC radiation detector with a 5×10 mm^2 area was fabricated. A 15 μm epitaxial SiC layer with an *n*-type doping density of 3.5×10^{14} cm^{-3} was deposited on an *n*-doped SiC substrate. A 1 μm buffer layer with a doping density of 1×10^{18} cm^{-3} was used between the epitaxial layer and substrate. A 500 nm SiO$_2$ dielectric layer was deposited between the Schottky electrode and SiC epitaxial layer. The Schottky electrode was made using a Ni/Au metal layer with each layer 30 nm thick and annealed at 400°C. In order to optimize ohmic electrode with graphene layer insertion based on MIS structure, four samples with different ohmic electrodes were fabricated. The first one was traditional ohmic electrode without graphene insertion, which was a Ti/Al/Ti/Au metal layer with each layer 30 nm

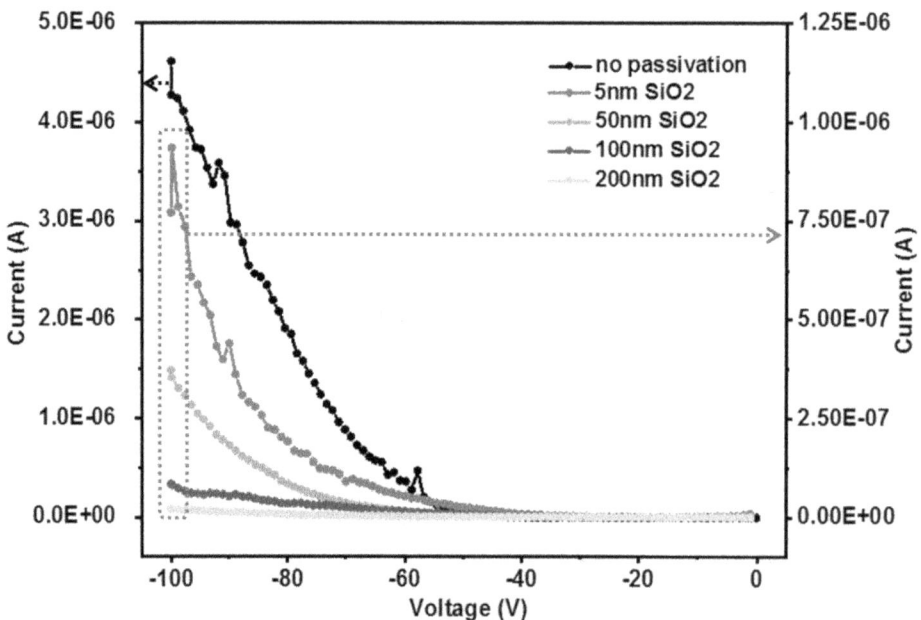

FIGURE 11

The dark current curves as a function of the bias with different thicknesses. From [22] Figure 3, reproduced with permission by Elsevier.

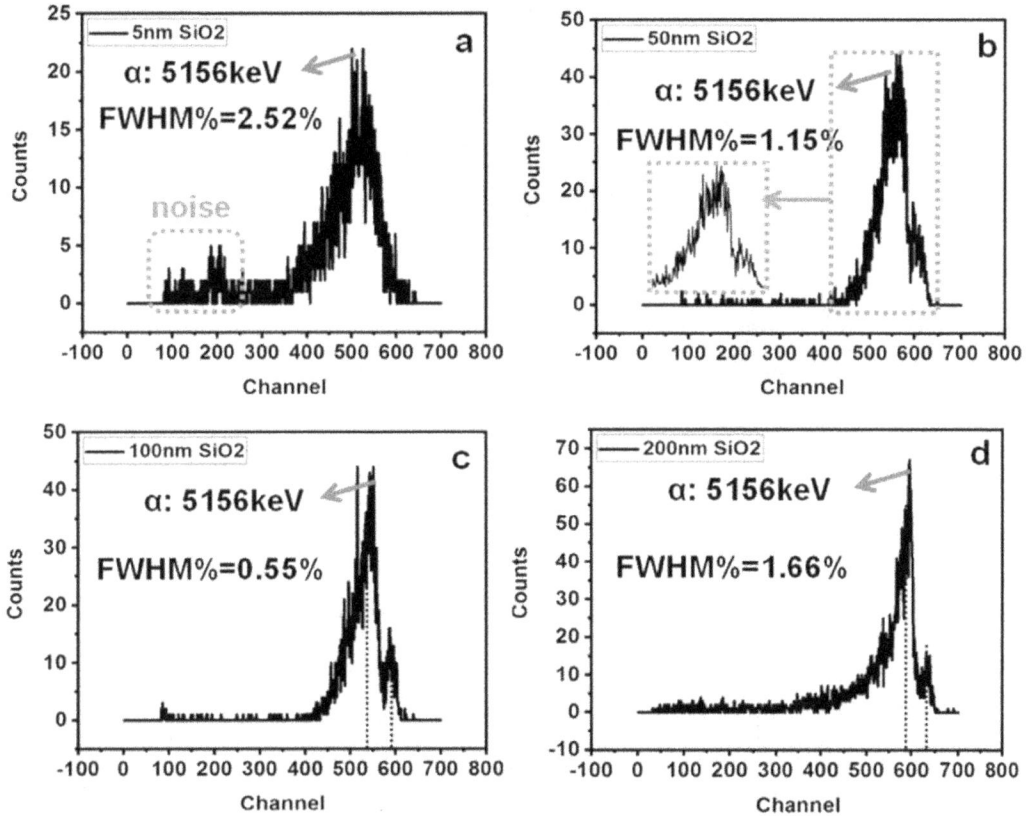

FIGURE 12

Pulse height spectra different thickness of SiO_2. From [22] Figure 5, reproduced with permission by Elsevier.

and annealed at 880°C; the second one was graphene inserted ohmic electrode between SiC and metal without annealing; the third one was graphene inserted ohmic electrode annealed at 400 °C; the last one was graphene inserted ohmic electrode annealed at 880°C. An illustration of the detector's structure was presented in Figure 13.

^{239}Pu and ^{241}Am sources were used to provide α particle irradiation. For all four devices, two peaks of the sources could be detected. The peak signals were related to the radiation energy, which was recorded as the channel number by a multichannel analyzer. As shown in Figure 14, the peak channel number of the conventional devices were less than those of graphene electrode devices. The peak position of the conventional device was reduced by around 150 channels compared with the graphene inserted device without annealing. For devices with graphene electrodes, the channel number of peaks decreased with the increasing annealing temperature. Since the number of channels in the multichannel analyzer corresponds to the signal voltage, the more channels, the greater the signal voltage. The larger signal voltage indicated the better charge collection ability of the device. Therefore, the graphene electrode could effectively improve the charge collection efficiency. After annealing, the crystallization quality of graphene decreases, and the higher the annealing temperature, the worse the crystallization quality. Therefore, the charge collection efficiency decreased, and the peak position moved to the low channel number. The peak position of annealing at 400°C moved only 16 channels, which indicating that annealing at 400 °C had little effect on the crystallization quality of graphene. The device

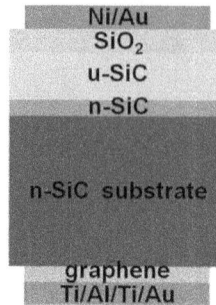

FIGURE 13

Illustration of the device's structure. From [21] Figure 1.b reproduced with permission by Elsevier.

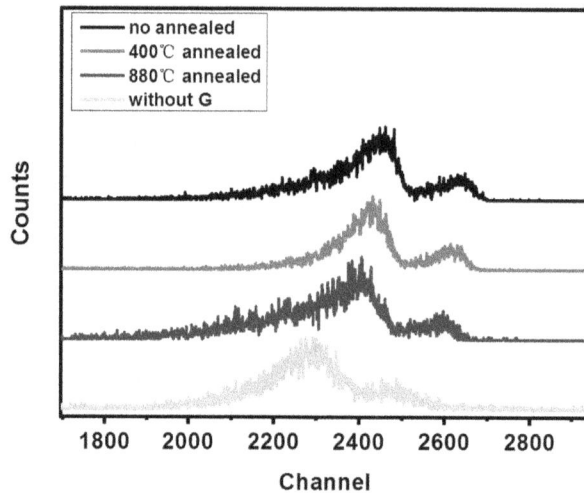

FIGURE 14

Pulse height spectra under irradiation measured in detectors at 40 V bias voltage. From [21] Figure 4 reproduced with permission by Elsevier.

annealed at 400°C achieved the best energy resolution of 3.9% under 40 V among the four devices.

4 Conclusion

SiC is one promising semiconducting material for ionizing radiation detection, particularly in hash environments [37-40]. So far, research on SiC-based materials' radiation detector have been performed. However, the defects present in SiC materials and unintentional doping, which lead to the high-leakage current, still present a main challenge in terms of improving device-level performance. The SiC radiation detectors with SBD or PIN structure, which have achieved good results for detecting alpha particles, electrons, X-rays, or neutrons usually needs high epitaxial layer quality and large thickness [41-42]. It is expensive and cannot be easily obtained domestically. A SiC radiation detector with

MIS structure reduces the surface trap charge density and reverse leakage current through surface passivation. Some experiments on alpha particle detection, which achieve good results, have been reported. But the thick dielectric layer of the MIS structure may affect the deposition of high-energy particles in detector space charge region and some of the particles may be deposited on the metal and dielectric layers, which may impede the carrier collection. Apart from improving SiC material crystal quality and decreasing the growth cost, we can invest in optimizing the structure of SiC MIS devices and choose the better insulation material for radiation detection such as electrons, X-rays, or neutrons and so on.

References

1. Nava F, Bertuccio G, Cavallini A, Vittone E, "Silicon carbide and its use as a radiation detector material", Measurement Science & Technology, 19(10), 102001, (2018). doi.org/10.1088/0957-0233/19/10/102001

2. Mandal KC, Krishna R, Muzykov PG, Laney Z, Das S, Sudarshan TS, "Radiation detectors based on 4H semi-insulating silicon carbide", Proceedings of SPIE – The International Society for Optical Engineering, 7805 (2010). doi.org/10.1117/12.863572

3. Nava F, Vanni P, Bruzzi M, Lagomarsino S, Sciortino S, Wagner G, Lanzieri C, "Minimum ionizing and alpha particles detectors based on epitaxial semiconductor silicon carbide", IEEE. T. Nucl. Sci., 51, 238–244 (2004). doi.org/10.1109/TNS.2004.825095

4. Chaudhuri SK, Zavalla KJ, Mandal KC, "High resolution alpha particle detection using 4H–SiC epitaxial layers: fabrication, characterization, and noise analysis", Nucl. Instrum. Meth. A., 728, 97–101 (2013). doi.org/10.1016/j.nima.2013.06.076

5. Zaťko B, Dubecký F, Šagátová A, Sedlačová K, Ryć L, "High resolution alpha particle detectors based on 4H-SiC epitaxial layer", J. Instrum., 10 (04), C04009 (2014). doi.org/10.1088/1748-0221/10/04/C04009

6. Bertuccio G, Casiraghi R, Nava F, "Epitaxial silicon carbide for X-ray detection", IEEE. T. Nucl. Sci, 48, 232–233 (2001). doi.org/10.1109/23.915369

7. Bozack MJ, "Surface studies on SiC as related to contacts", Phys. Status Solidi B, 202, 549–580 (1997). doi.org/10.1002/1521-3951(199707)202:1<549::AID-PSSB549>3.0.CO;2-6

8. Crofton J, Porter LM, Williams JR, "The physics of ohmic contacts to SiC", Phys. Status Solidi B, 202, 581–603 (2015). doi.org/10.1002/1521-3951(199707)202:1<581::AID-PSSB581>3.0.CO;2-M

9. Sciuto A, D'Arrigo G, Di Franco S, Mazzillo M, Franzo G, Torrisi L, Calcagno L, "4H-SiC detector in high photons and ions irradiation regime", IEEE Trans. Electron Devices, 65, 599–604 (2018). doi.org/10.1109/TED.2017.2785865

10. Margarone D, Krása J, Giuffrida L, Picciotto A, Torrisi L, Nowak T, et al., "Full characterization of laser-accelerated ion beams using Faraday cup silicon carbide and single-crystal diamond detectors", J. Appl. Phys., 109, 103302 (2011). doi.org/10.1063/1.3585871

11. Torrisi L, Sciuto A, Cannavò A, Di Franco S, Mazzillo M, Badalà P, Calcagno L, "SiC detector for sub-MeV alpha spectrometry", J. Electron. Mater., 46, 4242–4249 (2017). doi.org/10.1007/s11664-017-5379-y

12. Torrisi A, Wachulak PW, Fiedorowicz H, Torrisi L, "Monitoring of the plasma generated by a gas-puff target source", Phys. Rev. Accel. Beams, 22, 052901 (2019). doi.org/10.1103/PhysRevAccelBeams.22.052901

13. Itoh A, Matsunami H, "Analysis of Schottky barrier heights of metal/SiC contacts and its possible application to high-voltage rectifying devices", Phys. Status Solidi A, 162, 389–408 (2015). doi.org/10.1002/1521-396X(199707)162:1<389::AID-PSSA389>3.0.CO;2-X

14. Sciuto A, Torrisi L, Cannavò A, Mazzillo M, Calcagno L, "Advantages and limits of 4H-SIC detectors for high- and low-flux radiations", J. Electron. Mater., 46, 6403–6410 (2017). doi.org/10.1007/s11664-017-5675-6

15. La Via F, Roccaforte F, Makhtari A, Raineri V, Musumeci P, Calcagno L, "Structural and electrical characterisation of titanium and nickel silicide contacts on silicon carbide", Microelectron. Eng, 60(1–2), 269–282 (2002). doi.org/10.1016/S0167-9317(01)00604-9

16. Ruddy FH, Flammang RW, Seidel JG, "Low-background detection of fission neutrons produced by pulsed neutron interrogation", Nuclear Instruments & Methods in Physics Research, 598(2), 518–525 (2009). doi.org/10.1016/j.nima.2008.09.033

17. Tsuchida H, Kamata I, Ito M, Miyazawa T, Hoshino N, Fujibayashi H, et al., "Evolution of fast 4H-SiC CVD growth and defect reduction techniques", Materials Sci. Forum, 778–780, 85–90 (2014). https://doi.org/10.4028/www.scientific.net/MSF.778-780.85

18. Nguyen KV, Mannan MA, Mandal KC, "Improved n-Type 4H-SiC epitaxial radiation detectors by edge termination", IEEE Transactions on Nuclear Science, 62(6), 3199–3206 (2015). doi.org/10.1109/TNS.2015.2496902

19. Ohyu K, Ohkura M, Hiraiwa A, Watanabe K, "A mechanism and a reduction technique for large reverse leakage current in p-n junctions", IEEE Transactions on Electron Devices, 42(8), 1404–1412 (1995). doi.org/10.1109/16.398655

20. Sugimoto M, Kanechika M, Uesugi T, Kachi T, "Study on leakage current of pn diode on GaN substrate at reverse bias", Phys. Status Solidi C, 8, 2512–2514 (2011). doi.org/10.1002/pssc.201000935

21. Jia Y, Sun X, Shi Z, Jiang K, Wu T, Liang H, et al., "Improved performance of SiC radiation detectors due to optimized ohmic contact electrode by graphene insertion", Diamond and Related Materials, 115, 108355 (2021). doi.org/10.1016/j.diamond.2021.108355

22. Jia Y, Shen Y, Sun X, Shi Z, Jiang K, Wu T, et al., "Improved performance of SiC radiation detector based on metal–insulator-semiconductor structures", Nuclear Instruments and Methods in Physics Research Section A: Accelerators Spectrometers Detectors and Associated Equipment, 997(3), 165166 (2021). doi.org/10.1016/j.nima.2021.165166

23. Szweda R, "GaN and SiC detectors for radiation and medicine", III-Vs Review, 18(7), 40–41 (2005). doi.org/10.1016/S0961-1290(05)71301-6

24. Mandal KC, Muzykov PG, Krishna RM, Terry RJ, "Characterization of 4H-SiC epitaxial layers and high-resistivity bulk crystals for radiation detectors", IEEE Transactions on Nuclear Science, 59(4), 1591–1596 (2012). doi.org/10.1109/tns.2012.2202916

25. Wellmann PJ, "Review of SiC crystal growth technology", Semiconductor Science and Technology, 33(10), 103001 (2018). doi.org/10.1088/1361-6641/aad831

26. Eiting CJ, Krishnamoorthy V, Rodgers S, George T, "Demonstration of a radiation resistant, high efficiency SiC betavoltaic", Applied Physics Letters, 88(6), 1–38 (2006). doi.org/10.1063/1.2172411

27. Ramesh PD, Vaidhyanathan B, Ganguli M, Rao KJ, "Synthesis of β-SiC powder by use of microwave radiation", Journal of Materials Research, 9(12), 3025–3027 (1994). doi.org/10.1557/jmr.1994.3025

28. Casady JB, Johnson RW, "Status of silicon carbide (SiC) as a wide-bandgap semiconductor for high-temperature applications: A review", Solid-State Electronics, 39(10), 1409–1422 (1996). doi.org/10.1016/0038-1101(96)00045-7

29. Burk Jr. AA, O'Loughlin MJ, Siergiej RR, Agarwal AK, Sriram S, Clarke RC, "SiC and GaN wide bandgap semiconductor materials and devices", Solid-State Electronics 43(8), 1459–1464 (1999). doi.org/10.1016/S0038-1101(99)00089-1

30. Kaufmann IR, Pick A, Pereira MB, Boudinov HI, "Ni/Al_2O_3/4H-SiC structure for He^{++} energy detection in RBS experiments", Journal of Instrumentation, 11, 1748–0221: P10013 (2016). doi.org/10.1088/1748-0221/11/10/P10013

31. Kaufmann IR, Pick A, Pereira MB, Boudinov H, "Metal-insulator-SiC Schottky structures using HfO_2 and TiO_2 dielectrics", Thin Solid Films, 621, 184–187 (2017). doi.org/10.1016/j.tsf.2016.11.053

32. Kaufmann IR, Pereira MB, Boudinov HI, "Schottky barrier height of $Ni/TiO_2/4H$-SiC metal-insulator-semiconductor diodes", Semiconductor Science and Technology, 30(12), 125002 (2015). doi.org/10.1088/0268-1242/30/12/125002

33. Kaufmann IR, Pick AC, Pereira MB, Boudinov HI, "Characterization of a SiC MIS Schottky diode as RBS particle detector", Journal of Instrumentation, 13(02), P02017-P02017 (2018). doi.org/10.1088/1748-0221/13/02/P02017

34. Han SY, Kim KH, Kim JK, Jang HW, Lee KH, Kim N-K, et al., "Ohmic contact formation mechanism of Ni on *n*-type 4H–SiC", Appl. Phys. Lett., 79, 1816–1818 (2001). doi.org/10.1063/1.1404998

35. Zhang Y, Guo T, Tang X, Yang J, He Y, Zhang Y, "Thermal stability study of n-type and p-type ohmic contacts simultaneously formed on 4H-SiC", J. Alloys Compd., 731, 1267–1274 (2018). doi.org/10.1016/j.jallcom.2017.10.086

36. Cuong VV, Ishikawa S, Maeda T, Sezaki H, Yasuno S, Koganezawa T, et al., "High-temperature reliability of Ni/Nb ohmic contacts on 4H-SiC for harsh environment applications", Thin Solid Films, 669, 306–314 (2019). doi.org/10.1016/j.tsf.2018.11.014

37. Mandal KC, Muzykov PG, Chaudhuri SK, Terry JR, "Low energy x-ray and γ-ray detectors fabricated on n-type 4H-SiC epitaxial layer", IEEE Transactions on Nuclear, 60(4), 2888–2893 (2013). doi.org/10.1109/TNS.2013.2273673

38. Mandal KC, Muzykov PG, Terry JR, "Design, fabrication, characterization, and evaluation of x-ray detectors based on n-type 4H-SiC epitaxial layer", ECS Transactions, 45(7), 27–33 (2012). doi.org/10.1149/1.3701522

39. Addamiano A, Sprague JA, "'Buffer-layer' technique for the growth of single crystal SiC on Si", Appl. Phys. Lett., 44(5), 525–527 (1984). doi.org/10.1063/1.94820

40. Mandal KC, Krishna RM, Muzykov PG, Das S, Sudarshan TS, "Characterization of semi-insulating 4H silicon carbide for radiation detectors", IEEE Transactions on Nuclear Science, 58(4), 1992–1999 (2011). doi.org/10.1109/TNS.2011.2152857

41. Saxena V, Su JN, Steckl AJ, "High-voltage Ni- and Pt-SiC Schottky diodes utilizing metal field plate termination", IEEE Transactions on Electron Devices, 46(3), 456–464 (1999). doi.org/10.1109/16.748862

42. Tarplee MC, Madangarli VP, Zhang Q, Sudarshan TS, "Design rules for field plate edge termination in SiC Schottky diodes", IEEE Transactions on Electron Devices, 48(12), 2659–2664 (2001). doi.org/10.1109/16.974686

15

Internal Atomic Distortion and Crystalline Characteristics of Epitaxial SiC Thin Films Studied by Short Wavelength and Synchrotron X-ray Diffraction

Gu Xu

Department of Materials Science and Engineering, McMaster University, Hamilton, Ontario, Canada.

Zhe Chuan Feng, Jeffrey Yiin, Vishal Saravade, Benjamin Klein, and Ian T. Ferguson
Southern Polytechnic College of Engineering and Engineering Technology,
Kennesaw State University, Marietta, GA, USA.

1 Introduction

Silicon carbide (SiC) has been recognized as an important material for a wide variety of high-power and high-temperature electronic applications [1-5]. SiC exhibits a large number (>250) of polytypes with different structural and physical properties, as well as possesses a wide range of applications [6-14]. Research on SiC has attracted attention since the early stages of Si development [15]. However, due to the lack of proper growth technology for large size of wafers, development on SiC was hindered while technology development on Si has advanced to a higher degree. Interests on SiC have been renewed since 1980s because of two major technology breakthroughs. One is the successful growth of epitaxial cubic (3C-) SiC film on Si substrate by chemical vapor deposition [16]. The other is the development of a modified sublimation method for growing a large boule of bulk SiC single crystals [17]. Since the 1990s, large sizes of 6H- and 4H-SiC wafers have been commercially available, which greatly promote the research and development on SiC and applications [18]. For device applications, heterogeneous epitaxial 3C-SiC thin film, grown on top of single crystal silicon and 6H- or 4H-SiC, as well as homoepitaxial SiC on SiC (6H on 6H and 4H on 4H) substrate by chemical vapor deposition (CVD) are the most popular [15,16,18].

To characterize thin films, X-ray diffraction (XRD) is commonly used to measure the lattice constant from the Bragg diffraction peak position, and the crystal quality through the Bragg peak profile. Research and development activities on SiC materials and devices have widely used XRD to characterize the crystalline quality of cubic SiC on Si (100) [19,20]

DOI: 10.1201/9780429198540-19

and on Si (111) [21]. It can also be used to study the ion-implantation [22,23] and poly-morphism in SiC [24,25], monitor the alteration of polytypes during growth [26], investigate recrystallization and the orientation relationship [27], and explore heavy doping [28], and doping-induced mis-match through reciprocal maps [29].

Moving into the 21st century, XRD has continually played an important role in various research on SiC bulk materials, epitaxial films, and structures [30-44]. Also, synchrotron radiation (SR) XRD and SR X-ray topography have been applied to investigate the growth process and sample quality of SiC crystals [45-50].

XRD studies on SiC materials [5-29] are usually conducted using a copper source with 1.54 Å radiation. The analyses of the XRD patterns on the diffraction peak position, intensity, and width can lead to the information on sample structure, crystallinity, perfection, defects, strain, etc. The XRD line widths are related to the size (depth) and the mosaicity of the coherent region. Understanding of SiC crystal properties could be significantly improved and more accurately understood by tuning the X-ray source incident radiation. Xu and Feng presented, in Phys. Rev. Lett. [51], a research on Internal Atomic Distortion and Layer Roughness of Epitaxial SiC Thin Films Studied by Short Wavelength X-ray Diffraction. By using a short wavelength (0.71 Å) XRD measurement for 3C-SiC on Si (100), up to five order Bragg peaks and the crystallographic structure factors could be observed. Through this new method, a quantitative characterization of the internal atomic layer roughness and distortion of hetero-epitaxial 3C-SiC and homo-epitaxial 6H-SiC thin films could also be achieved. In this chapter, X-ray diffraction studies of SiC using this newly developed method are reviewed theoretically and experimentally. This provides a practical and useful method to investigate thin film materials and systems even beyond SiC. Further, synchrotron radiation (SR) XRD with a considerably short wavelength (0.443 Å) is also investigated to measure 3C-SiC hetero-epitaxial film on Si and homo-epitaxial 4H-SiC/4H-SiC and 6H-SiC/6H-SiC. Discussion on the internal SiC crystalline structural characteristics are systematically discussed.

2 Research Background

Usually no information from inside the unit cell, such as the internal atomic layer arrangement, can be extracted from the XRD investigation of SiC. Due to the lattice mismatch of 3C-SiC grown on Si, the internal Si and C atomic layers near the interface region could be distorted and have a complicated arrangement, as shown in Figure 1. It is important to understand these details in the crystalline properties of SiC. However, there are no details studied on these internal atomic layer arrangements in the literature.

Typical X-ray measurement using 1.54 Å can measure up to two orders of Bragg's diffraction peaks in SiC. However, SiC could exhibit additional peaks that can provide important information about its crystalline properties. Using short wavelength X-ray, a new method is presented here, to study the atomic internal layer distortion or roughness. A molybdenum anode is used to generate wavelength of 0.71 Å, less than one half of that from a copper anode. This enables measurement of Bragg reflections up to the 5th order, along the normal of the thin films in either (100) or (111) orientation. From the *peak intensities*, the absolute values of the crystallographic structure factors, $|F(hkl)|$, can be deduced and electron density of the thin film crystal can be calculated using Fourier transform (Eq. (1)).

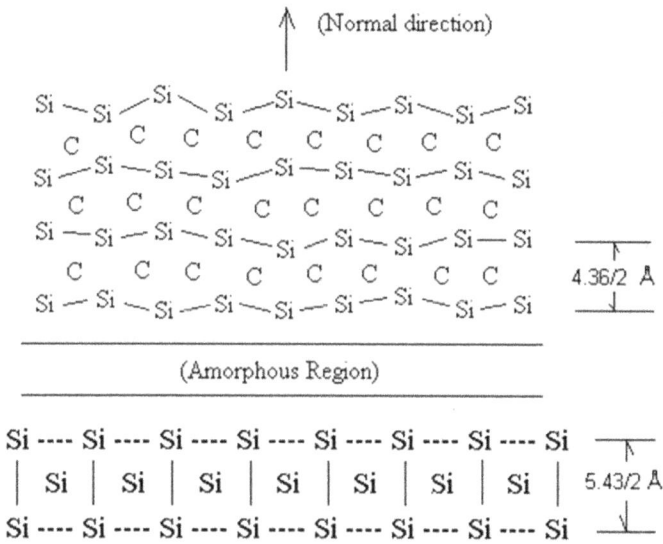

FIGURE 1
Cross-sectional view of 3C-SiC on single crystal silicon. From [51] Figure 1 with permission of reproduction by APS.

$$\rho(x,y,z) = \Sigma_{hkl} \ |F(hkl)| \ \cos(2\pi[hx/a+ky/b+lz/c] - \phi_{hkl}) \qquad (1)$$

where a, b, c are the lattice constants along x, y, z, and ϕ_{hkl} is the phase [52]. The results could be compared with the calculated electron distribution using documented atomic scattering factors. Atomic layer roughness can also be quantified from the broadening (flattening) of the electron distribution maxima for silicon and carbon atoms.

3 Experimental and Fundamental Details

SiC thin films as 3C-SiC grown on Si (100) and 6H-SiC grown on 6H-SiC using are investigated. Two 3C-SiC/Si with film thickness of ~6 μm for sample 1 and ~12 μm for sample 2, and one 6 μm 6H-SiC film homo-epitaxied on 6H-SiC substrate were deposited. Short wavelength X-ray was obtained from a Siemens D8 diffractometer using a molybdenum anode ($K\alpha$1 only). X-ray intensity data was collected with a scan step size of 0.004° per 2θ step.

In a usual diffractometer, an *intensity maximum* contains simultaneous diffraction contributions from a large volume surrounding a reciprocal lattice point. This is due to simple collimation and a broad spectrum ($K\alpha$1+$K\alpha$2) in the incident beam. Thus, this maximum does not equal the Bragg *peak intensity*, contributed only by the center of this volume [52]. A sample should be scanned continuously, to measure the "*integrated intensity*" over a range of 2θ (e.g., 0.5°). Assuming a standard peak profile, the *peak intensity* could be extracted from the *integrated intensity*. Lorentz factor must be used in the calculation, to compensate for the changing sweeping time of various reciprocal lattice points through the Ewald sphere [52]. However, such a scheme is not applicable because of the lattice

3C-SiC/Si (moly anode)

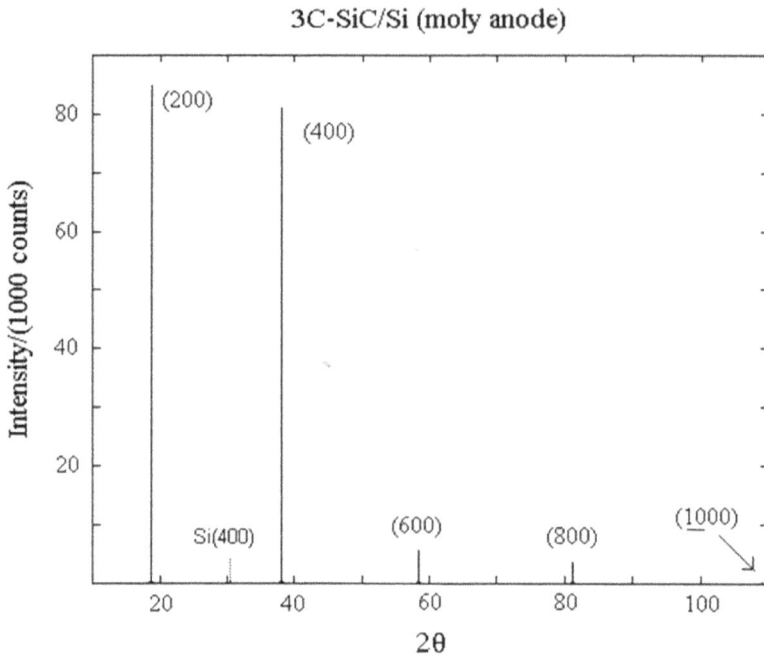

FIGURE 2
The first five Bragg peak intensities of 3C-SiC. From [51] Figure 2 with permission of reproduction by APS.

distortion, which changes the standard peak profile. To overcome these issues, a highly collimated and monochromatic beam ($K\alpha1$ only) is used. This ensures that only a tiny volume surrounding a reciprocal lattice point is covered at any time moment. (The 2θ resolution is <0.002°, which is much less than the Bragg peak width.) Using the step mode, intensity variation near the reciprocal lattice point was measured. The *peak intensity* can then be directly obtained by the *intensity maximum* without involving the Lorentz factor.

Peak intensities varied by the mosaic spreading along the transverse direction are measured and directly related to the variation of lattice constants of the blocks projected onto the film normal (the scattering vector).

Using molybdenum anode, peak intensities from ω–2θ scan along the surface normal of a 3C-SiC/Si thin film sample (growth time = 2 hr), are shown in Figure 2. Five Bragg peaks due to SiC (200, 400, ... , 1000) are observed along 2θ = 18.7–109°. Peaks due to the silicon substrate are excluded, although they are detectable. Instead of the detailed peak profile, from which the peak breadth is usually studied, peak intensity of various Bragg reflections is analyzed. Eq. (1) is used to investigate the atomic arrangement within the crystalline unit cell along the normal direction. Bragg peak intensities are proportional to the square of the structure factors $|F(hkl)|^2$, after corrections by the polarization factor and Debye factor for thermal vibration (because the scan was done by step mode rather than continuous mode, no Lorentz factor is involved):

$$I(hkl) \propto (1+\cos^2 2\theta)/2 \, \exp(-2M) \, |F(hkl)|^2 \qquad (2)$$

where $M = B(\sin\theta/\lambda)^2$, and B is the thermal Debye parameter measuring the thermal motion [53]. Using $B = 0.25$ as the average for silicon and carbon, $|F(200\text{–}1000)|$ for sample 1 obtained from Eq. (2) were 9.63, 8.6, 2.34, 2.03, and 0.60. Similarly, the structure factors for sample 2 were 9.47, 10.0, 3.02, 2.73, and 0.78 (in arbitrary units).

4 Theoretical Fourier Transform Calculation on 3C-SiC

To apply these $F(hkl)$ to reconstruct the atomic arrangement via Fourier transform using Eq. (1), experimentally measured structure factors are calculated with a theoretical situation, i.e., when the silicon atoms are stabilized on 0, ½, and 1, and the carbon atoms are in 1/4 and 3/4 of the relative co-ordinates of the unit cell. In this case, the internal atomic layer arrangement is not distorted and there is no internal atomic layer roughness involved. We have

$$F(h00) = \Sigma_j f_j(\sin\theta/\lambda)\exp(2\pi i h x_j/a) = f_{Si}(\sin\theta/\lambda) + f_C(\sin\theta/\lambda)\exp(2\pi i h/4) \qquad (3)$$

where the last step is only valid for h = even numbers. $f_{Si}(\sin\theta/\lambda)$ and $f_C(\sin\theta/\lambda)$, the atomic scattering factors for silicon and carbon atoms, respectively, are dependent on the scattering angle and can be obtained from the International Tables for X-ray Crystallography [53]. Calculated values are 5.83, 8.49, 3.05, 4.40, and 1.19 for $F(200, 400,..., 1000)$ (all ϕ_{h00} are zero), respectively.

Following Eq. (1), 1D electron densities along the film normal can now be constructed using Fourier transform. Because of the arbitrary proportional constant in Eq. (2), to compare the calculated and measured results, these structure factors were "normalized" by a constant, so that the total area under each electron distribution curve is the same. It is observed that the measured and calculated $F(h00)$ numbers happen to be very close with normalization factors in a range of 1.0–1.15. Experimental and theoretical results are listed together in Table 1.

From Figure 3, it is clear that the Si peak is wider for both hetero-epitaxial SiC thin films. This indicates roughening of the Si plane within the silicon-carbide layers, although there is an amorphous region between the thin film and the silicon substrate (Figure 1). The distortion or roughness of the Si planes inside the epitaxial SiC film can be estimated by the increase of the full widths at half maximum (FWHM) of the Si maxima from that of the ideal case. For example, the FWHM of sample 1 (6 μm), is found to be 41% larger than that of the theoretical value. Similarly, the FWHM of sample 2 (12 μm) is about 29% larger. The distortion of atomic layers or the internal roughness of the individual atomic plane is lesser for layers away from the substrate. Therefore, the averaged internal roughness, which is shown by the X-ray diffraction, is smaller when the films are thicker.

From Figure 3, it is also observed that within the epitaxial thin film, the electron density for carbon is decreased, with no apparent increase of the FWHM. This can only happen when some of the carbon atoms are relocated away from their original positions. This is not unexpected under the high degree of roughness and distortion for the silicon atoms. In addition, the influence of the cut-off at (1000) for the Fourier series in Eq. (1) must be small, since the 1/8th maxima predicted by Eq. (3) are not evident in the film samples.

TABLE 1

Theoretical and experimental structure factors (normalized) for 3C-SiC

	F(200)	F(400)	F(600)	F(800)	F(1000)
Experimental 3C-SiC film (6 μm)	9.54	8.53	2.32	2.01	0.59
Experimental 3C-SiC film (12 μm)	8.38	8.85	2.67	2.41	0.69
Theoretical values of 3C-SiC	5.83	8.49	3.05	4.40	1.19

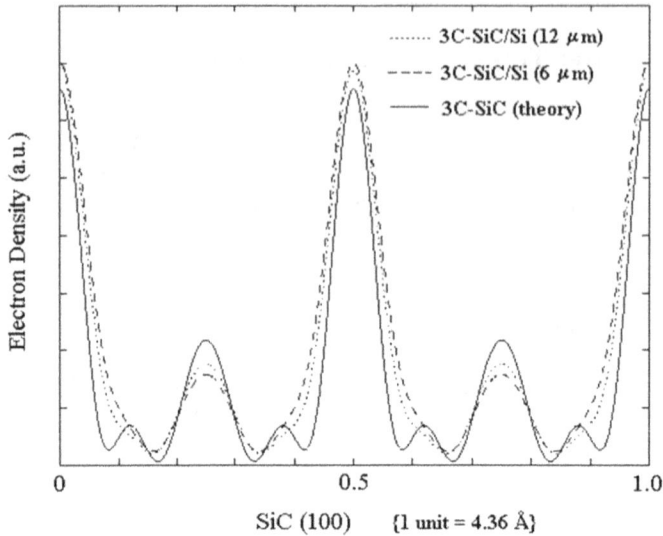

FIGURE 3

Measured and calculated electron density maps of 3C-SiC along the surface normal. From [51] Figure 3 with permission of reproduction by APS.

5 Theoretical Calculation of Homo-Epitaxial 6H-SiC and 4H-SiC

Homo-epitaxial silicon carbide thin film, where a 6H-SiC film of about 6 microns was grown on a single crystal 6H-SiC substrate is also investigated for a comparative study. The measured electron distribution is plotted together with those calculated using Eqs. (1–3) in Figure 4, where the two curves are almost identical to each other. It should be noted, however, that in this case there are non-vanishing phases that were acquired from the calculated values for both curves.

Another homo-epitaxial 4H-SiC/4H-SiC sample, 6 μm thick was grown on a single crystal 4H-SiC substrate, Figure 5. But in this case, the calculated electron distribution using Eqs. (1–3) in comparison with measured data plotted together in Figure 5 has deviation.

The scheme presented here is based on the kinematic theory, which is just an approximation from the dynamic theory, suitable for large and perfect crystals. However, in highly roughened thin films, due to the presence of lattice mismatch, the kinematic approach gives a fair description. This is confirmed by a good agreement between the experimental results and theoretical calculation.

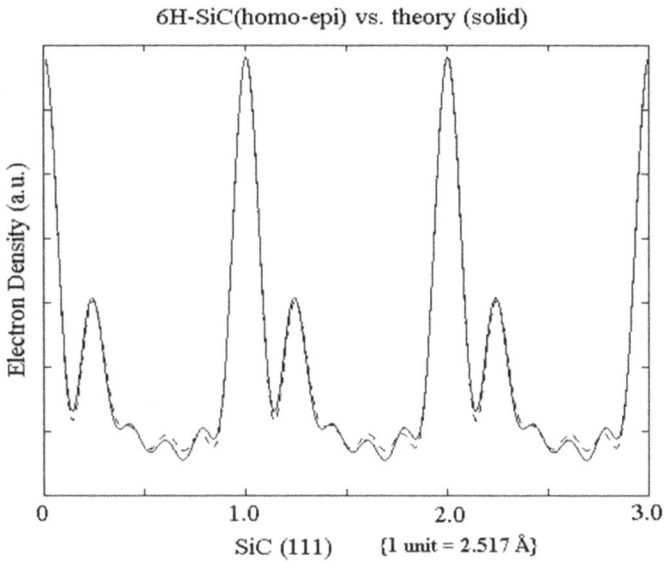

FIGURE 4
Measured and calculated electron density maps of 6H-SiC along the surface normal. From [51] Figure 4 with permission of reproduction by APS.

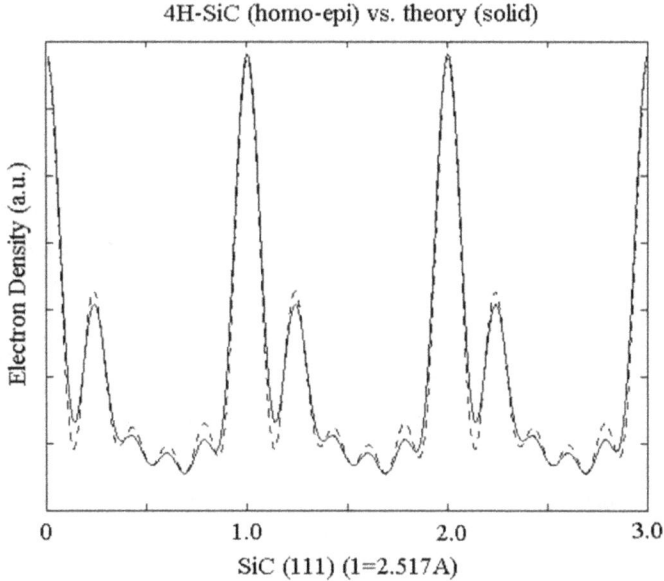

FIGURE 5
Measured and calculated electron density of homo-epitaxial 4H-SiC on single crystal 4H-SiC.

6 Synchrotron Radiation X-Ray Diffraction Measurements and Simulation

SR-XRD experiments were conducted at the National Synchrotron Radiation Research Center (NSRRC) in Hsinchu-Taiwan by the end station of BL01C2 with energy range of

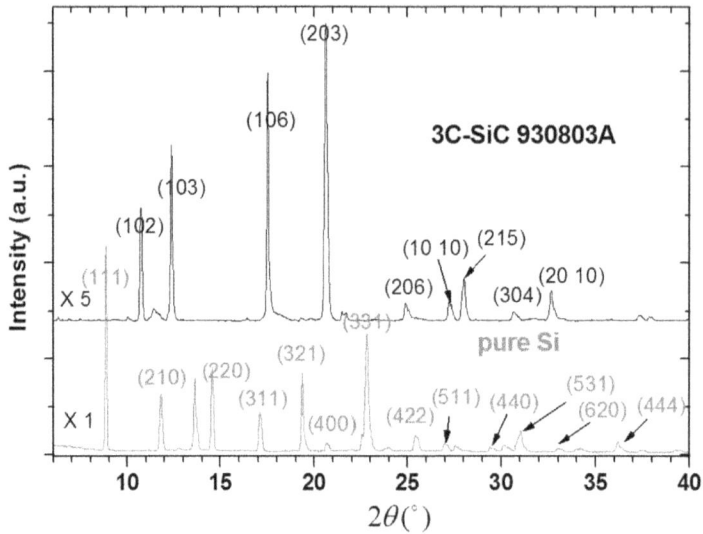

FIGURE 6
SR-XRD, using 28 keV (0.443 Å), scans for the Si substrate and a CVD-grown 3C-SiC/Si (100).

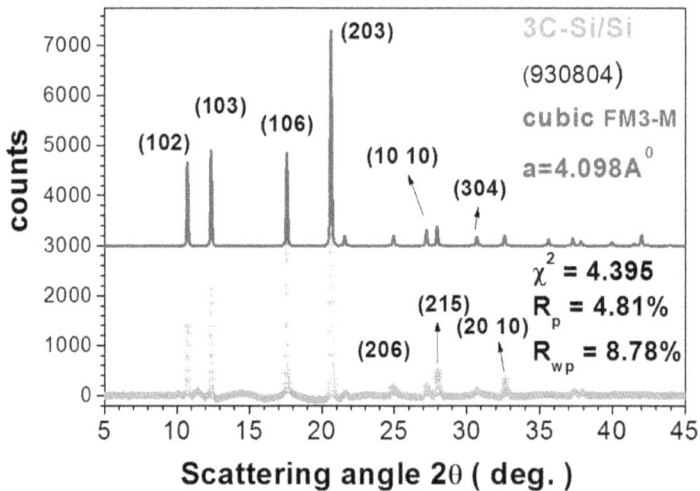

FIGURE 7
SR-XRD, using 28 keV (0.443 Å), scan for another CVD-grown 3C-SiC/Si (100) and comparative simulation result.

the beamline: 12–35 keV (1–0.35 Å) (www.nsrrc.org.tw/, http://tpsportal.nsrrc.org.tw). Beam energy of 28 keV, i.e., the short radiation wavelength of 0.443 Å was used. Figure 6 shows the synchrotron radiation X-ray diffraction (SR-XRD, $\lambda \sim 0.443$Å) patterns recorded at the room temperature (RT), for the Si substrate and a CVD-grown 3C-SiC/Si (100). Multiple diffraction peaks with a high resolution were detected due to the shorter X-ray wavelengths. The peaks are labeled in Figure 6.

    More 3C-SiC/Si (100) samples were measured by SR-XRD ($\lambda = 0.443$Å). Figure 7 exhibits another example and comparative simulation. All labels and assignments are matched well from the Inorganic Crystals Structure Database (ICSD) for cubic (3C) SiC.

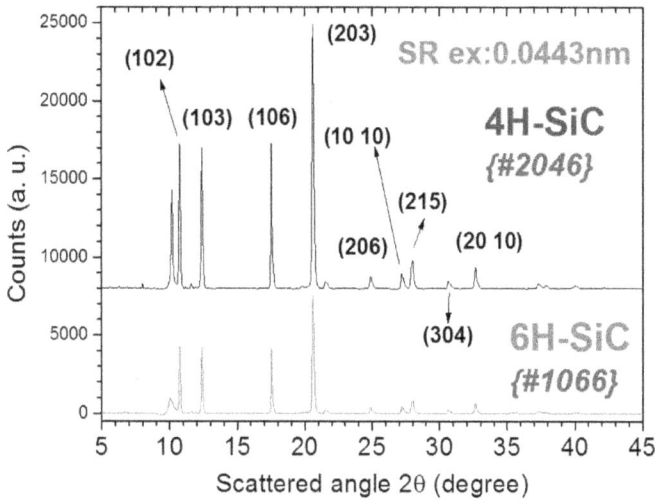

FIGURE 8
SR-XRD, using 28 keV (0.443 Å), scans for homo-epitaxial 4H-SiC/4H-SiC and 6H-SiC/6H-SiC.

High-energy SR-XRD ($\lambda = 0.443$Å) was further used to measure homo-epitaxial 4H-SiC and 6H-SiC samples. Figure 8 presents 2-scans for homo-epitaxial 4H-SiC/4H-SiC and 6H-SiC/6H-SiC. These SR-XRD spectra are similar to 3C-SiC, shown in Figures 5 and 6, as they are measured under the same excitation of 0.443Å. All labels and assigns in Figure 7 are matched for cubic (3C) SiC.

Atomic arrangements on 3C-SiC, 4H-SiC, and 6H-SiC, are analyzed to understand the crystalline properties. As predicted by Powell et al. [15], basic SiC tetrahedron forms the fundamental unit and different stacking of A, B, C sheets form SiC polytypes of ABC for 3C, ABAC for 4H and ABCACB for 6H SiC. As shown at Figure 1 of Chapter 3 in this book, i.e., Figure 1 of [54], Fissel presents the artificially layered heteropolytypic structures based on SiC polytypes, with clear pictures of these ABC stacking structures corresponding to 3C, 4H, and 6H-SiC. Our shortest wavelength (0.443 Å) excitation SR-XRD seems to detect the very basic SiC tetrahedron crystalline characteristics and common internal feature for all 3C, 4H, and 6H-SiC. These new results and knowledge base would promote further penetrative analyses and investigation to reveal more secrets inside the SiC polytypes.

7 Conclusion

A new method to accurately study the internal atomic arrangement of epitaxial materials with high resolution using short wavelength X-ray diffraction is developed. Cubic SiC films grown on Si (100) substrate experimentally exhibited up to five order Brag diffraction peaks along (100). This new method enables quantitative characterization of the internal atomic layer roughness and distortion of hetero-epitaxial 3C-SiC and homo-epitaxial 6H-SiC thin films. Electron density distribution can be reconstructed, using short wavelength X-ray to measure high order Bragg reflections. Compared with the calculated values from atomic scattering factors, the deviation of the unit cell structure from the ideal case can be identified. Many short wavelength (0.443Å) synchrotron radiation (SR) XRDs were also performed on hetero-epitaxial 3C-SiC on Si and homo-epitaxial 4H-SiC/4H-SiC and 6H-SiC/3C-SiC. Through a combination of theory and experiment, this study provides a practical and useful method to investigate thin film materials and systems.

We would like to acknowledge the support and help from NSRRC in Tsinchu-Taiwan.

References

1. W. J. Choyke, H. Matsunami, G. Pensl (Eds.), Silicon Carbide: Recent Major Advances, Springer, Berlin (2004).
2. Z. C. Feng (Ed.), SiC Power Materials – Devices and Applications, Springer, Berlin (2004).
3. Michael Shur, Sergey Rumyantsev and Michael Levinshtein (Eds.), SiC Materials and Devices, Vol. 1 and 2, World Scientific Publishing, Singapore (2006).
4. Peter Friedrichs, Tsunenobu Kimoto, Lothar Ley and Gerhard Pensl (Editors), Silicon Carbide: Volume 1: Growth, Defects, and Novel Applications (2009); Silicon Carbide: Volume 2: Power Devices and Sensors (2011), WILEY-VCH.
5. Tsunenobu Kimoto and James A. Co, Fundamentals of Silicon Carbide Technology: Growth, Characterization, Devices and Applications (Wiley – IEEE) (2014).
6. G. Pensl and W. J. Choyke, "Electrical and optical characterization of SiC", Physica B 185, 264–283 (1993).
7. J. B. Casady and R. W. Johnson, "Status of silicon carbide (SiC) as a wlde-bandgap semiconductor for high-temperature applications: a review", Solid-State Electronics 39, 1409–1422 (1996).
8. W. J. Choyke and G. Pensl, "Physical properties of SiC", MRS Bulletin, 22, 25–29 (1997).
9. N. G. Wright and A. B. Horsfall, "SiC sensors: a review", J. Phys. D: Appl. Phys. 40, 6345–6354 (2007).
10. S. E. Saddow, C. L. Frewin, M. Nezafatil, A. Oliveros I, S. Afroz, J. Register, M. Reyes, and S. Thomas, "3C-SiC on Si: A Bio- and Hemo-compatible Material for Advanced Nano-Bio Devices", IEEE Nanotechnology Materials and Devices Conference (NMDC), p.46–53 (2014).
11. Cai Chen, Fang Luo, and Yong Kang, "A Review of SiC Power Module Packaging: Layout, Material System and Integration", CPSS Transactions on power electronics and applications, 2, 170–186 (2017).
12. Kosuke Uchida, Toru Hiyoshi, Yu Saitoh, Takeyoshi Masuda, Tatsushi Kaneda, and Takashi Tsuno, "High Current SiC Transistors for Automotive Applications", SEI Technical Review, no. 88, 63–66 (2019).
13. W. M. Klahold, W. J. Choyke, and R. P. Devaty, "Band structure properties, phonons, and exciton fine structure in 4H-SiC measured by wavelength-modulated absorption and low-temperature photoluminescence", Physical Review B 102, 205203 (2020).
14. Lubin Han, Lin Liang, Yong Kang, and Yufeng Qiu, "A Review of SiC IGBT: Models, Fabrications, Characteristics and Applications", IEEE Transactions on Power Electronics, 36, 2080–2093 (2021).
15. J. A. Powell, P. Pirouz and W. J. Choyke, in *Semiconductor Interfaces, Microstructures and Devices: Properties and Applications*, ed. Z. C. Feng, 1993, Institute of Physics Publishing, Bristol, pp. 257.
16. S. Nishino, J. A. Powell and H. A. Will, Appl. Phys. Lett. 42, 460 (1983).
17. Y. M. Tairov and V. F. Tsvetkov, J. Cryst. Growth 43, 209 (1978); 52, 146 (1981).
18. W. J. Choyke, H. Matsunami and G. Pensel ed., *Fundamental questions and applications of SiC*, in Phys. Stat. Sol. (a) Vol. 162, No.1, and (b) Vol. 202, No. 1, 1997.
19. Th. Kunstmann and S. Veprek, Appl. Phys. Lett. 67, 3126 (1995).
20. C. A. Zorman, A. J. Fleischman, A. S. Dewa, M. Mehregany, C. Jacob, S. Nishino and P. Pirouz, J. Appl. Phys. 78, 5136 (1995).

21. K.-W. Lee, K.-S. Yu, J. W. Bae and Y. Kim, Mat. Sci. Forum, 264–268, 175 (1998).
22. S. Miyagawa, S. Nakao, K. Saitoh, M. Ikeyama, H. Niwa, S. Tanemura, Y. Miyagawa and K. Baba, J. Appl. Phys. 78, 7018 (1995).
23. H. Weishart, W. Matz and W. Skorupa, in S. Nakashima, H. Matsunami, S. Yoshida and H. Harima ed., Silicon Carbide and Related Materials 1995, Inst. Phys. Conf. Ser. No. 142, p. 541.
24. A. Ellison, J. Di. Persio and C. Brylinski, *ibid*, p. 441.
25. H. Romanus, G. Teichert and L. Spiess, Mat. Sci. Forum, 264–268, 437 (1998).
26. V. Ivantsov and V. Dmitriev, Mat. Sci. Forum, 264–268, 73 (1998).
27. Z. J. Zhang, K. Narumi, H. Naramoto, S. Yamamoto and A. Miyashita, J. Phys.: Condens. Matter 10, 11713 (1998).
28. S. Rendakova, V. Ivantsov and V. Dmitriev, Mat. Sci. Forum, 264–268, 163 (1998).
29. C. Hallin, A. Ellison, I. G. Ivanov, A. Henry, N. T. Son and E. Janzen, Mat. Sci. Forum, 264–268, 123 (1998).
30. A. A. Lebedev, G. N. Mosina, I. P. Nikitina, N. S. Savkina, L. M. Sorokin, and A. S. Tregubova, Investigation of the Structure of (p)3C-SiC–(n)6H-SiC Heterojunctions, Technical Physics Letters, 27, 1052–1054 (2001), https://doi.org/10.1134/1.1432347.
31. Robert S. Okojie, Thomas Holzheu, XianRong Huang, and Michael Dudley, X-ray diffraction measurement of doping induced lattice mismatch in n-type 4H-SiC epilayers grown on p-type substrates, Applied Physics Letters 83, 1971 (2003); doi: 10.1063/1.1606497.
32. M. D'angelo, H. Enriquez, V. Yu. Aristov, and P. Soukiassian, Atomic structure determination of the Si-rich β-SiC (001) 3x2 surface by grazing-incidence x-ray diffraction: A stress-driven reconstruction, Physical Review B 68, 165321 (2003), https://doi.org/10.1103/PhysRevB.68.165321.
33. Hun Jae Chung, Marek Skowronski, High-resolution X-ray diffraction and optical absorption study of heavily nitrogen-doped 4H–SiC crystals, Journal of Crystal Growth 259, 52–60 (2003), doi:10.1016/S0022-0248(03)01584-7.
34. X. R. Huang, M. Dudley, W. Cho, R. S. Okojie and P. G. Neudeck, Characterization of SiC Epitaxial Structures Using High-Resolution X-ray Diffraction Techniques, Materials Science Forum 457–460, 157–162 (2004), https://doi.org/10.4028/www.scientific.net/MSF.457-460.157.
35. M. A. Mastro, M. Fatemi, D. K. Gaskill, K.-K. Lew, B. L. Van Mil, and C. R. Eddy Jr.C. E. C. Wood, X-ray diffraction study of crystal plane distortion in silicon carbide substrates, Journal of Applied Physics 100, 093510 (2006); doi: 10.1063/1.2362918.
36. A. Audren, I. Monnet, D. Gosset, Y. Leconte, X. Portier, L. Thomé, F. Garrido, A. Benyagoub, Effects of electronic and nuclear interactions in 3C-SiC, Nuclear Instruments and Methods in Physics Research B 267, 976–979 (2009), doi:10.1016/j.nimb.2009.02.033.
37. C. Dupeyrat, A. Declémy, M. Drouet, A. Debelle, L. Thomé, Fe-implanted SiC as a potential DMS: X-ray diffraction and Rutherford backscattering and channelling study, Nuclear Instruments and Methods in Physics Research B, 268, 2863–2865 (2010), doi:10.1016/j.nimb.2010.03.018.
38. Dominique Gosset, Christian Colin, Aurelien Jankowiak, Thierry Vandenberghe, and Nicolas Lochet, X-ray Diffraction Study of the Effect of High-Temperature Heat Treatment on the Microstructural Stability of Third-Generation SiC Fibers, J. Am. Ceram. Soc., 96, 1622–1628 (2013), DOI: 10.1111/jace.12174.
39. Gerard Colston, Stephen D. Rhead, Vishal A. Shah, Oliver J. Newell, Igor P. Dolbnya, David R. Leadley, Maksym Myronov, Mapping the strain and tilt of a suspended 3C-SiC membrane through micro X-ray diffraction, Materials and Design, 103, 244–248 (2016), http://dx.doi.org/10.1016/j.matdes.2016.04.078.
40. L. B. Bayu Aji1, E. Stavrou,· J. B. Wallace,· A. Boulle,· A. Debelle,· S. O. Kucheyev, Comparative study of radiation defect dynamics in 3C-SiC by X-ray diffraction, Raman scattering, and ion channeling, Applied Physics A 125, 28 (2019), https://doi.org/10.1007/s00339-018-2325-7.
41. Nurul Hidayat, Abdulloh Fuad, Nandang Mufti, Ummu Kultsum, Anggun Amalia Fibriyanti, Sunaryono, Bambang Prihandoko, In-situ High-Resolution Transmission Electron

Microscopy and X-ray Diffraction Studies on Nanostructured β-SiC and Its Promising Feature for Photocatalytic Hydrogen Production, IOP Conf. Series: Materials Science and Engineering 515, 012012 (2019), doi:10.1088/1757-899X/515/1/012012.

42. Mojmír Meduňa, Thomas Kreiliger, Marco Mauceri, Marco Puglisi, Fulvio Mancarella, Francesco La Via, Danilo Crippa, Leo Miglio, Hans von Känel, X-Ray diffraction on stacking faults in 3C-SiC epitaxial microcrystals grown on patterned Si(001) wafers, Journal of Crystal Growth, 507, 70 (2019), doi.10.1016j.jcrysgro.2018.10.046.

43. Takaaki Koyanagi, David J. Sprouster, Lance L. Snead, YutaiKatoh, X-ray characterization of anisotropic defect formation in SiC under irradiation with applied stress, Scripta Materialia,197, 113785 (2021), https://doi.org/10.1016/j.scriptamat.2021.113785.

44. Deng Ya, Zhang Yumin, Zhou Yufeng. Measurement of residual stress in single-crystal sic by X-ray diffraction method. *Chinese Journal of Theoretical and Applied Mechanics*, 54, 147–153 (2022), doi: 10.6052/0459-1879-21-426

45. J. Chaudhuri, K. Ignatiev, J.H. Edgar, Z.Y. Xie, Y. Gao, Z. Rek, Low temperature chemical vapor deposition of 3C-SiC on 6H-SiC — high resolution X-ray diffractometry and synchrotron X-ray topography study, Materials Science and Engineering B76, 217–224 (2000), https://doi.org/10.1016/S0921-5107(00)00451-7.

46. H. Enriquez, M. D'angelo, V. Yu. Aristov, V. Derycke, P. Soukiassian, G. Renaud, A. Barbier, S. Chiang, and F. Semond, Silicon carbide surface structure investigated by synchrotron radiation-based x-ray diffraction, Journal of Vacuum Science & Technology B 21, 1881 (2003); doi: 10.1116/1.1588650.

47. Michael Dudley, XianRong Huang and William M Vetter, Contribution of x-ray topography and high-resolution diffraction to the study of defects in SiC, J. Phys. D: Appl. Phys. 36, A30–A36 (2003), DOI: 10.1088/0022-3727/36/10a/307.

48. George M. Amulele, Murli H. Manghnani, Baosheng Li, Daniel J. H. Errandonea, Maddury Somayazulu, and Yue Meng, High pressure ultrasonic and x-ray studies on monolithic SiC composite, Journal of Applied Physics 95, 1806 (2004); doi: 10.1063/1.1639141.

49. S. Milita, M. De Santis, D. Jones, A. Parisini, V. Palermo, Real time investigation of the growth of silicon carbide nanocrystals on Si(1 0 0) using synchrotron X-ray diffraction, Applied Surface Science 254, 2162–2167 (2008), doi:10.1016/j.apsusc.2007.09.004.

50. Nikolaos Baimpas, Alexander J.G. Lunt, Igor P. Dolbnya, Jiri Dluhos, Alexander M. Korsunsky, Nano-scale mapping of lattice strain and orientation inside carbon core SiC fibres by synchrotron X-ray diffraction, Carbon 79, 85–92 (2014), http://dx.doi.org/10.1016/j.carbon.2014.07.045.

51. G. Xu and Z. C. Feng, Internal atomic distortion and layer roughness of epitaxial SiC thin films studied by short wavelength x-ray diffraction, Phys. Rev. Lett. 84, 1926–1929 (2000). DOI: 10.1103/PhysRevLett.84.1926.

52. B.E. Warren, X-ray Diffraction, Addison-Wesley, New York (1969).

53. C.H. MacGillavry, K. Lonsdale and G.D. Rieck, International Table of X-ray Crystallography, The Kynoch Press, UK (1968).

54. A. Fissel, Artificially layered heteropolytypic structures based on SiC polytypes: molecular beam epitaxy, characterization and properties, Physics Reports, 2003, 379, pp. 149–255; DOI: 10.1016/S0370-1573(02)00632-4.

Index

Power materials, xiv
Precision control, 34, 173
Precursor, 32–35, 48, 63, 105, 147–153, 160, 175, 264, 331
Precursor chemistry, 142
Preliminary data, 16
Presolar SiC, 3–24
Pressure, 19, 136–153
Primitive cell, 235–236, 352
Primitive meteorites, 5–8
Principal planes, 13
Propagate, 67–71, 303
Propane, 142–152, 175
Protective layer, 329
Pseudo-temperature, 136
P_{si}/Delta, 291
Pulsed laser deposition (PLD), 30–34
Pumping, 126, 136, 255, 274, 279
Purity, 15, 23–24, 30–54, 63, 66, 88–92, 110, 128–152, 233–266, 321, 386
Pyroelectric coefficients, 356
Pyrometer, 37, 135–136
Pyrophoric, 152–153

Q

Qiu, Zhi Ren, 219
Quality, 14, 42, 47, 53–56, 63–92, 100–116, 126–153, 160–168, 174–188, 220–228, 250–261, 292–315, 329–362, 380–397, 403–415, 419–420
Quantitative description, 234–242
Quantitative measurement, 68
Quantum
    quantum bit (qubit), 249, 253
    quantum chemistry methods, 336–350
    quantum communications, 249–261
    quantum confinement effect, 202–205
    quantum dots (QDs), 249
    quantum efficiency, 275, 358, 376–397
    quantum well, 202
Quartz, 34, 128–150
Quasi
    quasi-Brewster angle, 315
    quasi-static approximation, 85
    quasicubic, 228, 250–274
Quenching, 22, 258, 281, 392

R

Rabi oscillation, 254–266
Radiation
    radiation damage, 5, 20

radiation defects, 20
radiation detector, 351, 403–415
radiation resistance, 113, 376
Radii, 9, 42–43
Radio-frequency (RF), 129
Raman
    Raman mapping, 327–338
    Raman microscopy, 327–363
    Raman scattering, 18, 65–68, 91, 219–243, 331–347
    Raman tensor, 235–243
Re-evaporation, 48
Reaction kinetics, 33
Reactive ion etching, 134
Reactor configuration, 136–142
Realistic rigid-ion model (RIM), 66
Reciprocal lattice, 180, 421–422
Reciprocal maps, 420
Recirculating cells, 136–142
Recombination, 15–23, 55, 106, 167, 200–210, 254–282, 378
Reflection coefficients, 292, 298, 346
Reflection electron energy loss spectroscope (REELS), 291
Reflectivity, 12, 36, 66, 76–92, 294
Reflection ellipsometry, 299
Refractive index, 294–316, 346
Reliability, 62, 100–116, 160–168, 376–381
Removal, 6, 46, 146, 209, 299, 334
Renishaw Series 1000 Raman microscope, 221
Research and development (R&D), xiii, 24, 57, 103–116, 126, 160–164, 419
Residual stress, 99, 318, 329–332, 430
Residue, 6–7, 205
Resonant
    resonant excitation, 254–282
    resonant modes, 20
    resonantly enhanced electronic Raman scattering, 228–229
Responsivity, 376–397
Reststrahlen band, 66–92
Reynolds, J.H., 4
Rhombic, 62
Rhombohedral lattices, 13
Richardton, 4
Rocking curve, 52–53
Room temperature, 15–22, 77–92, 110–116, 163, 175, 208–211, 221–235, 250–256, 273–281, 313, 335–336, 353–354, 381–395, 426
Rotation Raman, 220–243
Rough grinding, 300–320
Round-shaped steps, 39
Rubber, 37

For Product Safety Concerns and Information please contact our EU
representative GPSR@taylorandfrancis.com
Taylor & Francis Verlag GmbH, Kaufingerstraße 24, 80331 München, Germany